THE RISE OF
PLACENTAL
MAMMALS

George Gaylord Simpson (1902–1984) examining fossils at the American Museum in the late 1940s or early 1950s.

ORIGINS AND
RELATIONSHIPS
OF THE MAJOR
EXTANT CLADES

THE RISE OF PLACENTAL MAMMALS

Edited by KENNETH D. ROSE
and J. DAVID ARCHIBALD

THE JOHNS HOPKINS
UNIVERSITY PRESS
Baltimore and London

© 2005 The Johns Hopkins University Press
All rights reserved. Published 2005
Printed in the United States of America on acid-free paper
9 8 7 6 5 4 3 2 1

The Johns Hopkins University Press
2715 North Charles Street
Baltimore, Maryland 21218-4363
www.press.jhu.edu

Frontispiece courtesy of L. F. Laporte.

Library of Congress Cataloging-in-Publication Data

The rise of placental mammals : origins and relationships of the
major extant clades / edited by Kenneth D. Rose and J. David
Archibald.
      p.   cm.
   Includes bibliographical references (p.   ) and index.
   ISBN 0-8018-8022-X (hardcover : alk. paper)
      1. Mammals—Evolution.   2. Mammals—Classification.
I. Rose, Kenneth David, 1949–   II. Archibald, J. David.
QL708.5.R57   2005
599.138—dc22   2004014817

A catalog record for this book is available from the British
Library.

To George Gaylord Simpson,
twentieth-century giant of paleontology

CONTENTS

CONTRIBUTORS

J. David Archibald, Department of Biology, San Diego State University, San Diego, California

Robert J. Asher, Museum für Naturkunde, Institut für Systematische Zoologie, Berlin, Germany

Jonathan I. Bloch, Museum of Paleontology, University of Michigan, Ann Arbor, Michigan (*current affiliation:* Florida Museum of Natural History, University of Florida, Gainesville, Florida)

Douglas M. Boyer, Museum of Paleontology, University of Michigan, Ann Arbor, Michigan (*current affiliation:* Department of Anatomical Sciences, Stony Brook University, Stony Brook, New York)

Daryl P. Domning, Laboratory of Evolutionary Biology, Department of Anatomy, Howard University, Washington, D.C.

Eduardo Eizirik, Laboratory of Genomic Diversity, National Cancer Institute, Frederick, Maryland

Robert J. Emry, Department of Paleobiology, National Museum of Natural History, Smithsonian Institution, Washington, D.C.

Jörg Erfurt, Institut für Geologische Wissenschaften und Geiseltalmuseum, Martin-Luther-University, Halle, Germany

John J. Flynn, Department of Geology, The Field Museum, Chicago, Illinois

Timothy J. Gaudin, Department of Biological and Environmental Sciences, University of Tennessee, Chattanooga, Tennessee

Emmanuel Gheerbrant, Laboratoire de Paléontologie, UMR 8569 du CNRS, Département Histoire de la Terre, Muséum National d'Histoire Naturelle, Paris, France

Philip D. Gingerich, Museum of Paleontology and Department of Geological Sciences, University of Michigan, Ann Arbor, Michigan

Patricia A. Holroyd, Museum of Paleontology, University of California, Berkeley, California

Jeremy J. Hooker, Department of Palaeontology, The Natural History Museum, Cromwell Road, London, United Kingdom

Léo F. Laporte, Department of Earth Sciences, University of California, Santa Cruz, California

Jin Meng, Division of Paleontology, American Museum of Natural History, New York, New York

William J. Murphy, Laboratory of Genomic Diversity, National Cancer Institute, Frederick, Maryland

Jason C. Mussell, Center for Functional Anatomy and Evolution, The Johns Hopkins University School of Medicine, Baltimore, Maryland

Michael J. Novacek, Division of Paleontology, American Museum of Natural History, New York, New York

Stephen J. O'Brien, Laboratory of Genomic Diversity, National Cancer Institute, Frederick, Maryland

Kenneth D. Rose, Center for Functional Anatomy and Evolution, The Johns Hopkins University School of Medicine, Baltimore, Maryland

Guillermo W. Rougier, Department of Anatomical Sciences and Neurobiology, University of Louisville, Louisville, Kentucky

Eric J. Sargis, Department of Anthropology, Yale University, New Haven, Connecticut

Mary T. Silcox, Department of Anthropology, University of Winnipeg, Winnipeg, Manitoba, Canada

Nancy B. Simmons, Department of Mammalogy, American Museum of Natural History, New York, New York

Mark S. Springer, Department of Biology, University of California, Riverside, California

Gerhard Storch, Forschungsinstitut Senckenberg, Frankfurt am Main, Germany

Pascal Tassy, Laboratoire de Paléontologie, UMR 8569 du CNRS, Département Histoire de la Terre, Muséum National d'Histoire Naturelle, Paris, France

Jessica M. Theodor, Department of Geology, Illinois State Museum, Springfield, Illinois

Gina D. Wesley-Hunt, Committee on Evolutionary Biology, University of Chicago, Chicago, Illinois

John R. Wible, Section of Mammals, Carnegie Museum of Natural History, Pittsburgh, Pennsylvania

André R. Wyss, Department of Geological Sciences, University of California, Santa Barbara, California

ACKNOWLEDGMENTS

W E THANK THE CONTRIBUTORS FOR providing us with such comprehensive syntheses, and for doing so in an expeditious manner. We also thank our editor at the Johns Hopkins University Press, Vincent Burke, and his colleagues Juliana McCarthy, Ken Sabol, Martha Sewall, Sarah Shepke, and Janice Wheeler, for shepherding this book to publication. We are grateful to Peter Strupp, Cyd Westmoreland, and the staff of Princeton Editorial Associates for expediting the final stages of production. Léo Laporte generously provided the frontispiece. Finally, we are indebted to the following individuals, who provided prompt and constructive reviews of the chapter manuscripts: Marc Allard, Robert Asher, Michael Benton, Annalisa Berta, Percy Butler, Nicholas Czaplewski, Mary Dawson, Lawrence Flynn, Ewan Fordyce, Peter Giere, Philip Gingerich, Gregg Gunnell, Sue Hand, Jean-Louis Hartenberger, Paul Higgs, Luke Holbrook, Inés Horovitz, Christine Janis, Jason Lillegraven, Greg McDonald, Malcolm McKenna, Don Prothero, Eric Seiffert, Fred Szalay, Richard Tedford, Sergio Vizcaíno, David Webb, Lars Werdelin, Anne Yoder, and two anonymous reviewers.

J. DAVID ARCHIBALD
AND KENNETH D. ROSE

1

WOMB WITH A VIEW:
THE RISE OF PLACENTALS

ALTHOUGH EVOLUTIONARY SUCCESS IS A difficult if not impossible concept to define, we believe we know it when we see it. This is the case with the extant placentals, the clade of mammals to which we belong. Living placentals include only 1,050 genera and some 4,400 species (Wilson and Reeder, 1993)—compare this to just one of the other tetrapod clades, the birds, which has more than 9,000 species. Nevertheless, placentals range tremendously in their ecological diversity, from tiny shrews to the gargantuan blue whale, from pinnipeds swimming the frigid high-latitude oceans to the golden moles swimming the hot sands of southern Africa.

The evolutionary success of mammals is one of the few in evolutionary history for which we can offer an explanation. First, there is what Gould (1989) popularized as historical contingencies. For placental mammals, these were mode of reproduction, level of metabolism, and an ancestral, generalized quadrupedal stance (Archibald, 2001). Euviviparity, which includes lengthy *in utero* development of the embryo, requires that all support and sustenance come from the mother through the chorio-allantoic placenta. This mode of reproduction is unique to placentals. It allows the mother to continue normal activities while pregnant. Placentals, like other mammals, are endothermic, producing their heat through metabolic means. In small mammals, such as most rodents, between 80% and 90% of food goes toward maintaining endothermy (Vaughan et al., 2000). The common ancestor of all mammals, as well as that leading to eutherians, was a small, insectivorous quadruped that retained five digits on all four limbs. Such a generalized pattern permitted a wide diversity of stance and locomotion in later eutherians. For example, placentals have limbs modified for greatly varied activities, including swimming, flight, digging, running, hopping, climbing, brachiation, and capture of prey.

Second, placentals had spread to all continents except Australia and possibly Antarctica by the time dinosaurs became extinct some 65 million years ago. Thus, they were serendipitously poised to inherit the ecological space vacated by dinosaurs. They began almost immediately to speciate, although it was millions of years before placentals gained in size and ecological diversity (Kirchner and Weil, 2001). But even by about 10–15 million years after the Cretaceous/Tertiary (K/T) boundary, the vast majority of major placental clades that we call orders are recognizable.

The study of the evolutionary history of mammals, or any other taxon, requires a well-established, testable argument for the relationship of the included species. As with many plants and animals, our current ideas of systematic relationships for mammals trace their beginnings to Linnaeus (1758). Table 1.1 shows a sampling of some better-known, higher-level classifications emphasizing the ordinal-level trends in classification starting with Linnaeus. This table focuses on placentals, but as early classifications had not yet realized the higher relationships of placentals, marsupials, and monotremes, the last two taxa were confounded with placentals in earlier classifications. One obvious trend since Linnaeus is the increase in recognized placental orders (or equivalents). From eight extant orders recognized by Linnaeus in 1758 and Cuvier in 1817, the number has increased to 18 today (Wilson and Reeder, 1993; McKenna and Bell, 1997; Murphy et al., 2001). Unquestionably, what has been meant by an order has changed over time. Even accounting for this, mammals that originally had been grouped solely on a shared body plan were often recognized as lineages or clades once evolution was widely accepted in the mid-nineteenth century.

Although Simpson (1945) provided one of the best reviews of the various higher taxa of mammals, Gregory's (1910) older treatment remains a superlative narrative of the history of mammalian systematics, even though it was published almost a century ago. Gregory (1910: 87) pointed to a number of modifications that changed and improved our understanding of mammalian systematics. Three stand out: "The anthropocentric classification . . . gives way to the evolutionary classification," "[d]iscovery and development of the principles of the evolution of the feet . . . and of the teeth," and "[r]eunion and integration of results of mammalogy, comparative anatomy, embryology, paleontology." The first of these changes placed humans with other primates rather than in a separate order (Cuvier's Bimana for humans disappears). The second of these differentiated homologies from homoplasies found among mammalian teeth and feet (e.g., there are groups of "even-toed" and "odd-toed" ungulates). The third saw the better integration of "soft" and "hard" anatomy (e.g., monotremes and marsupials are recognized to be only distantly related to placentals).

From the time of Simpson's classification in 1945 onward, there was the general perception that orders represented

true evolutionary lineages or clades. The orders Macroscelidea (Butler, 1956) and Scandentia (Butler, 1972) were the last two to be recognized at the ordinal level, resulting in the standard 18 anatomically based orders of placental mammals as listed by Wilson and Reeder (1993): Xenarthra, Insectivora, Scandentia, Dermoptera, Chiroptera, Primates, Carnivora, Cetacea, Sirenia, Proboscidea, Perissodactyla, Hyracoidea, Tubulidentata, Artiodactyla, Pholidota, Rodentia, Lagomorpha, and Macroscelidea. Although there were many studies that tried to link various orders based on anatomical data and to find the origins of these orders, few well-supported results were forthcoming. In other words, with a few possible exceptions (see below), these 18 orders were the most inclusive groups of placental mammals for which we had good evidence for monophyly. In some ways, little had changed since the time of Gregory. In 1910, he noted that Linnaeus' classifications of 1758 and 1766 were "really an attempt to express relationship between distinct orders (as they are now accepted), an attempt that was certainly premature in Linné's time, since even now when the content of mammalogy is a hundred times greater, the interordinal connections are still either wholly unsettled or at best more a matter of probability than of demonstrated certainty" (Gregory, 1910: 30).

There are four superordinal groupings, however, that have long had anatomical support (Table 1.1). The oldest and generally most consistent is the grouping of rodents and lagomorphs under some common name, the most familiar being Glires, a name dating back to Linnaeus in 1758 (first used as an order). With the exception of the classification of McKenna and Bell (1997), it is still commonly accepted. The next oldest superordinal grouping of extant placentals that has had consistent support is Gregory's (1910) Archonta. Although two of the original members, macroscelidids and chiropterans, are now removed, primates, tupaiids, and dermopterans remain. The third is Simpson's (1945) proposed Paenungulata, whose extant members are the proboscideans, hyraxes, and sirenians. This name has found wide acceptance, again with the exception of McKenna and Bell (1997), who used the name Uranotheria for the same grouping. The fourth and final is Xenarthra, which has been recognized as a superordinal clade since 1975 by McKenna, although he referred to it as Edentata. This usage continued with McKenna and Bell (1997) and in various molecular studies (e.g., Murphy et al., 2001). With this consistent history, it should be no surprise that four chapters of the present volume deal with these four superordinal groupings.

A major change in mammalian systematics since Simpson (1945) has been the advent of powerful techniques that enable the study of ever-increasing portions of the genome. It is impossible to point to any one study that brought these techniques to maturity, but certainly the research of Murphy et al. (2001) demonstrates the trend. Such studies have provided strong evidence for four superordinal clades: Afrotheria, Xenarthra, Euarchontoglires, and Laurasiatheria (Table 1.1). Particularly notable is the recognition of an African clade, Afrotheria, including six previously recognized groups. Of these six groups, five were traditional orders, whereas

Table 1.1 Historical perspective of placental orders

LINNAEUS, 1758[a]
Mammalia
  Unguiculata[c]
    Order Primates (primates, dermopterans,
      chiropterans)
    Order Bruta (proboscideans, sirenians,
      bradypodids, myrmecophagids,
      pholidotans)
    Order Ferae (carnivorans)
    Order Bestiae (suids, tayassuids,
      dasypodids, erinaceids, soricids, talpids,
      didelphids)
    Order Glires (rhinocerotids, lagomorphs,
      rodents)
  Ungulata[c]
    Order Pecora (tylopods, ruminants)
    Order Belluae (equids, hippopotamids)
  Mutica[c]
    Order Cete (cetaceans)

CUVIER, 1817
Mammalia[b]
  Order Bimanes (humans)
  Order Quadrumanes (= primates, excluding
    humans)
  Order Carnassiers
    Cheiroptères (dermopterans, chiropterans)
    Insectivores (erinaceids, soricids, talpids,
      chrysochlorids, tenrecids)
  Order Carnivores
    Plantigrades (procyonids, some mustelids)
    Digitigrades (some mustelids, canids,
      viverrids, hyaenids, felids)
    Amphibes (pinnipeds)
    Marsupiaux (marsupials)
  Order Rongeurs (including lagomorphs)
    À clavicules (rodents with clavicles, the
      primate *Daubentonia*)
    San clavicules (rodents without clavicles,
      lagomorphs)
  Order Édentés
    Tardigrades (sloths)
    Édentés ordinaires (dasypodids, pangolins,
      myrmecophagids, tubulidentates)
    Monotrèmes (monotremes)
  Order Pachydermes
    Proboscidiens (elephants)
    Pachydermes ordinaries (hippopotamids,
      suiforms, hyracoids, ceratomorphs)
    Solipèdes (equids)
  Order Ruminans (tylopods, ruminants)
    Sans cornes (camelids, tragulids) avec
      cornes (ruminants except tragulids)
  Order Cétacés
    Herbivores (sirenians)
    Ordinaires (cetaceans)

FLOWER, 1883
Eutheria or Monodelphia
  Order Edentata (including xenarthrans,
    tubulidentates, pholidotes)
  Order Sirenia
  Order Cetacea
  Order Ungulata
    Suborder Artiodactyla
    Suborder Perissodactyla
    Suborder Hyracoidea
    Suborder Proboscidea

Order Rodentia
  Suborder Duplicidentata (= Lagomorpha)
  Suborder Simplicidentata (= Rodentia)
Order Chiroptera
Order Insectivora
  Suborder Dermoptera
  Suborder Insectivora (including Tupaiidae,
    Macroscelididae)
Order Carnivora
Order Primates

GREGORY, 1910
Eutheria
  Therictoidea
    Order Insectivora
    Order Ferae (including Fissipedia and
      Pinnipedia = Carnivora)
  Archonta
    Order Menotyphla (Tupaiidae and
      Macroscelididae)
    Order Dermoptera
    Order Chiroptera
    Order Primates
  Rodentia
    Order Glires
      Suborder Duplicidentata
        (= Lagomorpha)
      Suborder Simplidentata (= Rodentia)
  Edentata
    Order Tubulidentata
    Order Pholidota
    Order Xenarthra
  Paraxonia
    Order Artiodactyla
  Ungulata
    Order Sirenia
    Order Proboscidea
    Order Hyraces (=Hyracoidea)
    Order Mesaxonia (including Perissodactyla)
  Cetacea
    Order Odontoceti
    Order Mystacoceti

SIMPSON 1945
Eutheria
  Unguiculata
    Order Insectivora (including
      Macroscelidea)
    Order Dermoptera
    Order Chiroptera
    Order Primates (including Scandentia)
    Order Edentata (including Xenarthra)
    Order Pholidota
  Glires
    Order Lagomorpha
    Order Rodentia
  Mutica
    Order Cetacea
  Ferungulata
    Ferae
      Order Carnivora
    Protungulata
      Order Tubulidentata
    Paenungulata
      Order Proboscidea
      Order Hyracoidea
      Order Sirenia

Mesaxonia
  Order Perissodactyla
Paraxonia
  Order Artiodactyla

McKENNA AND BELL 1997
Placentalia
  Xenarthra
    Order Cingulata
    Order Pilosa
  Epitheria
    Anagalida
      Mirorder Macroscelidea
        Order Lagomorpha
        Order Rodentia
    Ferae
      Order Cimolesta (including
        Pholidota)
      Order Carnivora
    Lipotyphla
      Order Chrysochloridea
      Order Erinaceomorpha
      Order Soricomorpha
    Archonta
      Order Chiroptera
      Order Primates (including
        Dermoptera)
      Order Scandentia
    Ungulata
      Order Tubulidentata
      Eparctocyonia
        Order Cete (including Cetacea)
        Order Artiodactyla
      Altungulata
        Order Perissodactyla
        Order Uranotheria (including
          Hyracoidea, Sirenia, Proboscidea)

MURPHY ET AL. 2001
Placentalia
  Afrotheria
    Order Tubulidentata
    Order Macroscelidea
    Order Tenrecoidea
      (= Afrosoricida[d])
    Paenungulata
      Order Proboscidea
      Order Hyracoidea
      Order Sirenia
  Xenarthra (cingulates, pilosans)
  Boreoeutheria
    Euarchontoglires
      Archonta (= Euarchonta[d])
        Order Primates
        Order Scandentia
        Order Dermoptera
      Glires
        Order Lagomorpha
        Order Rodentia
    Laurasiatheria
      Order Lipotyphla (= Eulipotyphla[d])
      Order Chiroptera
      Order Carnivora
      Order Pholidota
      Order Perissodactyla
      Order Artiodactyla
        (= Cetartiodactyla[d])

---

[a] The taxa noted in parentheses for Linnaeus and other authors are the approximate equivalents recognized today.

[b] In this edition, Cuvier maintained marsupials and monotremes within placentals, but he did note that marsupials might belong in their own order (Gregory, 1910).

[c] The superordinal taxa are from Linnaeus (1766, not seen) as indicated by Gregory (1910).

[d] Taxa used by Murphy (2001) that were treated as synonyms by Archibald (2003).

the sixth, Tenrecoidea (McDowell, 1958), has been part of the established Lipotyphla. Its transfer to the African clade was not predicted based upon anatomy, but molecular evidence indicates that tenrecoids shared a more recent ancestor with elephants than they did with shrews, hedgehogs, or moles. The other three molecularly based superordinal clades held fewer surprises. In fact, the four anatomically based superordinal clades discussed earlier—Glires, Archonta (with the removal of Chiroptera), Paenungulata, and Xenarthra—are now supported by molecular evidence (e.g., Murphy et al., 2001). The greatest contribution of mammalian molecular systematic studies has been the strong support of superordinal clades that have not been recovered with any certainty based on anatomy. At the ordinal level, where the anatomical and molecular studies agree on 16 of 18 orders, the changes were less profound. Aside from the breakup of Lipotyphla, the most striking revision concerns Cetacea. As some anatomical studies had already showed, cetaceans were at least the sister taxon of artiodactyls, but the molecules are more radical in nesting Cetacea within Artiodactyla.

Before outlining the history and organization of this book on placental mammals, it is worthwhile to define what we mean by this term. By and large throughout this book, Placentalia is defined as the taxon including all extant placentals and their most recent common ancestor. A more inclusive taxon, Eutheria, is here retained to include all extinct mammals that share a more recent common ancestor than they do with Metatheria (including Marsupialia; Rougier et al., 1998).

The authors recount what we currently know of the initial radiation and ordinal relationships of placental mammals, primarily focusing on the anatomical evidence. The most recent volume in English dealing with an overview of the placental radiation was published more than 10 years ago (Szalay et al., 1993). Much has transpired since that time, most notably, a greater exchange between molecular and the more traditional anatomically based systematics. In this volume, 11 chapters examine all major clades or orders of extant placentals. Two chapters provide a wider and deeper perspective of the molecular and morphological evidence for placental origins and diversification. In addition, a chapter recounts George Gaylord Simpson's seminal contributions to the study of placentals.

Rather than charging the contributing authors with providing new data or new phylogenetic analyses, we asked them to summarize objectively the current state of knowledge and views about the origin and relationships of placental clades, presenting consensual views when possible, and recognizing significant minority viewpoints when not. Except for the overview chapter on molecular systematics, the authors were asked to focus on the morphological evidence and to note major points of agreement or discrepancy with molecular data. Although we are certainly not unbiased ourselves, we feel that the authors have admirably fulfilled our request.

The inception of this volume dates from the 2001 annual meeting of the Society of Vertebrate Paleontology in Bozeman, Montana, when the editors found they shared a mutual interest in hosting a symposium on the subject of placental evolution. David's interests are in the timing, biogeography, and relationships of stem placentals, whereas Ken's are more related to the question of the origin of extant placental orders. The dovetailing of our interests helped to bring this symposium to fruition at the annual meeting of the Society of Vertebrate Paleontology in Norman, Oklahoma, in 2002. This also coincided with the centenary of the birth of one of the greatest twentieth-century paleontologists, George Gaylord Simpson, who devoted much of his career to elucidating the early evolution and relationships of placental mammals. Thus, it was an easy choice to dedicate the symposium and its companion volume to this important scientific figure.

The chapters of this book, although generally based on the symposium, are more current, more comprehensive, and more detailed. Manuscripts were received in the spring or summer of 2003 and each was externally reviewed by at least two individuals. The authors then revised and returned final manuscripts in the autumn of 2003.

In chapter 2, Simpson biographer Léo Laporte recounts the major themes of G. G. Simpson's career, including the introduction of quantitative techniques to vertebrate paleontology and his influential use of paleontology in contributing to the Modern Synthesis of evolution. Laporte points out in a quote from Philip Gingerich that Simpson wrote about twice as many papers on systematics (mostly on mammals) than on broader evolutionary themes, although he is more widely known for the latter. It is practically impossible for mammalian paleobiologists to investigate any group of early mammals without referring to Simpson's work.

The next two chapters provide, respectively, anatomical and molecular overviews of what we know concerning earlier segments of eutherian evolution. In the past five to 10 years, there has been an explosion of new Cretaceous fossils and molecular studies dealing with the eutherian radiation. Chapter 3, by John Wible, Guillermo Rougier, and Michael Novacek, begins by providing a morphological characterization of eutherians based on dental, cranial, and postcranial anatomy. This discussion includes not only placentals, but more basal eutherians, as well as more distant outgroups. The authors next examine the interrelationships of Cretaceous eutherians and possible relationships to extant placental clades. The authors find no consensus among paleontologists on whether such taxa as the Cretaceous "zhelestids" and the zalambdalestids are closely related to extant superordinal placental clades. Issues surrounding the timing of the origin and initial radiation of placental mammals are also examined: here there is agreement that the origin and diversification of extant placental orders did not occur until after the K/T boundary.

In chapter 4, the second contribution on higher relationships, Mark Springer, William Murphy, Eduardo Eizirik, and

Stephen O'Brien examine the molecular evidence. They find that the current interpretation of the timing of the origin of extant placental orders, as deduced from molecular data, now shows more concordance with paleontological data than it has in the past. Of the 18 orders these authors recognize, eight appear after the K/T boundary, whereas 10 predate it by as much as 18 million years. Although this may at first seem a large difference between anatomy and molecules, the authors note that earlier molecular studies placed rodent origins well over 100 million years ago. Nonetheless, Springer et al. find that the molecular evidence supports a Cretaceous super- or interordinal radiation of placental mammals, which Wible and his coauthors (chapter 3) find problematic based on fossil evidence.

Unquestionably, the greatest contribution of these molecular studies has been the apparent untangling of superordinal relationships, which have eluded anatomical evidence for more than 100 years. As mentioned earlier, molecular studies have revealed four reasonably well-substantiated superordinal clades—Afrotheria, Xenarthra, Laurasiatheria, and Euarchontoglires. The most surprising of these is Afrotheria, which clusters previously disparate groups of largely African-centered taxa. Less surprising is Paenungulata (hyraxes, elephants, and sirenians), which both molecules and anatomy recover, albeit in differing combinations of included orders. The biogeography of these four clades is also discussed by Springer and his coauthors, with molecules usually supporting a Gondwanan center for placental origins and fossils supporting a Laurasian origin.

The next 11 chapters approximately follow the superordinal clades as recognized by molecular studies. The chapters deal with clusters of higher taxa conventionally considered "orders" of placental mammals. This hierarchical level is one of the more interesting in the history of the study of mammalian evolution (see Table 1.1). Until the advent of widespread molecular studies, all major taxa seemed to have a taxonomic rank that served as a barrier, beyond which one could not clearly argue for monophyly. Because such ranks are not biologically meaningful from one major taxon to another, it is not surprising that the rank of this barrier varies from taxon to taxon. For mammals—in particular, placental mammals—the barrier falls at the rank of the order.

Chapter 5, by Robert Asher, is the first of the chapters dealing with either superordinal- or ordinal-level groupings. Asher addresses the issue of what, if anything, are insectivores? More than any other group of placentals, insectivores, in any taxonomic guise, have had a checkered history. With the dissolution and exclusion of Menotyphla (tree and elephant shrews) from Insectivora more than 50 years ago, the latter term and Lipotyphla (hedgehogs, shrews, moles, *Solenodon*, *Nesophontes*, golden moles, and tenrecs) have often been used interchangeably. Asher explores a number of characters and character complexes that have been considered important in insectivoran systematics. Even when golden moles and tenrecs are removed, as argued by molecular data, the remaining Lipotyphla defies easy characterization, especially when sometimes incomplete fossil taxa are included in analyses. With the current allocation of insectivoran taxa to at least two (if not more) major clades of Placentalia, Asher notes that the older and now often rejected idea of insectivorans as basal placentals may in some form be correct.

In chapter 6, Patricia Holroyd and Jason Mussell tackle two of the least diverse but most enigmatic placental orders, Macroscelidea and Tubulidentata. Elephant shrews were first placed within or aligned with various insectivores. Their recognition as an ordinal-level clade did not occur until the 1950s. Holroyd and Mussell discuss four major hypotheses of macroscelidean relationships. For tubulidentates, the authors recognize three prevailing hypotheses of relationship. They could find no clear consensus on the origins of either of these two orders, but some of the hypotheses are more consistent with one another. First, a condylarthan/tethythere origin based on anatomical data is more consistent with an afrothere clade, which is recognized in molecular studies. Second, a close relationship of macroscelideans and tubulidentates with other, largely African, lineages fits the biogeographic picture for these taxa. As these authors note, however, there remain many problems with these hypotheses.

In chapter 7, Emmanuel Gheerbrant, Daryl Domning, and Pascal Tassy examine the superordinal placental clade Paenungulata—named in 1945 by George Gaylord Simpson. In addition to the extant Proboscidea, Hyracoidea, and Sirenia, various extinct taxa have been included in the clade. Molecular analyses strongly support this clade, although relationships within Paenungulata based on molecules have varied. Although Paenungulata is often supported on the basis of anatomical studies, the authors note the competing issue of possible hyracoid-perissodactyl relationships. The poor resolution may be the result of the lack of appropriate fossil African taxa of afrotheres. Recent fossil discoveries support an African origin of proboscideans and hyracoids by at least the late Paleocene, whereas the earliest known sirenian is an early Eocene terrestrial quadruped from the Western Hemisphere.

Kenneth Rose, Robert Emry, Timothy Gaudin, and Gerhard Storch in chapter 8 examine two orders that have often been linked, xenarthrans and pholidotans. They deem the evidence for a monophyletic Xenarthra, including sloths, armadillos, anteaters, and such extinct relatives as glyptodonts, to be compelling. As the name implies, all xenarthrans are characterized (except in quite derived taxa, such as glyptodonts) by having accessory articulations in parts of their vertebral column. The authors find little compelling morphological or molecular evidence for the more inclusive Edentata, which has included the extant orders Xenarthra and Pholidota (pangolins), often with a variety of extinct taxa as well. They find rather meager anatomical evidence linking Pholidota with Carnivora, a relationship more strongly supported by molecules. The best-known but most enigmatic taxa involved in the question of Edentata are the extinct palaeanodonts. Most of the anatomical evidence supports a palaeanodont and pangolin clade, but some is sufficiently equivocal to permit a possible xenarthran tie.

Chapter 9, by Mary Silcox, Jonathan Bloch, Eric Sargis, and Douglas Boyer, examines the superordinal clade Archonta, or Euarchonta, as they prefer. Archonta was named by Gregory (1910) for the orders we now recognize as Scandentia, Macroscelidea, Dermoptera, Chiroptera, and Primates. Acceptance of such a clade (in various guises) did not become widespread until the 1970s. The authors note that within Archonta, the clade Volitantia (for Dermoptera and Chiroptera) was strongly supported by anatomical data. Although their phylogenetic analysis supported Volitantia within Archonta, removal of bats does not alter the remaining topology, including a Scandentia and Dermoptera clade. Most recently, molecular studies remove Chiroptera from Archonta, but strongly retain a clade including Dermoptera, Scandentia, and Primates. It is this revised clade (Euarchonta) that concerns much of the chapter. Although the fossil record of scandentians in the early Tertiary remains poor, and that for dermopterans is not much better, the primate fossil record is quite good. The fossil record provides evidence bearing on the timing and place of origin of archontans. For both Dermoptera *sensu lato* and Primates *sensu lato*, the earliest representatives are known from the early Paleocene of North America, with primates, at least, in Asia and Africa by the late Paleocene. A discrepancy still remains between the timing of the origin of archontans based on fossils, which place it near the K/T boundary, versus that based on molecules, which place it about 85 million years ago.

In chapter 10, Jin Meng and André Wyss deal with the superordinal clade Glires, which includes the two extant orders Rodentia and Lagomorpha. As these authors point out, the question of whether Rodentia and Lagomorpha form a clade has been long enduring. Although this debate was, until recently, based only on anatomical studies, even with the advent of molecular studies, the question of gliran monophyly remained equivocal. This question now appears to be near resolution. Members of Glires share a number of specializations in the anterior dentition; notably, reduction to one pair of upper and lower incisors in rodents and two pairs of upper and one pair of lower incisors in lagomorphs. These incisors are evergrowing and have enamel restricted more or less to the anterior surface. Accompanying these modifications is the development of a large diastema between the incisors and cheek teeth. Although some other mammals show similar changes, modifications of the cheek teeth in basal rodents and lagomorphs further argue for their forming a clade. In addition, molecular studies now strongly support a Glires clade, with this clade being sister to Archonta (Euarchontoglires). There is now considerable confidence that the stem taxa of both rodents and lagomorphs can be traced to the early Paleocene, thus arguing that Glires dates back at least to the K/T boundary. Even the most recent molecular dates, however, still place this split at slightly more than 80 million years ago. Meng and Wyss do

not support recent paleontological studies arguing that stem glirans are known from some 85 million years ago.

Nancy Simmons reviews the most recent ideas on the evolutionary history of Chiroptera in chapter 11. Bats appear in the early Eocene fully volant and capable of echolocation. They clearly spread rapidly after their origin, as they are known from the early Eocene of North America, Europe, Africa, and Australia, and in the last case, are the earliest definite placentals known from that continent. Although bat monophyly has been questioned in the past, evidence from numerous organ systems, as well as molecular studies, now make it one of the most strongly supported ordinal clades of placentals. The same cannot be said for within-Chiroptera relationships. The idea of a megachiropteran clade and a microchiropteran clade has been the standard, based largely on anatomical evidence from fossil and recent forms. Microchiropteran monophyly has been challenged, however, mostly from molecular studies, which link some microchiropteran families with megachiropterans. This suggests that echolocation evolved in basal bats only to be lost in megachiropterans. One of the most intriguing results discussed by Simmons is that the diversification of extant families of bats occurred mostly in the Eocene, certainly one of the earliest such radiations among placental clades.

In chapter 12, the evolutionary history of Carnivora is reviewed by John Flynn and Gina Wesley-Hunt. Until relatively recently, any eutherian exhibiting a carnassial pair formed by the last upper premolar and first lower molar was considered to belong to Carnivora. Later analyses suggest that extant or crown-group Carnivora forms a clade to the exclusion of the more basal stem taxa, Viverravidae and some Miacidae, which also have this carnassial pair. The more inclusive Carnivoramorpha, which includes all of these taxa, is first seen in the early Paleocene of North America. Within crown-group Carnivora, molecular or combined data sets find strong support for all major clades except Viverridae and Mustelidae. Flynn and Wesley-Hunt also discuss the possible relationship of various creodonts to carnivoramorphans. Although the authors support the general view that creodonts may be sister to Carnivoramorpha, they find little evidence that creodonts form a monophyletic clade. They note that possible relations of Carnivora to other placentals based on morphology remain sketchy. A link between Carnivora and Pholidota has weak morphological but stronger molecular support.

In chapter 13, on Perissodactyla, Jeremy Hooker notes that there is unanimous agreement regarding the three extant clades (rhinos + tapirs and horses). There is, however, no overall consensus on the higher-level relationships among extant and extinct perissodactyls when the extinct chalicotherioids and brontotherioids are included. Hooker finds that some anatomical studies still conclude that there is a close relationship between perissodactyls and hyracoids, whereas molecular studies place these taxa far apart, the former in Laurasiatheria and the latter in Afrotheria. The question of the origins of Perissodactyla is, as for other placental orders, not confidently resolved, although phenacodontid condylarths are usually implicated. When more recently recognized Asian phenacodonts and perissodactyls are included in the phylogenetic analysis, a picture emerges in which brontotheres, not horses or tapiroids, are most basal in the order.

Artiodactyla has always seemed to be one of the most clearly delimited placental orders, characterized by its double-trochleated astragalus, which was long thought to be unique to the order. Similarly, within Artiodactyla, the tripartite groups of suiforms, tylopods, and ruminants appeared to be relatively stable. If recent molecular results continue to be supported, however, the monophyly of Artiodactyla can no longer be maintained, unless whales are included, the oldest of which are now known to have had a similar double-trochleated astragalus. Nor would suiforms be monophyletic, and the interrelationships of the three artiodactyl clades would change radically. In chapter 14, Jessica Theodor, Kenneth Rose, and Jörg Erfurt examine the traditional, anatomically based concept of Artiodactyla and also explore the ramifications of the changes argued by the molecules. They note that the issue of discerning the time and place of origin of artiodactyls remains ambiguous, as the oldest representatives first appear in North America, Europe, and southern Asia almost simultaneously in the earliest Eocene, without a clear ancestor or sister taxon. An archaic ungulate (condylarth) is implicated, but the possible candidates include arctocyonids, hyopsodontids, mioclaenids, or mesonychians. Moreover, some anatomically based analyses argue that whales are the sister taxon to artiodactyls rather than belonging within Artiodactyla. These studies suggest that the most likely ancestor for both artiodactyls and whales is an arctocyonid, but the evidence for this is not very strong. Artiodactyls and whales first appear in the fossil record within a few million years of each other—55 and 53.5 million years ago, respectively. Such similar dates seem concordant; however, if hippopotamids, which are first known from 15–16 million years ago, are the sister taxon to whales (as indicated by molecular data), then there is a gap in the fossil record of more than 37 million years between the first whales and hippopotamids.

No less then Charles Darwin in *Origin of Species* (1859) commented on the possible origin of cetaceans, speculating that they could have arisen from an aquatic bearlike creature snapping at insects in the water. Ridicule of this off-hand remark led to its exclusion in all later editions of this landmark volume. As Philip Gingerich recounts in chapter 15, the enigmatic origin of cetaceans has prompted considerable speculation. He reviews how the very rapid accumulation of data from both anatomical and molecular studies has resolved this enigma. Cetaceans are now clearly recognized at least as the sister taxon to Artiodactyla, if not sister to hippopotamids. Gingerich expands on the discussion begun in chapter 14, which suggests that if the anthracotheres are the closest extinct group to hippopotamids, then the former

are the possible sister taxon to cetaceans. The author concludes with a discussion of the environmental context of the origin and diversification of cetaceans, arguing that the origin not only of Cetacea but of many other orders of mammals that appear near the Paleocene-Eocene boundary is correlated with a thermal maximum. Cetaceans appeared along the shores of the warm Tethys Sea at about this time. These early forms, the archaeocetes, probably did not, however, survive the cooling event at the Eocene-Oligocene boundary. This is the earliest likely time for the origin of the two major extant cetacean clades, Odontoceti and Mysticeti.

As this collection of papers demonstrates, there is a consensus that nearly all of the 18 conventional placental orders are compellingly monophyletic. The only exceptions are Lipotyphla and Artiodactyla as commonly conceived, for which both morphological and molecular evidence now challenge the traditional taxonomic arrangements. Despite these findings, it must be admitted that we generally lack fossil evidence of the precise phylogenetic origins of the orders. In most cases, we can at best point to a family of archaic eutherians as a likely source (e.g., cimolestids for Carnivora, or more precisely, Carnivoramorpha; phenacodontids for Perissodactyla; arctocyonids for Artiodactyla). With regard to higher level relationships, the morphological evidence supports the monophyly of Glires, Archonta (or Euarchonta), Artiodactyla (or Cetartiodactyla, including whales), Tethytheria, and, to a lesser extent, Paenungulata and a Carnivora + Pholidota clade. The oldest fossils that can be definitively allocated to any of the 18 orders postdate the K/T boundary and in many cases (particularly crown groups) are no older than Eocene. In almost all cases, the authors conclude that the orders most likely did not originate until after the K/T boundary. Thus although probable divergence dates based on molecular evidence are becoming more compatible with those based on morphological evidence (e.g., for perissodactyls, artiodactyls, bats, crown-group Carnivora, and paenungulates), there remains significant discrepancies between molecular and morphological estimates for some other groups (e.g., primates, rodents, Glires).

REFERENCES

Archibald, J. D. 2001. Eutheria; pp. 1–4 *in* Encyclopedia of Life Sciences. www.els.net. Macmillan Publishers, Nature Publishing Group, London.

———. 2003. Timing and biogeography of the eutherian radiation: Fossils and molecules compared. Molecular Phylogeny and Evolution 28: 350–359.

Butler, P. M. 1956. The skull of *Ictops* and the classification of the Insectivora. Proceedings of the Zoological Society, London 126: 453–481.

———. 1972. The problem of insectivore classification; pp. 253–265 *in* K. A. Joysey and T. S. Kemp (eds.), Studies in Vertebrate Evolution. Winchester Press, New York.

Cuvier, G. 1817. La règne animal distribué d'après son organisation, pour servir de base l'histoire naturelle des animaux et d'introduction l'anatomie comparée. Les mammifères. Deterville, Paris.

Darwin, C. R. 1859. On the Origin of Species by Means of Natural Selection, or the Preservation of Favored Races in the Struggle for Life. John Murray, London.

Flower, W. H. 1883. On the arrangement of the orders and families. Proceedings of the Zoological Society, London 1883: 178–186.

Gould, S. J. 1989. Wonderful Life: The Burgess Shale and the Nature of History. W. W. Norton, New York.

Gregory, W. K. 1910. The orders of mammals. Bulletin of the American Museum of Natural History 27: 1–524.

Kirchner, J. W., and A. Weil. 2001. Delayed biological recovery from extinctions throughout the fossil record. Nature 404: 177–179.

Linnaeus, C. 1758. Systema Naturae per Regna tria Naturae, secundum Classes, Ordines, Genera, Species, cum Characteribus, Differentiis, Synonymis, Locis. Tenth edition. Impensis Direct, Salvius, Stockholm.

———. 1766. Systema Naturae per Regna tria Naturae, secundum Classes, Ordines, Genera, Species, cum Characteribus, Differentiis, Synonymis, Locis. Twelfth edition. Impensis Direct, Salvius, Stockholm.

McDowell, S. B., Jr. 1958. The Greater Antillean insectivores. Bulletin of the American Museum of Natural History 115: 113–214.

McKenna, M. C. 1975. Toward a phylogenetic classification of Mammalia; pp. 21–46 in W. P. Luckett and F. S. Szaly (eds.), Phylogeny of the Primates: A Multidisciplinary Approach. Plenum, New York.

McKenna, M. C., and S. K. Bell. 1997. Classification of Mammals above the Species Level. Columbia University Press, New York.

Murphy, W. J., E. Eizirik, S. J. O'Brien, O. Madsen, M. Scally, C. J. Douady, E. Teeling, O. A. Ryder, M. J. Stanhope, W. W. de Jong, and M. S. Springer. 2001. Resolution of the early placental mammal radiation using Bayesian phylogenetics. Science 294: 2348–2351.

Rougier, G. W., J. R. Wible, and M. J. Novacek. 1998. Implications of *Deltatheridium* specimens for early marsupial history. Nature 396: 459–463.

Simpson, G. G. 1945. The principles of classification and a classification of mammals. Bulletin of the American Museum of Natural History 85: 1–350.

Szalay, F. S., M. J. Novacek, and M. C. McKenna (eds.). 1993. Mammal Phylogeny: Placentals. Springer-Verlag, New York.

Vaughn, T. A., J. M. Ryan, and N. J. Czaplewski. 2000. Mammalogy. Fourth edition. Saunders College Publishing, Fort Worth.

Wilson D. E., and D. M. Reeder (eds.). 1993. Mammal Species of the World: A Taxonomic and Geographic Reference. Second edition. Smithsonian Institution Press, Washington D.C.

LÉO F. LAPORTE

LOOKING BACK AT THE RECORD: GEORGE GAYLORD SIMPSON AND PALEOMAMMALOGY

I N 1976, AT AGE 74, George Gaylord Simpson observed that he had been "unable to work steadily on one subject or even in a single field," yet he was clearly pleased by all that he had accomplished (Simpson, 1976: 1). A quarter of a century later, one can agree with that personal judgment for two reasons.

First, Simpson catalyzed the belated reconciliation of paleontology with contemporary biology. Paleontology at the beginning of the twentieth century had been something of an orphan in the biological sciences. Because of its emphasis on the older tradition of comparative morphology and the use of fossils by geologists in determining the relative ages of rocks, paleontology neither had a home in the new institutional centers established for biology, nor was it considered a formal discipline in biology. Worse yet, as the historian Ronald Rainger (1988: 220) has noted, paleontology did not demonstrate "much likelihood of becoming a foundation for serious programs of research in biology." Simpson himself remarked that "[t]he geneticists tended to consider that paleontology was incapable of rising above pure description and they did not even take the trouble to study descriptive paleontology for its bearing on genetics. It was easier to conclude that it had no such bearing. The paleontologists were, as a rule, quite willing to accept this stultifying conclusion, which also spared them the trouble of learning genetics" (1946: 53). Simpson's contributions, especially his book, *Tempo and Mode in Evolution,* published in 1944, brought paleontology into the mainstream of contemporary modern biology.

Secondly, Simpson's contributions had great heuristic value in serving as guides to subsequent research for the discipline of paleontology. Using the rubrics from

Simpson's own retrospective assessment, we can briefly summarize his contributions as follows in terms of methodology, evolution, biogeography, and systematics.

METHODOLOGY

In the late 1930s, Theodosius Dobzhansky's *Genetics and the Origin of Species* (1937) introduced Simpson to the work of population geneticists, and subsequently to the earlier writings of R. A. Fisher (esp. 1930), J.B.S. Haldane (esp. 1932), and Sewall Wright (esp. 1931, 1932), which motivated him to move from the typological to the populational approach in defining fossil species. And at the suggestion of his wife, Anne Roe, a clinical psychologist, Simpson began using formal small sample statistics in his paleontological research (Laporte, 2000: 100). This collaboration led to their book *Quantitative Zoology*, published in 1939, which was the first attempt to apply statistical methods to characterize faunas, whether extant or extinct, methods that were comprehensible to those not otherwise mathematically inclined. They claimed that statistics was the "best means of describing and interpreting what animals are and do" (1939: vii).

Early exemplars of Simpson's use of statistical reasoning occur in short and long versions, both published in 1937. In the one, initially presented orally at Harvard's tercentenary celebration in 1936 and published the following year as "Patterns of Phyletic Evolution," Simpson hypothesized a set of data showing a morphological character of a fossil increasing in dimensions within a vertical stratigraphic sequence and suggested the mutually exclusive interpretations that might result. The data might record the decline over stratigraphic time of one smaller species and the rise of descendant larger species; or simply a single species in which the character in question was increasing through time; or perhaps one larger species gradually diverging from the smaller ancestral species. Simpson (1937a: figs. 3–5, 7) then indicated graphically how one could determine the correct interpretation by simple statistical analysis of the data, stratum-by-stratum, through the rock sequence. Toward the end of his career, Simpson (1980: 112) credited this paper as marking his "abandonment of the typological thinking of my college teachers and started aiming me toward statistical biometry and the deeper investigation of evolutionary theory and taxonomic stance."

The other, longer 1937 publication was his monograph on the Paleocene mammals of the Fort Union Group of Montana, in which Simpson made more extensive use of statistical analysis for species discrimination. He argued that selecting just one character to determine a species might contradict species delineation using a different, single character. Instead, Simpson (1937b: 73–76) calculated a variety of statistics on 19 primary dental traits for 22 specimens of multituberculates:

After consideration of all these primary and secondary data, it was clear that of the eight groups finally achieved and checked

each represents a variable morphological unit, that the variation in each is not greater than commonly occurs in natural species, but that no two [groups] can be combined without producing a unit statistically heterogeneous and morphologically much more variable than a species. The biological conclusion is thus that eight species are present.

Thereafter, statistical methodology gave Simpson a quantitative basis for his treatment of organisms as "variable morphological units," rather than as idealized types. He thus placed himself squarely within the conceptual domain of the population geneticists, not incidentally bringing along with him his paleontological colleagues.

EVOLUTION

More broadly, Simpson's reading of Dobzhansky and the other geneticists stimulated his writing of *Tempo and Mode in Evolution,* which not only refuted those internal evolutionary mechanisms (such as aristogenesis, orthogenesis, inertia, and momentum) that were favored by most of his peers and predecessors, but more significantly, embraced modern population genetics and ecology to interpret rates and patterns of micro- and macroevolution (Laporte, 2000: 112 ff.). In particular, Simpson (1944: 180) wanted to show "how necessary it is to consider evolution in terms of the interaction [between] organisms and environment." His visual construct of the "adaptive grid," with its array of ecological zones of varying breadth, succinctly summarized his conceptual framework for explaining differing rates and patterns of evolution (1944: figs. 26, 28). To emphasize the significance of such variable rates and their resultant patterns, he coined the terms "bradytely," "horotely," and "tachytely" for unusually slow, broadly average, and unusually fast evolution, respectively.

Because the impact of *Tempo and Mode* was so great, at least among biologists, it is easy to ignore the steps that led up to it. In fact, it was in the Fort Union monograph, mentioned above, that the roots of several of the major concepts and approaches that Simpson articulated so cogently in *Tempo and Mode* can be found, including species as populations, organism-environment interactions, bias of the fossil record, and biometrical analysis. Two years after the completion of that monograph in the spring of 1936, Simpson began writing *Tempo and Mode in Evolution.*

Retrospectively, Simpson himself (1975: 3) realized the value of the Fort Union study, because, although he referred to it as "a more or less routine monograph," he acknowledged that "it did introduce some new ways of going at things." And what were those new ways? The answer is well expressed by the historian and philosopher of science Dudley Shapere (1980: 389):

And in the case of the synthetic theory of evolution, perhaps what we ought to be asking is not whether paleontology (or any other discipline) was brought into some tight deductive unification with Darwinian-Mendelian theory but rather to what extent the removal of barriers to seeing paleontological data in Darwinian-Mendelian terms, and the consequent application of the latter terms to those data, reoriented the kinds of questions asked, the kinds of research engaged in, the expectations about the sorts of things one should expect to find, and so forth, in paleontology and other fields.

BIOGEOGRAPHY

Simpson's work on the classification of mammals, and more particularly, his research on South American mammals, necessarily led him to consider their paleobiogeography. Trained as much as a geologist as a paleontologist, Simpson was naturally as interested in the temporal dimension of mammalian distribution as the spatial. Avoiding notions of all-or-nothing factors important in such distribution, Simpson explored the ways terrestrial vertebrates could disperse by what he termed "corridor," "filter," and "sweepstake" routes over geologic time, thereby explicating historical, as well as ecological, patterns of geographic distribution (e.g., Simpson, 1952, 1953). Simpson's treatment of historical biogeographic issues focused primarily on his own area of expertise, Cenozoic mammals, and he believed that mobile organisms dispersing across stable continents was fully adequate to explain their paleobiogeography. Simpson therefore regarded both continental drift theory and transoceanic land bridges to be neither sufficient nor necessary to account for his Cenozoic mammalian data (Laporte, 2000: 195 ff.). Not until the early 1970s did Simpson (1976: 8–11) accept plate tectonics, when he conceded that the theory could "enrich both in knowledge and in the several principles of biogeography." But, clearly, he did not believe that plate tectonics in any way invalidated the principles of historical biogeography he had previously set forth.

Nevertheless, what further insights might we gain about why Simpson held to his stabilist views as long as he did? First, and paradoxically, Simpson perhaps put too much confidence in the fossil record in determining past continental configurations. Because fossil mammals did not conclusively demonstrate drift, Simpson was therefore convinced that continental positions must have always been fixed. Ironically, even Simpson was seduced by his own incisive and inexorable logic. Second, it was not part of Simpson's research style to reconsider a problem once he had seriously thought about it and come up with a satisfying—to him, at least—answer. Simpson would attack a problem by careful evaluation of the data, by deliberate and logical formulation of his argument, and by construction of whatever theory was necessary to explain his conclusions. Once having done that—and the process might be extended over a number of years, as in the case of historical biogeography—Simpson moved on to new problems and new areas of research interest. He did not tinker, fuss over, or agonize about issues on which he had made a well-considered scientific judgment. Therefore, after the 1950s, Simpson did not again address the issue of drift, at least in the broad terms of his earlier work. Of course, he continued to make biogeographic statements as appropriate when writing about evolution, describing a fauna, or the like. But he did not reopen the whole question of his views of historical biogeography in the light of the newly established mobilist theory. Undoubtedly, Simpson viewed his theory of historical biogeography as essentially unaffected by continental drift.

SYSTEMATICS

As Philip Gingerich (1986: 5) has pointed out, "[although] Simpson is most widely known for his [publications] on evolution . . . , roughly twice the number of titles and pages . . . were devoted to morphology and systematics."

Simpson's Yale dissertation on American Mesozoic mammals and his postdoctoral research done at the British Museum on their European counterparts described these very small, hitherto neglected fossils and clarified their relationships to their hypothetical Cenozoic descendants. In his introduction to the British Museum publication, Simpson (1928: 1) noted:

[T]he "mammals were already well advanced at the beginning of the Eocene: it has been estimated that two-thirds of their total development from the reptilian to the recent mammalian type had already taken place by the end of the Mesozoic. This lost two-thirds, this mammalian prehistory, is much more basic in character . . . [and] must contain the answers to the most fundamental problems of mammalian classification and phylogeny.

Simpson further observed that the Mesozoic mammals had been long neglected because they are the smallest, the rarest, and the most fragile of fossils. Given that the last general review of the group had been made forty years before, in 1888 by H. F. Osborn, and before standard use of binocular microscopes, renewed study of these important fossils was most timely.

Simpson's study of Mesozoic mammals, however, was by no means preordained. Poking around in the basement of Yale's Peabody Museum, Simpson became aware of the Marsh Collection, which included such fossils. He then approached Richard Swann Lull, his graduate advisor, and suggested that he study them for his dissertation. Lull refused, saying that the fossils were "most important and most valuable, and you are the rawest student in this institution, so let's just skip that idea" (Simpson interview with the author, 18–19 August 1981, Tucson, Arizona). Nevertheless, displaying his inveterate persistence and determination, Simpson began working on the fossils, mostly from the Jurassic Morrison and Cretaceous Lance formations. As he recalled years later, "I moved into two big rooms in the tower of the new Peabody Museum. In the one next to the top I had an office, the top one a photographic studio where I did stereo photographs—the first time ever done, I think, with fossils —of the Mesozoic mammals, made some notes, and an outline. I showed all of this to Lull and he said, 'you're bit-ing off more than you can chew, but O.K., go ahead'" (Simp-son interview with the author, 18–19 August 1981, Tucson, Arizona). Figure 58 of the 1929 monograph presented what Simpson (1929: 143) called his "suggested relationships of the major groups of mammals," which indicates a poly-phyletic origin, an interpretation he favored for a number of years.

After his appointment to the American Museum as assis-tant curator of vertebrate paleontology, his duties included "an attempt to arrange a large collection [of mammals liv-ing and fossil] for cataloguing, storage, exhibition, and teach-ing, [in order to] present a workable synthesis. There is a real need for a general classification not wholly satisfied by those available, and no apology is necessary for trying to fill this need in some measure." Simpson (1931: 259) acknowledged critical examination of his classification by his American Mu-seum colleagues Henry Fairfield Osborn, Walter Granger, Barnum Brown, William King Gregory, Childs Frick, Wil-liam Diller Matthew, and Robert Hatt.

Simpson recorded 129 extinct and 113 living families of mammals, and compared his classification with an 1872 "arrangement of the families of mammals" by Theodore Gill. Simpson noted that his updated classification shows an almost threefold increase in fossil families, but less than 8% increase for living ones—a measure, according to Simpson (1931: 260–261), of "the progress of the [previous] sixty years . . . due almost entirely to paleontological discovery." As a result of such discovery, Simpson remarked that "pale-ontology has become the more important factor in dealing with questions related to classification . . . [and] the prob-lems of phylogeny and taxonomy become more and more the peculiar field of the paleomammalogist" (p. 261). To this non-taxonomist, one interesting innovation in Simpson's classification was his placing the families Hominidae and Pongidae into a new superfamily, the Hominoidea, distinct from the Old World monkeys. He stated that this arrange-ment follows "what seems a reasonable interpretation of the anatomical and paleontological facts actually known, however contrary to more theoretical or emotional inter-pretations" (p. 287).

The 1931 classification was, according to Simpson, "an outline" for the subsequent 1945 publication, *The Principles of Classification and a Classification of Mammals.* Simpson pointed out that his own knowledge, experience, and judg-ment had increased over the years since his initial appoint-ment as assistant curator in 1927, and also since the 1931 outline was published. He prefaced the work with the com-ment that "[d]oubtless waiting another 10 years would have resulted in further desirable changes, but as Dr. Johnson found, a whole life cannot be spent on one task and a whole life would not suffice. In the meantime a classification is needed" (1945: vii).

The 1945 monograph consists of three parts. Part 1 (33 pages) is an essay on the principles of taxonomy and in-cludes sections on the data of phylogeny, morphology and phylogeny, phylogeny and classification, monophyly and

polyphyly, vertical and horizontal classification, splitting and lumping, and nomenclature. In this last section, Simpson (1945: 30) discussed his ideas regarding types and hypo-digms, the latter a word he introduced several years before (in 1940):

The hypodigm of a genus, species, or other taxonomic group is a series of concrete specimens . . . the correct procedure of practi-cal nomenclature is to infer from a hypodigm the probable char-acter and limits of a natural population and then to decide the valid name of that inference by observing whether any name-bearers, types, occur in the hypodigm. . . . If the hypodigm does not include an established type, the group is new . . . and a new name is proposed and attached to a specimen in the hypodigm.

Of course, this distinction between type and hypodigm re-flected Simpson's acceptance of species as variable popula-tions of individuals and his rejection of species as types. Part I was a prelude to his 1961 book, *The Principles of Animal Taxonomy,* wherein he further codified his self-described "eclectic evolutionary taxonomy" (1976: 6).

In Part 2 (129 pages), Simpson presented the formal clas-sification of mammals, which went down to the generic level. Part 3 (110 pages) then offered comments on virtually all of the taxa he listed, giving reasons pro and con for a given taxonomic decision, including some of the relevant ambiguities. For example, in his discussion of the Homi-noidea, Simpson (1931: 259) agreed with W. K. Gregory in "the conviction that the gibbons, apes, and man are a unit, derived from common ancestry." Simpson (1931: 259) went on to say:

On the basis of usual diagnostic characters, such as the teeth, viewed with complete objectivity, this union seems warranted. I nevertheless reject . . . [including man] for two reasons: (a) men-tality is also a zoological character to be weighed in classification and evidently entitling man to some distinction, without leaning over backward to minimize our own importance, and (b) there is not the slightest chance that zoologists and teachers generally, however convinced of man's consanguinity with the apes, will agree on the didactic or practical use of one family embracing both.

Simpson (1945: 68) did place *Eoanthropus* ("Piltdown Man") among the Hominidae in Part 2, but he raised some indirect doubt about it in the commentary of Part III. "Almost none of [the] anthropological 'genera' has any zoological reason for being. All known hominids, recent and fossil, could well be placed in *Homo.* . . . [However], *Eoanthropus* (if the ape-like jaw belongs to it) may be given separate generic rank" (1945: 188).

One can judge the impact of Simpson's 1945 classification by the fact that in 1985, *Current Contents* called it a "citation classic, having been cited in over 565 publications since 1945," whereas over that same period *Tempo and Mode* had been cited 215 times, and *Major Features of Evolution*—the successor to *Tempo and Mode*—500 times. When informed of this, Simpson claimed that he was "somewhat baffled by the apparent evidence that this is my most cited publication. I have considered *Tempo and Mode in Evolution* and *Major Features of Evolution* were more often cited, sometimes adversely" (Current Contents, 1985: 12).

In his 1976 retrospective assessment, Simpson (1976: 8) volunteered the judgment that the 1945 classification was "decidedly out of date," and if he were to revise Simpson 1961, he would "retain essentially the same philosophy overall but would make extended use of the advanced numerical methods of the pheneticists and the advanced cladistic methods of the Hennigians."

Simpson (1940b: 200) himself reflected briefly on the inevitability of not always being right in his 1940 paper on the earliest primates:

Many errors are inevitably involved, but surely it is better to interpret the evidence actually available as best one can, accepting correction with good grace when it comes. . . . The [accretion] of interpretation and theory is as essential to the achievement of knowledge as is the accumulation of the raw materials from which they arise, and isolated observations of fact have little value until some progress has been made toward their interpretation.

This remark was no mere lip service to good scientific practice, for in his autobiography, Simpson (1978: 199) pointed out a "howling [taxonomic] mistake" he had made. In a 1967 publication on Tertiary primates of Africa, Simpson (1967: 50 ff.) named a new genus and species, *Propotto leakeyi,* and classified it as a lorisform primate. Soon after, Alan Walker wrote Simpson and told him, most cordially, that *Propotto* was a fruit bat and that he was preparing a manuscript to correct the error. In the ensuing correspondence Simpson complained that he had "slept little and worked not at all . . . a blunder for which I can least excuse myself. . . . I agree with you . . . that correction must be published promptly. . . . [my] embarrassment is in fact deserved and inescapable. . . . I appreciate your courtesy and effort to lighten the burden of my completely asinine blunder." Simpson also unequivocally refused Walker's generous offer to coauthor the correction "of work done entirely by you" (Walker to Simpson, July 3, 1968; February 4, 1969; February 25, 1969; and Simpson to Walker, undated circa February 5–6, 1969; February 7, 1969; May 13, 1969; I thank Professor Walker for bringing this correspondence to my attention).

Just as Darwin didn't get everything right in the nineteenth century, Simpson made his own share of mistakes in the twentieth, as present-day paleobiologists can no doubt attest. Nevertheless, Darwin blazed a fresh path for understanding the world of animate nature, and Simpson was able to illuminate a small portion of that world with similar originality, insight, and persistence.

REFERENCES

Current Contents. 1985. Volume 16, no. 17, p. 12. Institute for Scientific Information, Philadelphia.

Dobzhansky, T. 1937. Genetics and the Origin of Species. Columbia University Press, New York.

Fisher, R. A. 1930. The Genetical Theory of Natural Selection. Clarendon Press, London.

Gill, T. 1872. Arrangement of the families of mammals with analytical tables. Smithsonian Miscellaneous Collections 11: 1–98.

Gingerich, P. D. 1986. George Gaylord Simpson: Empirical theoretician. Contributions to Geology, University of Wyoming Special Paper 3, pp. 3–9.

Haldane, J.B.S. 1932. Causes of Evolution. Harper, New York.

Laporte, L. F. 2000. George Gaylord Simpson, Paleontologist and Evolutionist. Columbia University Press, New York.

———. 2003. G. G. Simpson web site http://people.ucsc.edu/~laporte/simpson/Index.html.

Rainger, R. 1988. Vertebrate paleontology as biology: Henry Fairfield Osborn and the American Museum of Natural History, pp. 219–256 *in* K. Benson, J. Maienschein, and R. Rainger (eds.), American Development of Biology. University of Pennsylvania Press, Philadelphia.

Shapere, D. 1980. The meaning of the evolutionary synthesis; pp. 388–398 *in* E. Mayr and W. Provine (eds.), The Evolutionary Synthesis. Harvard University Press, Cambridge.

Simpson, G. G. 1928. A Catalogue of the Mesozoic Mammalia in the Geological Department of the British Museum. British Museum [Natural History], London.

———. 1929. American Mesozoic Mammalia. Memoirs of the Peabody Museum of Natural History of Yale University, 3, part 1.

———. 1931. A new classification of mammals. American Museum of Natural History Bulletin 59: 259–293.

———. 1937a. Patterns of phyletic evolution. Geological Society of America Bulletin 48: 303–314.

———. 1937b. The Fort Union of the Crazy Mountain Field, Montana, and its mammalian faunas. United States National Museum Bulletin 169: 1–287.

———. 1940a. Types in modern taxonomy. American Journal of Science 238: 413–431.

———. 1940b. Studies on the earliest primates. American Museum of Natural History Bulletin 77: 185–212.

———. 1944. Tempo and Mode in Evolution. Columbia University Press, New York.

———. 1945. The principles of classification and a classification of mammals. American Museum of Natural History Bulletin 85: 1–350. (Completed in December, 1942, just before Simpson enlisted in the U.S. Army, but not published until three years later, owing to the necessary transcription from his original handwriting to typescript, as well as well as final checking and editorial preparation by his museum colleagues.)

———. 1946. Tempo and mode in evolution. New York Academy of Sciences Transactions, Series II, 8: 45–60.

———. 1952. Probabilities of dispersal in geologic time. American Museum of Natural History Bulletin 99: 163–176.

———. 1953. Evolution and Geography: An Essay on Historical Biogeography. Condon Lectures. Oregon State System of Higher Education, Eugene.

———. 1961. Principles of Animal Taxonomy. Columbia University Press, New York.

———. 1967. The Tertiary lorisiform primates of Africa. Bulletin of the Museum of Comparative Zoology, Harvard University 136(3): 39–62.

———. 1975. Transcription of Comments on His Bibliography. G. G. Simpson Papers, MS collection 31. American Philosophical Society, Philadelphia.

———. 1976. The compleat [sic] paleontologist? Annual Reviews of Earth and Planetary Science 4: 1–13.

———. 1978. Concession to the Improbable. Yale University Press, New Haven.

———. 1980. Why and How: Some Problems and Methods in Historical Biology. Pergamon, Oxford.

Simpson, G. G., and A. Roe. 1939. Quantitative Zoology. McGraw-Hill, New York.

Wright, S. 1931. Evolution in mendelian populations. Genetics 16: 97–159.

———. 1932. The roles of mutation, inbreeding, crossbreeding, and selection in evolution. Proceedings: 6th International Congress of Genetics 1: 356–366.

JOHN R. WIBLE,
GUILLERMO W. ROUGIER,
AND MICHAEL J. NOVACEK

3

ANATOMICAL EVIDENCE FOR SUPERORDINAL/ORDINAL EUTHERIAN TAXA IN THE CRETACEOUS

PLACENTAL MAMMALS REPRESENT THE VAST majority of the species of living mammals (4,354 of 4,629 reported in Wilson and Reeder, 1993). They occur on all continents except Antarctica and occupy a diverse array of niches, from the skies to the oceans to nearly all land masses in between (Nowak, 1991). Placentalia is a subgroup of Eutheria, the clade of therian mammals more closely related to placentals than to marsupials (Fig. 3.1; Rougier et al., 1998). Eutheria, in addition to crown-group Placentalia, include the extinct members of the placental stem lineage (Novacek et al., 1997). The crown-group Marsupialia, with 272 living species (Wilson and Reeder, 1993), and the extinct members of the marsupial stem lineage (Rougier et al., 1998) are subgroups of Metatheria, the sister group of Eutheria, the former being the clade of therian mammals more closely related to marsupials than to placentals.

Neontologists and paleontologists generally recognize 18 orders of living placentals (Fig. 3.1; Novacek, 1992b; Wilson and Reeder, 1993). The earliest members of these 18 orders postdate the extinction of the dinosaurs at the K/T boundary, with only a few possible (and controversial) exceptions (cf. Archibald et al., 2001; Meng et al., 2003a; O'Leary et al., 2004). Above the ordinal level, several superordinal clades (Fig. 3.1) are also recognized by neontologists and paleontologists (Novacek, 1992b, 1999; Shoshani and McKenna, 1998). Over the past several years, sequence-based studies have challenged not only the usefulness of fossils for phylogenetic reconstruction (see Smith, 1998, for a compilation of sources), but also the validity of most of these superordinal and a few ordinal groupings (Fig. 3.2; e.g., Eizirik et al., 2001;

Fig. 3.1. Relationships (thin lines) and geochronological ranges (horizontal bars) for the major clades of placental mammals and selected outgroups. The vertical line between the Mesozoic and Cenozoic represents the K/T boundary. Modified and redrawn from Novacek (1999: fig. 7). Redrawn by Ed Heck, American Museum of Natural History.

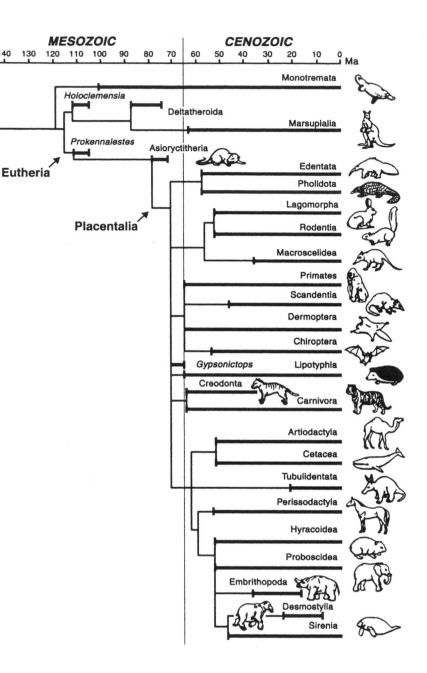

Madsen et al., 2001; Murphy et al., 2001a,b). Moreover, divergence dates for various cladogenetic events within Placentalia and Eutheria based on molecular data are generally much older than those based on the fossil record (compare Figs. 3.1 and 3.2; Springer et al., 2003).

Current paleontological data support a Cretaceous origin for Eutheria, with the earliest known representative, *Eomaia,* of Early Cretaceous age, from the middle Barremian of Liaoning Province, China, about 125 million years ago (Fig. 3.3; Ji et al., 2002). Three other genera of Early Cretaceous eutherians have been reported (but see below): *Murtoilestes* from the upper Barremian–middle Albian of eastern Russia (Averianov and Skutschas, 2001), and *Prokennalestes* (Kielan-Jaworowska and Dashzeveg, 1989) and *Montanalestes* (Cifelli, 1999) from the Albian-Aptian of Mongolia and North America, respectively. According to the

phylogenetic analysis by Luo et al. (2003), the earliest metatherian, *Sinodelphys,* is also from the middle Barremian of Liaoning. In contrast, recent molecular estimates for the Eutheria-Metatheria dichotomy range between extremes of 190 (Woodburne et al., 2003) and 130 (Janke et al., 2002) million years.

Over the past 25 years, our knowledge of eutherians during the Cretaceous has increased dramatically. In the 1979 edited volume on Mesozoic mammals, Kielan-Jaworowska et al. recognized 14 genera of Cretaceous eutherians (five Asian, eight North American, and one South American). In contrast, in their 1997 classification of mammals, McKenna and Bell recognized 32 genera of Cretaceous eutherians (20 Asian, nine North American, one South American, and two European), and since that publication, seven more Asian genera (Nessov, 1997; Novacek et al., 1997; Nessov et al.,

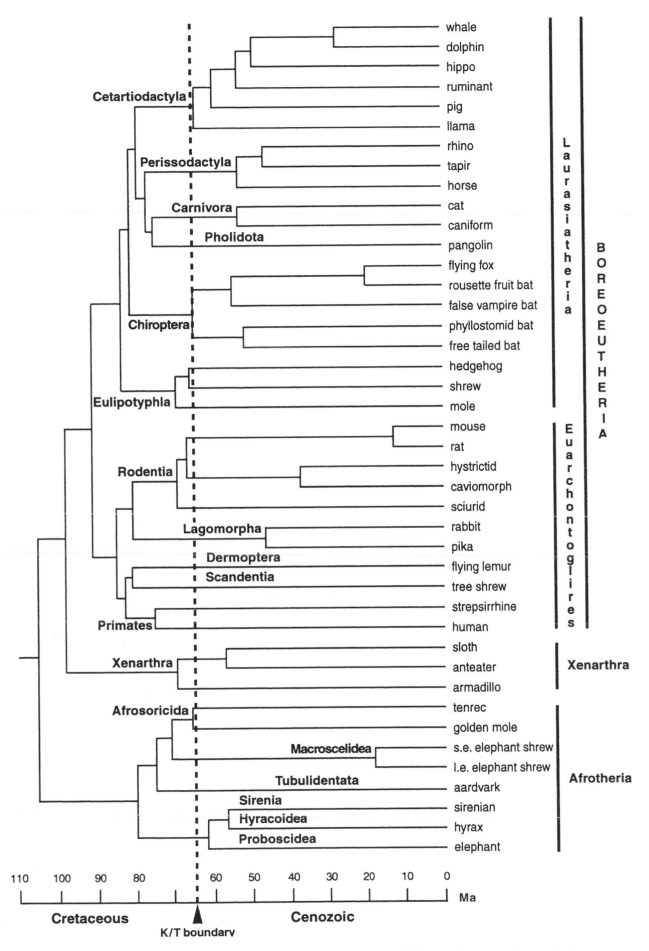

Fig. 3.2. Molecular time scale for the placental orders. Modified and redrawn from Springer et al. (2003: fig. 2). Redrawn by Gina Scanlon, Carnegie Museum of Natural History.

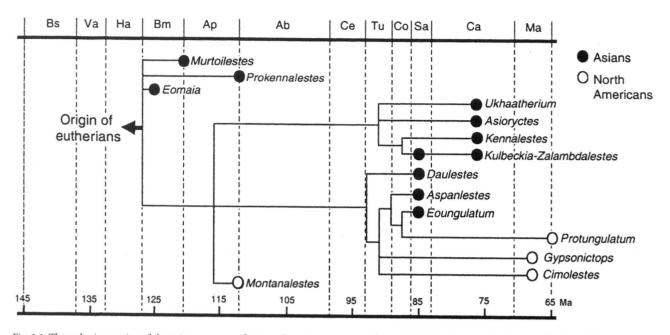

Fig. 3.3. The eutherian portion of the strict consensus of 50 equally parsimonious trees from the PAUP analyses in Ji et al. (2002: fig. 6b), which considered 268 skeletal and dental characters across these taxa plus 27 others. The zalambdalestid *Kulbeckia* was not included in the original PAUP analysis, but was added by hand to the figure by Ji et al. (2002). Abbreviations: Bs, Berriasian; Va, Valanginian; Ha, Hauterivian; Bm, Barremian; Ap, Aptian; Ab, Albian; Ce, Cenomanian; Tu, Turonian; Co, Coniacian; Sa, Santonian; Ca, Campanian; Ma, Maastrichtian. Redrawn from Ji et al. (2002: fig. 6b). Redrawn by Gina Scanlon, Carnegie Museum of Natural History.

1998; Averianov and Skutschas, 2001; Ji et al., 2002; Rana and Wilson, 2003) and one additional North American genus have been named (Cifelli, 1999). In addition to the increase in numbers of taxa and specimens, there has been an increase in our overall knowledge of anatomical systems in selected Cretaceous eutherians, with detailed descriptions of cranial (Wible et al., 2001, 2004) and postcranial (Horovitz, 2000, 2003) anatomy.

Our knowledge of relevant eutherian outgroups has also increased dramatically. Rougier et al. (1998) recently published the first comprehensive phylogenetic analysis of basal metatherians, and numerous studies (e.g., Rougier et al., 1996a; Luo et al., 2001, 2002) have addressed the wider relationships of therian mammals. Regarding metatherian anatomy, very complete skulls and skeletons of *Pucadelphys* and *Mayulestes* from the early Paleocene of Bolivia have been well documented (Marshall et al., 1995; Muizon, 1998), and new specimens of Late Cretaceous *Deltatheridium* from Mongolia have provided details of both cranial and postcranial anatomy (Rougier et al., 1998; Horovitz, 2000). The Early Cretaceous *Sinodelphys* is represented by a fairly complete skeleton, although few details of the skull are preserved (Luo et al., 2003). Very complete skulls and skeletons are described for the prototribosphenidan *Vincelestes* from the Early Cretaceous of Argentina (Rougier, 1993), and new specimens preserving fairly complete skeletons are known for a Late Jurassic dryolestoid (Krebs, 1991) and for Early Cretaceous "symmetrodonts" (Hu et al., 1997, 1998; Rougier et al., 2003).

These advances have not yet significantly improved our understanding of the phylogenetic relationships among Cre-

taceous eutherians or between Cretaceous and younger eutherians, which is needed to evaluate placental superordinal relationships. Constraining progress are incomplete data—most Cretaceous eutherians are known from isolated teeth and fragmentary jaws. In fact, only eight genera preserve upper and lower dentitions in association: *Zalambdalestes, Barunlestes, Asioryctes, Ukhaatherium, Kennalestes,* and *Hyotheridium* from the Campanian of the Gobi Desert, Mongolia (Gregory and Simpson, 1926; Kielan-Jaworowska and Trofimov, 1980; Kielan-Jaworowska, 1981, 1984; Novacek et al., 1997; Wible et al., 2004); *Daulestes* from the Coniacian of Uzbekistan (McKenna et al., 2000); and *Eomaia* (Ji et al., 2002). Only the first five Mongolian taxa and *Daulestes* have well-preserved dentitions and reasonably complete skulls. Fairly complete skeletons are known only for the first four Mongolian genera (Kielan-Jaworowska, 1977, 1978; Horovitz, 2000, 2003) and *Eomaia* (Ji et al., 2002).

Universal agreement on the taxa to be included in Eutheria does not exist. Currently, the most controversial forms are the Early Cretaceous ausktribosphenids, which include two lower-jaw genera, *Ausktribosphenos* and *Bishops,* from the Aptian of Australia. Rich et al. (1997) provisionally referred the former to their Placentalia (Eutheria of this chapter) (see also Woodburne et al., 2003). Subsequently, Rich et al. (1999) suggested it may be the sister group of Erinaceidae, which includes the living hedgehogs. Other opinions include affinities with more basal mammalian clades, including symmetrodonts (Kielan-Jaworowska et al., 1998) or with monotremes and a variety of other Gondwanan Mesozoic mammals united in Australosphenida (Luo et al., 2001, 2002).

MORPHOLOGICAL CHARACTERIZATION OF EUTHERIA

Lillegraven (1975: 718) lamented "that there are astonishingly few significant differences of likely phylogenetic antiquity between marsupials and placentals; the greatest differences are seen in the morphology of the reproductive tract arrangement, in reproductive endocrinology, and in brain structure." It remains true that the most fundamental differences between living placentals and marsupials are in the reproductive tract and the pattern of reproduction. Placentals possess unique trophoblastic tissues that form an active barrier between maternal and embryonic tissues, prolonging the period of intrauterine development and allowing for increased metabolic rates, brain development, karyotypic variability, and adult morphological divergence (Lillegraven et al., 1987). However, fossil discoveries since 1975 of basal eutherians and metatherians have revealed significant differences in skeletal morphology as well.

Cranial Anatomy

The oldest known eutherian skull is that of *Eomaia* (Ji et al., 2002), but it is badly damaged and preserved mostly as molds and impressions, providing few details. Of the remaining seven Late Cretaceous genera known from fairly complete skulls, the form that has been described and illustrated in the most detail and provides the most sutural information is *Zalambdalestes* from the Campanian of Mongolia (Figs. 3.4, 3.5; Kielan-Jaworowska, 1984; Wible et al., 2004). The skull of *Zalambdalestes* is unusual compared with other Cretaceous eutherians in its relatively large size (nearly 50 mm to more than 60 mm, whereas other genera range from less than 20 mm to nearly 40 mm) and its greatly elongated snout. *Zalambdalestes* is our major point of reference for the condition in basal eutherians in the following comparisons.

Therians are distinguished from Early Cretaceous *Vincelestes* and more distant outgroups in many aspects of cranial anatomy (see Rougier et al., 1992, 1996a, 1998; Hopson and Rougier, 1993; Rougier, 1993; Wible and Hopson, 1993; Rougier and Wible, in press). Perhaps the bones that are most affected are the paired petrosals, squamosals, and alisphenoids. In *Vincelestes* (Figs. 3.4, 3.5), the squamosal bone is confined to the posterolateral side wall of the braincase; it does not contribute to the primary braincase per se, but overlies the petrosal and parietal. The squamosal contains no foramina, and the glenoid fossa is positioned posteriorly, without a distinct postglenoid process. The petrosal bone, which houses the organs of hearing and balance, has an anterior lamina that forms the braincase in front of the squamosal and provides borders for foramina transmitting the second and third divisions of the trigeminal nerve. The petrosal's lateral flange, the ventral edge of the anterior lamina, forms the lateral wall of the roof of the middle-ear cavity and contributes to the floor beneath the trigeminal

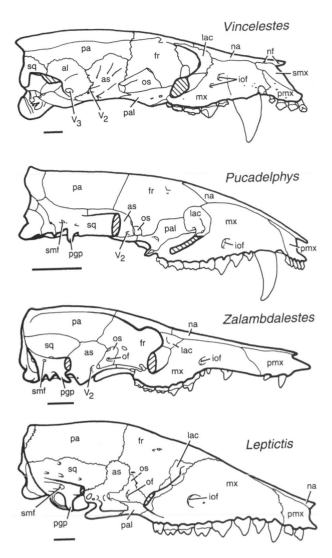

Fig. 3.4. Skulls in lateral view with zygomatic arches removed. Parallel lines represent cut edges. The prototribosphenidan *Vincelestes* from the Early Cretaceous of Argentina is redrawn from Rougier (1993: fig. 40); the metatherian *Pucadelphys* from the early Paleocene of Bolivia is redrawn from Marshall et al. (1995: figs. 12B, 15, © Publications Scientifiques du Muséum National d'Histoire Naturelle, Paris); the eutherian *Zalambdalestes* from the Late Cretaceous of Mongolia is redrawn from Wible et al. (2004: fig. 44A,B); and the placental *Leptictis* from the Oligocene of North America is redrawn from Novacek (1986a: figs. 1, 10). Scale = 5 mm. Abbreviations: al, anterior lamina of petrosal; as, alisphenoid; fr, frontal; iof, infraorbital foramen; lac, lacrimal; mx, maxilla; na, nasal; nf, nasal foramen; of, optic foramen; os, orbitosphenoid; pa, parietal; pal, palatine; pgp, postglenoid process; pmx, premaxilla; smf, suprameatal foranen; smx, septomaxilla; sq, squamosal; V_2, foramen for maxillary nerve; V_3, foramen for mandibular nerve. Redrawn by Gina Scanlon, Carnegie Museum of Natural History.

ganglion. In contrast, in therians (Figs. 3.4, 3.5), the squamosal is expanded forward, contacting the alisphenoid, and contributes to the primary braincase. The glenoid fossa is shifted anteriorly, and a large postglenoid process is found in the fossa's rear wall. The squamosal is pierced by two large vascular foramina, ventrally (postglenoid foramen) and laterally (suprameatal foramen), which indicate the development of new patterns of venous drainage. The petrosal's anterior lamina and lateral flange are essentially lost, and that bone no longer contributes to the side wall of the

Fig. 3.5. Skulls in ventral view. The proto-tribosphenidan *Vincelestes* from the Early Cretaceous of Argentina is redrawn from Rougier (1993: fig. 41); the metatherian *Mayulestes* from the early Paleocene of Bolivia is redrawn from Muizon (1998: figs. 6B, 15, © Publications Scientifiques du Muséum National d'Histoire Naturelle, Paris); the eutherian *Zalambdalestes* from the Late Cretaceous of Mongolia is redrawn from Wible et al. (2004: fig. 43); and the placental *Leptictis* from the Oligocene of North America is redrawn from Novacek (1986a: figs. 5, 14, 20). Scale = 5 mm. Abbreviations: ab, auditory bulla; al, anterior lamina of petrosal; as, alisphenoid; bs, basisphenoid; ecp, ectopterygoid process; enp, entoptery-goid process; gf, glenoid fossa; ju, jugal; lf, lateral flange of petrosal; M1, upper first molar; P2, upper second premolar; P3, upper third premolar; P4, upper fourth premolar; pal, palatine; pgf, postglenoid foramen; pr, promontorium of petrosal; ps, presphenoid; pt, pterygoid; V$_3$, foramen for mandibular nerve. Redrawn by Gina Scanlon, Carnegie Museum of Natural History.

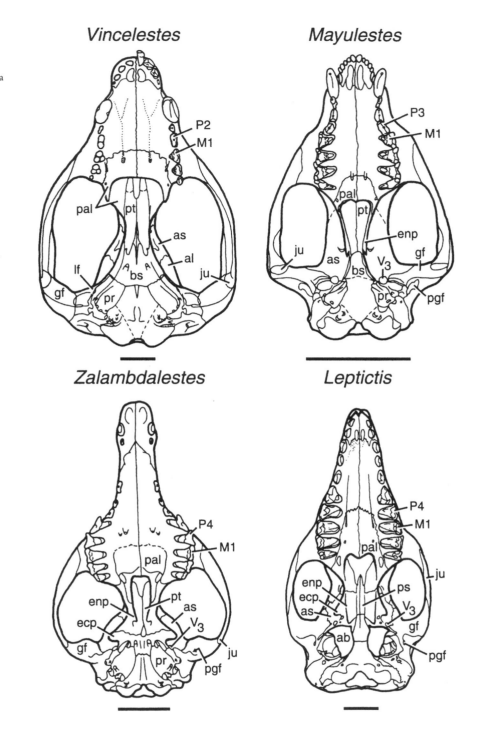

braincase, except in a petrosal referred to the Early Creta-ceous eutherian *Prokennalestes,* which has a remnant of the anterior lamina that was exposed on the braincase (Wible et al., 2001). The therian alisphenoid provides borders for the foramina transmitting the second and third divisions of the trigeminal nerve and for the floor beneath the trigemi-nal ganglion. Another noteworthy difference concerns the inner ear housed in the petrosal: in *Vincelestes,* the cochlear duct is coiled approximately 270°, whereas in therians, it is coiled at least 360°. There are also significant differences on the rostrum. As in more distant outgroups, *Vincelestes* has a well-developed septomaxilla bone in the side wall of

the external nares, vascular foramina in the nasal bone, and multiple exits in or along the maxilla for the infraorbital nerve and vessels (Fig. 3.4). In contrast, in basal therians, the infraorbital nerve and vessels have a single exit, and vascu-lar foramina in the nasal and the septomaxilla are absent.

Many of the osseous cranial features distinguishing metatherians from *Vincelestes* and eutherians are related to further changes in the vascular pattern (Wible, 1987, 1990, 2003; Rougier et al., 1992; Wible and Hopson, 1995; Rou-gier and Wible, in press). The metatherian internal carotid artery does not occupy a groove running the length of the promontorium of the petrosal (cochlear housing), but its

extracranial course is shifted anteromedially. The stapedial artery (the most significant extracranial branch of the internal carotid in the platypus and many living placentals) is absent in all living marsupials; and the absence of grooves, foramina, and canals on the petrosals of extinct metatherians indicates that the stapedial artery and its primary branches were also lacking. A prootic canal through the petrosal, by which the prootic sinus leaves the skull in monotremes, is ubiquitous among non-therian mammals. Rather than a substantial, vertical channel, the prootic canal is transformed into a narrow, horizontal structure in basal metatherians, which is retained in some living marsupials (Sánchez-Villagra and Wible, 2002). Another significant change is on the metatherian mandible, with the sharp medial inflection of the angular process (Sánchez-Villagra and Smith, 1997), although an inflected angle does not occur in *Sinodelphys* (Luo et al., 2003).

There are not many osseous cranial features that universally distinguish eutherians from metatherians and more distant outgroups (Novacek, 1986a; Rougier et al., 1998). For example, whereas the cranial vascular pattern and associated bony structures are relatively uniform among metatherians (Archer, 1976; Wible, 1987, 1990, 2003), a high degree of diversity exists among eutherians that has long been recognized as being systematically informative (Tandler, 1899; Bugge, 1974; Wible, 1987). Even among Cretaceous eutherians, there are forms that have well-developed grooves on the petrosal for the internal carotid and stapedial arteries (e.g., *Daulestes;* McKenna et al., 2000), for just the stapedial artery (e.g., *Zalambdalestes;* Wible et al., 2004), or for neither (e.g., *Zhelestes;* Archibald et al., 2001). The loss of the prootic canal had been proposed as a eutherian synapomorphy (Wible, 1990), but that structure has recently been reported in isolated petrosals referred to *Prokennalestes* (Wible et al., 2001) and to Late Cretaceous "zhelestids" (Archibald et al., 2001).

Moreover, the few osseous cranial features distinguishing eutherians do not pertain to a particular anatomical system, but are from various regions. For example, in the Cretaceous eutherian orbitotemporal region (Fig. 3.4), there is a separate optic foramen in the orbitosphenoid for the optic nerve, the palatine's contribution to the orbit is restricted, and the zygomatic arch is delicate (Rougier et al., 1998; Wible et al., 2004). In contrast, in *Vincelestes* and basal metatherians (Fig. 3.4; Rougier, 1993; Marshall et al., 1995; Muizon, 1998; Rougier et al., 1998), the exit for the optic nerve is behind the orbitosphenoid via the large sphenorbital fissure (except in *Deltatheridium*, which has a separate optic foramen; Rougier et al., in press); the palatine's orbital contribution is extensive, from the maxillary foramen to the sphenorbital fissure; and the zygomatic arch is robust. In the eutherian mesocranium (Fig. 3.5), between the choanae and ear regions, the paired pterygoid bones are greatly expanded and, with the alisphenoids, contribute to large, paired, neomorphic muscular processes, the ectopterygoid processes (Rougier et al., 1998; Wible et al., 2004), for the attachment of the external pterygoid muscle

(Turnbull, 1970). The pterygoids of *Vincelestes*, the Paleocene metatherian *Mayulestes*, and *Zalambdalestes* illustrate the differences (Fig. 3.5). In all three genera, the paired pterygoids contact on the midline. However, in *Vincelestes*, they are confined to the roof of the basipharyngeal canal (Rougier, 1993); in *Mayulestes*, they sport thin, vertical entopterygoid processes (Muizon, 1998) for the attachment of the internal pterygoid muscles (Turnbull, 1970); and in *Zalambdalestes*, they form the entopterygoid processes and contribute to the larger ectopterygoid processes (Wible et al., 2004).

There are three mandibular characters that are widely used in phylogenetic analyses of Cretaceous therians: the labial mandibular foramen, and the so-called "coronoid facet" and "Meckelian sulcus." Variability and lack (or poor knowledge) of modern analogs in all three lead us to caution against their usage and suggest closer scrutiny. The labial mandibular foramen is a large aperture in the anteroventral edge of the masseteric fossa that connects to the mandibular canal in, for example, *Prokennalestes* (Kielan-Jaworowska and Dashzeveg, 1989). Such a large aperture is variably present in *Zalambdalestes* (Wible et al., 2004), and tiny foramina in the same position are noted for *Daulestes* (McKenna et al., 2000) and *Vincelestes*, with the number and size varying among specimens of the latter (Rougier, 1993). A separate small coronoid bone is known for various Mesozoic mammaliaforms, and a distinct facet on the medial side of the ascending mandibular ramus has been interpreted as housing this bone in various primitive Mesozoic mammals (Wible, 1991). In *Vincelestes*, only one of the dozen specimens with mandibles has such a depression (Rougier, 1993). Metatherian mandibles are smooth in this region, but in many Cretaceous eutherians (e.g., *Prokennalestes*, Kielan-Jaworowska and Dashzeveg, 1989; *Asioryctes*, Kielan-Jaworowska, 1981) rather than a depression, there is a raised rugosity that has been posited as a remnant of a fused coronoid bone (Kielan-Jaworowska, 1981), for which we find no support. Many Mesozoic mammaliaforms have a distinct longitudinal sulcus that runs the length of the horizontal mandibular ramus (Simpson, 1928a,b, 1929), which has been interpreted variously as housing Meckel's cartilage (Bensley, 1902) or nervous and/or vascular structures (Simpson, 1928b). Despite this unresolved controversy, the term "Meckelian sulcus" is most often employed for an array of morphologies ranging from deep grooves clearly homologous with the condition in non-mammalian amniotes to barely marked longitudinal depressions on the lingual surface of the mandible. Recently, an Early Cretaceous gobiconodontid and "symmetrodont" have been described as preserving an ossified Meckel's cartilage that occupies only the broad posterior part of what might be identified as the Meckelian sulcus (Wang et al., 2001; Meng et al., 2003b). Some Cretaceous eutherians (e.g., *Prokennalestes;* Kielan-Jaworowska and Dashzeveg, 1989) and some living marsupials and placentals (Bensley, 1902) have fainter and much shorter versions of the Meckelian sulcus of more primitive taxa, yet our knowledge of what occupies these structures is severely limited.

The diversity of osseous cranial features present in the relatively small sample of Cretaceous eutherians is magnified many times over in placentals, resulting in the dramatic differences among living and extinct placentals, from the blue whale to the giant anteater to the pygmy shrew (Novacek, 1993; Starck, 1995). Among the features distinguishing early placentals, such as Oligocene *Leptictis* (Novacek, 1986a), from *Zalambalestes* are a posteriorly narrow nasal bone, broad contact between the frontal and maxilla on the rostrum, a reduced facial exposure of the lacrimal, a jugal that does not reach to the glenoid fossa, pterygoids that do not contact on the midline, and an osseous auditory bulla enclosing the middle ear (Figs. 3.4, 3.5).

Dental Anatomy

As stated earlier in the chapter, of the nearly 40 genera of Cretaceous eutherians, only eight are known from associated upper and lower dentitions, of which, only six preserve views of the occlusal surfaces. Our knowledge of the outgroups is even more scant. Associated upper and lower dentitions are known for only three Cretaceous metatherians, *Sinodelphys* from the middle Barremian of China (Luo et al., 2003), and *Deltatheridium* (Gregory and Simpson, 1926; Kielan-Jaworowska, 1975b; Rougier et al., 1998) and *Asiatherium* (Szalay and Trofimov, 1996), both from the Campanian of Mongolia. Among the post-dryolestoid outgroups to Theria, associated upper and lower dentitions are known only for *Vincelestes,* and its dentition is highly specialized (Fig. 3.5), with a reduced postcanine dental formula (two premolars and three molars) and low crowned molars (Butler, 1990; Rougier, 1993). Other than *Vincelestes,* the peramurid *Peramus* and the aegialodontid *Kielantherium* are the best-known dental taxa proximate to Theria. *Peramus* from the Late Jurassic of England is known from several partial mandibles and one partial maxilla; preserved is evidence of two distal, single-rooted lower incisors (more were likely present), the alveolus for a single-rooted lower canine, and eight upper and lower postcanine teeth (Clemens and Mills, 1971). The postcanine teeth have been interpreted as four premolars and four molars (Simpson, 1928a; Clemens and Mills, 1971) or five premolars and three molars (McKenna, 1975; Butler and Clemens, 2001), which we follow in this chapter. *Kielantherium* from the Albian-Aptian of Mongolia is known from one partial mandible preserving four molars, alveoli for four double-rooted premolars, and anteriorly, a broken alveolus either for the canine or a fifth premolar (Dashzeveg and Kielan-Jaworowska, 1984).

Few Cretaceous eutherians preserve complete upper and lower incisors. *Eomaia* has impressions for five upper and four lower, peglike incisors; the first upper and lower incisors are semiprocumbent, the remaining uppers nearly vertical, and the remaining lowers decreasing in procumbency (Ji et al., 2002). The same incisor formula with all uppers vertical and decreasing procumbency in the lowers occurs in Late Cretaceous *Asioryctes* (Kielan-Jaworowska, 1981) and *Ukhaatherium* (Novacek et al., 1997), as well as in various metatherians (Figs. 3.4, 3.5). Distinguishing the eutherians and metatherians are two significant features. In the uppers, the metatherian ultimate incisor is within the premaxilla, whereas in Late Cretaceous eutherians, it is either in the maxilla (as in the Coniacian zalambdalestid *Kulbeckia,* Archibald and Averianov, 2003; and in *Ukhaatherium*) or between the maxilla and premaxilla (as in *Asioryctes* and *Kennalestes,* Kielan-Jaworowska, 1981). The Late Cretaceous eutherian condition may be primitive, in that the ultimate incisor is in the maxilla in *Morganucodon, Haldanodon,* and an undescribed new genus of dryolestoid from the Late Jurassic (Hopson et al., 1999; Wible et al., 2004). In the lowers, the metatherian second incisor exhibits an unusual condition not found in eutherians or other outgroups; its root is wedged between that of its neighbors, producing the staggered condition described by Hershkovitz (1995), whereby the lateral side of the root is covered by a bony buttress that projects above the alveolar line of the adjacent teeth. The staggered condition does not occur in *Sinodelphys* (Luo et al., 2003).

Although some Cretaceous eutherians retain many simple incisors, others reduce and dramatically modify the teeth. Perhaps the most extreme example of the latter is the zalambdalestids, in which the lower mesial incisor is a greatly enlarged, procumbent, evergrowing tooth with enamel restricted to the buccal half and with a root extending below the premolars or even the first molar (Archibald et al., 2001; Fostowicz-Frelik and Kielan-Jaworowska, 2002; Wible et al., 2004). Unlike the condition in the placental superorder Glires (rodents, lagomorphs, and relatives), the zalambdalestid lower mesial incisor does not contact a similarly enlarged, procumbent, gliriform upper incisor. Regarding reduction in the number of incisors, the homologies proposed by some (e.g., Meng and Wyss, 2001; Meng et al., 2003a) follow a model whereby reduction occurs in a posterior-to-anterior direction. We are not aware of data supporting this, and in fact, reduction within the zalambdalestid lineage does not follow this model. Whereas the Campanian *Zalambdalestes* and *Barunlestes* have two small incisors posterior to the enlarged mesial incisor (Figs. 3.4, 3.5; Kielan-Jaworowska, 1984), the Coniacian *Kulbeckia* has three (Archibald and Averianov, 2003). According to Archibald and Averianov (2003), it is the second incisor of *Kulbeckia* that is likely missing in *Zalambdalestes* and *Barunlestes.*

The upper and lower canines are large teeth in basal metatherians and most Cretaceous eutherians; the zalambdalestids *Zalambdalestes* and *Barunlestes* are notable exceptions, with the lower canine resembling its adjacent incisor in size and semiprocumbent position, and the upper canine considerably reduced in the latter (Kielan-Jaworowska, 1984). What distinguishes the canines among Cretaceous eutherians is the number of roots. Whereas metatherians and some Cretaceous eutherians (e.g., *Ukhaatherium,* Novacek et al., 1997) have single-rooted upper and lower canines, other Cretaceous eutherians (e.g., *Kulbeckia,* Archibald and Averianov, 2003; *Asioryctes* and *Kennalestes,* Kielan-Jaworowska, 1981) have double-rooted canines in

both jaws. Yet others (e.g., *Zalambdalestes,* Kielan-Jaworowska, 1984; *Cimolestes,* Clemens, 1973) have a double-rooted upper and single-rooted lower, the condition occurring in *Vincelestes* (Rougier, 1993). The Late Jurassic "arguimurid" *Nanolestes* has a double-rooted lower canine (Martin, 2002), whereas that tooth is single-rooted in *Peramus* (Clemens and Mills, 1971).

A growing number of Cretaceous eutherians has been found with five premolars, as interpreted for Late Jurassic *Peramus* (McKenna, 1975; Butler and Clemens, 2001) and *Nanolestes* (Martin, 2002), suggesting that five may be the primitive number (McKenna, 1975; Novacek, 1986b; Rougier et al., 1998; Ji et al., 2002). In contrast, metatherians have three (or, in some instances, fewer) premolars (Thenius, 1989; Rougier et al., 1998), although *Sinodelphys* has four (Luo et al., 2003). Among the Cretaceous eutherians with five premolars are the three oldest known from relatively complete jaws: *Eomaia* has impressions for five premolars in the upper and lower jaws (Ji et al., 2002); *Prokennalestes* (Kielan-Jaworowska and Dashzeveg, 1989) and *Otlestes* from the Cenomanian of Uzbekistan (Nessov et al., 1994) have five lower premolars (their maxillary dentitions are incompletely known). The Cretaceous eutherians with the fewest premolars are *Barunlestes* and one specimen of *Zalambdalestes,* which have three upper and lower premolars (Kielan-Jaworowska, 1975a; Wible et al., 2004).

Living placentals range between a maximum of four premolars and a minimum of none (Thenius, 1989). A reduced number of premolars in different placental groups with sister taxa having a higher number of premolars suggests that premolar reduction is a common evolutionary event within Eutheria. At issue is whether patterns of reduction can be identified to support homologies for phylogenetic analysis. In at least two Late Cretaceous taxa, data from juvenile and adult specimens (*Kennalestes,* Kielan-Jaworowska, 1981) and from upper and lower dentitions (*Gypsonictops,* Lillegraven, 1969; Clemens, 1973) support a reduction from five to four premolars in the middle of the series. Luckett (1993) noted that this contrasts with the pattern known for some living placentals having four or fewer premolars, in which reduction proceeds in an anterior-posterior sequence. Luckett (1993) further suggested that the tooth lost in the middle of the premolar series in *Kennalestes* and *Gypsonictops* is not a distinct tooth position, but the deciduous second premolar. Lending some support to this suggestion is a juvenile *Daulestes* from the Coniacian of Uzbekistan with four upper and five lower premolars, interpreted as having the deciduous and replacement second premolar present at the same time in the lower jaw (McKenna et al., 2000). In any case, the Cretaceous eutherian record does not support a single pattern of premolar reduction. *Zalambdalestes* has different patterns in the maxilla and mandible: most specimens have four upper and lower premolars, but the first upper and second lower are variably present (Wible et al., 2004).

The three premolars occurring in metatherians are simple, double-rooted teeth with a tall primary cusp, and there is a sharp morphological break between the premolars and molars (Fig. 3.5; Thenius, 1989; Rougier et al., 1998). Whereas the anterior premolars are similarly simple (or even simpler, with only a single root) in Cretaceous eutherians, their ultimate and penultimate premolars exhibit varying degrees of molarization; that is, premolars that show all the main cusps of the molars, which results in the intergradation of the premolars and molars (Fig. 3.5). There is also a tendency to greater molarization in the posterior upper premolars than in the lowers. A molarized ultimate upper premolar with well-developed protocone, parastyle, and metastyle is nearly ubiquitous among Cretaceous eutherians, whereas a molarized lower ultimate premolar with well-developed trigonid and talonid is more restricted, occurring in zalambdalestids, *Gypsonictops,* and some "zhelestids" (Archibald et al., 2001). Molarization of the penultimate premolar is rarer, occurring only in the uppers of zalambdalestids and *Gypsonictops* (Archibald et al., 2001). In fact, the penultimate upper premolar is usually a tall, trenchant tooth, as in *Peramus* (Rougier et al., 1998). In contrast, in basal metatherians, the ultimate upper premolar is a tall, trenchant tooth (Rougier et al., 1998).

All sufficiently known Cretaceous eutherians have three upper and lower molars, as in *Vincelestes* (Rougier, 1993) and as interpreted for *Peramus* (McKenna, 1975; Butler and Clemens, 2001). In contrast, all sufficiently known basal metatherians have four molars, as in *Kielantherium* (Dashzeveg and Kielan-Jaworowska, 1984), except *Sinodelphys,* which has four upper molars, but only three lowers (Luo et al., 2003). One hypothesis accounting for the discrepancy is that the metatherian first molar is homologous to the eutherian deciduous, fully molarized ultimate premolar; typically, the latter is replaced by a semimolarized tooth in eutherians, but would not be replaced in metatherians (Owen, 1868; Luckett, 1993). Comparative studies of latest Cretaceous therians (e.g., Clemens, 1966, 1973; Lillegraven, 1969; Clemens and Lillegraven, 1986) have identified morphological criteria for distinguishing the molars of eutherians and metatherians. However, the discovery of more primitive taxa from the medial Cretaceous has weakened the utility of these criteria (Cifelli and Muizon, 1997; Cifelli, 1999), and recent phylogenetic analyses have identified few characters that distinguish the molars of basal eutherians and metatherians (Rougier et al., 1998; Luo et al., 2002, 2003).

Among Early Cretaceous eutherians, complete upper and lower molar series are known only for *Prokennalestes,* although these are not found in association (Kielan-Jaworowska and Dashzeveg, 1989); only impressions of the upper and lower molars are known for *Eomaia.* In the upper molars of *Prokennalestes* (Fig. 3.6), the stylar shelf is broad (roughly one-third of the crown width), with a prominent stylocone and well-developed parastylar and metastylar regions; the last is greatly reduced in the ultimate molar. The paracone and metacone bases are adjoined, and the former cusp is noticeably taller than the latter; the protocone is lower than both the paracone and metacone. Small, unwinged paraconule and metaconule are found proximal to the protocone, and pre- and postcingula are absent. A deep ectoflexus is found only on the penultimate molar; it is also present on the first molar in *Eomaia* (pers, observ.; contra Ji et al., 2002). Trends in the upper molars of Late Cretaceous eutherians (see, for example, Clemens and Lillegraven, 1986; Butler, 1990) include the reduction of the stylar shelf and stylocone; an increase in the width, length, and height of the protocone; separation of the paracone and metacone, with an increase in the size of the latter; strengthening of the conules, with the development of strong conular cristae; and development of pre- and postcingula, including the appearance of a hypocone from the latter (Fig. 3.6).

In the lower molars of *Prokennalestes* (Fig. 3.6), the trigonid is more acute with the paraconid more posteriorly positioned than in basalmost metatherians; the protoconid is the largest cusp on the trigonid, followed by the metaconid and the paraconid (the paraconid and metaconid are subequal in *Eomaia*, Ji et al., 2002, and *Montanalestes*, Cifelli, 1999). The talonid is narrower and much lower than the trigonid, with the hypoconulid roughly equidistant from the hypoconid and entoconid. The cristid obliqua extends up the posterior face of the metaconid. Trends in the lower molars of Late Cretaceous eutherians (see, for example, Clemens and Lillegraven, 1986; Butler, 1990) include further

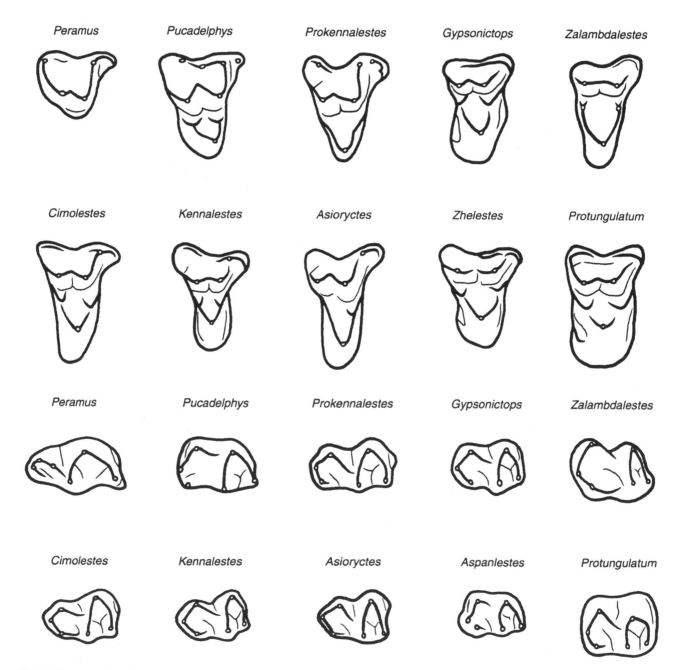

Fig. 3.6. Upper second molar (top two rows) and lower second molar (bottom two rows) for selected Mesozoic taxa. Not to scale. *Peramus, Prokennalestes, Gypsonictops, Cimolestes, Kennalestes, Asioryctes, Zhelestes, Aspanlestes,* and *Protungulatum* redrawn from Butler (1990: figs. 3, 8, 10); *Pucadelphys* redrawn from Marshall et al. (1995: fig. 7B, E, © Publications Scientifiques du Muséum National d'Histoire Naturelle, Paris); and *Zalambdalestes* redrawn from Wible et al. (2004: fig. 5). Redrawn by Gina Scanlon, Carnegie Museum of Natural History.

anteroposterior compression of the trigonid; an increase in the size of the metaconid; a decrease in the size of the paraconid; a change in the height differential between the trigonid and talonid; widening of the talonid; and a more labial position for the cristid obliqua (Fig. 3.6).

Another feature of the dentition distinguishing eutherians and metatherians is the pattern of replacement. In living placentals, all antemolar teeth undergo a single replacement (diphyodonty), with the possible exception of the first premolar position, which is rarely replaced (Luckett, 1993). The replacement of more than one premolar position has been recorded in some Cretaceous eutherians (e.g., *Daulestes*, McKenna et al., 2000) and is likely primitive for Theria, in light of similar patterns occurring in Early Cretaceous *Slaughteria* (Kobayashi et al., 2002), which falls outside Theria in the phylogenetic analysis by Rougier et al. (1998), and in Late Jurassic dryolestids (Martin, 1997). In contrast, in living marsupials, only one tooth, the ultimate premolar, is replaced in each jaw (Luckett, 1993), a derived pattern that is also recorded in some Cretaceous and Paleocene metatherians, including *Deltatheridium* (Rougier et al., 1998) and *Alphadon* (Cifelli et al., 1996; Cifelli and Muizon, 1998).

Postcranial Anatomy

Morphological approaches to higher-level eutherian phylogeny have traditionally been based on craniodental characters, with postcranial data, especially cruropedal characters, providing important evidence at selected placental nodes (Rose, 1999, 2001; Thewissen and Madar, 1999; Horovitz, 2000, 2003; Thewissen et al., 2001; Sargis, 2002). This is not surprising, as the postcranium is known in some detail in only a handful of Cretaceous eutherians, all from Asia (Kielan-Jaworowska, 1977, 1978; Novacek et al., 1997; Horovitz, 2000, 2003; Ji et al., 2002). Isolated tarsals and assorted other postcranial elements have been reported from a variety of localities and ages, including the latest Cretaceous of North America (Szalay and Decker, 1974) and of India (Godinot and Prasad, 1994; Prasad and Godinot, 1994).

Cretaceous eutherians are generalized in their postcranial anatomy; that is, they are plantigrade quadrupedal animals, occupying a terrestrial or scansorial milieu and lacking obvious specializations for extreme cursoriality or fossorial habits (Kielan-Jaworowska et al., 1979; Ji et al., 2002). Close similarities can be found between Cretaceous eutherians and living "insectivorans," in which facultative scansoriality and an uneven substrate locomotion are likely to be primitive (Jenkins, 1974; Jenkins and McClearn, 1984; Ji et al., 2002). The zalambdalestids *Zalambdalestes* and *Barunlestes* are exceptions; their skeleton is highly derived and specialized for a digitigrade, ricochetal mode of locomotion (Kielan-Jaworowska, 1978).

In the following overview (see Fig. 3.7), we compare the skeletons of basal eutherians, best represented by *Eomaia* (Ji et al., 2002; pers. observ.) and *Ukhaatherium* (Horovitz, 2000, 2003), *Zalambdalestes* (Kielan-Jaworowska, 1978; pers. observ.), and placentals with *Vincelestes* (Rougier, 1993), the

basal metatherians *Deltatheridium* (unpublished skeleton), *Mayulestes* and *Pucadelphys* (Marshall et al., 1995; Muizon, 1998; Argot, 2001, 2002, 2003), and marsupials (Szalay, 1994; Horovitz and Sánchez-Villagra, 2003).

In the vertebral column, *Vincelestes*, metatherians, *Eomaia*, and *Ukhaatherium* lack the transverse foramen in the atlas's transverse process, or at least, it is uncertain. In placentals (with only a few exceptions; Lessertisseur and Saban, 1967) and *Zalambdalestes*, this foramen is present; however, the phylogenetic importance of the feature is obscure. The composite nature of the atlas (unfused arch plus intercentrum) among basal mammals is retained in *Vincelestes*, basal metatherians, *Ukhaatherium*, and *Zalambdalestes*, but is apparently independently fused into a single complex in marsupials and placentals, with the atlas forming a complete ring. The spines of the posterior cervical vertebrae are ancestrally well developed, as shown by *Vincelestes*, metatherians, and some xenarthrans and afrotherians among placentals. In small, stem placentals, such as *Eomaia*, *Ukhaatherium*, *Asioryctes*, and *Zalambdalestes*, spinous processes are absent or vestigial. A common feature of the six cervical vertebrae among therians is the development of a ventral process (inferior lamella or lamina) that serves as a pivot for the attachment of the longus colli muscle. This process is already well developed in *Vincelestes*, in which there is a marked differentiation between the sixth cervical vertebra and the immediately succeeding vertebrae, which have smaller projecting lamellae. Differentiated inferior lamellae are present in *Ukhaatherium* and *Asioryctes*, which show a greater difference in the development of the lamellae between the fifth and sixth cervical vertebrae than is seen in *Vincelestes*. However, the inferior lamella is small or poorly differentiated in *Zalambdalestes*. Among placentals, there is substantial variability in the development of this feature, with some forms having the inferior lamella of the sixth vertebra as the only structure projecting ventrally and others having a distinct process on the more anterior cervical positions.

The number of dorsal vertebrae is relatively constant in therians, with a total count of 19 or 20 being common among generalized marsupials and placentals. The development of a distinctive lumbar region with five or six lumbar vertebrae is a derived feature already achieved by *Vincelestes*, probably by fusion of free ribs from the dorsal region. The eutherians *Eomaia* and *Ukaatherium* retain this generalized mammalian count.

In the pectoral girdle, *Vincelestes* retains a sagittal element fused to a paired component of the manubrium that is likely to represent the remnant of an endochondral component of the interclavicle (Rougier, 1993). The manubrium in the basal metatherian *Pucadelphys* has been described as an interclavicle (Marshall et al., 1995), but it does not seem to be any different from the manubria present among living marsupials. As a whole, the absence of an independent ossification in front of the manubrium occurs at the level of *Vincelestes* or before (Rougier, 1993; Hu et al., 1997; Ji et al., 1999). Metatherians and eutherians lack the extensive wing-like projections of the manubrium of *Vincelestes*, and the

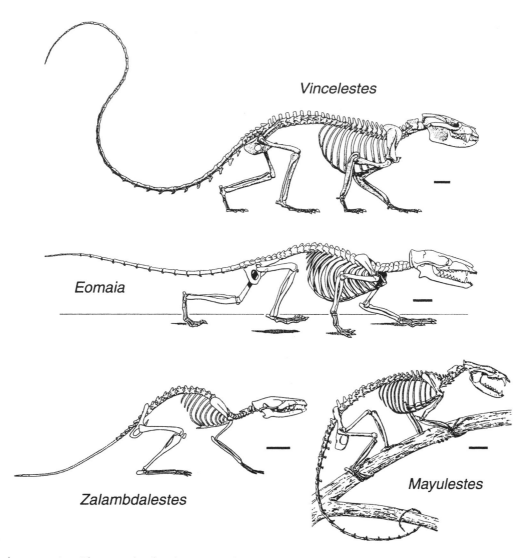

Fig. 3.7. Skeletal reconstructions. The prototribosphenidan *Vincelestes* from the Early Cretaceous of Argentina is from Rougier (1993: fig. 104); the metatherian *Mayulestes* from the early Paleocene of Bolivia is from Muizon (1998: fig. 57, © Publications Scientifiques du Muséum National d'Histoire Naturelle, Paris); the eutherian *Eomaia* is from Ji et al. (2002: fig. 1c); and the eutherian *Zalambdalestes* from the Late Cretaceous of Mongolia is modified from Kielan-Jaworowska (1978: fig. 17). Scale = 20 mm (except *Eomaia*, 10 mm). The presence or absence of epipubics is not known in *Mayulestes* (Muizon, 1998), but these bones are in the metatherian *Pucadelphys*, also from the early Paleocene of Bolivia (Marshall et al., 1995). Redrawn by Gina Scanlon, Carnegie Museum of Natural History.

putative endochondral remnant of the interclavicle becomes less conspicuous (if present at all) as an independent ossification.

In the scapulocoracoid of *Vincelestes,* the coracoid process is extremely prominent and the coracoid makes a substantial contribution (about one-third) to the glenoid cavity, a condition that is probably ancestral for therians. In marsupials and placentals, the coracoid contribution is reduced. There is a large angle between the glenoid and the scapular plane in *Vincelestes;* less so in *Ukhaatherium* and probably *Eomaia.* In *Zalambdalestes,* however, the scapula and the glenoid are essentially in the same plane, a feature likely to be ancestral for Placentalia. Basal metatherians present a condition similar to placentals, with essentially coplanar glenoid and scapular lamina. The relative torsion of the scapular lamina also affects the proportional size of the in-

fraspinous fossa and the orientation of the spinous process. In therian outgroups, the supraspinous fossa is narrow and, near the acromion, extends over the scapular spine to sandwich the infraspinous fossa between the lamina and the posteriorly extended spine. This condition is also present in *Ukhaatherium,* but with a more developed supraspinous fossa and a less posteriorly directed spine, so that the infraspinous fossa is not as enclosed between the lamina and the acromion/spine. *Zalambdalestes* does not show an extensive posterior projection of the spine, a condition similar to that seen in tenrecs and hedgehogs. Metatherians show a scapula with a spine nearly at a right angle to the plane determined by the lamina.

In the forelimb, therian humeri show a remarkable difference from the very stout and strongly twisted humerus, with a rounded radial condyle, of basal mammals (Sereno and

McKenna, 1995; Gambaryan and Kielan-Jaworowska, 1997) and *Vincelestes* (Rougier et al., 1996b). Metatherians and eutherians have gracile, relatively straight humeri with variously developed trochleae on their distal ends. The reduction in the angle between the plane formed by the greater and lesser tuberosities and the distal end of the humerus can be related to the more parasagittal position of the forearm from a more laterally adducted position present primitively among most mammaliaforms, with the possible exception of some multituberculates (Sereno and McKenna, 1995). Humeri and antebrachia become longer and more gracile in such basal eutherians as *Ukhaatherium,* compared with those of *Vincelestes* and metatherians, which commonly have an extensive deltopectoral crest and a conspicuous teres major process. Generalized placentals retain a gracile construction, with a relatively narrow and short deltopectoral crest, features that are profoundly modified among some of the more derived orders. The hand is generalized in its proportions among basal therians, with possible features indicating scansorial adaptations in *Eomaia* (Ji et al., 2002).

In the pelvis, the dorsal edge of the acetabulum is notched (open dorsally) in basal mammaliaforms; this persists in *Vincelestes* and possibly, to a lesser degree, in *Ukhaatherium.* However, this condition is not present in basal metatherians, *Zalambdalestes,* and placentals, in which the acetabulum is bordered by a distinct crest. In basal mammaliaforms, the obturator foramen is subequal in size to the acetabulum or slightly larger; *Vincelestes* and *Ukhaatherium* have a proportionally small circular obturator foramen, but this feature is larger and elongated in *Zalambdalestes,* most basal placentals, and metatherians.

Another distinctive feature of basal therians is the presence of epipubic bones, known to be present in a wide range of therians and therian stem taxa, ranging from "symmetrodonts" to metatherians and basal eutherians (Krebs, 1991; Rougier, 1993; Hu et al., 1997; Novacek et al, 1997; Springer et al., 1997; Horovitz and Sánchez-Villagra, 2003; Reilly and White, 2003). The fate of the epipubics in placentals is uncertain. No member of the crown-group has epipubics resembling those in non-placental mammals. However, given the widespread presence of a baculum (os penis) among "insectivorans," carnivorans, rodents, primates, dermopterans, and chiropterans (Saban, 1967), and given the similar developmental pattern of the epipubics of non-placental mammals and the baculum of placentals, it is possible that the epipubics and baculum are homologous (Jellison, 1945). Absence of paired epipubics is synapomorphic for the crown-group Placentalia, and their reduction and ultimate disappearance or transformation are likely related to locomotor changes, although some relationship to shorter gestation time, prolonged external suspension of poorly developed young, and pouch development cannot be ruled out (Novacek et al., 1997; Springer et al., 1997; Reilly and White, 2003).

The femur in *Vincelestes* and more basal mammals has greater and lesser trochanters of similar height, with the femoral head moderately inflected medially and a poorly differentiated femoral neck. The achievement of a more parasagittal posture in therians than that predicted for more basal forms results in a femoral head more medially directed, greater and lesser trochanters more unequal in size, and a lesser trochanter located more distally on the femoral shaft. A patella, which among living marsupials is present only as cartilage, occurs as a bone in *Eomaia, Ukhaatherium,* and *Zalambdalestes.* An ossified patella is absent in *Vincelestes* and metatherians, rendering the cartilaginous or absent patella as a primitive feature at the level of Theria. The presence of a patella in monotremes and multituberculates (Krause and Jenkins, 1983; Kielan-Jaworowska and Gambaryan, 1994) can be viewed as suggesting a loss of ossification for this structure in the lineage leading to Eutheria and a subsequent reossification.

The distal ends of the tibia and fibula are specialized among basal placentals for a tenon-mortise crural joint, in which dorsiflexion occurs primarily in the upper ankle joint and rotation is accommodated by the lower ankle joint (Szalay and Decker, 1974; Lewis, 1989; Szalay, 1994; Horovitz, 2000, 2003). The concave mortise is jointly formed by the styloid processes of the tibia and fibula, which trap between them the pulleylike dorsal surface of the astragalus (the tenon). A similar pattern is present among derived marsupials (Szalay, 1994), *Zalambdalestes* (Kielan-Jaworowska, 1978), and *Deccanolestes* (isolated tarsals from India; Prasad and Godinot, 1994), but is absent in basal metatherians and eutherians (Szalay, 1994; Horovitz, 2000, 2003). Among these basal therians, including *Ukhaatherium, Kennalestes, Deltatheridium,* and *Pucadelphys,* the astragalus lacks a prominent trochlea, and the astragalar surface is convex, with poorly separated individual facets that form an angle greater than 90° (Horovitz, 2000). In basal Mesozoic mammals and in basal metatherians, the distal end of the tibia is spiral, which accommodates a variety of movements without restricting the upper ankle joint to any particular plane. In *Ukhaatherium,* the distal end of the tibia is not well preserved (Horovitz, 2000: fig. 1.6), but the articulation does not seem to be spiral. However, the morphology of the astragalus suggests that the upper ankle joint had a larger degree of movement and was less stable than that of most placentals. Among therian stem groups, such as *Vincelestes* and dryolestoids, the astragalar head is not distinct; in contrast, in therians, a head is conspicuous, and it is further developed in such eutherians as *Eomaia, Ukhaatherium, Kennalestes, Protungulatum, Deccanolestes,* and *Zalambdalestes* (Szalay and Decker, 1974; Kielan-Jaworowska, 1978; Prasad and Godinot, 1994; Horovitz, 2000).

The calcaneum of *Vincelestes, Ukhaatherium, Deltatheridium,* and probably *Eomaia* shows a pronounced plantar curvature, which is absent in placentals and *Zalambdalestes,* and is also absent in the basal metatherians *Pucadelphys* and *Mayulestes.* The major axis of the foot of early mammals (as represented by the direction of the third metatarsal) seems to have formed an angle with the main axis of the calcaneum (represented by the tuber). This angulation is present in multituberculates (Kielan-Jaworowska and Gam-baryan, 1994), triconodonts (Ji et al., 1999), and probably, *Vincelestes*, based on its isolated tarsals. The primitive pes arrangement is likely retained in *Eomaia*, as shown by the left foot (pers. observ.; contra Ji et al., 2002), with an offset foot in which the calcaneum contacts the fifth metatarsal. *Zalambdalestes*, *Protungulatum*, and *Deccanolestes* show or suggest the placental condition with an in-line arrangement of the tuber and the pes. This feature is uncertain in *Ukhaatherium* and *Kennalestes*.

In general, postcranial evolution in early therians seems to illustrate a few general principles and trends:

1. Acquisition of more parasagittally located limbs, especially the forelimbs, which primitively are more adducted than the hind limbs;
2. Greater freedom of the shoulder girdle by reducing the stiffness of the sternoclavicular and scapuloclavicular joints;
3. Overall reduction of the number of elements involved in an articulation, as in the case of the sternum, shoulder girdle, knee, and tarsus; and
4. Cruropedal specialization, in which a given articulation has a dominant plane of movement, with strong restriction or minor function in alternative planes.

PHYLOGENETIC RELATIONSHIPS OF CRETACEOUS EUTHERIANS

Early researchers (e.g., Gregory and Simpson, 1926; Simpson, 1929, 1945) assigned the meager record of Cretaceous eutherians (e.g., *Gypsonictops*, *Batodon*, *Zalambdalestes*) to the wastebasket order Insectivora, in recognition of the overall primitive nature of these forms. The 1960s and 1970s saw a variety of hypotheses (e.g., Van Valen, 1964; Lillegraven, 1969; McKenna, 1975; Szalay, 1977) linking particular Cretaceous eutherians with living placental orders. A recent offering of similarly speculative hypotheses is in McKenna and Bell (1997), who assigned 24 of the 32 Cretaceous genera within crown-group Placentalia in the absence of specific diagnoses, description of character evidence, or phylogenetic analysis: 13 in Epitheria (non-edentaten placentals); three in Anagalida, along with elephant shrews, lagomorphs, and rodents; three in Ferae, along with carnivorans and pholidotans; three in Soricomorpha; one in Ungulata; and one in Meridiungulata, the South American ungulates. The past decade or so has seen a number of phylogenetic analyses with published taxon-character matrices that include Cretaceous eutherians (Novacek, 1992a; Archibald, 1996; Nessov et al., 1998; Rougier et al., 1998; Archibald et al., 2001; Meng and Wyss, 2001; Ji et al., 2002; Luo et al., 2003; Meng et al., 2003a).

In the few analyses that include Early Cretaceous eutherians, these chiefly Asian forms fall at or near the base of the clade (see Fig. 3.3; Archibald et al., 2001; Ji et al., 2002; Luo et al, 2003). Among Late Cretaceous eutherians, three

higher-level clades currently appear to be generally accepted (Nessov et al., 1998; Kielan-Jaworowska et al., 2000). Two are endemic Asian clades that include some very complete, well-preserved specimens: Zalambdalestidae of Archibald and Averianov (2003), with four genera from Mongolia, Uzbekistan, Tadjikistan, and Kazakhstan, between the late Turonian and Campanian (Archibald et al., 2001; Archibald and Averianov, 2003; Wible et al., 2004); and Asioryctitheria of Novacek et al. (1997), with three genera from Mongolia, Uzbekistan, and China, between the Coniacian and Campanian (Nessov, 1997; Novacek et al., 1997; Kielan-Jaworowska et al., 2003). The third clade, the paraphyletic "Zhelestidae" of Nessov et al. (1998), is the poorest known anatomically, represented largely by fragmentary upper and lower jaws. It is also the most diverse and has the widest distribution in time and space, with perhaps 11 genera, between the upper Cenomanian–lower Turonian of Japan, the Turonian and Coniacian of Uzbekistan, the Judithian and Lancian of North America, the Campanian and Maastrichtian of Spain and France, and possibly the Maastrichtian of Madagascar (Nessov et al., 1998; Setoguchi et al., 1999; Averianov et al., 2003). Currently, controversy surrounds the relationships among these groupings, and especially between these groupings and crown-group placentals.

"Zhelestidae"-Ungulata Relationships

To evaluate the relationships of six genera of Asian and North American "zhelestids," Archibald (1996) constructed a data matrix of 18 characters of the upper postcanine dentition across 18 Late Cretaceous eutherians. Included was *Protungulatum*, a North American Puercan form also reported from the latest Cretaceous of Saskatchewan (Johnson and Fox, 1984; Fox, 1989), which has been regarded as an archaic ungulate (Van Valen, 1978; Archibald, 1998). Ungulata is the superordinal grouping in Fig. 3.1 that includes the living artiodactyls, whales, aardvarks, perissodactyls, hyraxes, elephants, and sirenians. Archibald's (1996) PAUP analysis identified the "zhelestids" as a paraphyletic assemblage of basal members of the ungulate lineage (=*Protungulatum*), and he used Ungulatomorpha for the clade including "Zhelestidae" and Ungulata. Nessov et al. (1998) added two Tertiary taxa to the Archibald (1996) matrix, an early primatomorph *Purgatorius* and an early rodent *Tribosphenomys*. Interestingly, the addition of these two taxa did not support the results of Archibald (1996); *Protungulatum* and *Tribosphenomys* were sister taxa in a clade including *Purgatorius* and the North American "zhelestid" *Alostera*. Despite this discrepancy, Nessov et al. (1998) continued to accept "zhelestids" as primitive ungulatomorphs.

Archibald et al. (2001) reported an expanded analysis, with a matrix of 70 craniodental characters across the same 18 Late Cretaceous eutherians considered by Archibald (1996) and Nessov et al. (1998) plus seven additional ones (two from the Early Cretaceous, two from the Late Cretaceous, and three from the early Tertiary). Their principal goal was to test the affinities of zalambdalestids (discussed separately

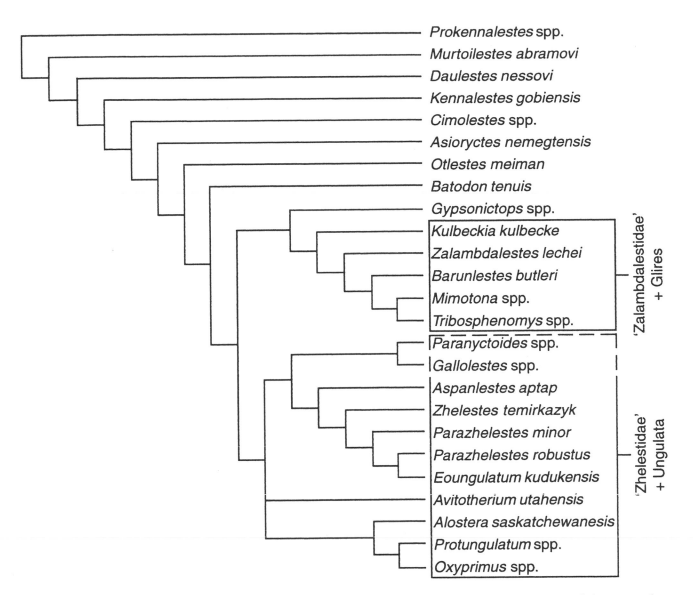

Fig. 3.8. Strict consensus of two equally most parsimonious trees from PAUP analysis in Archibald et al. (2001: fig. 3b). Data matrix includes 25 taxa and 70 craniodental characters. *Paranyctoides* and *Gallolestes* were not formally referred to "Zhelestidae"; hence, the dashed lines around these taxa by these authors. Redrawn from Archibald et al. (2001: fig. 3b). Redrawn by Gina Scanlon, Carnegie Museum of Natural History.

below). In the first PAUP analysis by Archibald et al. (2001), which included the Cretaceous taxa except *Protungulatum*, the seven genera of Asian and North American "zhelestids" fell in a monophyletic clade along with *Paranyctoides*, a Late Cretaceous form whose dentition resembles that of Tertiary nyctitheriid "insectivorans" (Fox, 1979). In their second PAUP analysis (Fig. 3.8), which contained all 25 taxa, the clade of "zhelestids" plus *Paranyctoides* also included the archaic ungulates *Protungulatum* and *Oxyprimus*, supporting the Ungulatomorpha (+ *Paranyctoides*) of Archibald (1996) and a Turonian origin for the ungulate lineage.

To date, the Ungulatomorpha hypothesis of Archibald and coworkers has not been tested by other researchers. We point out the following shortcomings in their hypothesis:

1. All the taxa that fall in Ungulatomorpha in the various analyses are poorly known, represented by fragmentary upper and lower jaws, with none of the uppers and lowers found in association. Isolated petrosals have been referred to the Uzbekistani "zhelestids" and *Protungulatum* (Archibald et al., 2001).

2. Although *Protungulatum* and *Oxyprimus* have been proposed as archaic ungulates (Van Valen, 1978; Archibald, 1998), this assignment has not yet been tested by phylogenetic analysis; and the concept of Ungulata is itself poorly defined, as recent studies that include molecular data (Murphy et al., 2001b; Asher et al., 2003) disperse ungulate-grade taxa throughout the eutherian tree.

3. Sufficient relevant taxa have not been included in any of the phylogenetic analyses supporting Ungulatomorpha. To test the ungulate nature of "zhelestids," additional basal ungulates, as well as taxa from other basal placental clades (e.g., "insectivorans"), are required.

4. Characters cited by other authors (Novacek et al., 1997; Meng and Wyss, 2001) as support for a monophyletic Placentalia that excludes other Eutheria, including such Cretaceous forms as *Ukhaatherium, Asioryctes,* and *Zalambdalestes,* were not incorporated into the data matrix of Archibald et al. (2001). A significantly expanded character matrix is needed that draws from the more complete cranioskeletal evidence in well-represented Cretaceous taxa, as well as in many living placental taxa.

Zalambdalestidae-Glires Relationships

As noted earlier in the chapter, the principal goal of the phylogenetic analyses by Archibald et al. (2001) was to test the affinities of the zalambdalestids *Kulbeckia, Zalambdalestes,* and *Barunlestes* to other Late Cretaceous eutherians and to two archaic members from both Ungulata and Glires. *Zalambdalestes* has been purported to have lagomorph affinities (Van Valen, 1964) or to be a member of a larger grouping, Anagalida, which includes lagomorphs, rodents, and elephant shrews (McKenna and Bell, 1997). In contrast, Novacek et al. (1997) postulated *Zalambdalestes* as a member of the placental stem lineage, because it retains primitive features, such as epipubic bones, that are absent in placentals. The first PAUP analysis by Archibald et al. (2001), which excluded the representatives from Ungulata and Glires, supported a monophyletic Zalambdalestidae placed at an octachotomy, including "Zhelestidae" and the remaining Late Cretaceous taxa. Their second PAUP analysis, including the members of Ungulata and Glires, identified Glires (=*Mimotona* + *Tribosphenomys*), and *Barunlestes, Zalambdalestes,* and *Kulbeckia* as successive outgroups to Glires (Fig. 3.8). By-products of this hypothesis are that the gliriform mesial lower incisor of zalambdalestids is homologous with that in rodents and lagomorphs, and the origin of the Glires lineage can be traced at least to the Coniacian.

Two published phylogenetic analyses contradict the Zalambdalestidae-Glires hypothesis proposed by Archibald et al. (2001). Meng and Wyss (2001) constructed a matrix of 82 osteological characters across 36 taxa (Late Cretaceous *Asioryctes, Kennalestes, Zalambdalestes,* and *Barunlestes,* plus 22 gliroid and 10 non-gliroid placentals). In their PAUP analyses, the zalambdalestids were far removed from Glires, either outside the placentals or as the second placental ingroup after the tree shrew *Tupaia.* Expanding the analyses by Meng and Wyss (2001), Meng et al. (2003a) constructed a matrix of 227 osteological characters across the same 36 taxa plus 14 more (three gliroid and 11 non-gliroid placentals). Their PAUP results mirrored those of Meng and Wyss (2001), with the zalambdalestids falling outside the placentals in all iterations. These two reports, which did not include some of the Cretaceous forms analyzed by Archibald et al. (2001), are much more comprehensive in their sampling of both Glires and other placentals and in their range of relevant morphological evidence. Thus, Meng and Wyss (2001) and Meng et al. (2003a) offer a more rigorous dis-

crimination of two alternatives; namely, whether Glires are more closely related to zalambdalestids than they are to incontrovertible crown-group placentals or the reverse. Meng and coauthors show that a number of placentals are more closely related to Glires than are zalambdalestids, concluding that the enlarged mesial lower incisors in these clades have been convergently acquired.

Asioryctitheria-Zalambdalestidae Relationships

Novacek et al. (1997) named Asioryctitheria for the Mongolian Late Cretaceous *Asioryctes, Ukhaatherium* and *Kennalestes,* based on basicranial similarities, including some reminiscent of those in some "insectivorans." Without discussion, Novacek (1997) figured *Zalambdalestes* at an unresolved trichotomy with *Asioryctes* and the crown-group Placentalia, offering the possibility of asioryctithere-zalambdalestid relationships. Wible et al. (2004) expanded on this possibility, noting that the basicranial resemblances used by Novacek et al. (1997) to characterize asioryctitheres, plus several more features, are also present in new specimens of *Zalambdalestes* (see below). This hypothesis was also supported in the phylogenetic analysis by Ji et al. (2002), which was designed to test the relationships of *Eomaia* among 13 other Cretaceous eutherians. They recovered an unresolved clade of *Zalambdalestes, Ukhaatherium, Asioryctes,* and *Kennalestes* (see Fig. 3.3).

PLACENTAL SUPERORDINAL RELATIONSHIPS

Since 1986, Novacek and collaborators have published a series of phylogenetic analyses evaluating the morphological bases for placental superordinal relationships (see Fig. 3.1; Novacek, 1992a,b). The character base includes both hard- and soft-tissue structures, but very few dental features, which makes it difficult to incorporate most Cretaceous eutherians into the analyses. We view the merging of data sets for largely Cretaceous taxa with largely Tertiary and/or living taxa as the next critical step for morphology-based phylogenies.

Some discrepancies notwithstanding, other key morphological analyses of placental relationships agree with many of the clades proposed by Novacek (e.g., MacPhee, 1994; Shoshani and McKenna, 1998). More extreme differences exist between Novacek's trees and those recently produced from large molecular data sets (e.g., Eizirik et al., 2001; Madsen et al., 2001; Murphy et al., 2001a,b; Árnason et al., 2002). These studies (see Fig. 3.2) have questioned some of the traditional 18 orders of extant placentals (e.g., cetaceans as the sister group of artiodactyl hippopotamids; tenrecs and golden moles allied with elephant shrews and not with hedgehogs, shrews, and moles in Lipotyphla) and have proposed new combinations of orders (e.g., uniting tenrecs, golden moles, elephant shrews, aardvarks, sirenians, hyraxes, and elephants in Afrotheria). Most of these recent molecular studies (e.g., Eizirik et al., 2001; Madsen et al., 2001; Murphy et al., 2001a,b; Delsuc et al., 2002) have identified three major superordinal groupings (see Fig. 3.2): two Southern Hemisphere clades, Afrotheria (Africa) and Xenarthra (South America); and a Northern Hemisphere clade, Boreoeutheria, composed of Euarchontoglires (Glires, Primates, Dermoptera, and Scandentia) and Laurasiatheria (Cetartiodactyla, including artiodactyls and whales; Perissodactyla; Carnivora; Pholidota; Chiroptera; and Eulipotyphla, including hedgehogs, shrews, and moles). Additionally, these studies support a Gondwanan origin for Placentalia, more often with Afrotheria at the base, followed by Xenarthra. However, a combined molecular-morphological study that does support Afrotheria (Asher et al., 2003) does not place this clade in a basal placental position. A Northern Hemisphere origin of Placentalia is consistent with the analyses presented by Archibald et al. (2001) and Ji et al. (2002; see Fig. 3.3).

These varying phylogenetic hypotheses also have implications for contrasting and controversial estimates of divergence times for major placental and eutherian clades (Alroy, 1999; Foote et al., 1999; Springer et al., 2003; O'Leary et al., 2004). Recent molecular-based studies generally support much earlier divergence dates for various cladogenetic events than are suggested by the fossil record (cf. Figs. 3.1, 3.2; Archibald and Deutschman, 2001; Archibald, 2003). The earliest known eutherian *Eomaia* (Ji et al., 2002) occurs well within the Cretaceous at 125 million years. However, the fossil date of 125 million years and a similar date for the earliest known metatherian *Sinodelphys* (Luo et al., 2003) are at stark odds with nearly all sequence-based dates for the eutherian-metatherian split (as reviewed in O'Leary et al., 2004). The suggestion that the early occurrence of *Eomaia* helps to close a gap between molecule- and fossil-based estimates of divergence for Eutheria (Ji et al., 2002) and Placentalia (for which the age of this taxon is irrelevant) is therefore erroneous (O'Leary et al., 2004). As noted, the argument by Archibald et al. (2001) that Placentalia originated prior to 85 million years has not been comprehensively tested. A study combining both molecular and morphological evidence for all the living mammal taxa sampled by Murphy et al. (2001a) and selected fossil taxa shows some results that place *Ukhaatherium* and *Zalambdalestes* outside or inside of Placentalia depending on the optimality criterion used to select the tree (Asher et al., 2003). A topological resolution index proposed in the same paper favors a stable outgroup position of these Cretaceous taxa to crown-group Placentalia, preserved in both of the two most parsimonious trees (MPT). The 11 MPTs favored by the Incongruence Length Difference (ILD) index, which measures congruence across data partitions (Farris et al., 1994; Wheeler, 1995), place *Ukhaatherium* and *Zalambdalestes* in an unstable position within Placentalia (either near chiropterans or near tenrecs + macroscelideans). Thus, there is the possibility, at least under the assumption of one optimality criterion (ILD), for a pre-Tertiary origin of Placentalia. In most of the 12 parameter sets for character weighting, MPTs either showed *Ukhaatherium* and *Zalambdalestes* as

excluded from Placentalia, or as forming a basal clade that also includes *Erinaceus* outside of other Placentalia (Asher et al., 2003).

Three models of placental origin and diversification have been characterized (Archibald and Deutschman, 2001: 113): the Explosive Model, with the origin and diversification in the early Tertiary (see, e.g., Fig. 3.1); the Long Fuse Model, with Cretaceous origin and early Tertiary diversification (see, e.g., Fig. 3.2); and the Short Fuse Model, with origin and diversification "before or shortly after the appearance of eutherians in the fossil record." Although many molecular studies (e.g., Kumar and Hedges, 1998) and at least one paleontological contribution (McKenna and Bell, 1997) have supported the Short Fuse Model, a recent recalibration based on molecular data that are constrained by several fossil-derived calibration dates by Springer et al. (2003) claimed to be more in line with the Long Fuse Model. The study of Springer et al. (2003), however, shows many deep-seated divergences for placentals, including splits within the orders Rodentia and Primates, as well as the early divergence dates for orders grouped within the proposed clade Afrotheria. Following the same methods, Roca et al. (2004) proposed an early divergence (76 million years) of the solenodon of Cuba and Hispaniola from other eulipotyphlans. The analysis of fossil taxa by Archibald et al. (2001) is consistent with the Long Fuse Model, whereas the studies of Meng and Wyss (2001) and Meng et al. (2003a) are more consistent with the Explosive Model.

SUMMARY

Twenty-five years ago, just over a dozen genera of Cretaceous eutherians were known, all but one from Asia and North America. Today, the number is at least 40, all but three from the same two continents. Although some very well-preserved, reasonably complete Asian specimens recently have been found, most Cretaceous eutherians are known from fragmentary jaws and teeth. The few well-preserved Cretaceous eutherians, along with similarly well-known basal metatherians and proximate therian outgroups (e.g., the Early Cretaceous prototribosphenidan *Vincelestes*) have provided considerable detail about the morphological changes in the earliest members of the placental and marsupial lineages. These more complete fossils have moved the discussion of eutherian origins from a debate about tooth morphology to one considering all aspects of morphological diversity: dental, cranial, and postcranial. One, perhaps surprising, observation is that Cretaceous eutherians retain more primitive characters in both the cranium and postcranium than previously appreciated, including the postcanine dental formula, the principal pattern of cranial arterial supply, and the presence of epipubic bones.

The fragmentary nature of most Cretaceous eutherians has constrained broad-scale phylogenetic analyses among these taxa or between them and younger eutherians, including crown-group Placentalia. Nevertheless, several recent analyses have supported close affinities between certain Cretaceous eutherians and placental superordinal groupings; namely, Asian zalambdalestids with Glires and "zhelestids" with Ungulata. These analyses suffer from deficiencies in taxonomic and morphological coverage, and more comprehensive analyses, including more relevant taxa and characters, have already been published that do not support the zalambdalestid-Glires clade. We think that current evidence is congruent with a position for all, or most, Cretaceous eutherians outside of the crown-group Placentalia. Even if some Cretaceous eutherians can eventually be interpreted as basal members of stem lineages of placental orders, it seems firmly established that some forms—such as *Prokennalestes,* asyoryctitheres, and zalambdalestids—represent stem lineages that are part of a preplacental diversification. Recovery of these immediate placental outgroups has to be hailed as one of the most important achievements for establishing character polarity within the diverse orders of placentals. This basal eutherian radiation, inaccessible to molecular studies, is one gem that morphological studies provide in the discussion of placental origins and radiation.

Divergence dates for various cladogenetic events within Placentalia, Eutheria, and Theria based on molecular data are generally much older than those based on fossils. However, the recent discovery of the eutherian *Eomaia* from the mid-Barremian of China, about 125 million years ago, has closed the gap at the Eutheria-Metatheria dichotomy a little, with molecular estimates for that event ranging between extremes of 190 and 130 million years.

ACKNOWLEDGMENTS

We thank the editors for inviting us to contribute to this volume. For comments on the manuscript, we thank Rob Asher, Inés Horovitz, and Jay Lillegraven. Funding for this article was provided by the National Science Foundation (DEB-0129031, DEB-0129061, and DEB-0129127) and the Antorchas Foundation.

REFERENCES

Alroy, J. 1999. The fossil record of North American mammals: Evidence for a Paleocene evolutionary radiation. Systematic Biology 48: 107–118.

Archer, M. 1976. The basicranial region of marsupicarnivores (Marsupialia), interrelationships of carnivorous marsupials, and affinities of the insectivorous peramelids. Zoological Journal of the Linnean Society 59: 217–322.

Archibald, J. D. 1996. Fossil evidence for a Late Cretaceous origin of "hoofed" mammals. Science 272: 1150–1153.

———. 1998. Archaic ungulates ("Condylarthra"); pp. 292–331 *in* C. Janis, K. Scott, and L. Jacobs (eds.), Tertiary Carnivores, Ungulates, and Ungulatelike Mammals. Cambridge University Press, Cambridge.

———. 2003. Timing and biogeography of the eutherian radiation: Fossils and molecules compared. Molecular Phylogenetics and Evolution 28: 350–359.

Archibald, J. D., and A. O. Averianov. 2003. The Late Cretaceous placental mammal *Kulbeckia.* Journal of Vertebrate Paleontology 23: 404–419.

Archibald, J. D., A. O. Averianov, and E. G. Ekdale. 2001. Late Cretaceous relatives of rabbits, rodents, and other extant eutherian mammals. Nature 414: 62–65.

Archibald, J. D., and D. H. Deutschman. 2001. Quantitative analysis of the timing of the origin and diversification of extant placental orders. Journal of Mammalian Evolution 8: 107–124.

Argot, C. 2001. Functional-adaptive anatomy of the forelimb in Didelphidae, and the paleobiology of the Paleocene marsupials *Mayulestes ferox* and *Pucadelphys andinus.* Journal of Morphology 247: 51–79.

———. 2002. A functional-adaptive analysis of the hindlimb anatomy of extant marsupials, and the paleobiology of the Paleocene marsupials *Mayulestes ferox* and *Pucadelphys andinus.* Journal of Morphology 253: 76–108.

———. 2003. Functional-adaptive anatomy of the axial skeleton of some extant marsupials and the paleobiology of the Paleocene marsupials *Mayulestes ferox* and *Pucadelphys andinus.* Journal of Morphology 255: 279–300.

Árnason, Ú., J. A. Adegoke, K. Bodin, E. W. Born, Y. B. Esa, A. Gullberg, M. Nilsson, R. V. Short, X. Xu, and A. Janke. 2002. Mammalian mitogenomic relationships and the root of the eutherian tree. Proceedings of the National Academy of Sciences USA 99: 8151–8156.

Asher, R. J., M. J. Novacek, and J. H. Geisler. 2003. Relationships of endemic African mammals and their fossil relatives based on morphological and molecular evidence. Journal of Mammalian Evolution 10: 131–194.

Averianov, A. O., J. D. Archibald, and T. Martin. 2003. Placental nature of the alleged marsupial from the Cretaceous of Madagascar. Acta Palaeontologica Polonica 48: 149–151.

Averianov, A. O., and P. P. Skutschas. 2001. A new genus of eutherian mammal from the Early Cretaceous of Transbaikalia, Russia. Acta Palaeontologica Polonica 46: 431–436.

Bensley, B. A. 1902. On the identification of Meckelian and mylohyoid grooves in the jaws of Mesozoic and Recent Mammalia. University of Toronto Studies, Biological Series 3: 75–81.

Bugge, J. 1974. The cephalic arterial system in insectivores, primates, rodents and lagomorphs, with special reference to the systematic classification. Acta Anatomica 87 (supplement 62): 1–159.

Butler, P. M. 1990. Early trends in the evolution of tribosphenic molars. Biological Reviews 65: 529–552.

Butler, P. M., and W. A. Clemens. 2001. Dental morphology of the Jurassic holotherian mammal *Amphitherium,* with a discussion of the evolution of mammalian post-canine dental formulae. Palaeontology 44: 1–20.

Cifelli, R. L. 1999. Tribosphenic mammal from the North American Early Cretaceous. Nature 401: 363–366.

Cifelli, R. L., and C. de Muizon. 1997. Dentition and jaw of *Kokopellia juddi,* a primitive marsupial or near-marsupial from the Medial Cretaceous of Utah. Journal of Mammalian Evolution 4: 241–258.

———. 1998. Tooth eruption and replacement pattern in early marsupials. Comptes Rendus Académie des Sciences Paris, Sciences de la Terre et des Planètes 326: 215–220.

Cifelli, R. L., T. B. Rowe, W. P. Luckett, J. Banta, R. Reyes, and R. I. Howes. 1996. Fossil evidence for the origin of the marsupial pattern of tooth replacement. Nature 379: 715–718.

Clemens, W. A., Jr. 1966. Fossil mammals of the type Lance Formation Wyoming. Part II. Marsupialia. University of California Publication in Geological Sciences 62: 1–122.

———. 1973. Fossil mammals of the type Lance Formation Wyoming. Part III. Eutheria and summary. University of California Publication in Geological Sciences 94: 1–102.

Clemens, W. A., Jr., and J. A. Lillegraven. 1986. New Late Cretaceous, North American advanced therian mammals that fit neither the marsupial nor eutherian molds; pp. 55–86 *in* K. M. Flanagan and J. A. Lillegraven (eds.), Vertebrates, Phylogeny, and Philosophy. Contributions to Geology, The University of Wyoming, Special Paper 3. University of Wyoming, Laramie, Wyoming.

Clemens, W. A., Jr., and J.R.E. Mills. 1971. Review of *Peramus tenuirostris* Owen (Eupantotheria, Mammalia). Bulletin of the British Museum (Natural History), Geology 20: 87–113.

Dashzeveg, D., and Z. Kielan-Jaworowska. 1984. The lower jaw of an aegialodontid mammal from the Early Cretaceous of Mongolia. Zoological Journal of the Linnean Society 82: 217–227.

Delsuc, F., M. Scally, O. Madsen, M. J. Stanhope, W. W. de Jong, F. M. Catzeflis, M. S. Springer, and E.J.P. Douzery. 2002. Molecular phylogeny of living xenarthrans and the impact of character and taxon sampling on the placental tree rooting. Molecular Biology and Evolution 19: 1656–1671.

Eizirik, E., W. J. Murphy, and S. J. O'Brien. 2001. Molecular dating and biogeography of the early placental mammal radiation. Journal of Heredity 92: 212–219.

Farris, J., M. Källersjo, A. G. Kluge, and C. Bult. 1994. Testing significance of incongruence. Cladistics 10: 315–319.

Foote, M., J. P. Hunter, C. M. Janis, and J. J. Sepkoski, Jr. 1999. Evolutionary and preservational constraints on origins of biologic groups: Divergence times of eutherian mammals. Science 283: 1310–1314.

Fostowicz-Frelik, Ł., and Z. Kielan-Jaworowska. 2002. Lower incisor in zalambdalestid mammals (Eutheria) and its phylogenetic significance. Acta Palaeontologica Polonica 47: 177–180.

Fox, R. C. 1979. Mammals from the Upper Cretaceous Oldman Formation, Alberta. III. Eutheria. Canadian Journal of Earth Sciences 16: 114–125.

———. 1989. The Wounded Knee local fauna and mammalian evolution near the Cretaceous-Tertiary boundary, Saskatchewan, Canada. Palaeontographica, Abteilung A 208: 11–59.

Gambaryan, P. P., and Z. Kielan-Jaworowska. 1997. Sprawling versus parasagittal stance in multituberculate mammals. Acta Palaeontologica Polonica 42: 13–44.

Godinot, M., and G.V.R. Prasad. 1994. Discovery of Cretaceous arboreal eutherians. Naturwissenschaft 81: 79–81.

Gregory, W. K., and G. G. Simpson. 1926. Cretaceous mammal skulls from Mongolia. American Museum Novitates 225: 1–20.

Hershkovitz, P. 1995. The staggered marsupial third lower incisor: Hallmark of cohort Didelphimorphia, and description of a new genus and species with staggered i3 from the Albian (Lower Cretaceous) of Texas. Bonner Zoologische Beiträge 45: 153–169.

Hopson, J. A., G. F. Engelmann, G. W. Rougier, and J. R. Wible. 1999. Skull of a paurodontid mammal (Holotheria, Dryolestoidea) from the Late Jurassic of Colorado. Journal of Vertebrate Paleontology 19 (supplement to 3): 52A–53A.

Hopson, J. A., and G. W. Rougier. 1993. Braincase structure in the oldest known skull of a therian mammal: Implications for mammalian systematics and cranial evolution. American Journal of Science 293A: 268–299.

Horovitz, I. 2000. The tarsus of *Ukhaatherium nessovi* (Eutheria, Mammalia) from the Late Cretaceous of Mongolia: An appraisal of the evolution of the ankle in basal therians. Journal of Vertebrate Paleontology 20: 547–560.

———. 2003. Postcranial skeletal morphology of *Ukhaatherium nessovi* (Eutheria, Mammalia) from the Late Cretaceous of Mongolia. Journal of Vertebrate Paleontology 23: 857–868.

Horovitz, I., and M. Sánchez-Villagra. 2003. A morphological analysis of marsupial mammal higher-level phylogenetic relationships. Cladistics 19: 182–212.

Hu, Y.-M., Y.-Q. Wang, C.-K. Li, and Z.-X. Luo. 1998. Morphology of dentition and forelimb of *Zhangheotherium*. Vertebrata PalAsiatica 36: 102–125.

Hu, Y.-M., Y.-Q. Wang, Z.-X. Luo, and C.-K. Li. 1997. A new symmetrodont mammal from China and its implications for mammalian evolution. Nature 390: 137–142.

Janke, A., O. Magnell, G. Wieczorek, M. Westerman, and Ú. Árnason. 2002. Phylogenetic analysis of 18S rRNA and the mitochondrial genomes of the wombat, *Vombatus ursinus*, and the spiny anteater, *Tachyglossus aculeatus*, increased support for the Marsupionta hypothesis. Journal of Molecular Evolution 54: 71–80.

Jellison, W. B. 1945. A suggested homolog of the os penis or baculum of mammals. Journal of Mammalogy 26: 146–147.

Jenkins, F. A., Jr. 1974. Tree shrew locomotion and the origins of primate arborealism; pp. 85–115 *in* F. A. Jenkins, Jr. (ed.), Primate Locomotion. Academic Press, New York.

Jenkins, F. A., Jr., and D. McClearn. 1984. Mechanisms of hind foot reversal in climbing mammals. Journal of Morphology 182: 197–219.

Ji, Q., Z.-X. Luo, and S.-A. Ji. 1999. A Chinese triconodont mammal and mosaic evolution of the mammalian skeleton. Nature 398: 326–330.

Ji, Q., Z.-X. Luo, C.-X. Yuan, J. R. Wible, J.-P. Zhang, and J. A. Georgi. 2002. The earliest known eutherian mammal. Nature 416: 816–822.

Johnson, P. A., and R. C. Fox. 1984. Paleocene and Late Cretaceous mammals from Saskatchewan, Canada. Palaeontographica, Abteilung A 186: 163–222.

Kielan-Jaworowska, Z. 1975a. Preliminary description of two new eutherian genera from the Late Cretaceous of Mongolia. Palaeontologia Polonica 33: 5–16.

———. 1975b. Evolution of the therian mammals in the Late Cretaceous of Asia. Part I. Deltatheridiidae. Palaeontologia Polonica 33: 102–132.

———. 1977. Evolution of the therian mammals in the Late Cretaceous of Asia. Part II. Postcranial skeleton in *Kennalestes* and *Asioryctes*. Palaeontologia Polonica 37: 65–83.

———. 1978. Evolution of the therian mammals in the Late Cretaceous of Asia. Part III. Postcranial skeleton in Zalambdalestidae. Palaeontologia Polonica 38: 5–41.

———. 1981. Evolution of the therian mammals in the Late Cretaceous of Asia. Part IV. Skull structure in *Kennalestes* and *Asioryctes*. Palaeontologia Polonica 42: 25–78.

———. 1984. Evolution of the therian mammals in the Late Cretaceous of Asia. Part V. Skull structure in Zalambdalestidae. Palaeontologia Polonica 46: 107–117.

Kielan-Jaworowska, Z., R. L. Cifelli, and Z.-X. Luo. 1998. Alleged Cretaceous placental from down under. Lethaia 31: 267–268.

Kielan-Jaworowska, Z., and D. Dashzeveg. 1989. Eutherian mammals from the Early Cretaceous of Mongolia. Zoologica Scripta 18: 347–355.

Kielan-Jaworowska, Z., J. G. Eaton, and T. M. Bown. 1979. Theria of metatherian-eutherian grade; pp. 182–191 in J. A. Lillegraven, Z. Kielan-Jaworowska, and W. A. Clemens (eds.), Mesozoic Mammals: The First Two-Thirds of Mammalian History. University of California Press, Berkeley.

Kielan-Jaworowska, Z., and P. P. Gambaryan. 1994. Postcranial anatomy and habits of Asian multituberculate mammals. Fossils and Strata 36: 1–92.

Kielan-Jaworowska, Z., J. H. Hurum, and D. Badamgarav. 2003. An extended range of the multituberculate Kryptobaatar and distribution of mammals in the Upper Cretaceous of the Gobi Desert. Acta Palaeontologica Polonica 48: 273–278.

Kielan-Jaworowska, Z., M. J. Novacek, B. A. Trofimov, and D. Dashzeveg. 2000. Mammals from the Mesozoic of Mongolia; pp. 573–626 in M. J. Benton, M. A. Shishkin, D. M. Unwin, and E. N. Kurochkin (eds.), The Age of Dinosaurs in Russia and Mongolia. Cambridge University Press, Cambridge.

Kielan-Jaworowska, Z., and B. A. Trofimov. 1980. Cranial morphology of the Cretaceous eutherian mammal Barunlestes. Acta Palaeontologica Polonica 25: 167–185.

Kobayashi, Y., D. A. Winkler, and L. L. Jacobs. 2002. Origin of the tooth-replacement pattern in therian mammals: Evidence from a 110 Myr old fossil. Proceedings of the Royal Society of London B269: 369–373.

Krause, D. W., and F. A. Jenkins, Jr. 1983. The postcranial skeleton of North American multituberculates. Bulletin of the Museum of Comparative Zoology 150: 199–246.

Krebs, B. 1991. Das Skelett von Henkelotherium guimarotae gen. et sp. nov. (Eupantotheria, Mammalia) aus dem Oberen Jura von Portugal. Berliner geowissenschaftliche, Abhandlungen A 133: 1–121.

Kumar, S., and B. Hedges. 1998. A molecular tree for vertebrate evolution. Nature 392: 917–919.

Lessertisseur, J., and R. Saban. 1967. Squelette axial; pp. 584–708 in P. Grassé (ed.), Traité de Zoologie. Volume 16, no. 1. Masson et Cie, Paris.

Lewis, O. J. 1989. Functional Morphology of the Evolving Hand and Foot. Clarendon Press, Oxford.

Lillegraven, J. A. 1969. Latest Cretaceous mammals of the upper part of the Edmonton Formation of Alberta, Canada, and review of marsupial-placental dichotomy in mammalian evolution. The University of Kansas Paleontological Contributions, Article 50 (Vertebrata 12): 1–122.

———. 1975. Biological considerations of the marsupial-placental dichotomy. Evolution 29: 707–722.

Lillegraven, J. A., S. D. Thompson, B. K. McNab, and J. L. Patton. 1987. The origin of eutherian mammals. Biological Journal of the Linnean Society 32: 281–336.

Luckett, W. P. 1993. An ontogenetic assessment of dental homologies in therian mammals; pp. 182–204 in F. S. Szalay, M. J. Novacek, and M. C. McKenna (eds.), Mammal Phylogeny: Mesozoic Differentiation, Multituberculates, Monotremes, Early Therians, and Marsupials. Springer-Verlag, New York.

Luo, Z.-X., R. L. Cifelli, and Z. Kielan-Jaworowska. 2001. Dual origin of tribosphenic mammals. Nature 409: 53–57.

Luo, Z.-X., Q. Ji, J. R. Wible, and C.-X. Yuan. 2003. An Early Cretaceous tribosphenic mammal and metatherian evolution. Science 302: 1934–1940.

Luo, Z.-X., Z. Kielan-Jaworowska, and R. L. Cifelli. 2002. In quest for a phylogeny of Mesozoic mammals. Acta Palaeontologica Polonica 47: 1–78.

MacPhee, R.D.E. 1994. Morphology, adaptations, and relationships of Plesiorycteropus, and a diagnosis of a new order of eutherian mammals. Bulletin of the American Museum of Natural History 220: 1–214.

Madsen, O., M. Scally, C. J. Douady, D. J. Kao, R. W. DeBry, R. Adkins, H. M. Armine, M. J. Stanhope, W. W. de Jong, and M. S. Springer. 2001. Parallel adaptive radiations in two major clades of placental mammals. Nature 409: 610–614.

Marshall, L. G., C. de Muizon, and D. Sigogneau-Russell. 1995. Pucadelphys andinus (Marsupialia, Mammalia) from the early Paleocene of Bolivia. Mémoires du Muséum National d'Histoire Naturelle 165: 1–164.

Martin, T. 1997. Tooth replacement in Late Jurassic Dryolestidae (Eupantotheria, Mammalia). Journal of Mammalian Evolution 4: 1–18.

———. 2002. New stem-lineage representatives of Zatheria (Mammalia) from the Late Jurassic of Portugal. Journal of Vertebrate Paleontology 22: 332–348.

McKenna, M. C. 1975. Toward a phylogenetic classification of the Mammalia; pp. 21–46 in W. P. Luckett and F. S. Szalay (eds.), Phylogeny of the Primates: A Multidisciplinary Approach. Plenum, New York.

McKenna, M. C., and S. K. Bell. 1997. Classification of Mammals above the Species Level. Columbia University Press, New York.

McKenna, M. C., Z. Kielan-Jaworowska, and J. Meng. 2000. Earliest eutherian mammal skull, from the Late Cretaceous (Coniacian) of Uzbekistan. Acta Palaeontologica Polonica 45: 1–54.

Meng, J., Y.-M. Hu, and C.-K. Li. 2003a. The osteology of Rhombomylus (Mammalia, Glires): Implications for phylogeny and evolution of Glires. Bulletin of the American Museum of Natural History 275: 1–247.

Meng, J., Y.-M. Hu, Y.-Q. Wang, and C.-K. Li. 2003b. The ossified Meckel's cartilage and internal groove in Mesozoic mammaliaforms: Implications to origin of the definitive mammalian middle ear. Zoological Journal of the Linnean Society 138: 431–448.

Meng, J., and A. R. Wyss. 2001. The morphology of Tribosphenomys (Rodentiaformes, Mammalia): Phylogenetic implications for basal Glires. Journal of Mammalian Evolution 8: 1–71.

Muizon, C. de. 1998. Mayulestes ferox, a borhyaenoid (Metatheria, Mammalia) from the early Palaeocene of Bolivia. Phylogenetic and palaeobiologic implications. Geodiversitas 20: 19–142.

Murphy, W. J., E. Eizirik, W. E. Johnson, Y. P. Zhang, O. A. Ryder, and S. J. O'Brien. 2001a. Molecular phylogenetics and the origins of placental mammals. Nature 409: 614–618.

Murphy, W. J., E. Eizirik, S. J. O'Brien, O. Madsen, M. Scally, C. J. Douady, E. Teeling, O. A. Ryder, M. J. Stanhope, W. W. de Jong, and M. S. Springer. 2001b. Resolution of the early placental mammal radiation using Bayesian phylogenetics. Science 294: 2348–2351.

Nessov, L. A. 1997. Cretaceous Nonmarine Vertebrates of Northern Eurasia. L. B. Golovneba and A. O. Averianov (eds.). Institute of the Earth's Crust, St. Petersburg University, St. Petersburg. (In Russian with English summary.)

Nessov, L. A., J. D. Archibald, and Z. Kielan-Jaworowska. 1998. Ungulate-like mammals from the Late Cretaceous of Uzbekistan and a phylogenetic analysis of Ungulatamorpha; *in* K. C. Beard and M. R. Dawson (eds.), Dawn of the Age of Mammals in Asia. Bulletin of Carnegie Museum of Natural History 34: 40–88.

Nessov, L. A., D. Sigogneau-Russell, and D. E. Russell. 1994. A survey of Cretaceous tribosphenic mammals from middle Asia (Uzbekistan, Kazakhstan and Tajikistan), of their geological setting, age and faunal environment. Palaeovertebrata Montpellier 23: 51–92.

Novacek, M. J. 1986a. The skull of leptictid insectivorans and the higher-level classification of eutherian mammals. Bulletin of the American Museum of Natural History 183: 1–112.

———. 1986b. The primitive eutherian dental formula. Journal of Vertebrate Paleontology 6: 191–196.

———. 1992a. Fossils, topologies, missing data, and the higher level phylogeny of eutherian mammals. Systematic Biology 41: 58–73.

———. 1992b. Mammalian phylogeny: Shaking the tree. Nature 356: 121–125.

———. 1993. Patterns of skull diversity in the mammalian skull; pp. 438–545 *in* J. Hanken and B. K. Hall (eds.), The Skull. Volume 2. Patterns of Structural and Systematic Diversity. University of Chicago Press, Chicago.

———. 1997. Mammalian evolution: An early record bristling with evidence. Current Biology 7: R489–R491.

———. 1999. 100 million years of land vertebrate evolution: The Cretaceous-Early Tertiary transition. Annals of the Missouri Botanical Garden 86: 230–258.

Novacek, M. J., G. W. Rougier, J. R. Wible, M. C. McKenna, D. Dashzeveg, and I. Horovitz. 1997. Epipubic bones in eutherian mammals from the Late Cretaceous of Mongolia. Nature 389: 483–486.

Nowak, R. M. 1991. Walker's Mammals of the World. Fifth Edition. The Johns Hopkins University Press, Baltimore.

O'Leary, M. A., M. Allard, M. J. Novacek, J. Gatesy, and J. Meng. 2004. Building the mammalian sector of the tree of life: Combining different data and a discussion of divergence times for placental mammals; pp. 490–516 *in* J. Cracraft and M. Donoghue (eds.), Assembling The Tree of Life. Oxford University Press, Oxford.

Owen, R. 1868. On the Anatomy of the Vertebrates. Volume III. Mammals. Longmans, Green, and Co., London.

Prasad, G.V.R., and M. Godinot. 1994. Eutherian tarsal bones from the Late Cretaceous of India. Journal of Paleontology 68: 892–902.

Rana, R. S., and G. P. Wilson. 2003. New Late Cretaceous mammals from the Intertrappean beds of Rangapur, India and paleobiogeographic framework. Acta Palaeontologica Polonica 48: 331–348.

Reilly, S. M., and T. D. White. 2003. Hypaxial motor patterns and the function of epipubic bones in primitive mammals. Science 299: 400–402.

Rich, T. H., P. Vickers-Rich, A. Constantine, T. F. Flannery, L. Kool, and N. van Klavern. 1997. A tribosphenic mammal from the Mesozoic of Australia. Science 278: 1438–1442.

———. 1999. Early Cretaceous mammals from Flat Rocks, Victoria, Australia. Records of the Queen Victoria Museum 106: 1–35.

Roca, A. L., G. K. Bar-Gal, E. Eizirik, K. M. Helgen, R. Maria, M. S. Springer, S. J. O'Brien, and W. J. Murphy. 2004. Mesozoic origin for West Indian insectivores. Nature 429: 649–651.

Rose, K. D. 1999. Postcranial skeleton of Eocene Leptictidae (Mammalia), and its implications for behavior and relationships. Journal of Vertebrate Paleontology 19: 355–372.

———. 2001. The ancestry of whales. Science 293: 2216–2217.

Rougier, G. W. 1993. *Vincelestes neuquenianus* Bonaparte (Mammalia, Theria) un primitivo mamífero del Cretácico Inferior de la Cuenca Neuquina. Ph.D. Dissertation, University of Buenos Aires.

Rougier, G. W., Q. Ji, and M. J. Novacek. 2003. A new symmetrodont mammal with fur impressions from the Mesozoic of China. Acta Geologica Sinica 77: 7–14.

Rougier, G. W., and J. R. Wible. In press. Major changes in the mammalian ear region and basicranium; *in* M. T. Carrano, T. J. Gaudin, R. W. Blob, and J. R. Wible (eds.), Amniote Paleobiology. University of Chicago Press, Chicago.

Rougier, G. W., J. R. Wible, and J. A. Hopson. 1992. Reconstruction of the cranial vessels in the Early Cretaceous mammal *Vincelestes neuquenianus:* Implications for the evolution of the mammalian cranial vascular system. Journal of Vertebrate Paleontology 12: 188–216.

———. 1996a. Basicranial anatomy of *Priacodon fruitaensis* (Triconodontidae, Mammalia) from the Late Jurassic of Colorado, and a reappraisal of mammaliaform interrelationships. American Museum Novitates 3183: 1–38.

Rougier, G. W., J. R. Wible, and M. J. Novacek. 1996b. Multituberculate phylogeny. Nature 379: 406.

———. 1998. Implications of *Deltatheridium* specimens for early marsupial history. Nature 396: 459–463.

———. In press. New specimen of *Deltatheroides cretacicus* (Metatheria, Deltatheroida) from the Late Cretaceous of Mongolia; *in* M. R. Dawson and J. A. Lillegraven (eds.), Fanfare for an Uncommon Paleontologist: Papers in Honor of Malcolm C. McKenna. Bulletin of Carnegie Museum of Natural History.

Saban, R. 1967. Enderostes; pp. 1079–1087 *in* P. Grassé (ed.), Traité de Zoologie. Volume 16, Number 1. Masson et Cie, Paris.

Sánchez-Villagra, M. R., and K. K. Smith. 1997. Diversity and evolution of the marsupial mandibular angular process. Journal of Mammalian Evolution 4: 119–144.

Sánchez-Villagra, M. R., and J. R. Wible. 2002. Patterns of evolutionary transformation in the petrosal bone and some basicranial features in marsupial mammals, with special reference to didelphids. Journal of Zoological Systematics and Evolutionary Research 40: 26–45.

Sargis, E. J. 2002. A multivariate analysis of the postcranium of tree shrews (Scandentia, Tupaiidae) and its taxonomic implications. Mammalia 66: 579–598.

Sereno, P. C., and M. C. McKenna. 1995. Cretaceous multituberculate skeleton and the early evolution of the mammalian shoulder girdle. Nature 377: 144–147.

Setoguchi, T., T. Tsubamoto, H. Hanamura, and K. Hachiya. 1999. An early Late Cretaceous mammal from Japan, with reconsideration of the evolution of tribosphenid molars. Paleontological Research 3: 18–28.

Shoshani, J., and M. C. McKenna. 1998. Higher taxonomic relationships among extant mammals based on morphology, with selected comparisons of results from molecular data. Molecular Phylogenetics and Evolution 9: 572–584.

Simpson, G. G. 1928a. A Catalogue of the Mesozoic Mammalia in the Geological Department of the British Museum. British Museum (Natural History), London.

————. 1928b. Mesozoic Mammalia. XII. The internal mandibular groove of Jurassic mammals. American Journal of Science 15: 461–470.

————. 1929. American Mesozoic Mammalia. Memoirs of the Peabody Museum of Yale University 3: 1–235.

————. 1945. The principles of classification and a classification of mammals. Bulletin of the American Museum of Natural History 85: 1–350.

Smith, A. B. 1998. What does paleontology contribute to systematics in a molecular world? Molecular Phylogenetics and Evolution 9: 437–447.

Springer, M. S., J.A.W. Kirsch, and J. A. Case. 1997. The chronicle of marsupial evolution; pp. 129–161 in T. J. Givinish and K. J. Systsma (eds.), Molecular Evolution and Adaptive Radiations. Cambridge University Press, Cambridge.

Springer, M. S., W. J. Murphy, E. Eizirik, and S. J. O'Brien. 2003. Placental mammal diversification and the Cretaceous-Tertiary boundary. Proceedings of the National Academy of Sciences USA 100: 1056–1061.

Starck, D. 1995. Lehrbuch der speziellen Zoologie. Volume II. Wirbeltiere. Part 5/1. Säugetiere. Gustav Fischer, Jena.

Szalay, F. S. 1977. Phylogenetic relationships and a classification of the eutherian Mammalia; pp. 315–374 in M. K. Hecht, P. C. Goody, and B. M. Hecht (eds.), Major Patterns in Vertebrate Evolution. Plenum, New York.

————. 1994. Evolutionary History of the Marsupials and an Analysis of Osteological Characters. Cambridge University Press, New York.

Szalay, F. S., and R. L. Decker. 1974. Origins, evolution, and function of the tarsus in Late Cretaceous Eutheria and Paleocene Primates; pp. 224–258 in F. A. Jenkins, Jr. (ed.), Primate Locomotion. Academic Press, New York.

Szalay, F. S., and B. A. Trofimov. 1996. The Mongolian Late Cretaceous Asiatherium, and the early phylogeny and paleobiogeography of Metatheria. Journal of Vertebrate Paleontology 16: 474–509.

Tandler, J. 1899. Zur vergleichenden Anatomie der Kopfarterien bei den Mammalia. Denkschriften Akademie der Wissenschaften, Wien, mathematisch-naturwissenschaftliche Klasse 67: 677–784.

Thenius, E. 1989. Zähne und Gebiß der Säugetiere; pp. 1–513 in J. Niethammer, H. Schliemann, and D. Starck (eds.), Handbook of Zoology. Volume VIII. Mammalia. Part 56. Walter de Gruyter, Berlin.

Thewissen, J.G.M., and S. I. Madar. 1999. Ankle morphology of the earliest cetaceans and its implications for the phylogenetic relations among ungulates. Systematic Biology 48: 21–30.

Thewissen, J.G.M., E. M. Williams, L. J. Roe, and S. T. Hussain. 2001. Skeletons of terrestrial cetaceans and the relationships of whales to artiodactyls. Nature 413: 277–281.

Turnbull, W. D. 1970. Mammalian masticatory apparatus. Fieldiana, Geology 18: 149–356.

Van Valen, L. 1964. A possible origin for rabbits. Evolution 18: 484–491.

————. 1978. The beginning of the Age of Mammals. Evolutionary Theory 4: 45–80.

Wang, Y.-Q., Y.-M. Hu, J. Meng, and C.-K. Li. 2001. An ossified Meckel's cartilage in two Cretaceous mammals and origin of the mammalian middle ear. Science 294: 357–361.

Wheeler, W. C. 1995. Sequence alignment, parameter sensitivity, and the phylogenetic analysis of molecular data. Systematic Biology 44: 321–331.

Wible, J. R. 1987. The eutherian stapedial artery: Character analysis and implications for superordinal relationships. Zoological Journal of the Linnean Society 91: 107–135.

————. 1990. Late Cretaceous marsupial petrosal bones from North America and a cladistic analysis of the petrosal in therian mammals. Journal of Vertebrate Paleontology 10: 183–205.

————. 1991. Origin of Mammalia: The craniodental evidence reexamined. Journal of Vertebrate Paleontology 11: 1–28.

————. 2003. On the cranial osteology of the short-tailed opossum Monodelphis brevicaudata (Didelphidae, Marsupialia). Annals of Carnegie Museum 72: 137–202.

Wible, J. R., and J. A. Hopson. 1993. Basicranial evidence for early mammal phylogeny; pp. 45–62 in F. S. Szalay, M. J. Novacek, and M. C. McKenna (eds.), Mammal Phylogeny: Mesozoic Differentiation, Multituberculates, Monotremes, Early Therians, and Marsupials. Springer-Verlag, New York.

————. 1995. Homologies of the prootic canal in mammals and non-mammalian cynodonts. Journal of Vertebrate Paleontology 15: 331–356.

Wible, J. R., M. J. Novacek, and G. W. Rougier. 2004. New data on skull structure in the Mongolian Late Cretaceous eutherian mammal Zalambdalestes. Bulletin of the American Museum of Natural History 281: 1–144.

Wible, J. R., G. W. Rougier, M. J. Novacek, and M. C. McKenna. 2001. Earliest eutherian ear region: A petrosal referred to Prokennalestes from the Early Cretaceous of Mongolia. American Museum Novitates 3322: 1–44.

Wilson, D. E., and D. M. Reeder (eds.). 1993. Mammal Species of the World: A Taxonomic and Geographic Reference. Smithsonian Institution Press, Washington, D.C.

Woodburne, M. O., T. H. Rich, and M. S. Springer. 2003. The evolution of tribosepheny and the antiquity of mammalian clades. Molecular Phylogenetics and Evolution 28: 360–385.

MARK S. SPRINGER,
WILLIAM J. MURPHY,
EDUARDO EIZIRIK,
AND STEPHEN J. O'BRIEN

MOLECULAR EVIDENCE FOR MAJOR PLACENTAL CLADES

SUCH FEATURES OF MAMMALIAN ANATOMY as the skeleton, muscular system, and reproductive system have long supplied systematists with evidence for proposing and testing hypotheses of supraordinal relationships. The ongoing challenge is to distinguish between homologous and homoplastic characters. Morphological phylogenies often support competing hypotheses (e.g., the placement of pholidotans as the sister taxon to either Xenarthra or Carnivora), but there remain recurrent hypotheses that pervade morphological studies of placental mammal phylogeny. Examples include Tethytheria (Proboscidea + Sirenia), Paenungulata (Tethytheria + Hyracoidea), Altungulata (Paenungulata + Perissodactyla), Ungulata (Paenungulata + Perissodactyla + Artiodactyla + Cetacea + Tubulidentata), Glires (Lagomorpha + Rodentia), Anagalida (Glires + Macroscelidea), Volitantia (Chiroptera + Dermoptera), Archonta (Volitantia + Primates + Scandentia), and Epitheria (all living placentals except Xenarthra). Against this backdrop of a priori hypotheses, the advent of molecular data provides an opportunity to evaluate morphologically based groups that reflect homology vs. homoplastic similarity.

Early molecular data sets included immunological distances (Goodman, 1975; Shoshani, 1986) and amino acid sequences for one or more proteins (Goodman, 1975; de Jong et al., 1981; Shoshani et al., 1985; Kleinschmidt et al., 1986; Miyamoto and Goodman, 1986). These studies failed to provide a well-resolved tree for the orders of placental mammals, but in some cases, provided support for a paenungulate clade, with or without the inclusion of aardvarks (de Jong et al., 1981; Kleinschmidt et al., 1986; Miyamoto and Goodman, 1986; Shoshani, 1986). Subsequently, protein sequences supported an association of elephant shrews with aardvarks and paenungulates (de Jong et al., 1993). In addition, early studies challenged the naturalness of such groups as Archonta (Goodman, 1975), Edentata (Shoshani, 1986), and Ungulata (Shoshani, 1986).

In the past decade, DNA studies have been at the forefront of molecular investigations into placental phylogeny. Initially, these studies were based on segments of single genes and met with limited success (Stanhope et al., 1992, 1993, 1996; Springer and Kirsch, 1993; Lavergne et al., 1996; Porter et al., 1996; Springer et al., 1997a). Nevertheless, they provided robust support for Paenungulata, as well as for an expanded clade that joined aardvarks and elephant shrews with paenungulates. DNA data sets that increased taxon sampling and included both mitochondrial and nuclear genes suggested that the paenungulate-aardvark-elephant shrew clade also included the insectivore families Chrysochloridae and Tenrecidae (Springer et al., 1997b; Stanhope et al., 1998a,b). Stanhope et al. (1998b) proposed the name Afrotheria for the clade that includes Hyracoidea, Proboscidea, Sirenia, Macroscelidea, Tubulidentata, and Afrosoricida—the latter a newly proposed order for chrysochlorids and tenrecids. This hypothesis met resistance among morphologists, both because morphological synapomorphies for Afrotheria were not forthcoming (Asher, 1999) and because this hypothesis challenged the monophyly of lipotyphlan insectivores (MacPhee and Novacek, 1993), rendering this group polyphyletic.

Large data sets, including both longer gene segments of individual genes (*BRCA1;* Madsen et al., 2001; Scally et al., 2001) and concatenated data sets incorporating segments from multiple genes (Eizirik et al., 2001; Madsen et al., 2001; Murphy et al., 2001a,b; Scally et al., 2001; Delsuc et al., 2002; Amrine-Madsen et al., 2003; Springer et al., in press), provided for increased resolution and divided placental mammals into four major groups: Afrotheria, Xenarthra, Laurasiatheria, and Euarchontoglires. The largest concatenations of DNA sequences, which range from 16.4 to 17.3 kb (Murphy et al., 2001b; Amrine-Madsen et al., 2003; Springer et al., in press), allow for only local rearrangements in the placental tree (e.g., the paenungulate trifurcation is not satisfactorily resolved) when they are analyzed with maximum likelihood and Bayesian methods (Fig. 4.1). It is worth noting that these data sets are dominated by nuclear exons, which have more resolving power than do mitochondrial protein-coding genes for investigating deep-level placental relationships (Springer et al., 2001a). The emerging view of higher-level placental relationships that has resulted from nuclear data sets, with or without the addition of mitochondrial rRNA genes, is partially corroborated by other lines of evidence, including rare genomic changes (see below), analyses of complete rRNA + tRNA gene sequences from the mitochondrial genome (Hudelot et al., 2003), and to a lesser extent, by mitochondrial protein-coding sequences (Árnason et al., 2002; Lin et al., 2002). At the same time, there have been important challenges to the growing consensus that there are four major clades of placental mam-

mals. The analysis by Árnason et al. (2002) of mitochondrial protein-coding sequences suggests that Laurasiatheria is diphyletic and that Euarchontoglires, Glires, and Rodentia are all paraphyletic taxa near the base of the placental tree. Another feature of these authors' topology is that Afrotheria and Xenarthra are deeply nested in the placental tree rather than basal or near-basal, as we have recovered in our analyses (e.g., Madsen et al., 2001; Murphy et al., 2001a,b; Amrine-Madsen et al., 2003). Asher et al. (2003) performed maximum parsimony analysis with a molecular concatenation that included 19 nuclear segments and three mitochondrial genes under 12 different optimization alignment settings that employed different character transformation weights. They recovered paraphyletic Rodentia, Glires, and Euarchontoglires at the base of Placentalia in six of 12 analyses. Rodent paraphyly at the base of Placentalia, with a basal split between murids and other placentals, also resulted in three of 12 analyses in the total evidence analyses by Asher et al. (2003) that included both molecular and morphological data. In analyses with depauperate taxon sampling, Misawa and Janke (2003) have suggested that Glires is paraphyletic and that lagomorphs are more closely related to primates and artiodactyls than to rodents. The possibility that rodents, Glires, and Euarchontoglires are paraphyletic at the base of Placentalia has profound consequences for the early biogeographic history of Placentalia and the deployment of morphological and genomic changes in this group.

For paleontologists who are primarily concerned with morphological data, choosing between disparate molecular views of placental relationships may seem daunting. We argue later in the chapter that a robust solution for placental relationships is already in place and allows for only local rearrangements. Some molecular studies that challenge this view, principally by altering the placement of the placental root, are compromised by limited taxon sampling and/or inadequate methods of phylogenetic analysis. Below, we review evidence for the major features of the phylogeny depicted in Figure 4.1, with the view that congruence from fundamentally different types of data is a guiding principle in systematics (Patterson, 1982). We also review problems associated with analyses that place rodents at the base of the placental tree, often as a paraphyletic taxon. Next, we provide a molecular timescale for placental evolution and discuss implications of placental phylogeny and divergence times for understanding the biogeographic history of Placentalia. Finally, we offer a brief prospectus on integrating molecular and morphological data, including data for fossil taxa.

MAJOR CLADES OF PLACENTAL MAMMALS

Maximum likelihood and Bayesian analyses of data sets that emphasize concatenated nuclear genes provide robust support for four major groups of placental mammals:

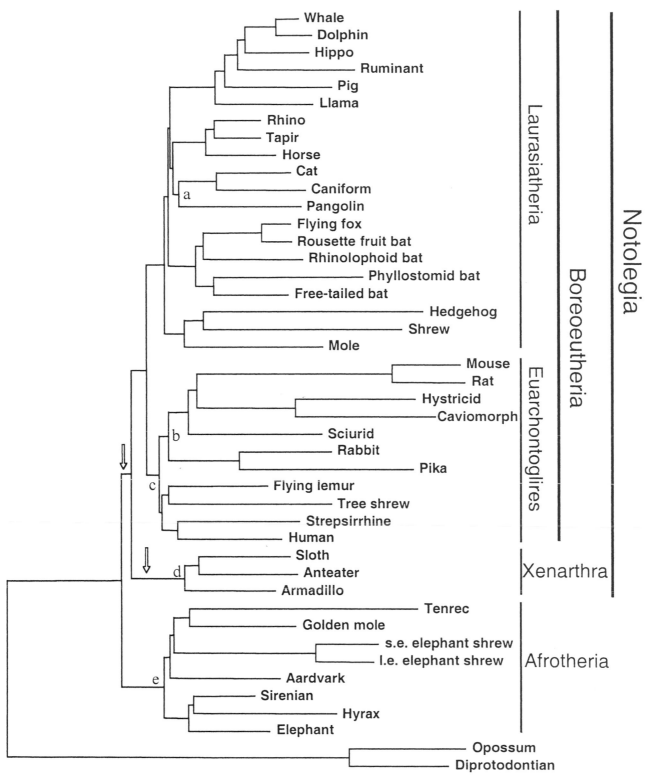

Fig. 4.1. Molecular phylogeny of placental mammals, based on the concatenated DNA sequence data set of Murphy et al. (2001b). Branch lengths are drawn to scale, indicating the proportional magnitude of molecular change per lineage. Major clades are indicated on the right. Arrows denote the two alternative positions for the root. Letters indicate additional lines of evidence supporting adjacent nodes: a, Carnivora + Pholidota (deletion in *apolipoprotein B* gene; osseus tentorium); b, Glires (morphological synapomorphies); c, Euarchontoglires (deletions in two different genes); d, Xenarthra (morphological synapomorphies; deletion in the *aA-crystallin* gene); e, Afrotheria (deletions in two different genes; presence of unique SINE elements).

Xenarthra, Afrotheria, Laurasiatheria, and Euarchontoglires (Madsen et al., 2001; Murphy et al., 2001a,b; Scally et al., 2001; Delsuc et al., 2002; Amrine-Madsen et al., 2003; Springer et al., in press) (Table 4.1). Furthermore, there is robust support for Boreoeutheria; that is, a clade comprised of Laurasiatheria and Euarchontoglires. Of these clades, only Xenarthra is supported by morphological data. The absence of morphological evidence for Afrotheria, Laurasiatheria, Euarchontoglires, and Boreoeutheria suggests that synapomorphies for these clades were never present, have been eroded in the subsequent evolutionary history of these taxa, or remain to be discovered, possibly among morphological characters that are not confederated with diet and locomotion. Even in the absence of morphological evidence for these clades, rare genomic changes have been discovered that support Xenarthra, Afrotheria, Euarchontoglires, and possibly Laurasiatheria. Xenarthra is supported by a three-amino-acid deletion in the eye lens protein aA-crystallin (van Dijk et al., 1999). Rare genomic changes supporting Afrotheria include a 9 bp deletion in *BRCA1* (Madsen et al., 2001; Scally et al., 2001), a 237–246 bp deletion in the *apolipoprotein B* alignment of Amrine-Madsen et al. (2003), and a family of short interspersed nuclear elements (SINE) called AfroSINES that is unique to this group (Nikaido et al., 2003). In addition, chromosome painting studies now reveal two potential genomic characters (associations or shared syntenies) supporting the monophyly of representative afrotherians (elephants and aardvarks; Froenicke et al., 2003). Rare genomic changes supporting Euarchontoglires include an 18-amino-acid deletion in exon 8 of the *SCA1* gene, and a 6 bp deletion in the *PRNP* gene (Poux et al., 2002; de Jong et al., 2003). Thomas et al. (2003) identified several indels, including transposon insertions, that occur in primates and rodents but not in carnivores, artiodactyls, or non-mammalian vertebrates that were sampled. These indels are consistent with the monophyly of Euarchontoglires, but will require additional taxon sampling within Placentalia to confirm that they are true synapomorphies. Finally, and with less certainty because of alternate alignment possibilities, we have discovered a putative deletion in *PLCB4* that supports Laurasiatheria (Fig. 4.2).

Given the monophyly of Xenarthra, Afrotheria, and Boreoeutheria, there are only three possible locations for the root of the placental tree. The first is between Xenarthra

Table 4.1 Bootstrap percentages and posterior probabilities for the major clades of placental mammals

Method	Data Set	Reference	Clade				
			Xenarthra	Afrotheria	Euarchontoglires	Laurasiatheria	Boreoeutheria
ML[a]	BRCA1 (2,947 bp)	Madsen et al. (2001)	100, 100	100, 100	100, 100	100, 100	100, 100
ML[a]	6 gene concatenation (5,708 bp)	Madsen et al. (2001)	Not applicable	100, 100	44, 66	99, 99	45, 64
ML	18 gene concatenation (9,779 bp)	Murphy et al. (2001a)	100	99	85	99	79
ML	22 gene concatenation (16,397 bp)	Murphy et al. (2001b)	100	100	100	100	100
ML	3 gene concatenation (4,350 bp)	Delsuc et al. (2002)	100	100	98	100	98
ML	APOB (1,342 bp)	Amrine-Madsen et al. (2003)	98	77	100	100	94
ML	23 gene concatenation (17,736 bp)	Amrine-Madsen et al. (2003)	100	100	100	100	100
Bayesian	22 gene concatenation (16,397 bp)	Murphy et al. (2001b)	1.00	1.00	1.00	1.00	1.00
Bayesian	3 gene concatenation (4,350 bp)	Delsuc et al. (2002)	1.00	1.00	1.00	1.00	1.00
Bayesian	APOB (1,342 bp)	Amrine-Madsen et al. (2003)	1.00	1.00	1.00	1.00	1.00
Bayesian	23 gene concatenation (17,736 bp)	Amrine-Madsen et al. (2003)	1.00	1.00	1.00	1.00	1.00
Bayesian	Mt tRNA + rRNA genes (3,571 bp)	Hudelot et al. (2003)	1.00	1.00	1.00	0.96	0.96
Bayesian	Concatenation of UTRs from 4 genes (1,762 bp)	Springer et al. (in press)	1.00	1.00	1.00	1.00	1.00
Bayesian	Concatenation of protein-coding segments from 15 genes (12,988 bp)	Springer et al. (in press)	1.00	1.00	1.00	1.00	1.00

[a] Duplicate maximum likelihood (ML) values in cells across these rows are for analyses without an allowance for rate-heterogeneity (first value in double-valued clade column entries) and with a gamma distribution of rates (second value).

```
Didelphis     CATGGAGAATAGAT--GAGTTC-CACATTTCAGTTTTAACATTTT
Macropus      CATGGAGAACACGTAAGAGTTA-GATATTTCAGTTTTAACTTTTT
Choloepus     CACAGAGACTTAGAATGA-TAA-CATAC-TCCTTTTGAGCATTT-
Amblysomus    CATAGAGACCTGGAATGACTTA-CATGC-TTCTATTTAACATTTT
Trichechus    CATAGACACTTGGAATGACTTA-CATAT-TTCTGTTTAACACTTC
Orycteropus   CATAGAGACTTGGAATGACTTA-CATA--TTCTATTTAACATTTT
Tamias        CACAAAGACGTGGAATGAC--A-CATAC-TTCTATTTATCAGTTT
Mus           CACATGGACTGGGGATGACCAAACGCCTCTTATATTTAACAATTT
Cavia         CACAGAGACTGGGGATGACTTA-TGTAC-TTCGATTTAACGGTTC
Sylvilagus    CGGAGAGACTTGTA-TGACCTA-CGTAC-ATATCTTTACCAGTTG
Cynocephalus  CACAGAGACTTGGGATGACTTA-CATAC-TTCAGTTTAACAGTTT
Tupaia        CACAGCGACATGGAACAATTTA-TATGC-TTCTCTTTAGCAGTTT
Homo          CAGAGAGACTTGGAATGTCTGA-CTGAC-TTCTATTTAACAGCTT
Erinaceus     CACAGAAAC----------TTA-CATAC-TTCTATTTAACATTTT
Sorex         CACCGAAAC----------GTA-CATAC-TTCTATTTAATGTTTT
Artibeus      CACAGA-C-----------TAA-CATGC-TTCTACTTCANGT---
Rousettus     CACAGA-T-----------TTA-CATACCTTTTGTTTCACATTTT
Megaptera     CACAGAGAC----------TTA-CGTAC-TTCTCTTAAATGTTTT
Hippopotamus  CACAGAGAC----------TTA-CGTAC-TTCTCTTAAACATTTT
Tragelaphus   CACAGAGAT---------------GTAC-TTCTCTTAAACATTCT
Equus         CACAGAGAT---------TTA---------------ACCTTTG
Ceratotherium CACAGAGAT---------TTA-CATAC-TTCTCTTTAACATCTT
Tapirus       CACAGAGAT---------TTA-CATAC-TTCTCTTTAACATTTT
Felis         CACCGAGAC----------TTA-CATAT-TTCTATTTAACATTTT
Manis         CACAGAGAC----------TTA-CATAC-TTCTACTTAACATTTC
```

Fig. 4.2. Putative deletion for Laurasiatheria in the 3′ UTR of the *PLCB4* gene. Representative sequences are from Murphy et al. (2001a,b). The sequence for *Homo* corresponds to positions 165–207 of GenBank sequence AY011788.

and Epitheria, which is consistent with some morphological studies (McKenna, 1975). The second possibility, which is favored by several molecular studies (Murphy et al., 2001b; Amrine-Madsen et al., 2003), is between Afrotheria and other placental mammals (i.e., Notolegia; Springer et al., in press). The final possibility is between Atlantogenata (i.e., Xenarthra + Afrotheria; Waddell et al., 1999) and Boreoeutheria. Swofford-Olsen-Waddell-Hillis (SOWH) tests (Swofford et al., 1996) reported by Murphy et al. (2001b) rejected the first and third possibilities in favor of a rooting between Afrotheria and other placentals, However, Buckley (2002) has shown that SOWH tests can give overconfidence in a topology when the assumptions of a model of sequence evolution are violated. Given the small likelihood differences that separate these three rooting positions, resolving the trifurcation at the base of Placentalia will require additional data, including more genes and improved taxon sampling to mitigate against long branches. Delsuc et al. (2002) have shown that locating the placental root is sensitive to taxon sampling among the basal groups. Improved models of sequence evolution may also prove important in resolving the placental root. Other a priori positions for the placental root, including those on the erinaceomorph and murid edges, are firmly rejected by both SOWH and Kishino and Hasegawa (KH; 1989) tests (Scally et al., 2001). The KH test is more conservative than the SOWH test in rejecting alternate hypotheses.

RELATIONSHIPS WITHIN THE MAJOR CLADES

Within Xenarthra, molecular analyses suggest an association of anteaters and sloths, to the exclusion of armadillos (see Fig. 4.1). This result agrees with morphological evidence

and with a recent molecular study that included most living xenarthran genera (Delsuc et al., 2002).

The basal split in Afrotheria is between Fossoromorpha (i.e., aardvarks, elephant shrews, afrosoricidans; Springer et al., in press) and Paenungulata, with elephant shrews and afrosoricidans clustering together in the former group. The paenungulate trifurcation is not satisfactorily resolved by our data. Other molecular studies also attest to the difficulty of resolving relationships within Paenungulata (Amrine and Springer, 1999). In contrast, the total evidence of Asher et al. (2003), which combines molecular and morphological data, finds strong support for Tethytheria. Arguing against Tethytheria is a newly discovered AfroSINE that hyracoids and sirenians share to the exclusion of proboscideans (Nishihara et al., 2003), consistent with the maximum likelihood and Bayesian results of Murphy et al. (2001b).

The basal split within Laurasiatheria is between Eulipotyphla and other taxa (i.e., Variamana = Chiroptera + Perissodactyla + Cetartiodactyla + Pholidota + Carnivora; Springer et al., in press). The name Variamana reflects the highly divergent manus that occurs in members of this group. Within Variamana, there is some support for a monophyletic Fereuungulata (i.e., Perissodactyla + Cetartiodactyla + Pholidota + Carnivora; Waddell et al., 1999). Within Fereuungulata, an association of Carnivora and Pholidota is strongly supported by primary sequence analyses (see Fig. 4.1) and a 363 bp deletion in the *apolipoprotein B* alignment of Amrine-Madsen et al. (2003). Morphological evidence for this clade includes an osseous tentorium (Shoshani and McKenna, 1998). Within Chiroptera, microbats are paraphyletic and Yinpterochiroptera (i.e., megabats + non-nycterid rhinolophoids; Springer et al., 2001b; Teeling et al., 2002) is supported.

In contrast to some mitochondrial studies (Árnason et al., 2002; Hudelot et al., 2003), which support an association of

Scandentia and Lagomorpha, our data divide Euarchonto-glires into Glires and Euarchonta. Molecular support for Glires adds to morphological evidence (e.g., Meng et al., 2003) favoring this hypothesis (e.g., Asher et al., 2003). Within Euarchontoglires, chromosome data also support Glires monophyly (Murphy et al., 2001c; Stanyon et al., 2003). Similarly, strong molecular support for rodent monophyly corroborates morphological evidence for this clade (Luckett and Hartenberger, 1993). Euarchonta includes Primates, Dermoptera, and Scandentia and is similar to the morphological Archonta hypothesis but with bats now excluded. With the removal of bats from Archonta, presumed morphological synapomorphies for Volitantia (Simmons and Quinn, 1994; Simmons and Geisler, 1998) must now be viewed as homoplastic in bats and flying lemurs. At the same time, ad hoc explanations for the absence of archontan tarsal specializations in bats (e.g., Szalay and Drawhorn, 1980) are no longer necessary.

LONG BRANCHES AND THE PATHOLOGIC BEHAVIOR OF PARSIMONY

Long branch attraction occurs when convergent changes on long branches outnumber synapomorphic changes on shorter, internal branches. This problem can be most acute when outgroup sequences are highly divergent relative to ingroup sequences (Swofford et al., 1996). Felsenstein (1978) demonstrated that parsimony is more susceptible to long branch attraction than is maximum likelihood. As mentioned above, some molecular analyses root the placental tree within Rodentia or between representative rodents and other taxa. Rooting within Rodentia is common in studies with limited taxon sampling, especially with parsimony, and this result is a candidate for long branch attraction. An example is provided by one of our own data sets—*BRCA1*. Fig. 4.3 shows an unrooted parsimony tree based on the *BRCA1* data set of Madsen et al. (2001). It is evident from this tree that long branches include terminal branches leading to *Hystrix, Elephantulus, Erinaceus, Lepus, Rhynchocyon, Scalopus, Tonatia,* and *Tupaia,* and internal branches leading to hyraxes, murids, and tenrecids. In analyses that included a marsupial outgroup (*Vombatus*), the most parsimonious trees (four at 10,867 steps) all rooted on the branch leading to *Hystrix,* which renders Rodentia, Glires, and Euarchontoglires paraphyletic. When *Hystrix* is removed from the analysis, the most parsimonious trees (four at 10,376 steps) root on the murid branch. Again, Rodentia, Glires, and Euarchontoglires are all rendered paraphyletic. Continued pruning of the tree, now with deletion of both murids (*Mus, Rattus*) from the analysis, results in four trees (9,965 steps), all of which root on the internal branch leading to the tenrecids (*Echinops, Tenrec*). This renders Rodentia, Glires, and Euarchontoglires monophyletic, but Afrotheria paraphyletic. Subsequent deletion of the tenrecids results in a single most parsimonious tree (8,958 steps) that roots on

Tupaia. When *Tupaia* is subsequently deleted, two most parsimonious trees (8,664 steps) are recovered, both of which root on *Erinaceus.* Deletion of *Erinaceus* results in two most parsimonious trees (8,304 steps) that root on *Scalopus.* Finally, deletion of *Scalopus* results in rooting the placental tree on *Tonatia* (two trees at 8,004 steps), rendering Chiroptera paraphyletic and suggesting that the ancestral placental mammal was capable of powered flight! Clearly, the placental root is highly unstable, sensitive to taxon sampling, and jumps from one long branch to another in parsimony analyses with the *BRCA1* data set of Madsen et al. (2001). Parsimony analyses of the Murphy et al. (2001a) data set produced similar rooting artifacts that were not observed with other tree building methodologies. This behavior is a hallmark feature of long branch attraction. Asher et al. (2003) performed analyses with a molecular data set that included 49 extant taxa and 22 genes from Murphy et al. (2001b). As noted above, the resulting root was between murid rodents and other taxa in six of 12 analyses. In the remaining analyses, the root was attracted to *Erinaceus* or elephant shrews. Murids, hedgehogs, and elephant shrews derive from three of the four major clades of placental mammals. In addition, these taxa consistently exhibit long branches on molecular topologies (e.g., Fig. 4.1), which may explain the attraction of the marsupial outgroup to these taxa in parsimony analyses. In view of these observations, as well as statistical tests that corroborate the occurrence of long branch attraction in parsimony analyses of such gene segments as *BRCA1* (Madsen et al., 2001; Scally et al., 2001), we dismiss these anomalous results as phylogenetic artifacts.

In contrast, maximum likelihood and Bayesian analyses with taxonomically diverse data sets and multiple nuclear genes consistently place the placental root between Xenarthra and Epitheria, Afrotheria and Notolegia, or Atlantogenata and Boreoeutheria (Madsen et al., 2001; Murphy et al., 2001a,b; Scally et al., 2001; Delsuc et al., 2002; Huchon et al., 2002; Amrine-Madsen et al., 2003; Springer et al., in press). Waddell et al. (2001) have shown that posterior probabilities of clade support may be inflated in some cases, but bootstrap analyses with maximum likelihood also constrain the root of the placental tree to one of these three locations. Furthermore, rare genomic changes are incompatible with rooting the placental tree within Rodentia, Glires, Euarchontoglires, Laurasiatheria (contingent on the *PLCB4* deletion), or Afrotheria.

TIMESCALE FOR PLACENTAL MAMMAL EVOLUTION

A timescale for placental diversification is of considerable importance for unraveling the early biogeographic history of this taxonomic group. Archibald and Deutschman (2001) reviewed three competing models for placental diversification. The Explosive Model postulates that both inter- and intraordinal divergences occurred after the K/T boundary.

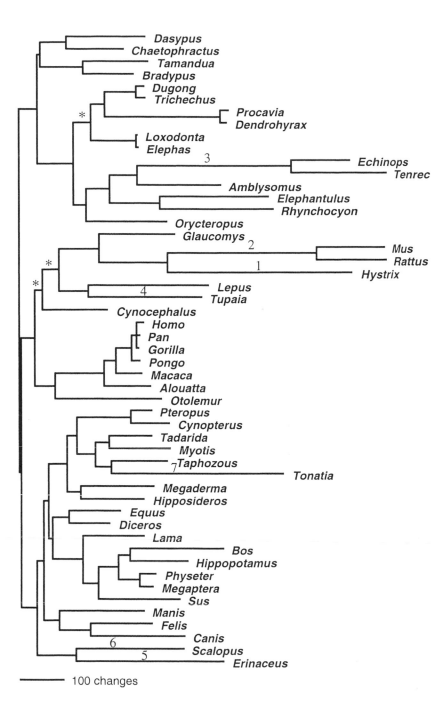

Fig. 4.3. One of four most parsimonious unrooted trees (accelerated transformation branch lengths) for the *BRCA1* data of Madsen et al. (2001). Asterisks denote branches that collapse on the strict consensus tree. The basal split is arbitrarily shown between Atlantogenata and Boreoeutheria. When the outgroup taxon *Vombatus* is included in the analysis, the placental root occurs on the terminal branch leading to *Hystrix*, which is labeled 1 in the figure. Numbers 2–7 correspond to successive rooting positions as placental taxa are sequentially deleted from maximum parsimony analyses.

100 changes

This model is preferred by some paleontologists (Gingerich, 1977; Carroll, 1997; Benton, 1999; Foote et al., 1999). At the other extreme, the Short Fuse Model postulates inter- and some intraordinal cladogenesis well back in the Cretaceous, including splits within Rodentia as old as 112–125 million years (Janke et al., 1997; Springer, 1997; Kumar and Hedges, 1998; Huelsenbeck et al., 2000). The Long Fuse Model is intermediate between these hypotheses and postulates Cretaceous interordinal and Cenozoic intraordinal divergences, although with an allowance for limited intraordinal diversification near the end of the Cretaceous (e.g., Eulipotyphla). In their analysis of the placental record, Archibald and Deutschman (2001) rejected the Short Fuse Model but could not discriminate between the Explosive and Long

Fuse Models. Some cladistic analyses support the Long Fuse Model by including 85- to 90-million-year-old zalambdalestids and zhelestids in crown group Eutheria (Archibald et al., 2001).

Virtually all molecular studies agree that interordinal diversification began well back in the Cretaceous (Springer, 1997; Kumar and Hedges, 1998; Penny et al., 1999; Eizirik et al., 2001). In our own recent work, we have employed both linearized trees and quartet dating and have estimated the base of Placentalia at approximately 103 million years (Murphy et al., 2001b). Our recent analysis using the relaxed clock method of Thorne et al. (1998) and Kishino et al. (2001) recovered dates in the range of 97–109 million years for the base of Placentalia and is generally consistent with

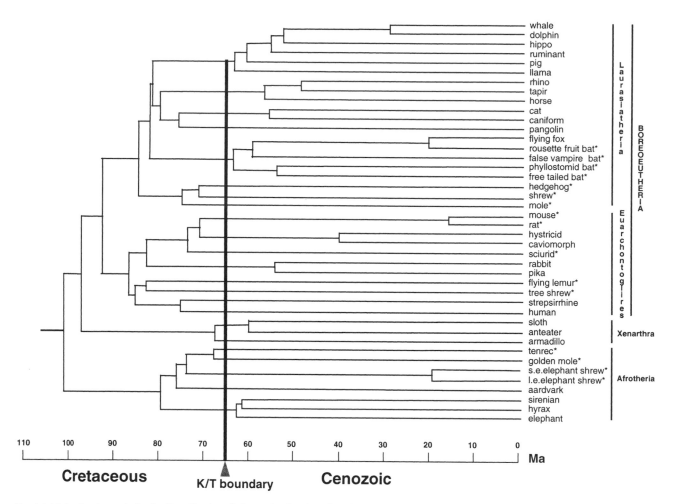

Fig. 4.4. Molecular timescale for the diversification of placental orders, based on the Murphy et al. (2001b) data set (*PNOC* segment excluded) and the *multidivtime* program of Thorne and Kishino (2002). Each of 18 different gene segments (*A2AB, ADORA, ADRB2, APP, ATP7A, BDNF, BMI1, BRCA1, CNR1, CREM, EDG1, IRBP, PLCB4, RAG1, RAG2, TYR, vWF, ZFX*) was allowed to have its own parameters under the F84 model of sequence evolution, with an allowance for a gamma distribution of nucleotide substitution rates among sites. We used 13 fossil constraints following Springer et al. (2003). It was necessary to exclude *PNOC* because marsupial sequences are lacking for this segment. Asterisks denote placental taxa included in the "K/T body size" taxon set of Springer et al. (2003), which was used to test the hypothesis that molecular estimates of Cretaceous divergence times are an artifact of increased body size subsequent to the K/T boundary. In those analyses, Springer et al. (2003) found that interordinal divergences remained in the Cretaceous.

the Long Fuse Model of diversification (Springer et al., 2003). Similar results were obtained by Douady and Douzery (2003). Although the original implementation of the Thorne/ Kishino method did not allow for distinct process partitions of heterogeneous molecular data, the multidivtime program of Thorne and Kishino (2002) allows different models of sequence evolution for individual gene segments. We employed this methodology with the Murphy et al. (2001b) data set and recovered divergence estimates that were generally within one to two million years of the values reported in Springer et al. (2003). These results are shown in Fig. 4.4 and Table 4.2. Ninety-five percent credibility intervals for all of the nodes in Figure 4.4 are given in Table 4.2.

We have further suggested that the separation of Xenarthra and Afrotheria at approximately 100–110 million years may have resulted from the vicariant separation of South America and Africa at 100–120 million years ago (Smith et al., 1994; Hay et al., 1999). This hypothesis ascribes an important role for both plate tectonic events and Gondwana in the early history of placental mammals.

Following the Murphy et al. (2001b) phylogeny, which roots between Afrotheria and other placentals, we suggested the possibility of a Gondwanan origin for Placentalia. According to this model, the basal split between Afrotheria and Notolegia corresponds to the vicariant event that separated South America and Africa. This hypothesis also requires subsequent dispersal to Laurasia, which had separated from Gondwana by 160–170 million years ago (Smith et al., 1994). In contrast, Archibald (2003) suggested that the agreement of molecular dates for the Xenarthra-Afrotheria split and geologic dates for the South America–Africa separation is mere coincidence, and that Xenarthra and Afrotheria represent separate dispersals to these continents from the Northern Hemisphere. Archibald (2003) further argued that a Laurasian origin for Placentalia, with subsequent dispersals to South America and Africa, is more parsimonious than a Gondwanan origin. Such exercises are sensitive to how higher-level taxa are coded for their place of origin. In contrast to Archibald (2003), we coded Xenarthra and Afrotheria as Gondwanan, and Laurasiatheria and Euarchontoglires

Table 4.2 Bayesian estimates of divergence dates for placental nodes

Node	Date (My)	95% Credibility Intervals
Sirenian to Hyrax	60.8	56.3, 63.9
Base of Paenungulata	62.9	58.9, 64.9
Macroscelides to *Elephantulus*	18.2	15.0, 21.9
Base of Afrosoricida	67.9	62.2, 73.2
Afrosoricida to Macroscelidea	73.9	68.7, 78.7
Base of Fossoromorpha	75.6	70.6, 80.3
Base of Afrotheria	78.8	74.0, 82.9
Strepsirrhine to human	75.2	70.3, 80.0
Dermoptera to Scandentia	83.1	78.5, 88.0
Base of Euarchonta	85.2	80.9, 89.8
Rabbit to pika	52.9	47.9, 58.0
Hystricid to Caviomorph	40.1	35.5, 44.9
Mouse to rat	16.7	14.2, 19.6
Mouse-Rat to Hystricid-Caviomorph	71.6	66.8, 76.7
Base of Rodentia	74.9	70.3, 79.8
Base of Glires	82.3	77.8, 87.1
Base of Euarchontoglires	86.5	82.4, 91.0
Cat to Caniform	53.5	50.3, 57.7
Base of Ostentoria	74.6	71.3, 78.1
Rhino to tapir	48.3	44.6, 52.0
Base of Perissodactyla	56.1	54.1, 57.9
Ostentoria to Perissodactyla	77.9	75.1, 81.0
Mysticete to Odontocete	27.3	24.0, 30.6
Hippo to Cetacea	52.3	52.0, 53.3
Hippo-Cetacea to ruminant	55.2	53.5, 56.9
Pig to hippo-Cetacea-ruminant	59.7	57.6, 61.9
Base of Cetartiodactyla	62.3	59.9, 64.6
Base of Fereuungulata	79.4	76.5, 82.5
Flying fox to Rousette fruit bat	19.7	16.3, 23.1
Base of Yinpterochiroptera	59.5	58.2, 60.0
Phyllostomid to free-tailed bat	53.7	50.2, 57.4
Base of Chiroptera	63.1	61.2, 65.3
Base of Variamana	80.3	77.3, 83.4
Hedgehog to shrew	70.6	66.0, 75.4
Hedgehog-shrew to mole	75.6	71.7, 79.6
Base of Laurasiatheria	83.1	79.7, 86.8
Base of Boreoeutheria	91.8	87.7, 96.2
Sloth to anteater	59.0	53.6, 64.9
Base of Xenarthra	66.8	61.6, 72.4
Notolegia to Afrotheria	96.9	92.3, 101.8
Base of Placentalia	101.3	96.2, 106.5

Note: See also Fig. 4.4.

as Laurasian (Madsen et al., 2001). If we ignore presumed stem eutherian outgroups to Placentalia, the most recent common ancestor of placental mammals is reconstructed as having resided in Gondwana, with a single step representing dispersal to Laurasia (Fig. 4.5A). One of the other viable positions for the placental root, between Xenarthra and Epitheria, also recovers a Gondwanan placental root (Fig. 4.5B). The final alternative for the root, between Atlantogenata and Boreoeutheria, is equivocal for the geographic origin of Placentalia, as it allows for either Laurasia or Gondwana (Fig. 4.5C). However, such a root would still imply a strong biogeographic component for the early cladogenesis of Placentalia, suggesting an initial split be-

tween Gondwanan and Laurasian clades, followed in the former group by the divergence between Afrotheria (African) and Xenarthra (South American). In reconstructions that recognize the Laurasian fossil *Eomaia* as the oldest eutherian mammal (Ji et el., 2002), two of three reconstructions (Xenarthra-Epitheria, Afrotheria-Notolegia) are equivocal for the placental root (Figs. 4.5D,E) and the third (Atlantogenata-Boreoeutheria) favors a Laurasian origin (Fig. 4.5F). An additional complication for these reconstructions is the possibility that the oldest stem eutherians are from Gondwana rather than from Laurasia (Woodburne et al., 2003).

PROSPECTUS FOR FUTURE STUDIES

We stand on the threshold of a well-resolved molecular phylogeny for the extant orders of placentals mammals (see Fig. 4.1). The major groups have been defined, and only local rearrangements remain to be resolved. Of these, resolving the trifurcation at the base of Placentalia is most significant and merits attention and resources. Beyond the living eutherian orders, the challenge of integrating molecular and morphological data (fossils included) into a comprehensive and accurate phylogeny for all eutherian orders is formidable. The total evidence analysis by Asher et al. (2003) is the first step in this direction, but, in our view, suffers from the methodological limitations of parsimony. Other methodological approaches should also be explored, including molecular scaffolds (Springer et al., 2001b) and Bayesian methods (Ronquist and Huelsenbeck, 2003) that allow molecular and morphological data partitions to have their own models of evolution.

SUMMARY

Maximum likelihood and Bayesian analyses of DNA sequences, principally derived from nuclear genes, provide a well-resolved phylogeny for the orders of placental mammals, with only local rearrangements resisting resolution. Placentalia is divided into four major groups: Afrotheria (Afrosoricida, Hyracoidea, Macroscelidea, Proboscidea, Sirenia, and Tubulidentata), Xenarthra, Laurasiatheria (Carnivora, Cetartiodactyla, Chiroptera, Eulipotyphla, Perissodactyla, and Pholidota), and Euarchontoglires (Dermoptera, Lagomorpha, Primates, Rodentia, and Scandentia). Of these, Laurasiatheria and Euarchontoglires are sister taxa that together make up Boreoeutheria. Rare genomic events corroborate Xenarthra, Afrotheria, Euarchontoglires, and possibly, Laurasiatheria. Molecular estimates of divergence times are generally consistent with the Long Fuse Model of diversification and place the root of the placental tree in the range of 97–109 million years. There are only three viable positions for this root: (1) between Xenarthra and Epitheria; (2) between Afrotheria and Notolegia; or (3) between Atlantogenata and Boreoeutheria. Each of these possibilities is consistent with a strong biogeographic component for

Fig. 4.5. Biogeographic reconstructions of the origin of Placentalia, as inferred from alternative rooting schemes given the relationships of extant major eutherian clades. Abbreviations: G., Gondwana; L., Laurasia; La., Laurasiatheria; Eu., Euarchontoglires; Xe., Xenarthra; Af., Afrotheria. Groups in (D), (E), and (F) are equivalent to those in (A), (B), and (C), respectively. The letters L and G on branches indicate the most parsimonious reconstruction of the geographic location at the origin of that lineage. Arrowhead indicates inferred geographic location at the placental root. Schemes in (D)–(F) include the fossil eutherian *Eomaia*.

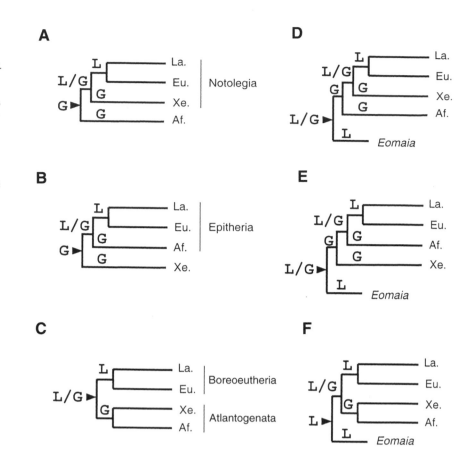

the early history of Placentalia. Beyond the living eutherian orders, a major challenge ahead is integrating neontological and paleontological data into a comprehensive and robust phylogeny for all eutherian orders.

ACKNOWLEDGMENTS

We thank J. David Archibald and Kenneth D. Rose for inviting us to contribute to this volume. Michael Benton, Paul Higgs, Anne Yoder, and an anonymous reviewer made helpful comments on an earlier draft of this chapter.

REFERENCES

Amrine, H. M., and M. S. Springer. 1999. Maximum likelihood analysis of the tethythere hypothesis based on a multigene data set and a comparison of different models of sequence evolution. Journal of Mammalian Evolution 6: 161–176.

Amrine-Madsen, H., K.-P. Koepfli, R. K. Wayne, and M. S. Springer. 2003. A new phylogenetic marker, *apolipoprotein B*, provides compelling evidence for eutherian relationships. Molecular Phylogenetics and Evolution 28: 225–240.

Archibald, J. D. 2003. Timing and biogeography of the eutherian radiation: Fossils and molecules compared. Molecular Phylogenetics and Evolution 28: 350–359.

Archibald, J. D., A. O. Averianov, and E. G. Ekdale. 2001. Late Cretaceous relatives of rabbits, rodents, and other extant eutherian mammals. Nature 414: 62–65.

Archibald, J. D., and D. Deutschman. 2001. Quantitative analysis of the timing of origin of extant placental orders. Journal of Mammalian Evolution 8: 107–124.

Árnason, Ú., J. A. Adegoke, K. Bodin, E. W. Born, Y. B. Esa, A. Gullberg, M. Nilsson, R. V. Short, X. Xu, and A. Janke. 2002. Mammalian mitogenomic relationships and the root of the eutherian tree. Proceedings of the National Academy of Sciences USA 99: 8151–8156.

Asher, R. J. 1999. A morphological basis for assessing the phylogeny of the "Tenrecoidea" (Mammalia, Lipotyphla). Cladistics 15: 231–252.

Asher, R. J., M. J. Novacek, and J. H. Geisler. 2003. Relationships of endemic African mammals and their fossil relatives based on morphological and molecular evidence. Journal of Mammalian Evolution 10: 131–194.

Benton, M. J. 1999. Early origins of modern birds and mammals: Molecules vs. morphology. BioEssays 21: 1043–1051.

Buckley, T. R. 2002. Model misspecification and probabilistic tests of topology: Evidence from empirical data sets. Systematic Biology 51: 509–523.

Carroll, R. L. 1997. Patterns and Processes of Vertebrate Evolution. Cambridge University Press, Cambridge.

de Jong, W. W., J.A.M. Leunissen, and G. J. Wistow. 1993. Eye lens crystallins and the phylogeny of placental orders: Evidence for a macroscelidid-paenungulate clade? pp. 5–12 in F. S. Szalay, M. J. Novacek, and M. C. McKenna (eds.), Mammal Phylogeny. Volume 2. Placentals. Springer-Verlag, New York.

de Jong, W. W., M.A.M. van Dijk, C. Poux, G. Kappe, T. van Rheede, and O. Madsen. 2003. Indels in protein-coding sequences of Euarchontoglires constrain the rooting of the eutherian tree. Molecular Phylogenetics and Evolution 28: 328–340.

de Jong, W. W., A. Zweers, and M. Goodman. 1981. Relationship of aardvark to elephants, hyraxes and sea cows from alpha-crystallin sequences. Nature 292: 538–540.

Delsuc, F., M. Scally, O. Madsen, M. J. Stanhope, W. W. de Jong, F. M. Catzeflis, M. S. Springer, and E.J.P. Douzery. 2002. Molecular phylogeny of living xenarthrans and the impact of character and taxon sampling on the placental tree rooting. Molecular Biology and Evolution 19: 1656–1671.

Douady, C. J., and E.J.P. Douzery. 2003. Molecular estimation of eulipotyphlan divergence times and the evolution of "Insectivora." Molecular Phylogenetics and Evolution 28: 285–296.

Eizirik, E., W. J. Murphy, and S. J. O'Brien. 2001. Molecular dating and biogeography of the early placental mammals. Journal of Heredity 92: 212–219.

Felsenstein, J. 1978. Cases in which parsimony and compatibility methods will be positively misleading. Systematic Zoology 27: 401–410.

Foote, M., J. P. Hunter, C. M. Janis, and J. J. Sepkoski, Jr. 1999. Evolutionary and preservational constraints on origins of biologic groups: Divergence times of eutherian mammals. Science 283: 1310–1314.

Froenicke L., J. Wienberg, G. Stone, L. Adams, and R. Stanyon. 2003. Towards the delineation of the ancestral eutherian genome organization: Comparative genome maps of human and the African elephant (*Loxodonta africana*) generated by chromosome painting. Proceedings of the Royal Society of London B 270: 1331–1340.

Gingerich, P. D. 1977. Patterns of evolution in the mammalian fossil record; pp. 469–500 *in* A. Hallam (ed.), Patterns of Evolution as Illustrated by the Fossil Record. Elsevier, Amsterdam.

Goodman, M. 1975. Protein sequence and immunological specificity: Their role in phylogenetic studies of primates; pp. 219–248 *in* W. P. Luckett and F. S. Szalay (eds.), Phylogeny of the Primates: A Multidisciplinary Approach. Plenum, New York and London.

Hay, W. W., R. M. DeConto, C. N. Wold, K. M. Wilson, S. Voigt, M. Schulz, A. R. Wold, W.-C. Dullo, A. B. Ronov, A. N. Balukhovsky, and E. Soding. 1999. Alternative global Cretaceous paleogeography; pp. 1–48 *in* E. Barrera and C. C. Johnson (eds.), Evolution of the Cretaceous Ocean-Climate System. Special Paper 332. Geological Society of America, Boulder, Colorado.

Huchon, D., O. Madsen, M.J.J.B. Sibbald, K. Ament, M. J. Stanhope, F. Catzeflis, W. W. de Jong, and E.J.P. Douzery. 2002. Rodent phylogeny and a timescale for the evolution of Glires: Evidence from an extensive taxon sampling using three nuclear genes. Molecular Biology and Evolution 19: 1053–1065.

Hudelot, C., V. Gowri-Shankar, H. Jow, M. Rattray, and P. G. Higgs. 2003. RNA-based phylogenetic methods: Application to mammalian mitochondrial RNA sequences. Molecular Phylogenetics and Evolution 28: 241–252.

Huelsenbeck, J. P., B. Larget, and D. Swofford. 2000. A compound process for relaxing the molecular clock. Genetics 154: 1879–1892.

Janke, A., X. Xu, and Ú. Árnason. 1997. The complete mitochondrial genome of the wallaroo (*Macropus robustus*) and the phylogenetic relationship among Monotremata, Marsupialia, and Eutheria. Proceedings of the National Academy of Sciences USA 94: 1276–1281.

Ji, Q., Z.-X. Luo, C. X. Yuan, J. R. Wible, J. P. Zhang, and J. A. Georgi. 2002. The earliest known eutherian mammal. Nature 416: 816–822.

Kishino, H., and M. Hasegawa. 1989. Evaluation of the maximum likelihood estimate of the evolutionary tree topologies from DNA sequence data, and the branching order in Hominoidea. Journal of Molecular Evolution 29: 170–179.

Kishino, H., J. L. Thorne, and W. J. Bruno. 2001. Performance of a divergence time estimation method under a probabilistic model of rate evolution. Molecular Biology and Evolution 18: 352–361.

Kleinschmidt, T., J. Czelusniak, M. Goodman, and G. Braunitzer. 1986. Paenungulata: A comparison of the hemoglobin sequences from elephant, hyrax, and manatee. Molecular Biology and Evolution 3: 427–435.

Kumar, S., and S. B. Hedges. 1998. A molecular timescale for vertebrate evolution. Nature 392: 917–920.

Lavergne, A., E. Douzery, T. Stichler, F. M. Catzeflis, and M. S. Springer. 1996. Interordinal mammalian relationships: Evidence for paenungulate monophyly is provided by complete mitochondrial 12S rRNA sequences. Molecular Phylogenetics and Evolution 6: 245–258.

Lin, Y.-H., P. A. McLenachan, A. R. Gore, M. J. Phillips, R. Ota, M. D. Hendy, and D. Penny. 2002. Four new mitochondrial genomes and the increased stability of evolutionary trees of mammals from improved taxon sampling. Molecular Biology and Evolution 19: 2060–2070.

Luckett, W. P., and J.-L. Hartenberger. 1993. Monophyly or polyphyly of the order Rodentia: Possible conflict between morphological and molecular interpretations. Journal of Mammalian Evolution 1: 127–147.

MacPhee, R.D.E., and M. J. Novacek. 1993. Definition and relationships of Lipotyphla; pp. 13–31 *in* F. S. Szalay, M. J. Novacek, and M. C. McKenna (eds.), Mammal Phylogeny. Volume 2. Placentals. Springer-Verlag, New York.

Madsen, O., M. Scally, C. J. Douady, D. J. Kao, R. W. DeBry, R. Adkins, H. Amrine, M. J. Stanhope, W. W. de Jong, and M. S. Springer. 2001. Parallel adaptive radiations in two major clades of placental mammals. Nature 409: 610–614.

McKenna, M. C. 1975. Toward a phylogenetic classification of the Mammalia; pp. 21–46 *in* W. P. Luckett and F. S. Szalay (eds.), Phylogeny of the Primates: A Multidisciplinary Approach. Plenum, New York and London.

Meng, J., Y. Hu, and C. Li. 2003. The osteology of *Rhombomylus* (Mammalia, Glires): Implications for phylogeny and evolution of Glires. Bulletin of the American Museum of Natural History 275: 1–247.

Misawa, K., and A. Janke. 2003. Revisiting the Glires concept—phylogenetic analysis of nuclear sequences. Molecular Phylogenetics and Evolution 28: 320–327.

Miyamoto, M. M., and M. Goodman. 1986. Biomolecular systematics of eutherian mammals: Phylogenetic patterns and classification. Systematic Zoology 35: 230–240.

Murphy, W. J., E. Eizirik, W. E. Johnson, Y. P. Zhang, O. A. Ryder, and S. J. O'Brien. 2001a. Molecular phylogenetics and the origins of placental mammals. Nature 409: 614–618.

Murphy, W. J., E. Eizirik, S. J. O'Brien, O. Madsen, M. Scally, C. J. Douady, E. Teeling, O. A. Ryder, M. J. Stanhope, W. W. de Jong, and M. S. Springer. 2001b. Resolution of the early placental mammal radiation using Bayesian phylogenetics. Science 294: 2348–2351.

Murphy, W. J., R. Stanyon, and S. J. O'Brien. 2001c. Evolution of mammalian genome organization inferred from comparative gene mapping. Genome Biology 2: 1–8.

Nikaido, M., H. Nishihara, Y. Hukumoto, and N. Okada. 2003. Ancient SINEs from African endemic mammals. Molecular Biology and Evolution 20: 522–527.

Nishihara, H., M. Nikaido, Y. Fukumoto, M. J. Stanhope, and N. Okada. 2003. Phylogenetic analysis among afrotherian mammals based on SINE insertions. Society for Molecular Biology and Evolution Abstracts, p. 84.

Patterson, C. 1982. Morphological characters and homology; pp. 21–74 in K. A. Joysey and A. E. Friday (eds.), Problems of Phylogenetic Reconstruction. Systematics Association Special Volume 21. Academic Press, London and New York.

Penny, D., M. Hasegawa, P. J. Waddell, and M. D. Hendy. 1999. Mammalian evolution: Timing and implications from using the logdeterminant transform for proteins of differing amino acid composition. Systematic Biology 48: 76–93.

Porter, C. A., M. Goodman, and M. J. Stanhope. 1996. Evidence on mammalian phylogeny from sequences of exon 28 of the von Willebrand factor gene. Molecular Phylogenetics and Evolution 5: 89–101.

Poux, C., T. Van Rheede, O. Madsen, and W. W. de Jong. 2002. Sequence gaps join mice and men: Phylogenetic evidence from deletions in two proteins. Molecular Biology and Evolution 19: 2035–2037.

Ronquist, F., and J. P. Huelsenbeck. 2003. MRBAYES 3: Bayesian phylogenetic inference under mixed models. Bioinformatics 19: 1572–1574.

Scally, M., O. Madsen, C. J. Douady, W. W. de Jong, M. J. Stanhope, and M. S. Springer. 2001. Molecular evidence for the major clades of placental mammals. Journal of Mammalian Evolution 8: 239–277.

Shoshani, J. 1986. Mammalian phylogeny: Comparison of morphological and molecular results. Molecular Biology and Evolution 3: 222–242.

Shoshani, J., M. Goodman, J. Czelusniak, and G. Braunitzer. 1985. A phylogeny of Rodentia and other eutherian orders: Parsimony analysis utilizing amino acid sequences of alpha and beta hemoglobin chains; pp. 191–210 in W. P. Luckett and J. L. Hartenberger (eds.), Evolutionary Relationships among Rodents: A Multidisciplinary Analysis. Plenum, New York.

Shoshani, J., and M. C. McKenna. 1998. Higher taxonomic relationships among extant mammals based on morphology, with selected comparisons of results from molecular data. Molecular Phylogenetics and Evolution 9: 572–584.

Simmons, N. B., and J. H. Geisler. 1998. Phylogenetic relationships of Icaronycteris, Archaeonycteris, Hassianycteris, and Palaeochiropteryx to extant bat lineages, with comments on the evolution of echolocation and foraging strategies in Microchiroptera. Bulletin of the American Museum of Natural History 235: 1–182.

Simmons, N. B., and T. H. Quinn. 1994. Evolution of the digital tendon locking mechanism in bats and dermopterans: A phylogenetic perspective. Journal of Mammalian Evolution 2: 231–254.

Smith, A. G., D. G. Smith, and B. M. Funnell. 1994. Atlas of Cenozoic and Mesozoic Coastlines. Cambridge University Press, Cambridge.

Springer, M. S. 1997. Molecular clocks and the timing of the placental and marsupial radiations in relation to the Cretaceous-Tertiary boundary. Journal of Mammalian Evolution 4: 285–302.

Springer, M. S., A. Burk, J. R. Kavanagh, V. G. Waddell, and M. J. Stanhope. 1997a. The interphotoreceptor retinoid binding protein gene in therian mammals: Implications for higher level relationships and evidence for loss of function in the marsupial mole. Proceedings of the National Academy of Sciences USA 94: 13754–13759.

Springer, M. S., G. C. Cleven, O. Madsen, W. W. de Jong, V. G. Waddell, H. M. Amrine, and M. J. Stanhope. 1997b. Endemic African mammals shake the phylogenetic tree. Nature 388: 61–64.

Springer, M. S., R. W. DeBry, C. Douady, H. Amrine, O. Madsen, W. W. de Jong, and M. J. Stanhope. 2001a. Mitochondrial versus nuclear gene sequences in deep-level mammalian phylogeny reconstruction. Molecular Biology and Evolution 18: 132–143.

Springer, M. S., and J.A.W. Kirsch. 1993. A molecular perspective on the phylogeny of placental mammals based on mitochondrial 12S rDNA sequences, with special reference to the problem of Paenungulata. Journal of Mammalian Evolution 1: 149–168.

Springer, M. S., W. J. Murphy, E. Eizirik, O. Madsen, M. Scally, C. J. Douady, E. C. Teeling, M. J. Stanhope, W. W. de Jong, and S. J. O'Brien. In press. A molecular classification for the living orders of placental mammals and the phylogenetic placement of primates; in M. Dagosto and M. Ravosa (eds.), Primate Origins and Adaptations. Plenum, New York.

Springer, M. S., W. J. Murphy, E. Eizirik, and S. J. O'Brien. 2003. Placental mammal diversification and the Cretaceous-Tertiary boundary. Proceedings of the National Academy of Sciences USA 100: 1056–1061.

Springer, M. S., E. C. Teeling, O. Madsen, M. J. Stanhope, and W. W. de Jong. 2001b. Integrated fossil and molecular data reconstruct bat echolocation. Proceedings of the National Academy of Sciences USA 98: 6241–6246.

Stanhope, M. J., W. J. Bailey, J. Czelusniak, M. Goodman, J.-S. Si, J. Nickerson, J. G. Sgouros, G.A.M. Singer, and T. K. Kleinschmidt. 1993. A molecular view of primate supraordinal relationships from the analysis of both nucleotide and amino acid sequences; pp. 251–292 in R.D.E. MacPhee (ed.), Primates and Their Relatives in Phylogenetic Perspective. Plenum, New York.

Stanhope, M. J., J. Czelusniak, J.-S. Si, J. Nickerson, and M. Goodman. 1992. A molecular perspective on mammalian evolution from the gene encoding interphotoreceptor retinoid binding protein, with convincing evidence for bat monophyly. Molecular Phylogenetics and Evolution 1: 148–160.

Stanhope, M. J., O. Madsen, V. G. Waddell, G. C. Cleven, W. W. de Jong, and M. S. Springer. 1998a. Highly congruent molecular support for a diverse superordinal clade of endemic African mammals. Molecular Phylogenetics and Evolution 9: 501–508.

Stanhope, M. J., M. R. Smith, V. G. Waddell, C. A. Porter, M. S. Shivig, and M. Goodman. 1996. Mammalian evolution and the interphotoreceptor retinoid binding protein (IRBP) gene: Convincing evidence for several superordinal clades. Journal of Molecular Evolution 43: 83–92.

Stanhope, M. J., V. G. Waddell, O. Madsen, W. W. de Jong, S. B. Hedges, G. C. Cleven, D. Kao, and M. S. Springer. 1998b. Molecular evidence for multiple origins of Insectivora and for a new order of endemic African insectivore mammals. Proceedings of the National Academy of Sciences USA 95: 9967–9972.

Stanyon, R., G. Stone, M. Garcia, and L. Froenicke. 2003. Reciprocal chromosome painting shows that squirrels, unlike murid rodents, have a highly conserved genome organization. Genomics 82: 245–249.

Swofford, D. L., G. P. Olsen, P. J. Waddell, and D. M. Hillis. 1996. Phylogenetic inference; pp. 407–492 *in* D. M. Hillis, C. Moritz, and B. K. Mable (eds.), Molecular Systematics. Sinauer, Sunderland, Massachusetts.

Szalay, F. S., and G. Drawhorn, 1980. Evolution and diversification of the Archonta in an arboreal milieu; pp. 133–169 *in* W. P. Luckett (ed.), Comparative Biology and Evolutionary Relationships of Tree Shrews. Plenum, New York.

Teeling, E. C., O. Madsen, R. A. Van Den Bussche, W. W. de Jong, M. J. Stanhope, and M. S. Springer. 2002. Microbat paraphyly and the convergent evolution of a key innovation in Old World rhinolophoid microbats. Proceedings of the National Academy of Sciences USA 99: 1431–1436.

Thomas, J. W., J. W. Touchman, R. W. Blakesley, et al. (68 other authors). 2003. Comparative analyses of multi-species sequences from targeted genomic regions. Nature 424: 788–793.

Thorne, J. L., and H. Kishino. 2002. Divergence time and evolutionary rate estimation with multilocus data. Systematic Biology 51: 689–702.

Thorne, J. L., H. Kishino, and I. S. Painter. 1998. Estimating the rate of evolution of the rate of molecular evolution. Molecular Biology and Evolution 15: 1647–1657.

van Dijk, M.A.M., E. Paradis, F. Catzeflis, and W. W. de Jong. 1999. The virtues of gaps: Xenarthran (edentate) monophyly supported by a unique deletion in aA-crystallin. Systematic Biology 48: 94–106.

Waddell, P. J., H. Kishino, and R. Ota. 2001. A phylogenetic foundation for comparative mammalian genomics. Genome Informatics 12: 141–154.

Waddell, P. J., N. Okada, and M. Hasegawa. 1999. Towards resolving the interordinal relationships of placental mammals. Systematic Biology 48: 1–5.

Woodburne, M. O., T. H. Rich, and M. S. Springer. 2003. The evolution of tribospheny and the antiquity of mammalian clades. Molecular Phylogenetics and Evolution 28: 360–385.

ROBERT J. ASHER

5

INSECTIVORAN-GRADE PLACENTALS

THE MAMMALIAN ORDER LIPOTYPHLA WAS named by Ernst Haeckel in 1866, and following Nowak (1999), consists of seven Recent taxa: Chrysochloridae (golden moles), Erinaceidae (hedgehogs and moonrats), Nesophontidae (the recently extinct Caribbean genus *Nesophontes*), Solenodontidae (the Caribbean genus *Solenodon* or "almiqui"), Soricidae (shrews), Talpidae (moles and desmans), and Tenrecidae (tenrecs and otter-shrews). Haeckel considered lipotyphlans to be a subset of a larger Insectivora, along with a menotyphlan group consisting of macroscelideans (elephant shrews) and scandentians (tree shrews). Menotyphla has long since been abandoned as a cohesive taxonomic unit (cf. Simpson, 1945); for many decades, the terms "Insectivora" and "Lipotyphla" have been used interchangeably. In this chapter, I use "Insectivora" as a descriptive, gradistic term referring to the seven Recent families listed in Nowak (1999), and "Lipotyphla" as a clade consisting of erinaceids, soricids, talpids, *Solenodon,* and *Nesophontes*.

Unlike menotyphlans, insectivorans have until recently been regarded as a natural, monophyletic group. Following Butler (1972, 1988), several anatomical characters are congruent with their monophyly. However, as summarized by MacPhee and Novacek (1993), none uniquely characterizes insectivorans to the exclusion of all other eutherians.

Since MacPhee and Novacek's overview of the group in 1993, much has changed regarding our understanding of mammalian phylogeny. For example, the molecular diversity of insectivoran mammals is much better understood now than it was a decade ago (McNiff and Allard, 1998; Stanhope et al., 1998; Emerson et al., 1999;

Mouchaty et al., 2000a,b; Douady et al., 2002a,b; Malia et al., 2002; Roca et al., 2004). In terms of our understanding of the evolutionary history of the various insectivoran lineages, the impact of these molecular data has been tremendous. Specifically, sequence data have been interpreted by a number of investigators (e.g., op. cit. and Madsen et al., 2001; Murphy et al., 2001a,b; Árnason et al., 2002) to support the polyphyly of Insectivora, and the association of African insectivorans (i.e., Tenrecidae and Chrysochloridae) with other African mammalian groups, including Macroscelidea, Tubulidentata, Proboscidea, Hyracoidea, and Sirenia, in the novel taxon Afrotheria (Stanhope et al., 1998). Significantly, a signal favoring an endemic African clade was supported in a recent combined data analysis that included morphological and molecular data, plus several fossil taxa (Asher et al., 2003).

From an anatomical perspective, the polyphyly of the Insectivora leads to several questions: What is the distribution of characters formerly regarded as diagnostic of Insectivora? Is there anatomical support for the various groups into which former insectivorans now belong? What extinct groups of placental mammals are related to modern insectivorans? Addressing these questions is the focus of this chapter. To clarify the definition of several insectivoran characters, I use the morphological data set sampled at the generic level by Asher et al. (2003) and map their distribution onto one of the placental phylogenies identified as optimal in that paper.

INSTITUTIONAL ABBREVIATIONS

The following zoological collections contributed to this study and are referred to in the text: American Museum of Natural History, New York (AMNH); Max Planck Institut für Hirnforschung, Stephan Collection now located at the Vogt Institut für Hirnforschung, Düsseldorf (MPIH); University of California Museum of Paleontology, Berkeley (UCMP); Zoologisches Institut, Universität Tübingen (ZIUT); Zoologisches Museum, Humboldt Universität zu Berlin (ZMB).

PHYLOGENY OF EUTHERIAN MAMMALS

Asher et al. (2003) combined a morphological character matrix with the 22-gene (19 nuclear and three mitochondrial) data set summarized by Murphy et al. (2001b). The morphological data were coded based on observations of 60 taxa from all extant and eight extinct eutherian orders, including seven extant insectivorans plus the extinct *Leptictis* and *Anagale,* sampling all extant insectivoran families except for Caribbean taxa, which until the study of Roca et al. (2004) were still poorly known genetically. DNA data were available only for the 49 extant taxa that are accessioned on GenBank (National Institutes of Health, 2004). Asher et al.

(2003) performed distinct phylogenetic analyses of these combined data under 12 parameter sets, each of which varied weights assigned to morphological characters, transversions, and transitions; gaps were weighted equally with morphological characters. Direct optimization (Wheeler, 1996, 1998), as applied in the program POY (Wheeler et al., 2003) run on a parallel supercomputing cluster (Janies and Wheeler, 2001), was used to reconstruct topologies. Direct optimization reconstructs trees simultaneously with making hypotheses of molecular homology (Felsenstein, 1988: 525), incorporates morphological and gap characters, skips an intermediate "alignment" stage, and uses parsimony as an optimality criterion. Using congruence across partitions for each parameter set as an optimality criterion, we identified two of the 12 total parameter sets as maximally congruent —one favored by the Incongruence Length Difference (ILD) index (Wheeler, 1995) and one favored by a new, topological index (Asher et al., 2003; original matrices, trees, and data are available at ftp.amnh.org/pub/people/asher).

With these data and methods, we found consistent support among the 12 combined analyses for the inclusion of tenrecs and golden moles within the Afrotheria, and a consistent lack of support for any association between African and northern insectivorans. Rather, maximally congruent analyses supported two insectivoran clades: a northern group, containing *Erinaceus, Sorex,* and talpids; and a southern group, with *Tenrec, Echinops,* and *Amblysomus* occurring among other endemic African mammals. Both trees favored by the topological index (depicted in Fig. 5.1) supported a tenrec–golden mole clade within the Afrotheria; however this arrangement did not appear in other parameter sets, including that favored by the ILD index. Besides support for Paenungulata, intra-afrotherian relationships remain controversial. However, insectivoran polyphyly occurs in all of the combined analyses presented by Asher et al. (2003). Optimizations of characters presented in Fig. 5.1 are therefore representative of the other topologies recovered by Asher et al. (2003), with the important caveat that a tenrec–golden mole clade is not consistently supported.

MORPHOLOGICAL SIMILARITIES AMONG INSECTIVORAN MAMMALS

To organize the following discussion, I focus on the six characters reviewed by MacPhee and Novacek (1993). These characters and others proposed by Butler (1988) and McDowell (1958) are listed in Table 5.1.

Reduced Pubic Symphysis and Simplified Gut

Following MacPhee and Novacek (1993), two of the six characters outlined by Butler (1972) are most consistent with the hypothesis of insectivoran monophyly: a reduced pubic symphysis and a simplified intestinal tract, typified by absence of the cecum. Asher et al. (2003) defined these

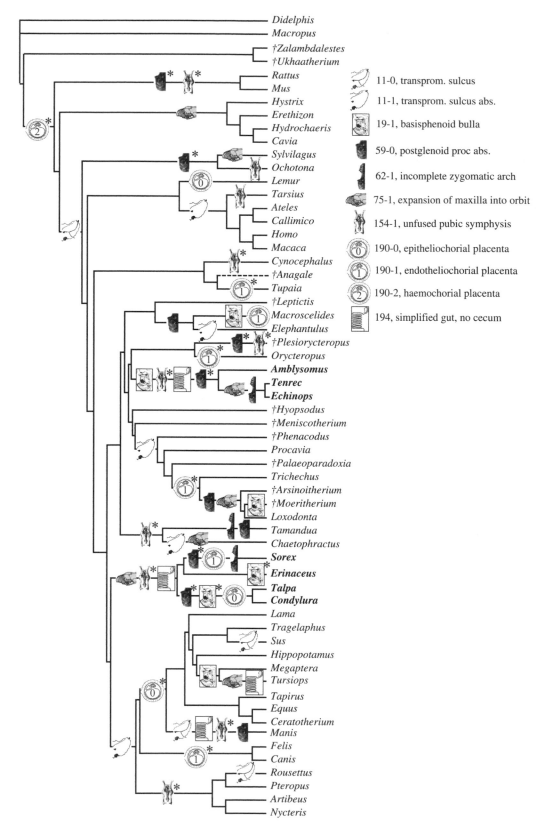

Fig. 5.1. Topology mapping the evolution of characters commonly recognized as diagnostic of Insectivora, based on 196 morphological characters, 19 nuclear, and three mitochondrial genes discussed in Asher et al. (2003), produced by weighting morphology ×2, transversions ×2, and transitions ×1, with gaps weighted the same as morphological characters. Topological congruence across genetic and morphological partitions was maximized with this weighting scheme. This was one of two most parsimonious trees (MPT; 75,150 steps); in the other MPT, *Anagale* (identified with a dotted line) shifts away from *Tupaia* to form a clade with macroscelidids. Otherwise, relationships are identical in the two trees. Character symbols are identified at right; asterisks indicate character changes arbitrarily optimized to delay transformations. Numbers adjacent to character symbols correspond to the morphological character list given in Asher et al. (2003; see also ftp.amnh.org/pub/people/asher). Insectivorans are in boldface; fossil taxa are identified with a dagger. Abbreviations: abs., absent; transprom., transpromontorial.

Table 5.1 Status of literature-based morphological similarities among insectivorans

Character (number-state) (from Asher et al., 2003)	References	Synapomorphy for Insectivoran Clades	Distribution among Insectivorans	Distribution among Noninsectivorans
Reduced pubic symphysis (154-1)	Butler (1972, 1988); MacPhee and Novacek (1993)	Ambiguous optimization	All Recent families	Chiroptera, Xenarthra
Simplified gut, absent cecum (194-1)	Butler (1972, 1988); MacPhee and Novacek (1993)	Lipotyphla Tenrecoidea	All Extant families (unknown in Nesophontidae)	Chiroptera, Xenarthra
Expansion of maxilla into orbital mosaic (75-1)	Butler (1972, 1988); MacPhee and Novacek (1993); Giere (2002)	Lipotyphla	Erinaceids, *Solenodon*, most tenrecids; slight in talpids and soricids; absent in *Potamogale* and chrysochlorids	*Procavia, Macropus*
"Mobile proboscis" (not included in Asher et al., 2003)	Butler (1972, 1988); MacPhee and Novacek (1993); Maier (2002); Whidden (2002)	See text	Unique insertion pattern of rostral muscles identified in all extant families by Whidden, 2002; cartilaginous nasal skeleton unique in soricids	Proboscideans and macroscelideans have elongate, complex proboscii, but do not share muscular or cartilaginous patterns with insectivorans
Incomplete zygomatic arch (62-1)	Butler (1972, 1988); MacPhee and Novacek (1993)	Autapomorphy for tenrecids and soricids	Zygomatic arch incomplete in soricids, *Solenodon*, *Nesophontes*, and tenrecids; composition of complete arch in talpids and chrysochlorids unclear	*Tamandua*
Hemochorial placenta (190-2)	Butler (1972, 1988); Luckett (1977); Mossman (1987); MacPhee and Novacek (1993); Carter (2001)	Changes on branches more basal than either tenrecoids or lipotyphlans	Erinaceids, tenrecids, chrysochlorids; epitheliochorial in talpids, endotheliochorial in soricids and *Solenodon*	Most other placental groups, including rodents, lagomorphs
Horizontal orientation of tympanic membrane (not included by Asher et al., 2003)	Butler (1988); MacPhee (1981)	See text	Horizontal in juvenile specimens of *Erinaceus*, *Microgale*, soricids, and talpids	Overlap between *Solenodon*, macroscelidids, and tupaiids
Basisphenoid contribution to anterior bulla (19-1)	McDowell (1958); MacPhee (1981)	Tenrecoidea, ambiguous in Lipotyphla	Present in erinaceids, tenrecids, talpids absent in soricids, *Solenodon*, *Nesophontes*	Present in *Macroscelides*, *Tarsius*, cetaceans, *Moeritherium*
Transpromontorial groove on ventrum of petrosal, usually but not exclusively associated with internal carotid artery (11-0)	Butler (1988); Wible (1986)	Changes on branches more basal than either tenrecoids or lipotyphlans	Present in all insectivoran-grade taxa	Present in lagomorphs, *Lemur*, *Tupaia*, *Anagale*, *Tamandua*, *Leptictis*, *Macroscelides*, *Orycteropus*, *Plesiorycteropus*, tenrecoids, *Hyopsodus*, *Meniscotherium*, *Rousettus*, *Manis*, and *Sus*
Absent postglenoid process (59-0)	McDowell (1958)	Changes on branches more basal than either tenrecoids or lipotyphlans	Absent in soricids, *Nesophontes*, *Solenodon*, talpids, tenrecids and chrysochlorids, present in *Erinaceus*	Absent in Manis, macroscelidids, *Plesiorycteropus*, proboscideans, *Arsinoitherium*, *Tamandua*, lagomorphs, and murids

Sources: Based on Asher et al. (2003). Additional information is provided here on character distribution in taxa not sampled by Asher et al. (2003), such as Caribbean insectivorans.

characters as follows (with the character number used by Asher et al. in parentheses):

Pubic symphysis (154): Insectivorans have been characterized as having either narrow or no contact between each pubis at the symphysis (e.g., *Tenrec*, state 1). Most other mammals have a craniocaudally broad pubic symphysis, similar in length to ischiopubic rami framing the obturator foramina of each os coxae (e.g., *Tupaia*, state 0).

Hindgut (194): The word "Lipotyphla" derives from the simplified intestinal tract found in insectivoran-grade mammals. Shrews, moles, hedgehogs, tenrecs, golden moles, and *Solenodon*, in addition to certain other mammals (e.g., many bats), lack a cecum and do not have well-differentiated proximal and distal parts of the gut tube (state 1). Most other mammals (e.g., *Sylvilagus*) possess a cecum, plus large and small intestines with distinct anatomical and functional properties (state 0).

A reduced pubic symphysis also appears in chiropterans, *Manis*, *Plesiorycteropus*, xenarthrans, *Cynocephalus*, *Tarsius*, *Ochotona*, *Mus*, and in some specimens of *Rattus* (Fig. 5.1). Because of the position of northern insectivorans (i.e., erinaceids, soricids, and talpids) between branches leading to Chiroptera and a xenarthran-afrotherian clade, with xenarthrans basal, and because of the position of tenrecs–golden moles adjacent to a clade containing *Plesiorycteropus*, reduction of the pubic symphysis cannot be optimized unambiguously as a synapomorphy for either insectivoran clade. Both possess the reduced state, which is derived relative to the condition in most other placental mammals, but not relative to their immediate sister taxa.

A simple gut tube with a reduced cecum is found in insectivorans, *Manis*, and *Tursiops*. Following Simmons and Geisler (1998: character 195), the cecum is present in the chiropterans sampled by Asher et al. (2003; i.e., nycterids, phyllostomids, and pteropodids), but several chiropterans not included in that data set, such as *Myotis* and *Vespertilio*, do have simple gut tubes lacking ceca. Despite these homoplastic occurrences, simplification of the gut tube is optimized as an independently occurring synapomorphy in both northern and southern insectivoran groups.

EXPANSION OF MAXILLA INTO ORBITAL MOSAIC

MacPhee and Novacek (1993: fig. 3.4) noted that the expansion of the maxilla into the orbital mosaic is widely distributed among insectivorans. Asher et al. (2003: 180) defined this character as:

Maxillary contribution to orbital mosaic (75): Many insectivoran-grade taxa have been noted to possess an expanded maxillary contribution to the medial orbital mosaic, extending posteriorly past the maxillary toothrow and superiorly to a level even with the upper margin of the infraorbital canal, and contributing to the separation of the palatine from more dorsal components of the orbital mosaic such as the frontal and lacrimal (state 1; see Butler, 1988; MacPhee and Novacek, 1993; Giere, 2002). Other taxa (e.g., *Felis*) show an orbital mosaic without a major contribution from the maxilla, consisting instead of the palatine, lacrimal, frontal, and/or ethmoid bones, with contact between the palatine and more dorsal orbital bones (state 0).

An expansive orbital wing of the maxilla appears not only among most insectivorans, but also in caviomorph rodents, *Sylvilagus*, *Chaetophractus*, proboscideans, and *Tursiops* (Fig. 5.1). I have previously (Asher, 1999, 2000) coded this character with three states, based on the presence of an intermediate morphology recognized in *Leptictis* by Novacek (1986), in which the maxilla is slightly expanded into the orbit, but does not separate the frontal and palatine. With the larger taxon sample of Asher et al. (2003), recognizing the gradations that define an intermediate state for *Leptictis* (Novacek, 1986) becomes more difficult. Hence, only the two states described above are recognized in Asher et al. (2003), with *Leptictis* coded as lacking an expanded maxilla (state 0). Furthermore, in Asher (1999, 2000), I mistakenly coded golden moles as having a posteriorly expansive maxilla. MacPhee and Novacek (1993: 18) have correctly noted that, at least for *Amblysomus*, the maxilla does not contribute significantly to the orbital mosaic, and that there is a small articulation between the palatine and frontal. *Amblysomus* has accordingly been coded in Asher et al. (2003) as state 0.

Using a sample of histologically prepared embryonic and juvenile specimens, Giere (2002) determined that, as in *Erinaceus*, the palatine is excluded from the orbital mosaic in soricids and talpids; however, the maxilla in these groups is less caudally extensive than in *Erinaceus*. In *Sorex*, the maxilla has a small caudal extension that separates the palatine from frontal and lacrimal (Giere, 2002: fig. 9), but the posterior and lateral expansion of the nasal capsule contributes more to the ventral restriction of the palatine bone than does the posterior expansion of the maxilla (Giere, 2002: 66). Thus, although insectivoran-grade placentals generally possess a large maxilla that excludes the palatine from dorsal connections to the lacrimal and frontal, this expansion is not as marked in soricids and talpids; nor are the sutures visible in adult skulls of either group. Notably, the African tenrecid *Potamogale*, not included in Asher et al. (2003) due to a lack of sequence data, exhibits no posterior expansion of the maxilla into the orbit, unlike all other tenrecids (Asher, 1999, 2000). Future analyses that incorporate the histological data described by Giere (2002) should be able to refine the orbital mosaic into one or more characters that better represent the variability of this region. Nevertheless, as summarized by Giere (2002) and defined by Asher et al. (2003), a posterior expansion of the maxilla associated with ventral restriction of the palatine comprises a synapomorphy for northern insectivorans, but not for tenrecs–golden moles, due to the absence of this character state in the latter group.

Mobile Proboscis

Perhaps due to its somewhat vague definition, "mobile proboscis" has been proposed not only as a synapomorphy for Insectivora, but also for the competing phylogenetic hypothesis of Afrotheria. Hedges (2001) suggested this as an afrotherian synapomorphy, based on the general observation that the trunk of an elephant shows a similar, if extreme, form of mobility as that seen in the snout of, for example, elephant shrews and sirenians.

Whidden (2002) explored the myological basis for this proposal, and concluded that no afrotherian pattern of snout musculature exists. Rather, he found that tenrecs and golden moles share a common pattern with northern insectivorans; namely, all insectivorans possess a series of five distinct muscles that move the snout, at least some of which are characterized by long tendinous insertions. In soricids and talpids, four of these muscles originate posterior to the squamosal root of the zygomatic arch. Macroscelideans have three elongate tendinous snout muscles that originate just anterior to the orbit. Hyraxes and aardvarks have two muscles, also originating from the antorbital region, but lacking tendons and broadly inserting into the distal rhinarium. Whidden (2002) concluded that the pattern of muscular support for the cartilaginous nasal skeleton was more congruent with the placement of tenrecs and golden moles with northern insectivoran taxa than with afrotherians.

From a structural standpoint, "mobile proboscis" is a vague designation because there are several different ways to increase the flexibility of the nasal skeleton. As reconstructed by Maier (2002), the proboscis of *Neomys* and *Sorex* (Soricidae) owes its mobility in part to three separate nasal roof cartilages: the tectum nasi anterioris, intermedium, and profundus (Fig. 5.2A). With muscular support and a flexible, cylindrical nasal septum, these cartilages are capable of telescoping into one another, providing considerable anteroposterior and dorsoventral mobility (Maier, 2002). Roux (1947: figs. 24–26) reconstructed the cartilaginous nose of another soricid, *Suncus,* and figured what appears to be a tectum nasi intermedium, distinct from the tectum nasi anterioris. Its design is essentially the same as that documented by Maier (2002) in *Neomys* and *Sorex.* Roux did not label the cartilaginous elements themselves, but rather, referred to the gaps separating them as "resorption deficiencies of the tectum nasi," perhaps not realizing that these structures persist into adulthood and are part of a specific functional regime.

This kind of mobile proboscis in shrews appears to be unique. Although the tenrecids *Micropotamogale, Geogale,* and *Tenrec* have a substantial cartilaginous nasal skeleton (Asher, 2001), none shows a proximally overlapping series of nasal tecti. Nor are these present in the cartilaginous ethmoidal nasal skeleton of an embryonic sirenian (Matthes, 1912). Non-soricids exhibit other specializations in their cartilaginous nasal skeleton, potentially related to increased mobility of the proboscis. For example, *Solenodon* has an incomplete zona annularis (Menzel, 1979; Asher, 2001); that is, the nasal sidewall is separated from the nasal floor, due to an anteroposteriorly elongate fissure extending the length of the cartilaginous nose (Fig. 5.2B). *Elephantulus* shows a series of dorsoventrally oriented fenestrae in the nasal side wall (Fig. 5.2C) that adds to mediolateral flexibility. This structure represents a morphology similar to that of *Rhynchocyon,* described by Parker (1885: 246–247, plate 36) as consisting of a series of "narrow rings," separated by regularly occurring dorsoventrally elongate fenestrae.

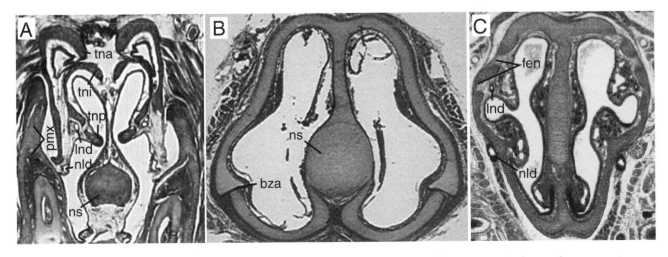

Fig. 5.2. (A) Coronal slice through the premaxilla of *Sorex araneus* (crown rump length [CRL] 40 mm, slice 10.2.5, ZIUT), showing the tectum nasi anterioris, tectum nasi intermedium, and tectum nasi profundus dorsal to the nasal septum in an individual that has retracted its mobile proboscis (see Maier, 2002). (B) Coronal slice through the cartilaginous external nose in *Solenodon paradoxus* (slice 59.3.2, ZIUT), showing the broken zona annularis; that is, the discontinuous side wall and floor of the nasal skeleton. (C) Coronal slice through the cartilaginous external nose in *Elephantulus myurus* (CRL 30 mm, slice 8.1.5, ZIUT) showing fenestrae in the nasal sidewall. Also labeled are the nasolacrimal duct and lateral nasal duct. Abbreviations: bza, broken zona annularis; fen, fenestrae; lnd, lateral nasal duct; nld, nasolacrimal duct; ns, nasal septum; pmx, premaxilla; tna, tectum nasi anterioris; tni, tectum nasi intermedium; tnp, tectum nasi profundus.

Thus, perhaps like "ossified bulla," "mobile proboscis" is by itself too broad an anatomical designation to convey much phylogenetic or functional information. Details such as those given by Maier (2002) about the mobility of a proboscis and by Whidden (2002) about its muscular control, are necessary to construct an informative character complex. Because such details are not widely available across mammalian genera, Asher et al. (2003) did not include characters of the nasal skeleton or musculature in their morphological matrix. However, Asher (2001) optimized 20 characters from the nasal skeleton and cranial arterial supply, not including those from the soricid nasal roof, discussed by Maier (2002), on insectivoran vs. afrotherian hypotheses of placental relationships. Among the nasal characters, none presented a distribution that is congruent with insectivoran monophyly, although the distribution of, for example, the lateral cartilaginous cover of the nasolacrimal duct in erinaceids and soricids (Asher, 2001: character 16), was consistent with the erinaceid-soricid clade recovered in some of the analyses discussed in Douady et al. (2002b), Malia et al. (2002), and Asher et al. (2003).

Placentation

As noted by MacPhee and Novacek (1993), a "hemochorial placenta," in which there is minimal tissue separating maternal and fetal blood supply, is not an accurate description of placentation in all insectivoran-grade mammals. Luckett (1977) characterized erinaceids, tenrecids, and chrysochlorids as hemochorial; soricids as endotheliochorial; and talpids as epitheliochorial; that is, both soricids and talpids show more layers of tissue separating maternal and fetal blood supply than do hemochorial taxa (Luckett, 1977: fig. 25). Wislocki (1940) described the fetal membranes of *Solenodon*, the morphology of which was summarized by Mossman (1987: 169) as strongly resembling that of the soricid *Crocidura*. Mossman further noted that there may be differences within some families (e.g., Talpidae). In fact, Malassiné and Leiser (1984) and A. M. Carter (pers. comm.) have indicated that *Talpa europaea* possess an endotheliochorial placenta, and that reports of an epitheliochorial placenta in *Scalopus* may result from observations on the placenta of a non-term individual. In addition, placentation in several insectivoran groups remains poorly understood. New data on placentation has been published for marsupials (Zeller and Freyer, 2001), elephant shrews (Cutler et al., 1998), and rodents (Mess, 2002); but according to Mossman (1987), who was very critical of the study of *Echinosorex* by Meister and Davis (1953), erinaceids, for example, remain well studied only for the single genus *Erinaceus*.

The removal of African insectivorans from Lipotyphla makes some details of insectivoran placentation summarized by Mossman (1987: 174–183) phylogenetically more interesting. First, the northern insectivorans *Scalopus, Talpa, Sorex, Suncus,* and *Erinaceus* show a "permanent, bilaminar, juxtauterine, omphalopleure"; that is, their yolk sac persists throughout gestation, is walled by both trophoblast and

endoderm, and is situated against the wall of the uterus. This condition differs from that seen in the chrysochlorid *Eremitalpa*, the tenrecid *Hemicentetes*, and the macroscelidid *Elephantulus*, which shows a juxtauterine, bilaminar omphalopleure during blastocyst stages only. Second, among northern insectivorans, the orientation of the yolk sac is consistently mesometrial (i.e., "at or towards the side of the uterus" Mossman, 1987: 310); whereas among the African taxa, it is described as mesometrial only in the tenrecid *Setifer*. In another tenrec, *Hemicentetes*, Mossman describes it as "lateral" and in the golden mole *Eremitalpa*, he describes it as "?variable." Thus, as discussed by Carter (2001), some features of placentation are consistent with the placement of northern insectivorans, or "eulipotyphlans," in their own clade apart from tenrecids and chrysochlorids. However, although the distribution of several placentation characters has been documented across most placental families (Luckett, 1977, 1985, 1993), most have yet to be incorporated into a broadly sampled, low-level mammalian phylogeny. Asher et al. (2003) incorporated four characters regarding placentation into their morphological character set: presence of a trophoblast (character 188), type of maternal-fetal nutrient exchange (189), type of placental membrane (190), and permanence of the yolk sac (191). None of these characters optimizes as a synapomorphy for either insectivoran group. The proposed hemochorial placenta synapomorphy (Butler, 1988) was a state in the placental membrane character defined as follows:

Placental membrane (190): One area of variation in the eutherian placenta is the way in which maternal blood reaches the trophoblast of the interhemal membrane. Contact between uterine epithelium and trophoblast is called epitheliochorial (e.g., *Equus*, state 0); contact between a maternal capillary endothelium and trophoblast is endotheliochorial (e.g., *Felis*, state 1); and direct contact between maternal blood and the trophoblast is called hemochorial (e.g., *Procavia*, state 2). Marsupials are coded as inapplicable for this character, as the function served by the eutherian trophoblast is carried out in marsupials by the shell membrane, combined with an abbreviated intra-uterine period of development.

This character is mapped onto the placental phylogeny in Fig. 5.1. In fact, contrary to the status of hemochorial placenta as an insectivoran synapomorphy, the only "core insectivoran" sampled by Asher et al. (2003) that shows a hemochorial placenta is *Erinaceus*. African insectivorans also possess a hemochorial placenta, as do rodents, lagomorphs, haplorhine primates, xenarthrans, *Elephantulus, Procavia,* and most chiropterans. Accordingly, a hemochorial placenta characterizes the basalmost branches on the topology depicted in Fig. 5.1, and is therefore reconstructed as primitive for both insectivoran groups.

Another aspect of reproductive anatomy not yet incorporated into a broadly sampled phylogeny of insectivorans, or mammals generally, concerns gamete and reproductive tract morphology. Both shrews (Phillips and Bedford, 1984; Bedford et al., 1994, 1997) and moles (Bedford et al., 1999) possess spermatozoa with very large, flattened heads, which exhibit barbs along their exterior. In addition, female shrews and moles show a highly convoluted, rugose isthmus of the fallopian tube, the walls of which may form crypts that function to store spermatozoa and/or regulate access of sperm to more distally located unfertilized eggs (Bedford et al., 1997). A barbed sperm head has also been observed among certain tenrecids (Bedford, pers. commun.), as well as elephant shrews (Woodall, 1991) and the erinaceid *Atelerix* (Bedford et al., 2000).

Reduction of the Jugal

Solenodon and *Nesophontes* (not sampled by Asher et al., 2003), as well as soricids and tenrecids, lack a complete zygomatic arch. MacPhee and Novacek (1993) summarized the suggestion of Butler (1972) that taxa without a complete arch, which correspondingly lack a jugal element, share the character state "reduced jugal" with talpids and chrysochlorids that possess a complete zygomatic arch arguably comprised of the maxilla alone. I follow MacPhee and Novacek in regarding the evidence on the composition of the zygomatic arch in talpids and chrysochlorids as too ambiguous to definitively homologize their complete zygomatic arch and lack of clear jugal sutures with the incomplete arch and missing jugal in soricids, tenrecids, *Solenodon*, and *Nesophontes*. Hence, Asher et al. (2003: 178) defined a zygomatic arch character based on the presence or absence of an ossified connection linking the ventrolateral aspect of the squamosal with the facial skeleton as follows:

Zygomatic arch (62): Many insectivorans have an incomplete zygomatic arch lateral to the mandibular coronoid process (e.g., *Tenrec*, state 1). Most other mammals have a robust zygomatic process (e.g., *Canis*, state 0). In *Megaptera*, the maxilla does not participate in the zygomatic arch; rather, the frontal comprises the anterior component of the arch (state 2).

Thus defined, with reference only to its anteroposterior continuity rather than to its suspected bony composition (except for the autapomorphic frontal arch in *Megaptera*), an incomplete arch is still rare among placentals—found only in myrmecophagids, in addition to tenrecids, soricids, and *Solenodon*. Lack of a complete zygomatic arch does not constitute a synapomorphy for either insectivoran group (see Fig. 5.1).

Other Morphological Similarities

Further similarities among insectivoran groups noted by Butler (1988) include the presence of a basisphenoid contribution to the auditory bulla, the transpromontorial course of the internal carotid artery, and the horizontal orientation of the tympanic membrane. McDowell (1958) suggested the absence of a true postglenoid process, the presence of a caudal tympanic process of the petrosal, and the presence of interradicular crests on the basal surface of cheek teeth

as defining characters for Insectivora. As morphological characters, molar interradicular crests and the orientation of the tympanic membrane have not yet been incorporated into a cladistic analysis that includes insectivorans.

Among these characters, only the basisphenoid contribution to the auditory bulla constitutes an unambiguous synapomorphy of one of the insectivoran clades depicted in Fig. 5.1, joining tenrecs and golden moles. This character state also occurs among talpids and erinaceids, but it cannot be unambiguously optimized on the topology depicted in Fig. 5.1 until more northern insectivoran taxa are added. Similarly, the absence of a postglenoid process is shared in both African and most northern insectivorans (present in *Erinaceus* and absent in murids, lagomorphs, elephant shrews, proboscideans, *Arsinoitherium*, *Tamandua*, and *Manis*), but does not unambiguously define a character change on the branches leading to either insectivoran group (see Fig. 5.1).

A transpromontorial sulcus on the ventrum of the pars cochlearis (Asher et al., 2003: character 11), typically associated with the internal carotid artery (Wible, 1986), is widely distributed across eutherian mammals and is never reconstructed in Fig. 5.1 as changing on branches leading to either the northern or southern insectivoran group.

The caudal tympanic process of the petrosal (CTPP; see MacPhee, 1981) was defined as "large" by Asher et al. (2003) when it contributes to a posterior wall of an ossified auditory bulla, in addition to shielding the fenestra cochleae ventrally (see also MacPhee et al., 1988). Based on this definition, the CTPP was coded as "small" in *Erinaceus*, but as contributing to the posterior bulla in *Sorex* and talpids. Hence, in Fig. 5.1, it does not optimize as a synapomorphy for northern insectivorans; nor does the CTPP character change on the branch leading to tenrecs and golden moles, due to its "large" state in *Amblysomus* but "small" state in *Tenrec* and *Echinops*. However, within the Tenrecidae, only the spiny tenrecs (including *Tenrec* and *Echinops*) exhibit a relatively small CTPP; potamogalines, *Geogale*, *Limnogale*, *Microgale*, and *Oryzorictes* (not included in Asher et al., 2003) show a larger, more shrewlike structure (Asher, 1999). Hence, future studies with a larger tenrecid sample may support the optimization of this character as a tenrec–golden mole synapomorphy.

MacPhee (1981) provided measurements of tympanic orientation in embryonic and juvenile specimens, as well as one adult *Tupaia* specimen, across several primates, insectivorans, tupaiids, and macroscelidids. Although the tympanic membrane becomes more vertically oriented as an individual ages, there appear to be consistent differences across similarly-aged specimens in the sample of MacPhee (1981). *Erinaceus*, *Hemicentetes*, *Microgale*, and *Setifer* show a relatively more horizontal tympanum than most other taxa. However, there is overlap between *Erinaceus* and *Elephantulus*, and a juvenile *Solenodon* specimen fits neatly into the range of embryonic-to-juvenile specimens of *Tupaia*. Thus, again it is apparent that considerable variation exists in the distribution of a character once hypothesized to unite insectivorans. However, variable distribution of a given character is by itself insufficient reason to exclude that character a priori from a phylogenetic analysis. Hence, assuming information on the tympanic angle across comparable ontogenetic stages of many placental taxa can be collected and satisfactorily categorized into discrete character states, such information may yet provide valuable resolution to some part(s) of the mammalian tree.

FOSSIL INSECTIVORANS

McKenna and Bell (1997) listed approximately 250 living and extinct genera of insectivoran-grade mammals. About 75% of these genera are extinct, and few of these are represented by material good enough to preserve the characters summarized above. Nevertheless, many do retain anatomical data that indicate taxonomic affinity with one or more Recent groups; in fact, extinct relatives of modern insectivoran taxa have been recognized for many decades (e.g., Gaillard, 1915; Grassé, 1955; Butler, 1956, 1984; Butler and Hopwood, 1957; Repenning, 1967; Schmidt-Kittler, 1973; Hutchinson, 1974; Engesser, 1975; Barnosky, 1981; Novacek, 1985; MacPhee et al., 1988; Gould, 1995; Wojcik and Wolsan, 1998; Ottenwalder, 2001; Lopatin, 2002).

As indicated by these and other authors, characters that permit recognition of fossil representatives of Recent groups include the craniomandibular complex of soricids and the forelimb and shoulder girdle of talpids. As a group, erinaceids have a distinctive dentition that, in combination with such middle ear characters as a basisphenoid bulla and a small or absent piriform fenestra, permits recognition of their extinct relatives (e.g., Butler, 1956). Similarly, tenrecids are characterized by zalambdodont teeth (see Asher et al., 2002: 14–16), a posteriorly extensive maxillary component to the ventral pterygoid plate (McDowell, 1958: fig. 31), and a basisphenoid bulla (Butler and Hopwood, 1957). Golden moles and *Solenodon* are also zalambdodont. The former have many apomorphies (Hickman, 1990) and can be identified based on several unique anatomical characters, such as the mandibular-hyoid articulation (Bronner, 1991), the presence of three longbones in the forearm (Grassé, 1955), a hypertrophied medial epicondyle of the humerus, a reduced number of manual digits, and a reduced mandibular coronoid process. *Solenodon* is not as derived postcranially, but shows a unique combination of zalambdodont teeth, a large piriform fenestra, and a lack of any ventrally ossified bullar components. Following Ottenwalder (2001), the extinct *Solenodon arrendondoi* from Cuba and *S. marcanoi* from Hispaniola (both from Pleistocene or younger deposits) are close relatives of the two extant species, also limited geographically to Cuba (*S. cubanus*) and Hispaniola (*S. paradoxus*).

Fragmentary Insectivoran Fossils

More taxonomically problematic than close relatives of modern families are several extinct taxa that, over the years, have been proposed as related to the higher-level insectivoran-grade taxa "Soricomorpha" and/or "Erinaceomorpha." The majority of these extinct taxa (e.g., the early Tertiary taxa *Leptacodon, Saturninia, Scenopagus*) are known solely from fragmentary dental material. Obviously, it is more difficult to evaluate the systematic position of taxa represented by poor material, but paleontologists must work with the material they have available. Phylogenetic hypotheses have been presented for these taxa, supporting, for example, the position of *Scenopagus* as part of an extinct radiation at the stem of extant erinaceids (Novacek, 1985), and of the nyctitheres *Leptacodon* and *Saturninia* as part of a radiation on the stem leading to "soricomorphs" (Butler, 1988), a group which following Butler includes soricids and solenodontids (including *Nesophontes*), in addition to talpids, tenrecids, and chrysochlorids. With the recovery of more material, these hypotheses will become subject to more exhaustive tests. For example, based on new postcranial remains, Hooker (2001) suggested that extinct nyctitheres may actually be related to primates and/or scandentians rather than to insectivoran-grade taxa. Such a relationship has been substantiated by Hooker's (2001) cladistic analyses of both pedal and craniodental characters, including pedal remains attributed by Hooker to ?*Cryptotopos* (Nyctitheriidae), and provides a provocative hypothesis amenable to future testing.

Better-Known Insectivoran Fossils

Several extinct insectivorans discussed below are known by well-preserved cranial and postcranial material. Unfortunately, this has not necessarily made identification of their taxonomic affinities easy. In three previous analyses, two focusing on the living family Tenrecidae (Asher, 1999, 2000) and one on extinct, dentally zalambdodont "apternodontids" from North America (Asher et al. 2002), I have presented phylogenies that include several taxa known from complete skulls and/or skeletons. In addition, Fig. 5.1 is based on a combined data analysis (Asher et al., 2003) focusing on the affinities of African insectivorans and samples the morphological and molecular diversity of all eutherian orders, including two skeletally well-known insectivoran-grade fossils, *Anagale* and *Leptictis*. I will defer definitive statements on the affinities of fossil insectivoran groups, pending the completion of a character matrix that samples all of these extinct groups and includes the morphological and sequence diversity of members of all modern placental orders. Nevertheless, for the purposes of this review, I present here an overview of previously published results, summarized in Table 5.2.

Insectivorans from Messel, Germany

Leptictidium (Maier, 1986), *Macrocranion* (MacPhee et al., 1988), and *Pholidocercus* (von Koenigswald and Storch, 1983) from the middle Eocene Messel locality near Darmstadt, Germany, are represented by fairly complete skeletons, although these are preserved in a somewhat flattened state.

Table 5.2 Summary of topologies for fossil insectivorans from previous studies

	Asher (1999)	Asher (2000)	Asher et al. (2002)	Asher et al. (2003)
Anagale	Not included	Not included	Not included	1. Sister to *Tupaia* (7/12) 2. Unresolved (3/12) 3. Unresolved with *Tupaia* at base of Rodentia to (1/12) 4. Lagomorph stem (1/12)
Apternodus	1. Sister to tenrecoid clade (5/8) 2. Sister to tenrecoids (1/8) 3. Sister to *Solenodon* (1/8) 4. Unresolved at base of tenrecid clade with chrysochlorids. talpids, soricids *Solenodon,* and *Nesophontes* (1/8)	1. Sister to *Solenodon,* in clade with *Nesophontes* and *Centetodon* (10/12) 2. Unresolved within tenrecoid clade (2/12)	1. Sister to soricid-parapternodontid-*Oligoryctes* clade (1/4) 2. Unresolved within soricid-parapternodontid-*Oligoryctes* clade (1/4) 3. Unresolved (2/4)	Not included
Centetodon	1. Sister to tenrecoid-*Nesophontes-Solenodon-Apternodus* clade (3/8) 2. Sister to [1] plus soricids and talpids (3/8) 3. Sister to [1] plus soricids (1/8) 4. Unresolved (1/8)	1. Sister to *Nesophontes-Solenodon-Apternodus* (10/12) 2. Sister to tenrecoid-soricid-talpid clade (2/12)	1. Sister to *Nesophontes-*tenrecid-*Solenodon-Apternodus-Oligoryctes-*parapternodontid-soricid clade (2/4) 2. Sister to *Solenodon-*tenrecid-*Apternodus-Oligoryctes-*parapternodontid-soricid clade (1/4) 3. Unresolved (1/4)	Not included
Gymnurechinus	Not included	1. Sister to *Erinaceus* (7/12) 2. Unresolved within Erinaceidae (1/12) 3. Sister to soricid-talpid clade among paraphyletic erinaceomorphs (2/12) 4. Unresolved (1/12) 5. Sister to *Echinosorex-Macrocranion* (1/12)	Not included	Not included
Leptictidium	Not included	1. Sister to *Leptictis-*macroscelidids (3/12) 2. Unresolved (3/12) 3. Sister to [1] plus *Macrocranion* and *Pholidocercus* (2/12) 4. Sister to *Leptictis* plus erinaceomorphs (2/12) 5. Sister to *Leptictis-*macroscelidid-*Orycteropus-Procavia-Tupaia* clade (1/12) 6. Sister to [5] plus erinaceomorphs (1/12)	Not included	Not included
Leptictis	1. Sister to *Tupaia* (4/8) 2. Sister to *Procavia-Tupaia* clade (1/8) 3. Sister to erinaceids (1/8) 4. Sister to Lipotyphla (1/8) 5. Unresolved (1/8)	1. Sister to macroscelidids (4/12) 2. Sister to [1] plus *Macrocranion* (2/12) 3. Unresolved (3/12) 4. Sister to erinaceomorphs (2/12) 5. Sister to erinaceomorph-macroscelidid-*Orycteropus-Procavia-Tupaia* clade (1/12)	Not included	1. Sister to macroscelidids (5/12) 2. Sister to [1] plus tenrecoids (1/12) 3. Unresolved (4/12) 4. Lipotyphlan stem (1/12) 5. Lagomorph stem (1/12)

continued

Table 5.2   continued

	Asher (1999)	Asher (2000)	Asher et al. (2002)	Asher et al. (2003)
Macrocranion	Not included	1. Sister to *Echinosorex* (6/12) 2. Unresolved (2/12) 3. Sister to soricid-talpid clade among paraphyletic erinaceomorphs (2/12) 4. Sister to macroscelidids (2/12)	Not included	Not included
Nesophontes	1. Sister to *Solenodon-Apternodus*-tenrecoid clade (3/8) 2. Sister to *Solenodon-Apternodus*-tenrecid clade (1/8) 3. Sister to Soricidae (1/8) 4. Sister to [1] plus talpids and soricids (2/8) 5. Unresolved (1/8)	1. Sister to *Apternodus-Solenodon* clade (8/12) 2. As in [1] but unresolved with respect to Centetodon (2/12) 3. Sister to tenrecoid-*Solenodon-Apternodus* clade (1/12) 4. As in [3] but unresolved with respect to soricids-talpids (1/12)	1. Sister to tenrecid-*Solenodon-Apternodus-Oligoryctes*-parapternodontid-soricid clade (1/4) 2. Sister to [1] plus *Centetodon* (1/4) 3. Unresolved within [1] (1/4) 4. Unresolved (1/4)	Not included
Oligoryctes	Not included	Not included	1. Sister to parapternodontids and soricids (2/4) 2. Unresolved (2/4)	Not included
Pholidocercus	Not included	1. Sister to other erinaceomorphs (5/12) 2. Unresolved within erinaceomorphs (1/12) 3. Sister to soricid-talpid-paraphyletic erinaceomorph clade (2/12) 4. Sister to *Leptictis-Macrocranion*-macroscelidid clade (2/12) 5. Unresolved (2/12)	Not included	Not included

Notes: Only fossils with (at least) complete cranial material are listed. Ratios in parentheses following clade identifies portion of analyses supporting that clade; for example, *Apternodus* was sister taxon to *Solenodon* in 10 of 12 analyses performed by Asher, (2000). *Apternodus* was coded as a single chimaeric taxon in Asher (1999, 2000) and as seven distinct species in Asher et al. (2002). Asher (1999) consists of 35 terminals and 71 osteological characters, focusing on tenrecids; Asher (2000) is an expanded version of Asher (1999) and consists of 42 terminals and 105 osteological, 20 soft tissue, and 12S rRNA sequence data for 17 of the extant terminals, focusing on tenrecids; Asher et al. (2002) consists of 30 terminals and 118 osteological characters, focusing on fossil zalambdodonts; and Asher et al. (2003) consists of 60 terminals, 196 morphological characters, and up to 19 nuclear and three mitochondrial genes for all 49 extant taxa, including all placental orders. I use the term Erinaceomorpha to indicate living erinaceids (*Echinosorex* and *Erinaceus*) plus the extinct *Gymnurechinus*, *Macrocranion*, and *Pholidocercus*. I use Tenrecoidea *sensu* McDowell (1958) to indicate a tenrecid-chrysochlorid clade.

I have previously included all three in a phylogenetic analysis of cranial, dental, and postcranial characters, plus 12S rRNA sequence data for several living taxa (Asher, 2000), but obtained little resolution in consensus topologies across analyses that varied such phylogenetic variables as relative character weighting and hypotheses of order. In topologies that were resolved, *Macrocranion* and *Pholidocercus* most often appeared adjacent to extant erinaceids. Characters used by Asher (2000) that support this arrangement include a distally fused tibia-fibula and a prominent rostral tympanic process of the petrosal. *Macrocranion* shares with erinaceids a small lower third molar and a prominent lacrimal flange; *Pholidocercus* and erinaceids share paired ectopterygoid processes and three vertebrae contributing to the sacroiliac

articulation. *Leptictidium* most often appeared at the base of a clade containing elephant shrews; potential synapomorphies include a molariform P3, similarly-sized anterior incisors, and five upper antemolars (i.e., four putative premolars plus a canine).

Centetodon

The early–middle Tertiary taxon *Centetodon* is also known by relatively well-preserved cranial material. Lillegraven et al. (1981) focused primarily on the intrageneric diversity of this animal, but suggested in closing their monograph that, like nyctitheres, *Centetodon* was part of a radiation possibly on the stem leading to modern soricids and *Solenodon*, encapsulated in the "Soricomorpha" of McKenna and Bell (1997). I have included *Centetodon* as a terminal taxon in three previous studies (Asher, 1999, 2000; Asher et al., 2002), but in this case have also encountered considerable topological variability, depending on the starting assumptions of the particular analysis. In the most taxonomically broadly sampled of these three studies (Asher, 2000), which included 32 insectivorans plus ten other mammalian taxa, *Centetodon* is most frequently reconstructed in a clade with the Caribbean genera *Solenodon* and *Nesophontes*, as well as the Tertiary North American taxon *Apternodus*. These taxa all share a derived anatomy of the craniomandibular articulation; as in soricids, they lack a postglenoid process and show a glenoid fossa positioned anterior, as opposed to anterolateral, to the pars cochlearis of the petrosal. The mandibular condyle is consequently supported posteriorly by a ventral extension of the alisphenoid, or the entoglenoid process (McDowell, 1958: fig. 25).

In contrast, a more detailed study of the seven species of the genus *Apternodus* (Asher et al., 2002) plus five additional dentally zalambdodont North American fossils (but fewer tenrecids and non-insectivoran mammals), did not support the position of *Centetodon* in a clade with *Apternodus*. Instead, *Centetodon* was reconstructed outside of an *Apternodus*-soricid-*Solenodon*-*Nesophontes*-tenrecid clade, a signal also recovered under some assumption sets by Asher (1999). Synthesis of these data sets so as to represent the diversity of both African insectivorans and North American fossils is necessary prior to favoring one of these competing topologies.

Leptictis and *Palaeoryctes*

Among fossil "Insectivora," MacPhee and Novacek (1993) focused primarily on the status of two fossil groups with possible close affinities to one or more of the six extant insectivoran families: palaeoryctids and leptictids. Their review of these fossils groups is still timely, and I do not repeat their discussion here, other than to note that they regarded *Leptictis* as a viable candidate for the sister taxon of Insectivora. In contrast, the most recent analyses in which I have included *Leptictis* as a terminal taxon (Asher, 2000; Asher et al., 2003) more frequently reconstruct *Leptictis* with elephant shrews than with insectivorans. Resemblances consistent with such a relationship include the presence of a large entotympanic component of the ossified auditory bulla, an optic foramen not considerably smaller than exit foramina of the trigeminal nerve, and an internally concave acetabular region of the pelvis.

Palaeoryctes remains less well known than *Leptictis*. It is known from a semicomplete skull (McDowell, 1958), but has not been included in recent, character-based phylogenetic analyses. Two other palaeoryctid genera, *Eoryctes* (Thewissen and Gingerich, 1989) and *Pararyctes* (Bown and Schankler, 1982) are also known from relatively complete cranial material, although cranial specimens of the latter genus have yet to be fully described. Asher et al. (2002) sampled these two taxa and noted convergence in the develop-

ment on the posterior skull of boxlike, masticatory muscle attachments (or "lambdoid plates") in some palaeoryctids and *Apternodus,* plus several important differences in the middle ear and jaw articulation. Specifically, palaeoryctids have an extensively ossified auditory bulla, consisting of a broad rostral tympanic process of the petrosal and a large entotympanic. They also show a prominent postglenoid process and a glenoid fossa anterolateral to the pars cochlearis of the petrosal. In the phylogeny focusing on *Apternodus* and other zalambdodont mammals presented by Asher et al. (2002), these character states contributed to the placement of palaeoryctids outside of a clade consisting of soricids, *Solenodon, Nesophontes, Centetodon,* tenrecids, and fossil zalambdodonts, but not including erinaceids. How palaeoryctids will fare in a more broadly sampled analysis remains to be demonstrated.

"Apternodontids" and Soricids

Until recently, any small fossil mammal from the North American Tertiary with triangular upper molars and a reduced, unbasined talonid was placed in the extinct family "Apternodontidae." This group was classified by McKenna and Bell (1997) in the Soricoidea, along with *Solenodon,* soricids, and the extinct *Nesophontes,* plesiosoricids, micropternodontids, and nyctitheriids.

Asher et al. (2002) reviewed the morphology and relationships of apternodontids, and agreed with McKenna and Bell (1997) that the majority of its constituent taxa are closely related to modern soricids and *Solenodon.* However, following Asher et al. (2002), the "Apternodontidae" as previously construed is paraphyletic: three of its constituent genera, *Oligoryctes, Koniaryctes,* and *Parapternodus,* are more closely related to soricids than to members of the genus *Apternodus.* Such a close relationship between dentally zalambdodont fossils and modern soricids had not previously been suggested.

With a fossil history extending from the middle Eocene through the early Miocene, *Domnina* has long been regarded as the basalmost soricid, along with other heterosoricines, such as *Trimylus* (Repenning, 1967; Engesser, 1975). Recently, Lopatin (2002) described *Soricolestes,* a middle Eocene fossil soricid from central Mongolia that most likely predates the earliest North American record of *Domnina* from the late Uintan of central Wyoming (Krishtalka and Setoguchi, 1977) and southern Saskatchewan (Storer, 1984). Like *Domnina* (Fig. 5.3A) and *Trimylus* (Fig. 5.3B), but unlike extant soricids (Fig. 5.3C), *Soricolestes* lacks a pocketed coronoid process. No specimen of *Soricolestes* preserves a mandibular condyle, but Lopatin (2002) has inferred for it a single unpaired cranial articulation, due to the non-pocketed coronoid and masseteric fossa on the type specimen. This inference is not justified, however, as *Trimylus* lacks a pocketed coronoid, shows a slight masseteric fossa, and yet possesses a widely divergent pair of articular facets on the mandibular condyle (Fig. 5.3B). These facets are much less divergent in *Domnina,* but there is nevertheless a distinct separation of

Fig. 5.3. Mandibles in internal (left) and posterior (right) views. (A) *Domnina thompsoni*, AMNH 32647. (B) *Trimylus dakotensis*, UCMP 37270. (C) *Crocidura olivieri*, AMNH 48490. The posterior view of mandibular condyle at right is AMNH 236229. and (D) *Solenodon paradoxus*, AMNH 185012. Scale bars = 2 mm.

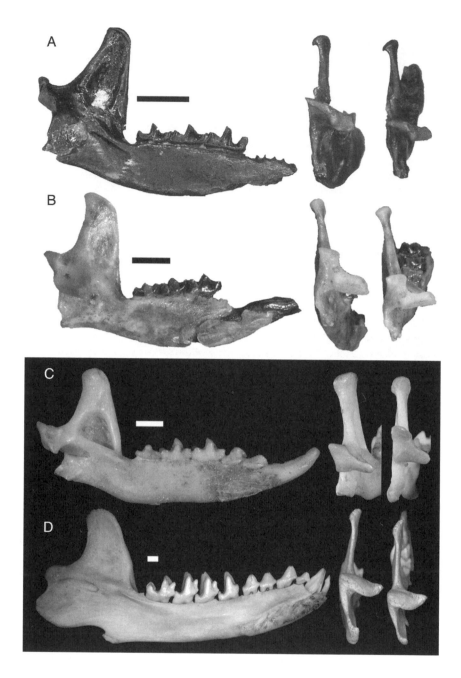

the condyle into upper and lower facets in this taxon as well (Fig. 5.3A).

McDowell (1958: 142) argued that both *Solenodon* and *Nesophontes* possessed a "divided" mandibular condyle, "intermediate between that of the Soricidae and that of other lipotyphlans." In *Solenodon*, McDowell (1958: fig. 8D–F) figured an "external capitular facet" on the dorsolateral aspect of the squamosal glenoid fossa, and a "postglenoid capitular facet" on its ventromedial aspect. On the mandibular condyle itself, he illustrated a marked division between these facets (1958: fig. 8F). Such a division is not present in the specimens of *Solenodon* available to me (e.g., ZMB 2761; AMNH 185012; see Fig. 5.3D). Furthermore, a histologically sectioned specimen of *Solenodon* (MPIH 6863; discussed in Mac-Phee, 1981, and Asher, 2001) that preserves several coronal slices through the craniomandibular joint shows no sign of

multiple articular capsules, as McDowell (1958) claimed to have observed in macerated skulls preserving dried bits of soft tissue. Admittedly, this histological series does not preserve the jaw joint in its entirety, but only several posterior slices. Nevertheless, I concur with MacPhee and Novacek (1993) that osteologically, *Solenodon* has only a single craniomandibular articulation (Fig 5.3D). It is difficult to prove that extinct shrews possessed separate synovial capsules, as demonstrated in living taxa (Fernhead et al., 1954); but based on its widely spaced articular facets (Fig. 5.3B), dual synovial capsules were probably present in *Trimylus,* and possibly also in *Domnina* (Fig. 5.3A). Confirmation of the condition in *Domnina* will require better cranial material that documents the temporal side of the temporomandibular joint.

If the hypothesis of Asher et al. (2002) is correct that elements of the paraphyletic "Apternodontidae" (i.e., *Oligoryctes* and the less well-known early Eocene *Parapternodus* and *Koniaryctes*) constitute soricid sister taxa, then the mosaic nature of craniomandibular evolution in these taxa becomes even more marked. Like modern shrews, but unlike fossil heterosoricines, *Oligoryctes* and *Parapternodus* possess an internally pocketed coronoid process. (Remains of the coronoid process are as yet unknown in *Koniaryctes*.) Characters from other anatomical regions supporting this zalambdodont-soricid clade include the absence of an alisphenoid canal and a tightly packed toothrow without diastemata, and in *Koniaryctes-Parapternodus*-soricids, a non-molarized p4 and a mesiodistally short m3 (Asher et al., 2002).

Based on a topology from Asher et al. (2002: fig. 60), two characters that represent the derived craniomandibular articulation in shrews exhibit several changes. The pocketed coronoid appears first in the soricid sister taxon *Oligoryctes* and is reversed in heterosoricines. Barring its occurrence in *Parapternodus* and *Koniaryctes* (for which the relevant anatomy is unknown), the double jaw articulation does not appear until the base of the Soricidae, with some reduction in the degree of separation between the two facets in the branch leading to *Domnina*. Variability in these craniomandibular joint characters may be indicative of clade-specific homoplasy, a concept discussed by Simpson (1945: 9–10) and later referred to as "underlying synapomorphy" (Saether, 1979). Other characters that may represent clade-specific homoplasy include tooth pigmentation, present in soricines, heterosoricines, and some *Apternodus* specimens (but not crocidurines); and the bulbous enlargement of the anterior dentition, present sporadically in soricids (*Drepanosorex*; see Reumer, 1984), the genus *Apternodus* (*A. baladontus* and *A. mediaevus*; see Asher et al. 2002), and the genus *Solenodon* (*S. cubanus* and *S. arredondoi*; see Ottenwalder, 2001).

TAXONOMY

Table 5.3 depicts the classifications of Simpson (1945), Van Valen (1967), and McKenna and Bell (1997; see also Novacek, 1986: table 1). Two newer synoptic classifications that account for recent molecular work are presented in Table 5.4. Interestingly, proposals for a polyphyletic Insectivora are not solely the product of sequence-based analyses. As shown in Table 5.3, Van Valen (1967) placed African tenrecs and golden moles in a distinct order from northern erinaceids, soricids, and talpids, although in this case, he placed *Solenodon* close to the African taxa in his "Deltatheridia," along with fossil palaeoryctids, micropternodontids, didymoconids, hyaenodontids, and oxyaenids. However, even Van Valen's novel classification did not foreshadow the extent to which insectivorans are today recognized as polyphyletic. In order to apply a complete, hierarchical nomenclature encompassing all Recent insectivoran groups and account for the recent influence of molecular data, it is necessary to include every single order of extant placental mammals; whereas classifications supporting insectivoran

monophyly could, obviously, be much more focused within Placentalia.

The choice of higher-level taxa listed in Table 5.4 follows the guidelines set out by Simpson (1945: 12–33), with the significant exception that I do not explicitly assign Linnean ranks to any of these hierarchical levels. Otherwise, as recommended in both Linnean (Simpson, 1945) and more recent (deQueiroz and Cantino, 2001) classification schemes, priority and novelty play major roles in determining which taxa are used. New names are applied reluctantly, and only for groups with no precedent in the literature (e.g., Afrotheria Stanhope et al. 1998). As pointed out by Archibald (2003), older taxa that remain at least partially intact, such as Lipotyphla Haeckel (1866) minus tenrecs and golden moles, may still be used in favor of new variants on those names, such as "Eulipotyphla." As discussed below, names that duplicate existing taxa, such as "Afrosoricida" Stanhope et al. (1998), should be discarded in favor of their senior synonyms, which, in the case of "Afrosoricida," is Tenrecoidea McDowell (1958).

The first of the two classifications in Table 5.4 integrates the work of Árnason et al. (2002) and Asher et al. (2003); the second is based on Murphy et al. (2001b) and Waddell et al. (2001). These classifications differ most conspicuously in the location of the placental root; according to the former, it is near or within rodents; following the latter, it is between afrotherians and xenarthrans. In fact, Árnason et al. (2002) favored a more basal position for erinaceids than did Asher et al. (2003), as well as a paraphyletic Primates and a more nested position for Lagomorpha. However, Árnason et al. (2002) agreed with Asher et al. (2003) in three important respects: 1) the position of tenrecs within Afrotheria (Árnason et al. did not sample golden moles), 2) an Afrotheria-Xenarthra clade as the sister taxon to laurasiatheres, and 3) a near-basal position of rodents. Based on these topologies, there are five novel clades that do not yet have a taxonomic designation. The first four result from the successive branching of murids, hystricognaths, lagomorphs, and primates from the remainder of Placentalia. For these nodes, I have applied the following names, following the format of "Exafroplacentalia" from Waddell et al. (2001): Exmuridplacentalia for all placentals excluding murid rodents, Exrodentplacentalia excluding all rodents, Exgliresplacentalia excluding rodents and lagomorphs, and Exprimateplacentalia excluding primates, rodents, and lagomorphs. The fifth novel taxon is a clade consisting of xenarthrans and afrotheres, here dubbed the Xenafrotheria. Significantly, should the placental root of Murphy et al. (2001b) prove correct, priority for higher-level nomenclature would then rest with the terminology summarized by Waddell et al. (2001), as the unrooted trees from both Asher et al. (2003) and Murphy et al. (2001b) are very similar.

In the first paper to formally recognize the Afrotheria, Stanhope et al. (1998) also introduced the taxon "Afrosoricida" to designate a tenrec–golden mole clade. This term has since been used in several publications (e.g., Madsen et al., 2001; Waddell et al., 2001; Douady et al., 2002a,b), whereas

Table 5.3 Summaries of previous synoptic classifications relevant to insectivoran mammals

Simpson (1945; order, superfamily, family, subfamily)	Van Valen (1967; order, suborder, superfamily, family)	McKenna and Bell (1997; grandorder, order, superfamily, family)
Insectivora	Insectivora	Lipotyphla
†Deltatheroidea	Proteutheria	Chrysochloridea
†Deltatheridiidae	†Endotherioidea	**Chrysochloridae**
†Deltatheridiinae	Tupaioidea	Erinaceomorpha
†Didelphodontinae	†Leptictidae	†Sespidectidae
Tenrecoidea	†Zalambdalestidae	†Litocherinae
†Palaeoryctidae	†Anagalidae	†Diacodontinae
Solenodontidae	†Paroxyclaenidae	†Amphilemuridae
†Apternodontinae	Tupaiidae	†Adapisoricidae
Solenodontinae	†Pantolestidae	†Creotarsidae
Tenrecidae	†Ptolemaiidae	Erinaceoidea
Tenrecinae	†Pentacodontidae	**Erinaceidae**
Oryzorictinae	†Apatemyoidea	Talpoidea
Potamogalidae	†Apatemyidae	†Proscalopidae
Chrysochloroidea	Macroscelidea	**Talpidae**
Chrysochloridae	Macroscelididae	†Dimylidae
Erinaceoidea	Dermoptera	Soricomorpha
†Zalambdalestidae	†Mixodectoidea	†Otlestidae
†Leptictidae	†Mixodectidae	†Geolabididae
Erinaceidae	Galeopithecoidea	Soricoidea
Echinosoricinae	†Plagiomenidae	**†Nesophontidae**
Erinaceinae	Galeopithecidae	†Micropternodontidae
†Dimylidae	Erinaceota	†Apternodontidae
Macroscelidoidea	Erinaceoidea	**Solenodontidae**
Macroscelididae	†Adapisoricidae	†Plesiosoricidae
Soricoidea	**Erinaceidae**	†Nyctitheriidae
†Nyctitheriidae	†Dimylidae	**Soricidae**
Soricidae	**Talpidae**	Tenrecoidea
Soricinae	Soricoidea	**Tenrecidae**
Crocidurinae	†Plesiosoricidae	
Scutisoricinae	**†Nesophontidae**	
Talpidae	**Soricidae**	
Uropsilinae	Deltatheridia	
Desmaninae	†Hyaenodonta	
Talpinae	†Palaeoryctoidea	
Scalopinae	†Palaeoryctidae	
Condylurinae	†Micropternodontidae	
†Nesophontidae	†Didymoconidae	
†Pantolestoidea	†Hyaenodontoidea	
†Pantolestidae	†Hyaenodontidae	
†Pantolestinae	†Oxyaenoidea	
†Pentacodontinae	†Oxyaenidae	
†Mixodectoidea	Zalambdodonta	
†Mixodectidae	Tenrecoidea	
	Tenrecidae[a]	
	Solenodontidae	
	Chrysochloroidea	
	Chrysochloridae	

Note: Linnean ranks intended by the original authors are indicated in parentheses, and correspond with nested indentations. Recent insectivoran families are in bold; daggers indicate extinct taxa.

[a] Van Valen placed "apternodontids" within the Tenrecidae.

other authors (e.g., Mouchaty, 1999; Asher, 2000; Malia et al., 2002; Archibald, 2003; Asher et al., 2003) have recognized that Tenrecoidea McDowell (1958), identical in content to Afrosoricida Stanhope et al. (1998), has priority. Several authors have used the term "tenrecoid" descriptively to identify a wide array of dentally zalambdodont eutherian mammals (e.g., Lillegraven et al., 1981; Asher, 1999), or as a higher-level taxon to identify clades dissimilar to that of McDowell (e.g., a clade made up of apternodontids, palaeoryctids, solenodontids, and tenrecids; Simpson, 1931, 1945), or for a solenodontid-soricid-talpid-tenrecid-chrysochlorid clade (Butler, 1956). The term "Tenrecomorpha" was not used by Simpson in either of his synoptic classifications (1931, 1945), but has been used by Butler (1972: 261) to group Malagasy

Table 5.4 Competing synoptic classifications of high-level mammalian groups incorporating recent proposals influenced by molecular data

Placentalia	Placentalia
Muridae	Afrotheria
Exmuridplacentalia*	Tubulidentata
Hystricognatha	Paenungulata
Exrodentplacentalia*	Hyracoidea
Lagomorpha	Sirenia
Exgliresplacentalia*	Proboscidea
Primates	Afroinsectiphilia
Exprimateplacentalia*	Macroscelididae
Scandentia	Tenrecoidea
Dermoptera	**Tenrecidae**
Laurasiatheria	**Chrysochloridae**
Lipotyphla	Exafroplacentalia
†Apternodontidae	Xenarthra
†Geolabididae	Boreotheria
†Nesophontidae	Laurasiatheria
Erinaceidae	Lipotyphla
Soricidae	**Erinaceidae**
Talpidae	**Soricidae**
Solenodon	**Talpidae**
Scrotifera	*Solenodon*
Chiroptera	Scrotifera
Carnivora	Chiroptera
Pholidota	Ferungulata
Perissodactyla	Euungulata
Cetartiodactyla	Cetartiodactyla
Xenafrotheria*	Perissodactyla
Xenarthra	Ferae
Afrotheria	Carnivora
Macroscelidea	Pholidota
Tubulidentata	Euarchontoglires
Uranotheria[a]	Glires
†Hyopsodontidae	Rodentia
†Meniscotheriidae	Lagomorpha
Paenungulata	Archonta
†Desmostylia	Primatomorpha
†Embrithopoda	Primates[b]
Proboscidea	Dermoptera
Sirenia	Scandentia
Hyracoidea	
Tenrecoidea	
Chrysochloridae	
Tenrecidae	

Sources: Classification at left is adapted from Árnason et al. (2002) and Asher et al. (2002, 2003); classification at right is adapted from Murphy et al. (2001a,b) and Waddell et al. (2001).

Notes: Following Simpson (1945), novel terms for high-level taxa are used only for clades not previously named in the literature. Nested clades indicate phylogenetic structure, but this does not necessarily correspond with Linnean ranks. Notation is as in Table 5.3. The position of *Solenodon* in both classifications is based on Roca et al. (2004). Higher-level taxa followed by an asterisk have not previously been named in the literature.

[a] McKenna and Bell (1997) first used Uranotheria as a senior synonym for Paenungulata, encompassing taxa descended from the most recent common ancestor of embrithopods, sirenians, proboscideans, and hyracoids. Here it is used as a stem taxon, with Paenungulata reserved for the crown.

[b] Murphy et al. (2001a) and Árnason et al. (2002) depict primates as paraphyletic, with dermopterans as the sister taxon to Anthropoidea.

tenrecs with African potamogalines, taxa which have occasionally been placed in separate families (e.g., Simpson, 1945; Eisenberg, 1975). Butler (1972) recognized "Chrysochloromorpha" as a suborder distinct from his "Tenrecomorpha."

In fact, the rules of nomenclature as maintained by the International Commission on Zoological Nomenclature (1999) do not apply above the level of the family. However, following the guidelines of Simpson (1945), which are applicable at higher levels, temporal precedent must still be respected by zoologists in order to preserve a stable system of nomenclature. Bronner et al. (2003) cited Simpson (1945) in their discussion of Afrotherian taxonomy, and noted the confusion engendered by the term "Afrosoricida," which has nothing to do with soricids or the African shrew *Afrosorex* (Hutterer, 1986). They retained "Afrosoricida," based on part of paragraph 19 of Simpson (1945: 32), who recommended that the common suprafamilial designation "oidea" not be used at an ordinal level. However, Bronner et al. (2003) did not fully cite Simpson (1945: 25) who states that "there is no standard termination for names of groups higher than superfamilies, and these may be but do not need to be derived from names of included genera. It is best to avoid using for these groups names [reserved for superfamilies through subtribes], but some names in -oidea, the superfamily ending, are in common use for groups higher than superfamilies are are not rejected on that account." In addition, Simpson (1945: 32) states that "there are no standard or required terminations for [categories above the superfamily]. Names similar in ending and therefore liable to confusion with those of lower categories, such as names ending in -oidea, are to be avoided as far as convenient but are not to be rejected on this account alone." Furthermore, the status of a tenrec–golden mole clade as an order, as opposed to a superfamily, is without solid empirical footing. Indeed, the Afrotheria was itself first proposed as an order (Stanhope et al., 1998); it would therefore follow that if Linnean ranks are to be used at all, clades nested within Afrotheria, such as Tenrecoidea, be categorized at a lower level, such as superfamily. In any event, the term "Tenrecoidea" of McDowell (1958) was the first taxonomic designation encompassing both African insectivoran families to the exclusion of other relevant taxa, does not contain a misleading root and should therefore be the preferred name for a taxon consisting of tenrecs and golden moles, if indeed such a clade proves stable in future phylogenetic analyses.

CONCLUSIONS

The variable affinities of fossil insectivorans with modern groups evident in Table 5.2 is partly because previous character-based phylogenies have focused on either tenrecids (Asher, 1999, 2000), North American zalambdodonts (Asher et al., 2002), or endemic African mammals (Asher et al., 2003); none to date has combined morphological and molecular matrices that sample broadly both fossil and living insectivorans at a low level (i.e., genus or species), with representatives of all placental orders. Complicating the mat-

ter, Insectivora appears to be at least diphyletic (Murphy et al., 2001a,b; Asher et al., 2003) or even triphyletic (Mouchaty et al., 2000a,b; Árnason et al., 2002). Hence, studies of insectivoran phylogeny require a very broad sample across all groups of placental mammals; and the hypothesis often credited to Huxley (1880) that "insectivorans" represent the ur-placental from which other groups descended (Butler, 1972; Wyss, 1987) may, in some form, be correct. For example, when fossil remains of their respective stem lineages are identified, basal afrotheres and/or laurasiatheres might fit neatly into the insectivoran grade (see, e.g., Seiffert and Simons, 2001).

Even when a data matrix is available that samples a large number of extinct and extant insectivorans plus representatives of all other placental orders, it is entirely possible that topological variability may yet occur, or that support indices may remain "low" by popular standards. Anatomical and genetic uniqueness, which can translate into high branch and/or bootstrap support, is artifactual for many living groups, because extinction has eliminated from our pool of terminal taxa the intermediates that would have otherwise blurred the distinction between that group and its living sister taxon (Horovitz, 1999). As summarized above, living soricids are easily recognizable; but their jaw-joint autapomorphies evolved in a mosaic fashion and have reversed in certain lineages, making it more challenging to define a "soricid" or recognize their stem relatives, given the diversity of this clade throughout time.

From a methodological standpoint, adding fossil taxa to a sequence-rich character matrix may drastically reduce character support, simply because the pool of positive information for a given fossil terminal is relatively small (O'Leary, 2000). By itself, this is not a sufficient reason to exclude taxa with missing data (Kearny, 2002); but commonly used support indices (Felsenstein, 1985; Bremer, 1994) in a combined-data context should not be uncritically compared with support indices applied to phylogenies of extant taxa, for which all of the terminals have a more evenly distributed character sample. A low support index for a given node containing a fossil taxon may simply reflect the uncertainty that paleontologists have long recognized regarding the phylogeny of certain extinct taxa.

Despite qualifications about the putative topological stability that may come with both the improvement in the fossil record of insectivoran taxa and the construction of character matrices to represent our understanding of their anatomy, applying an optimality criterion to as much data as possible regarding these animals remains an essential task. Accounting for the phylogenetic information content of fossils is the only means by which we can reconstruct a historically informative tree of life.

SUMMARY

Recent data that bear on placental interrelationships indicate that the "Insectivora," a group once believed to comprise living hedgehogs, shrews, moles, solenodons, tenrecs, and golden moles, is polyphyletic. Specifically, African tenrecs and golden moles form a clade with other African mammals, and holarctic hedgehogs, moles, shrews, and solenodons form an independent northern radiation of insectivoran mammals. The implications of this polyphyly for the evolution of characters once believed to constitute insectivoran synapomorphies are considerable. These characters are here mapped onto a phylogeny of placental mammals, based on a combined morphological-molecular dataset. One of these characters, the "mobile proboscis," is present among both insectivorans and African mammals and is, in fact, a complex that varies greatly both in skeletal support and muscular control across mammalian groups. Characters related to the craniomandibular joint in soricids have evolved in a mosaic fashion, and can help identify extinct soricid relatives. Although occasionally including wellpreserved fossil insectivoran taxa, phylogenies published to date have thoroughly sampled the diversity of neither living nor extinct insectivoran mammals; hence, considerable ambiguity remains, particularly regarding the affinities of extinct taxa. This ambiguity may be unavoidable, given the possible status of extinct insectivoran groups as stem taxa to multiple modern radiations.

ACKNOWLEDGMENTS

I thank Ken Rose and David Archibald for inviting me to participate in the 2002 Society of Vertebrate Paleontology Symposium on Placental Phylogeny in honor of G. G. Simpson. For helpful comments on the chapter, I thank Marc Allard, Peter Giere, and Anthony Carter. Jean Spence provided access to mammalogy collections at the American Museum of Natural History, and Bill Clemens and Pat Holroyd provided access to collections at the University of California Museum of Paleontology. I thank the American Museum of Natural History, Grant DEB 9800908 of the National Science Foundation (USA), and the Deutscher Akademischer Austauschdienst for support of the research that made this review possible.

REFERENCES

Archibald, J. D. 2003. Timing and biogeography of the eutherian radiation: Fossils and molecules compared. Molecular Phylogenetics and Evolution 28: 350–359.

Árnason, Ú., J. A. Addegoke, K. Bodin, E. W. Born, Y. B. Esa, A. Gullberg, M. Nilsson, M. Short, X. Xu, and A. Janke. 2002. Mammalian mitogenomic relationships and the root of the eutherian tree. Proceedings of the National Academy of Sciences USA 99: 8151–8156.

Asher, R. J. 1999. A morphological basis for assessing the phylogeny of the "Tenrecoidea" (Mammalia, Lipotyphla). Cladistics 15: 231–252.

———. 2000. Phylogenetic History of Tenrecs and Other Insectivoran Mammals. Ph.D. Dissertation, State University of New York, Stony Brook.

———. 2001. Cranial anatomy in tenrecid insectivorans: Character evolution across competing phylogenies. American Museum Novitates 3352: 1–54.

Asher, R. J., M. C. McKenna, R. J. Emry, A. R. Tabrum, and D. Kron. 2002. Morphology and relationships of Apternodus and other extinct, zalambdodont, placental mammals. Bulletin of the American Museum of Natural History 273: 1–118.

Asher, R. J., M. J. Novacek, and J. H. Geisler. 2003. Relationships of endemic African mammals and their fossil relatives based on morphological and molecular evidence. Journal of Mammalian Evolution 10: 131–194.

Barnosky, A. D. 1981. A skeleton of Mesoscalops (Mammalia, Insectivora) from the Miocene Deep River Formation, Montana, and a review of the proscalopid moles: Evolutionary, functional, and stratigraphic relationships. Journal of Vertebrate Paleontology 1: 285–339.

Bedford, J. M., G. W. Cooper, D. M. Phillips, and G. L. Dryden. 1994. Distinctive features of the gametes and reproductive tracts of the Asian musk shrew, Suncus murinus. Biology of Reproduction 50: 820–834.

Bedford, J. M., O. B. Mock, S. K. Nagdas, V. P. Winfrey, and G. E. Olson. 1999. Reproductive features of the eastern mole (Scalopus aquaticus) and star-nose mole (Condylura cristata). Journal of Reproduction and Fertility 117: 345–353.

———. 2000. Reproductive characteristics of the African pygmy hedgehog, Atelerix albiventris. Journal of Reproduction and Fertility 120: 143–150.

Bedford, J. M., O. B. Mock, and D. M. Phillips. 1997. Unusual ampullary sperm crypts, and behavior and role of the cumulus oophorus, in the oviduct of the least shrew, Cryptotis parva. Biology of Reproduction 56: 1255–1267.

Bown, T. M., and D. Schankler. 1982. A review of the Proteutheria and Insectivora of the Willwood Formation (lower Eocene), Bighorn Basin, Wyoming. Bulletin of the United States Geological Survey 1523: 1–79.

Bremer, K. 1994. Branch support and tree stability. Cladistics 10: 295–304.

Bronner, G. N. 1991. Comparative hyoid morphology of nine chrysochlorid species (Mammalia: Chrysochloridae). Annals of the Transvaal Museum 35: 295–311.

Bronner, G. N., M. Hoffman, P. J. Taylor, C. T. Chimimba, P. B. Best, C. A. Matthee, and T. J. Robinson. 2003. A revised systematic checklist of the extant mammals of the southern African subregion. Durban Museum Novitates 28: 56–95.

Butler, P. M. 1956. Erinaceidae from the Miocene of east Africa. British Museum (Natural History), London. Fossil Mammals of Africa 11: 1–75.

———. 1972. The problem of insectivore classification; pp. 253–265 in K. A. Joysey and T. S. Kemp (eds.), Studies in Vertebrate Evolution. Winchester Press, New York.

———. 1984. Macroscelidea, Insectivora, and Chiroptera from the Miocene of East Africa. Palaeovertebrata 14: 117–200.

———. 1988. Phylogeny of the insectivores; pp. 117–141 in M. J. Benton (ed.), The Phylogeny and Classification of the Tetrapods. Volume 2. Clarendon Press, Oxford.

Butler, P. M., and A. T. Hopwood. 1957. Insectivora and chiroptera from the Miocene rocks of Kenya colony. British Museum (Natural History), London. Fossil Mammals of Africa 13: 1–35.

Carter, A. M. 2001. Evolution of the Placenta and fetal membranes seen in the light of molecular phylogenetics. Placenta 22: 800–807.

Cutler, J. B., A. Dumas, and R. M. Wynn. 1998. Immunohistochemical analysis of the interhemal membrane of the elephant shrew. Gynecologic and Obstetric Investigation 46: 46–48.

deQueiroz, K., and P. D. Cantino. 2001. Phylogenetic nomenclature and the PhyloCode. Bulletin of Zoological Nomenclature 58: 254–271.

Douady, C. J., F. Catzeflis, D. J. Kao, M. S. Springer, and M. J. Stanhope. 2002a. Molecular evidence for the monophyly of Tenrecidae (Mammalia) and the timing of the colonization of Madagascar by Malagasy tenrecs. Molecular Phylogenetics and Evolution 22: 357–363.

Douady, C. J., P. I. Chatelier, O. Madsen, W. W. de Jong, F. Catzeflis, M. S. Springer, and M. J. Stanhope. 2002b. Molecular phylogenetic evidence confirming the Eulipotyphla concept and in support of hedgehogs as the sister group to shrews. Molecular Phylogenetics and Evolution 25: 200–209.

Eisenberg, J. F. 1975. Phylogeny, behavior, and ecology in the mammalia: Lessons from the Tenrecidae; pp. 47–68 in W. P. Luckett and F. S. Szalay (eds.), Phylogeny of the Primates: A Multidisciplinary Approach. Plenum, New York.

Emerson, G. L., C. W. Kilpatrick, B. E. McNiff, J. Ottenwalder, and M. W. Allard. 1999. Phylogenetic relationships of the order Insectivora based on complete 12s rRNA sequences from mitochondria. Cladistics 15: 221–230.

Engesser, B. 1975. Revision der europäischen Heterosoricinae (Insectivora, Mammalia). Eclogae Geologicae Helvetiae 68: 649–671.

Felsenstein, J. 1985. Confidence limits on phylogenies: An approach using the bootstrap. Evolution 39: 783–791.

———. 1988. Phylogenies from molecular sequences: Inference and reliability. Annual Review of Genetics 22: 521–565.

Fernhead, R. W., C.C.D. Shute, and A. A. Bellairs. 1954. The temporomandibular joint of shrews. Proceedings of the Zoological Society of London 125: 795–806.

Gaillard, C. 1915. Nouveau genre de musaraignes dans les dépôts Miocènes de la Grive-Saint-Alban (Isère). Annales de la Société Linnéene de Lyon 62: 83-98.

Giere, P. 2002. Grundplan Rekonstruktion und Ontogenese der Orbitalregion der "Eulipotyphla" (Mammalia). Ph.D. Thesis, Humboldt University, Wissenschaft und Technik, Berlin.

Gould, G. C. 1995. Hedgehog phylogeny (Mammalia, Erinaceidae) —the reciprocal illumination of the quick and the dead. American Museum Novitates 3131: 1–45.

Grassé, P. P. 1955. Traité de Zoologie: Anatomie, Systématique, Biologie. Masson, Paris.

Haeckel, E. 1866. Generelle Morphologie der Organismen; Allgemeine Grundzuege der Organischen Formen-Wissenschaft, Mechanisch Begruendet durch die von Charles Darwin Reformierte Descendenz-Theorie. Two volumes. Georg Reimer, Berlin.

Hedges, S. B. 2001. Afrotheria: Plate tectonics meets genomics. Proceedings of the National Academy of Sciences USA 98: 1–2.

Hickman, G. C. 1990. The Chrysochloridae: Studies toward a broader perspective of adaptation in subterranean mammals; pp. 23–48 in O. Reig and E. Nevo (eds.), Evolution of Subterranean Mammals at the Organismal and Molecular Levels. Alan R. Liss, New York.

Hooker, J. J. 2001. Tarsals of the extinct insectivoran family Nyctitheriidae (Mammalia): Evidence for archontan relationships. Zoological Journal of the Linnean Society of London 132: 510–529.

Horovitz, I. 1999. A phylogenetic study of living and fossil platyrrhines. American Museum Novitates 3269: 1–40.

Hutchinson, J. H. 1974. Notes on type specimens of European Miocene Talpidae and a tentative classification of Old World Tertiary Talpidae (Insectivora, Mammalia). Géobios 7: 211–256.

Hutterer, R. 1986. African shrews allied to Crocidura fischeri: Taxonomy, distribution and relationships. Cimbebasia, Series A 8: 24–35.

Huxley, T. H. 1880. On the application of the laws of evolution to the arrangement of the Vertebrata and more particularly of the Mammalia. Proceedings of the Zoological Society of London 1880: 649–662.

International Commission on Zoological Nomenclature. 1999. International Code of Zoological Nomenclature, 4th edition, adopted by the International Union of Biological Sciences. London, International Trust for Zoological Nomenclature.

Janies, D. A., and W. C. Wheeler. 2001. Efficiency of parallel direct optimization Cladistics 17: S71–S82.

Kearny, M. 2002. Fragmentary taxa, missing data, and ambiguity: Mistaken assumptions and conclusions. Systematic Biology 51: 369–381.

Krishtalka, L., and T. Setoguchi. 1977. Paleontology and geology of the Badwater Creek area, Central Wyoming. Part 13. The late Eocene Insectivora and Dermoptera. Annals of the Carnegie Museum 46: 71–99.

Lillegraven, J. A., M. C. McKenna, and L. Krishtalka. 1981. Evolutionary relationships of middle Eocene and younger species of Centetodon (Mammalia, Insectivora, Geolabididae) with a description of the dentition of Ankylodon (Adapisoricidae). University of Wyoming Publications 45: 1–115.

Lopatin, A. V. 2002. The earliest shrew (Soricidae, Mammalia) from the middle Eocene of Mongolia. Paleontological Journal 36: 650–659.

Luckett, W. P. 1977. Ontogeny of amniote fetal membranes and their application to phylogeny; pp. 439–516 in M. K. Hecht, P. C. Goody, and B. M. Hecht (eds.), Major Patterns in Vertebrate Evolution. Plenum, New York.

———. 1985. Superordinal and intraordinal affinities of rodents: Developmental evidence from the dentition and placentation; pp. 227–276 in W. P. Luckett and J.-L. Hartenberger (eds.), Evolutionary Relationships among Rodents. Plenum, New York.

———. 1993. Uses and limitations of mammalian fetal membranes and placenta for phylogenetic reconstruction. Journal of Experimental Zoology 266: 514–527.

MacPhee, R.D.E. 1981. Auditory regions of primates and eutherian insectivores: Morphology, ontogeny, and character analysis. Contributions to Primatology 18: 1–282.

MacPhee, R.D.E., and M. J. Novacek. 1993. Definition and relationships of Lipotyphla; pp. 13–31 in F. S. Szalay, M. J. Novacek, and M. C. McKenna (eds.), Mammal Phylogeny. Volume 2. Placentals. Springer-Verlag, New York.

MacPhee, R.D.E., M. J. Novacek, and G. Storch. 1988. Basicranial morphology of early Tertiary erinaceomorphs and the origin of primates. American Museum Novitates 2921: 1–42.

Madsen, O., M. Scally, C. J. Douady, D. J. Kao, R. W. DeBry, R. Adkins, H. M. Amrine, M. J. Stanhope, W. W. de Jong, and M. S. Springer. 2001. Parallel adaptive radiations in two major clades of placental mammals. Nature 409: 610–614.

Maier, W. 1986. *Leptictidium nasutum*—ein archaisches Säugetier aus Messel mit außergewöhnlichen biologischen Anpassungen. Natur und Museum 116: 1–19.

———. 2002. Zur funktionellen Morphologie der rostralen Nasenknorpel bei Soriciden. Mammalian Biology 67: 1–17.

Malassiné, A., and R. Leiser. 1984. Morphogenesis and fine structure of the near-term of placenta of Talpa europaea: 1. Endotheliochorial labyrinth. PLACENTA 5(2): 145-158.

Malia, M. J., R. M. Adkins, and M. W. Allard. 2002. Molecular support for Afrotheria and the polyphyly of Lipotyphla based on analyses of the growth hormone receptor gene. Molecular Phylogenetics and Evolution 24: 91–101.

Matthes, E. 1912. Die Regio ethmoidalis des Primordialcraniums von *Manatus latirostris*. Jenaische Zeitschrift 48: 1–31.

McDowell, S. B. 1958. The Greater Antillean insectivores. Bulletin of the American Museum of Natural History 115: 115–213.

McKenna, M. C., and S. K. Bell. 1997. Classification of Mammals Above the Species Level. Columbia University Press, New York.

McNiff, B., and M. W. Allard. 1998. A test of archontan monophyly and the phylogenetic utility of the mitochondrial gene 12S rRNA. American Journal of Physical Anthropology 107: 225–241.

Meister, W. and D. D. Davis. 1953. Placentation of a primitive insectivore *Echinosorex gymnura*. Fieldiana: Zoology 35(2): 1–31.

Menzel, K. H. 1979. Morphologische Untersuchungen an der vorderen Nasenregion von *Solenodon paradoxus* (Insectivora). Doctorate in Dentistry, Johann-Wolfgang-Goethe University Frankfurt am Main.

Mess, A. 2002. Evolutionary differentiation of placental organisation in hystricognath rodents; pp. 279–292 in C. Denys, L. Granjon, and A. Poulet. (eds.), African Small Mammals: Proceedings of the 8th International Symposium on African Small Mammals, Paris. IRD Editions, Paris.

Mossman, H. W. 1987. Vertebrate Fetal Membranes. Rutgers University Press, New Brunswick, New Jersey.

Mouchaty, S. K. 1999. Mammalian Molecular Systematics with Emphasis on the Insectivore Order Lipotyphla. Ph.D. Thesis, Lund University, Sweden.

Mouchaty, S. K., A. Gullberg, A. Janke, and Ú. Árnason. 2000a. The phylogenetic position of the Talpidae within Eutheria based on analysis of complete mitochondrial sequences. Molecular Biology and Evolution 17: 60–67.

———. 2000b. Phylogenetic position of the tenrecs (Mammalia: Tenrecidae) of Madagascar based on analysis of the complete mitochondrial genome sequence of *Echinops telfairi*. Zoologica Scripta 29: 307–317.

Murphy, W. J., E. Eizirik, W. E. Johnson, Y. P. Zhang, O. A. Ryder, and S. J. O'Brien. 2001a. Molecular phylogenetics and the origin of placental mammals. Nature 409: 614–618.

Murphy, W. J., E. Eizirik, S. J. O'Brien, O. Madsen, M. Scally, C. J. Douady, E. Teeling, O. A. Ryder, M. J. Stanhope, W. W. de Jong, and M. S. Springer. 2001b. Resolution of the early placental mammal radiation using Baysian phylogenetics. Science 294: 2348–2351.

National Institutes of Health. 2004. GenBank website www.ncbi.nlm.nih.gov.

Novacek, M. J. 1985. The Sespidectinae, a new subfamily of hedgehog-like insectivores. American Museum Novitates 2822: 1–24.

———. 1986. The skull of leptictid insectivorans and the higher level classification of eutherian mammals. Bulletin of the American Museum of Natural History 183: 1–111.

Nowak, R. M. 1999. Walker's Mammals of the World. Sixth Edition. The Johns Hopkins University Press, Baltimore.

O'Leary, M. A. 2000. Operational obstacles to total evidence analyses considering that 99% of life is extinct. Journal of Vertebrate Paleontology 20(supplement): 61A.

Ottenwalder, J. A. 2001. Systematics and biogeography of the West Indian genus *Solenodon;* pp. 253–329 in C. A. Woods and F. Sergile (eds.), Biogeography of the West Indies, Patterns and Perspectives. Second Edition. CRC Press, Boca Raton.

Parker, W. K. 1885. On the structure and development of the skull in the Mammalia. Part 3, Insectivora. Philosophical Transactions of the Royal Society of London 176: 121–276.

Phillips, D. M., and J. M. Bedford. 1984. Unusual features of sperm ultrastructure in the musk shrew *Suncus murinus*. Journal of Experimental Zoology 235: 119–126.

Repenning, C. A. 1967. Subfamilies and genera of the Soricidae. United States Geological Survey Professional Paper 565: 1–74.

Reumer, J.W.F. 1984. Ruscinian and early Pleistocene Soricidae (Insectivora, Mammalia) from Tegelen (The Netherlands) and Hungary. Scripta Geologica 73: 1–173.

Roca, A. L., G. K. Bar-Gal, E. Eizirik, K. M. Helgen, R. Maria, M. S. Springer, S. J. O'Brien, and W. J. Murphy. 2004. Mesozoic origin for West Indian insectivores. Nature 429: 649–651.

Roux, G. H. 1947. The cranial development of certain Ethiopian "insectivores" and its bearing on the mutual affinities of the group. Acta Zoologica 28: 165–397.

Saether, O. A. 1979. Underlying synapomorphies and anagenetic analysis. Zoologica Scripta 8: 305–312.

Schmidt-Kittler, N. 1973. *Dimyloides* Neufunde aus der oberoligozänen Spaltenfüllung "Ehrenstein 4" (Süddeutschland) und die systematische Stellung der Dimyliden (Insectivora, Mammalia). Mitteilungen der Bayerischen Staatssammlung für Paläontologie und Historische Geologie 13: 115–139.

Seiffert E. R., and E. L. Simons. 2001. *Widanelfarasia,* a diminutive new placental from the late Eocene of Egypt. Proceedings of the National Academy of Sciences USA 97: 2646–2651.

Simmons, N. B., and J. H. Geisler. 1998. Phylogenetic relationships of *Icaronycteris, Archaeonycteris, Hassianycteris,* and *Palaeochiropteryx* to extant bat lineages, with comments on the evolution of echolocation and foraging strategies in Microchiroptera. Bulletin of the American Museum of Natural History 235: 1–182.

Simpson, G. G. 1931. A new classification of mammals. Bulletin of the American Museum of Natural History 59: 259–293.

———. 1945. The principles of classification and a classification of mammals. Bulletin of the American Museum of Natural History 85: 1–350.

Stanhope, M. J., V. G. Waddell, O. Madsen, W. W. de Jong, S. B. Hedges, G. C. Cleven, D. Kao, and M. S. Springer. 1998. Molecular evidence for multiple origins of the Insectivora and for a new order of endemic African mammals. Proceedings of the National Academy of Sciences USA 95: 9967–9972.

Storer, J. E. 1984. Mammals of the Swift Current Creek local fauna (Eocene): Uintan, Saskatchewan. Saskatchewan Museum of Natural History Contributions 7: 1–158.

Thewissen, J.G.M., and P. D. Gingerich. 1989. Skull and endocranial cast of *Eoryctes melanus,* a new palaeoryctid (Mammalia:

Insectivora) from the early Eocene of western North America. Journal of Vertebrate Paleontology 9: 459–470.

Van Valen, L. 1967. New Paleocene insectivores and insectivore classification. Bulletin of the American Museum of Natural History 135: 217–284.

von Koenigswald, W., and G. Storch. 1983. *Pholidocercus hassiacus,* ein Amphilemuride aus dem Eozän der "Grube Messel" bei Darmstadt (Mammalia: Lipotyphla). Senckenberg Lethaia 64: 447–459.

Waddell, P. J., H. Kishino, and R. Ota. 2001. A phylogenetic foundation for comparative mammalian genomics. Genome Informatics 12: 141–154.

Wheeler, W. C. 1995. Sequence alignment, parameter sensitivity, and the phylogenetic analysis of molecular data. Systematic Biology 44: 321–331.

———. 1996. Optimization alignment: The end of multiple sequence alignment in phylogenetics? Cladistics 12: 1–9.

———. 1998. Alignment characters, dynamic programming and heuristic solutions; pp. 243–251 *in* R. DeSalle and B. Schierwater (eds.), Molecular Approaches to Ecology and Evolution. Birkhäuser-Verlag, Basel.

Wheeler, W. C., D. Gladstein, and J. DeLaet. 2003. POY direct optimization computer program, available at ftp.amnh.org/pub/molecular/poy.

Whidden, H. P. 2002. Extrinsic snout musculature in Afrotheria and Lipotyphla. Journal of Mammalian Evolution 9: 161–184.

Wible, J. R. 1986. Transformations in the extracranial course of the internal carotid artery in mammalian phylogeny. Journal of Vertebrate Paleontology 6: 313–325.

Wislocki, G. B. 1940. The placentation of *Solenodon paradoxus.* American Journal of Anatomy 66: 497–531.

Wojcik, J. M., and M. Wolsan. 1998. Evolution of Shrews. Mammal Research Institute, Bialowieza, Poland.

Woodall, P. F. 1991. An untrastructural study of the spermatozoa of elephant shrews (Mammalia: Macroscelidea) and its phylogenetic implications. Journal of Submicroscopy and Cytology 23: 47–58.

Wyss, A. R. 1987. Notes on Proteutheria, Insectivora, and Thomas Huxley's contribution to mammalian systematics. Journal of Mammalogy 68: 135–138.

Zeller, U., and C. Freyer. 2001. Early ontogeny and placentation of the grey short-tailed opossum, *Monodelphis domestica* (Didelphidae: Marsupialia): Contribution to the reconstruction of the marsupial morphotype. Journal of Zoological Systematics and Evolutionary Research 39: 137–158.

PATRICIA A. HOLROYD
AND JASON C. MUSSELL

MACROSCELIDEA AND TUBULIDENTATA

MACROSCELIDEA, THE ELEPHANT SHREWS OR sengis, and Tubulidentata, the aardvarks, are two of the least speciose and most phylogenetically perplexing orders of extant mammals. Both exclusively African in distribution today, their fossil records are sparse, leaving our understanding of their diversification frustratingly incomplete. Both have been postulated to have origins within extinct, archaic orders of mammals based on their morphology, whereas recent molecular analyses suggest affinities with other African mammals, united in the Afrotheria. Here we examine and summarize these various lines of evidence and suggest future avenues for elucidating the phyletic position of these enigmatic animals.

MACROSCELIDEA

Historically, the extant Macroscelidea were linked with a diverse array of mammals in the "Insectivora," and some fossil taxa were originally described as marsupials (Stromer, 1932), hyracoids (Andrews, 1914), or mixodectid insectivorans (Schlosser, 1911). The history of their eventual recognition as a distinct order has been reviewed in detail by Patterson (1965) and Butler (1995). The entire order is currently classified in a single family, having six subfamilies that contain 14 genera (Table 6.1). Phylogenetic relationships within the order (see, e.g., Novacek, 1984; Tabuce et al., 2001) indicate that the extant subfamilies (Macroscelidinae and Rhynchocyoninae) plus

Table 6.1 Classification of elephant shrews

Order Macroscelidea[a] Butler, 1956: 479
  Family Macroscelididae Bonaparte, 1838: 113
    Subfamily Herodotiinae Simons, Holroyd and Bown, 1991: 9734
      Chambius[b] Hartenberger, 1986: 247
      Herodotius Simons, Holroyd, and Bown, 1991: 9734
      Nementchatherium Tabuce, Coiffait, Coiffait, Mahboubi, and
        Jaeger, 2001: 535
    Subfamily Metoldobotinae Simons, Holroyd and Bown, 1991: 9736
      Metoldobotes Schlosser, 1910
    Subfamily Macroscelidinae Bonaparte, 1838: 113
      Hiwegicyon Butler, 1984: 137
      Pronasilio Butler, 1984: 135
      Palaeothentoides Stromer, 1932: 185
      Macroscelides Smith, 1829: 435
      Elephantulus Thomas and Schwann, 1906: 577 (non *E. rozeti*[c])
      Petrodromus Peters, 1846: 258
    Subfamily Myohyracinae Andrews, 1914: 171
      Myohyrax Andrews, 1914: 171
      Protypotheroides Stromer, 1922: 333
    Subfamily Mylomygalinae Patterson, 1965: 310
      Mylomygale Broom, 1948: 6–8
    Subfamily Rhynchocyoninae Gill, 1872: 19
      Rhynchocyon Peters, 1847: 36

[a] McKenna and Bell (1997) recently ranked Macroscelidea as a mirorder within the grandorder Anagalida. As no phylogenetic analysis has strongly demonstrated a relationship among zalambdalestids, anagalids, mimotonids, lagomorphs, rodents, mixodonts, and elephant shrews, we conservatively leave Macroscelidea at the rank of order.

[b] McKenna and Bell (1997) removed *Chambius* to the louisinine hyopsodontid condylarths. We follow Simons et al. (1991) and Tabuce et al. (2001) in placing this genus within the subfamily Herodotiinae.

[c] Douady et al. (2003) have demonstrated that the single species *"Elephantulus" rozeti* from North Africa is the sister taxon of *Petrodromus*. As of this writing, this species has not been formally assigned to either *Petrodromus* or a new genus.

Myohyracinae and Mylomygalinae form the crown group, and a monophyletic Herodotiinae (comprising three genera) and the monotypic Metoldobotinae are stem taxa forming the basal branchings of the clade.

Macroscelidea ranges from the Eocene to the Recent, and both fossil and recent macroscelideans are exclusively African in distribution. Their distribution today (Fig. 6.1) is primarily sub-Saharan Africa, with a disjunct distribution of *"Elephantulus" rozeti* (sister taxon of *Petrodromus*; Douady et al., 2003) also occurring along the north African coast, suggesting a more extensive undocumented range of Macroscelidinae in the past. Plio-Pleistocene taxa representing three of the six subfamilies are best known from South African cave sites (Macroscelidinae and Mylomygalinae) and Tanzanian sites (Macroscelidinae and Rhynchocyoninae; Butler and Greenwood, 1976; Butler, 1987). Additionally, the presence of Pliocene macroscelidines has been noted in Morocco (Jaeger, pers. comm. in Butler, 1995). Of the four extant genera, two are known from fossil sites within their extant range, and *Rhynchocyon* is known from Kenyan sites somewhat north of its present range. Only *Petrodromus* has not yet been reported as a fossil. A moderate diversity of Miocene elephant shrews has been recovered from sites in Uganda, Namibia, and Kenya, representing Macroscelidinae, Myohyracinae, and Rhynchocyoninae, and indicating that the two extant subfamilies had diverged by at least the early Miocene. Most of the Miocene to Pleistocene records are exclusively dental and gnathic specimens, although a rhynchocyonine skull (Butler and Hopwood, 1957) has been described, and additional cranial and undescribed postcranial fossils are known (Butler, 1987).

Although virtually all Neogene and Quaternary records occur in sub-Saharan Africa and within the geographic range of extant taxa, the Paleogene record is known exclusively from North Africa. Ranging in age from early or middle Eocene to early Oligocene, four monospecific genera have been described. The three Eocene taxa (*Chambius, Nementchatherium,* and *Herodotius*) form a monophyletic Herodotiinae, and the early Oligocene *Metoldobotes* represents a derived lineage more closely related to the crown group (Tabuce et al., 2001). All four of these taxa are quite poorly known at present, and our understanding of these taxa is based only on dental and/or gnathic remains, although *Metoldobotes* crania have been recovered but are not yet described (Simons, pers. comm. in Butler, 1995).

Characterization of the Macroscelidea

Macroscelidea was originally proposed as an order by Butler (1956). After adding a number of previously misattributed fossil taxa to the Macroscelididae, Patterson (1965) defined the order based on a number of hard and soft tissue characters. Those that are most important to our current understanding of the order and its possible relationships are the presence of large molariform fourth premolars, reduced-to-absent third molars, fused or closely appressed radius and ulna, and fused tibia-fibula. Although Patterson (1965) provided a fairly lengthy list of characters for the Macroscelidea, he wrote that many of the taxa were only known from dental and gnathic remains, and their attribution to the Macroscelidea was based on recognition of dental characters that were similar to those seen in the extant macroscelidids. In practice, paleontologists have used order Macroscelidea as a stem-based taxon name for all fossil taxa that are unequivocally more closely related to living macroscelideans than to other extant orders.

In particular, Paleogene fossils are assigned to the Macroscelidea almost entirely based on the presence of enlarged molariform premolars and reduced third molars. Like later macroscelideans, the Paleogene forms (Fig. 6.2C–E) are characterized by quadritubercular upper molars; molariform fourth premolars; reduced third molars; and the development of two short, anterobuccally directed lophs arising from the lingual cusps (Tabuce et al., 2001). Unlike the later occurring forms (Fig. 6.2A–B), which show trends toward increasing molarization of the fourth premolar, higher crowned teeth, and loss of the third molar, the Eocene taxa are typified by lower-crowned, more bunodont dentitions and less reduced third molars.

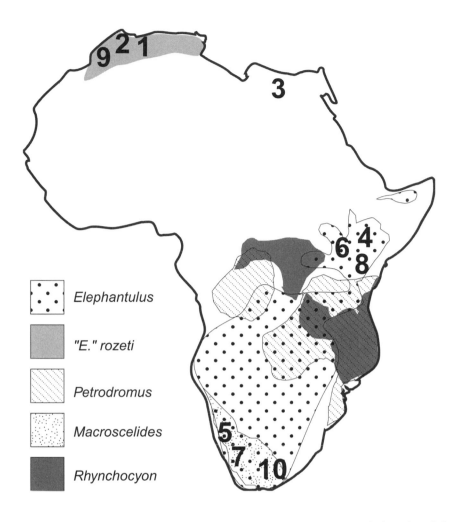

Fig. 6.1. Distribution of recent and fossil macroscelideans. The distribution of extant taxa, is indicated by patterned areas. Numbered fossil sites and recovered taxa, in order of increasing age, are 1: early-middle Eocene Jebel Chambi (*Chambius*), Tunisia; 2: late middle–late Eocene Bir El Ater (*Nementchatherium*), Algeria; 3: late Eocene–early Oligocene Fayum Province localities (*Herodotius* and *Metoldobotes*), Egypt; 4: assorted early Miocene (*Rhynchocyon, Hiwegicyon,* and *Myohyrax*) and middle Miocene (*Pronasilio* and *Myohyrax*) sites in Kenya; 5: early Miocene sites in the Sperrgebiet (*Myohyrax* and *Protypotheroides*) and middle Miocene–Pleistocene sites of Berg Aukas (undescribed Macroscelidinae), Namibia; 6: early Miocene Napak (*Myohyrax*), Uganda; 7: late Pliocene Klein Zee (*Palaeothentoides*), South Africa; 8: late Pliocene Laetolil (*Rhynchocyon*) and early Pleistocene Olduvai (*Macroscelides*), Tanzania; 9: late Pliocene Irhoud-Ocre (undescribed Macroscelidinae), Morocco; 10: multiple Pleistocene cave sites in the Transvaal region (*Elephantulus, Macroscelides,* and *Mylomygale*), South Africa. The extant taxa distribution is based on maps in the African Mammal Databank (Boitani et al., 1999); fossil occurrences are primarily from Butler (1995).

Legend:
- *Elephantulus*
- *"E." rozeti*
- *Petrodromus*
- *Macroscelides*
- *Rhynchocyon*

Position of Macroscelidea within Eutheria

The recognition of elephant shrews as a distinct clade of eutherian mammals and the discovery of their earliest representatives has not simplified the problem of their higher-order affinities. Currently, there are four prevailing ideas about their possible position within Eutheria, each of which presents a different set of circumstances for the timing and place of origin of the group and is based on different sets of data. For the purposes of this chapter, we call them the "Anagalida hypothesis," the "Afrotheria hypothesis," the "Condylarthran hypothesis," and the "Tethytheria hypothesis," and discuss the extent to which they are consistent with one another and with the known fossil record of Macroscelidea.

ELEPHANT SHREWS AS ANAGALIDANS. The hypothesis of a clade composed of Glires, Anagalidae, and Macroscelidea as part of a larger grouping called "Anagalida" originated with Simpson's (1931) and Evans's (1942) attempts to discern the position of the early Oligocene Mongolian form *Anagale* and their comparisons with macroscelideans. Simpson (1931), comparing the skull of *Anagale* to tupaiids and macroscelidids, found that *Anagale* shared a number of characters with both, but thought tupaiids and *Anagale* to be closer. Evans (1942) undertook a more detailed study of three representative taxa (*Tupaia, Anagale,* and *Rhynchocyon*) and found that *Anagale* was intermediate between the two. Although tupaiids were later moved elsewhere (see, e.g., Novacek, 1980), this hypothesized relationship between *Anagale* and elephant shrews was formalized by McKenna (1975), who united them in an emended Anagalida (McKenna, 1975: fig. 3) and averred that "anagalids now appear to me to be relatively plesiomorphous early Asian macroscelideans that had not yet fused the tibia with the fibula or yet strongly reduced or lost M3/3" (McKenna, 1975: 35). He hypothesized that the time of origin of the order was Late Cretaceous and that early macroscelideans dispersed between southwest Asia and Africa, where they were subsequently isolated, a notion he had earlier espoused when he considered them derived from leptictid-like ancestors (McKenna, 1969). Although emending the Anagalida, he did not provide any specific characters considered to unite the two groups to the exclusion of the other taxa considered. Szalay (1977: 356) included macroscelideans, anagalids, pseudictopids, eurymylids, and duplicidentates in an emended Lagomorpha united by a "characteristic restriction of upper and lower ankle movements . . . as well as cheek tooth hypsodonty for some form of terrestrial herbivory." He also observed that macroscelideans and duplicidentates share cranial similarities, but only noted crestlike

Fig. 6.2. Dental morphology in selected taxa. All taxa have been scaled to equal first molar length and aligned along the anterior edge of the first molar, to compare the relative differences in premolar molarization and M3 reduction. (A) Extant *Rhychocyon*. (B) Miocene *Rhynchocyon clarki*. (C) Late Eocene *Herodotius pattersoni*. (D) Early-middle Eocene *Chambius kasserinensis*. (E) Middle-late Eocene *Nementchatherium senarhense*. (F) Early Eocene *Haplomylus speirianus*. All drawings are after Butler (1995: figs. 2–3).

A

B

C

D

E

F

sculpturing of the orbital edges. However, he refrained from specifically uniting macroscelideans with any particular group within his emended Lagomorpha, as he did not think there was adequate knowledge to make a more precise determination of relationships.

Numerous detailed analyses of morphologic data sets undertaken in subsequent years consistently found a relationship between Glires and Macroscelidea, although the characters used and confidence expressed by the authors varied among studies. In cladistic analyses of extant orders (Novacek and Wyss, 1986; Novacek et al., 1988; Novacek, 1989, 1990, 1992), the presence of canals in the external alisphenoid for branches of the maxillary nerve, the relationship of the inferior ramus of the stapedial artery and inferior petrosal nerve to the tegmen tympani, and embryonic disc orientation were used to unite the group, but it was noted that the number of shared features was unfortunately few. The most recent analysis of extant orders and fossil taxa that has recovered a relationship between Glires, macroscelideans, and anagalids (=Anagalida) is that of Shoshani and McKenna (1998). In their analysis, Anagalida is united by 18 characters that describe aspects of the placenta, conditions of the supinator and semitendinous muscles, fused tibia-fibula, the presence of a long acromion process of the scapula with a metacromion process, and bony features of the orbit and basicrania. Studies have been based on the scoring of presence/absence data in a data set that grew from 66 (Novacek and Wyss, 1986) to 88 (Novacek, 1992), to 260 characters in Shoshani and McKenna (1998) and included both osteological and soft-tissue characters. In each data set, modern orders were treated as a single operational taxonomic unit (OTU) and were scored based solely on the condition in extant taxa (e.g., Novacek, 1992) or based on a composite of extant and fossil taxa (e.g., Shoshani and McKenna, 1998).

The only analysis that has focused on fossil Glires and also includes elephant shrews is Meng and Wyss's (2001) attempt to better assess relationships among Glires. In addition to basal genera assigned to Glires, they included the extant taxa *Elephantulus* and *Rhynchocyon*, fossil anagalids, selected Paleogene primates, *Leptictis*, *Asioryctes*, and *Kennalestes*. The two extant macroscelideans grouped together, but were otherwise part of a polytomy that included fossil members of Glires, fossil anagalids, and *Pseudictops*. This group was united by a series of dental characters and one character of the bulla: p4 trigonid with strong ridges, p4 talonid fully crested, unilaterally hypsodont cheek teeth, upper molar labial cingulum absent, complete protoloph between paracone and protocone on upper molars, upper molar protocone ridged and V-shaped, upper molar conules ridged, posterior wall of trigonid strongly crested, and auditory bulla formed primarily by the ectotympanic. They also recognized macroscelidean monophyly on the basis of the most extensive list of characters yet presented for the order: reduction of lower canine size; P3 becoming bilobed and elongated; P4 molarized; loss of upper and lower third molars; upper molar crowns longer than wide, having weak conules and anteroposteriorly elongate paracone and metacone; flat rather than concave mandibular glenoid fossa, with rudimentary postglenoid process; premaxilla with long posterior process contacting frontal narrowly; antorbital fossa prominent, associated with the development of nasolabial muscles; reduced mandibular coronoid process; auditory bulla formed by several elements; distal fusion of tibia and fibula; and the reversion of the calcaneal facet to the primitive eutherian condition.

Evaluating these studies in light of the fossil record of elephant shrews, few, if any, of the characters used in these analyses can be assessed in any of the extinct taxa, due to the absence of crania for most taxa. For example, in the data set of Novacek (1992), only three of the 88 characters would be scorable for any of the fossil elephant shrew taxa (Novacek, 1992: table 1, characters 54, 58, and 64), and all would be scored as the primitive condition. Among the dental characters cited by Meng and Wyss (2001) as synapomorphies for a Glires + Macroscelidea + Anagalidae + *Pseudictops* clade, most are features associated with increasing crown height and shearing capability and are only present in Neogene-to-Recent elephant shrews. Even among the characters that were found to unite *Rhynchocyon* and *Elephantulus,* the only one that is also known to be present in Paleogene elephant shrews is the molarization of P4; the others are characteristics only of the Neogene-to-Recent forms. Thus, there appear to be few dental characteristics that might unite Glires and Macroscelidea, and other possible synapomorphies of the crania and postcrania cannot be evaluated in the earliest members of the Macroscelidea.

If accurate, a sister taxon relationship with Glires and anagalids implies that the elephant shrew lineage had diverged by at least the late Paleocene, when the basal gliroid *Tribosphenomys* is known, and that this divergence probably occurred in Asia, where both anagalids and early gliroids first appear. By extension, this hypothesized relationship implies the existence of an Asian ghost lineage of early macroscelideans that would have crossed the Tethys by at least the early-middle Eocene, when *Chambius* is recorded. Alternatively, if we consider stem Glires *sensu* Archibald et al. (2001) and Archibald (2003), recognizing the Asian Late Cretaceous

Kulbeckia as the oldest member of Glires, the timing of this divergence would be even earlier.

ELEPHANT SHREWS AS AFROTHERES. By contrast with the morphological studies, molecular studies consistently fail to return an Anagalida clade comprising Glires and elephant shrews. Beginning with studies of eye-lens crystallins (e.g., de Jong, 1985; McKenna, 1987; de Jong, 1993), a possible relationship with taxa typically African in origin (e.g., tubulidentates, hyracoids, proboscideans) has been strongly suggested. As more molecular studies recovered this clade (e.g., Porter et al., 1996; Madsen et al., 1997; Stanhope et al., 1998a), Stanhope et al. (1998b) proposed the superordinal nomen Afrotheria to include Macroscelidea, Tubulidentata, Afrosoricida (=Tenrecoidea *sensu* McDowell, 1958) and paenungulates (hyracoids, sirenians, and proboscidians). These recent analyses of mitochondrial and nuclear DNA commonly recover elephant shrews as sister taxon to Tubulidentata or Afrosoricida near the base of Afrotheria (e.g., Stanhope et al., 1998a,b; Murphy et al., 2001a,b; van Dijk et al., 2001; Malia et al., 2002; Amrine-Madsen et al., 2003). Although some authors consider the molecular evidence "overwhelming" (e.g., Douady et al., 2002), no unambiguous morphological synapomorphies have been identified in support of this clade.

This hypothesis of relationships implies that elephant shrews had diverged by perhaps as early as the middle Cretaceous (Eizirik et al., 2001) or at least by the Late Cretaceous–earliest Paleogene (69 ± 5 million years ago; Douady and Douzery, 2003), based on different molecular clock methods. Also, if the basal placement of Afrotheria within eutherian mammals is correct, Africa may have been the area of origin for Eutheria.

ELEPHANT SHREWS AS CONDYLARTHS. Most discussions of elephant shrew higher-order relationships have focused on evidence from extant taxa and have not incorporated the fossil evidence. Historically, most attention was focused on their possible relationships with other "Insectivora," although Frechkop (1931) had suggested a possible relationship with ungulates, based on the masticatory apparatus of *Rhynchocyon* and *Elephantulus*. Examinations of the Paleogene elephant shrews have led to alternative hypotheses of relationships. Both Hartenberger (1986) and Simons et al. (1991) hypothesized a possible ancestor-descendant relationship between certain hyopsodontid condylarthrans and elephant shrews, based on dental morphological comparisons. Hartenberger (1986), in describing early-middle Eocene *Chambius,* drew comparisons with the European early-middle Eocene Louisininae, noting the reduction of the third molars and the resemblances of the overall molar morphology to that group. However, he concluded that a more comprehensive analysis was needed to test this relationship. Simons et al. (1991), in describing late Eocene *Herodotius,* also compared it favorably with louisinines, but thought greater cladistic similarity lay with the North

American early Eocene *Haplomylus,* listing seven shared derived features of that genus with the Macroscelidea: molariform p4 with enlarged, midline paraconid and talonid with two cusps; p2–3 with well-developed paraconids; m1–2 trigonids wider than talonids and reduced hypoconulids; P4–M2 postprotocristae directed toward the hypocone rather than toward the metaconule; P4 transversely widened and bearing enlarged parastyles; and the reduction of M3/m3. They also observed that these premolar characters are similar to the condition observed in phenacodontids, and considered that these shared characters indicated macroscelideans arose from within the condylarthrans.

Although both of these studies of the Paleogene taxa posited an origin for elephant shrews within the Condylarthra and, therefore, suggested relationships with ungulate or paenungulate taxa by extension, neither study specifically addressed the identification of the modern orders to which Macroscelidea might be related. Nor did either study examine the implications of these proposed relationships for the time and place of origin. An origin by the early Eocene might be suggested by a relationship with Louisininae, and a late Paleocene divergence at the latest would be required by a sister-group relationship with *Haplomylus.* No particular area of origin is favored by either hypothesized relationship, as hyopsodontids are Holarctic in distribution in the early Paleogene. The possibility of African phenacodontid condylarths has also been noted (Gheerbrant et al., 2001), and it may be that other condylarthran groups were present in early Paleogene Africa.

ELEPHANT SHREWS AS TETHYTHERES.   Tabuce et al. (2001) formally tested condylarthran and afrotherian hypotheses in the context of the description of middle-late Eocene *Nementchatherium.* Based on an analysis of dental characters, they recovered a monophyletic Macroscelidea whose closest sister taxon is *Microhyus* (previously classed as a louisinine hyopsodontid), followed successively by the proboscidean *Numidotherium,* the hyracoids and/or Perissodactyla, and phenacodontids. This relationship between Macroscelidea and the Tethytheria was supported by dental characters, including enlarged P4/p4 metacone and metaconid, alignment of the M1–2 protocone and paracone, hypocone and protocone subequal in height, presence of a hypoloph, and crestiform m1–2 hypoconulids. They also recovered the long-suspected paraphyly of condylarthrans, particularly hyopsodontids, relative to modern orders. As noted by the authors, their analysis is consistent with some molecular studies (e.g., de Jong, 1993), but the lack of homologous dental characters prevented them from scoring tenrecids and chrysochlorids in their study. The analysis of Tabuce et al. (2001) strongly suggests an African origin of the group, although the inclusion of the European *Microhyus* as sister taxon to macroscelideans implies that trans-Tethyan dispersal occurred, although the direction is unclear. A minimum time of divergence would be prior to the late Paleocene, when the earliest proboscidean *Phosphatherium* appears in Morocco (Gheerbrant et al., 1996).

TUBULIDENTATA

Today, only a single extant aardvark taxon, *Orycteropus afer,* is known from sub-Saharan Africa (Fig. 6.3). *Orycteropus* and two additional genera are known from the Miocene and/or Plio-Pleistocene: *Leptorycteropus* and *Myorycteropus* (Table 6.2). A fourth genus, *Plesiorycteropus,* from subfossil sites in Madagascar, was long classed among the tubulidentates (e.g., Patterson, 1975), but has recently been reexamined in detail by MacPhee (1994). He found *Plesiorycteropus* to be strongly convergent on aardvarks but to share no derived characters, and reassigned them to the new order Bibymalagasia. Other fossil taxa have also been suggested as possible tubulidentates, but have been reassigned upon closer examination (see discussions in Patterson, 1975; MacPhee, 1994).

Unequivocal Plio-Pleistocene records of aardvarks are known from Laetoli, Tanzania (Leakey, 1987); Omo, Ethiopia (Howell and Coppens, 1974); and various sites in Kenya (MacInnes, 1956; Patterson, 1975; Milledge, 2003), the Cape Province of South Africa (Hendey, 1973), and Algeria (*Orycteropus,* Patterson, 1975; Fig. 6.3). Outside Africa, tubulidentates have been reported from the Pliocene (Ruscinian) of France (Aymar, 1992).

The smaller forms *Myorycteropus* and *Leptorycteropus* have been found in the early Miocene of Rusinga (MacInnes, 1956), and the middle-late Miocene of Namibia and Uganda (Pickford, 1994, 1996). In the Miocene, the range of tubulidentates expanded into Turkey (late Miocene, Sen, 1994; middle Miocene, Fortelius, 1990), Greece (late Miocene, de Bonis et al., 1994; *Orycteropus,* Colbert, 1941), Italy (Rook and Masini, 1994), and Pakistan (e.g., Pickford, 1978). *Orctyeropus* is also known from northwestern Iran and Ukraine (Patterson, 1975).

Overall, the morphologies of the fossil and living taxa are quite similar, to the extent that Pickford (1975) even considered them all to belong to a single genus. The principal difference between the living form and the fossil genera is the smaller size of the latter. Additionally, *Leptorycteropus* is more gracile, and *Myorycteropus* more robust, with the bony features associated with digging accentuated.

No unquestioned Paleogene tubulidentates have been described, although Thewissen (1985) has suggested that a single specimen, the type of *Leptomanis,* from the late Eocene or early Oligocene of Quercy, France, may represent a Paleogene tubulidentate. The specimen is the dorsal part of a skull with elongate nasals, frontals, and parts of the orbits, temporal fossae, and ethmoids and is generally considered a junior synonym of the manid *Necromanis* (e.g., Emry, 1970; McKenna and Bell, 1997). Thewissen noted a number of similarities, especially the lateral position of the root of the zygomatic arch, which were more similar to *Orycteropus* and/or *Plesiorycteropus* than to manids. Until this specimen is reexamined in light of MacPhee's (1994) work on *Plesiorycteropus,* its affinities to *Leptomanis* and its impact on our understanding of tubulidentate differentiation will remain unclear.

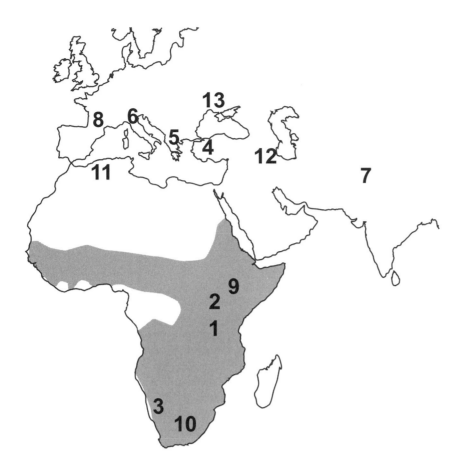

Fig. 6.3. Distribution of fossil and recent tubulidentates. The distribution of extant *Orycteropus afer* is indicated by the shaded area. All fossils sites are records of *Orcyteropus*, unless otherwise indicated, and references are those in the text. 1: Numerous sites in Kenya and Tanzania, including early Miocene Songhor (*Orycteropus* and *Myorycteropus*), Rusinga, and Mfwangano, Kenya; late Miocene *Leptorycteropus* and Pliocene *Leptorycteropus* and *Orycteropus*, Lothagam, Kenya; early Pleistocene Kanapoi, Koobi Fora, Rusinga, Kanjera, and Laetoli, Tanzania. 2: Middle-late Miocene, Uganda. 3: Middle-late Miocene, Namibia. 4: Middle-late Miocene, Turkey. 5: Late Miocene, Greece. 6: Late Miocene, Italy. 7: Late Miocene, Pakistan. 8: Pliocene, France. 9: Plio-Pleistocene, Ethiopia. 10: Pleistocene, South Africa. 11: Pleistocene, Algeria. 12: Age uncertain, Iran. 13: Age uncertain, Ukraine. The distribution of extant *Orycteropus afer* is based on maps in the African Mammal Databank (Boitani et al., 1999).

Table 6.2   Classification of Tubulidentata

Order Tubulidentata Huxley, 1872: 288
    Family Orycteropodidae Gray, 1821: 305
        Leptorycteropus Patterson, 1975: 186
        Myorycteropus MacInnes, 1956: 1
        Orycteropus Geoffroy Saint-Hilaire, 1796: 102

Notes: McKenna and Bell (1997) also included *Palaeorycteropus* Filhol, 1893 and *Archaeorycteropus* Ameghino, 1905 in the family Orycteropodidae. Both taxa are based on single bones from the Quercy Phosphorites; here we follow Simpson (1945) and Patterson (1975) in regarding these as indeterminate.

Characterization of the Tubulidentata

The teeth are the most distinctive characteristic of the order and inspired the ordinal name. The cheek teeth are rootless (allowing them to grow continually) and in *Orycteropus*, contain up to 1,500 hexagonal prisms of dentine, each of which is built around a thin, tubelike pulp cavity. Each tooth is columnar and is encircled in cementum rather than enamel. In all forms for which the snout is known, it is slightly to significantly elongate, a feature associated with myrmecophagy. The postcrania also show a number of features that readily differentiate these taxa from contemporaneous mammals—for example, prominent lateral process of the ischium, presence of a third trochanter on the femur, keeled metapodials, and a ball-like astragalar head. However, no postcranial characters stand out as unique, a fact

that has led to difficulty and misunderstandings regarding the composition of the order (see, e.g., discussions in Mac-Phee, 1994). Rather, it is the combination of these postcranial characters and their absence in contemporaneous forms that permit identification of postcranial remains as tubulidentate. In general, tubulidentates are distinctive among extant (and even Neogene) taxa in retaining so many primitive eutherian characters (e.g., the presence of a clavicle, lunar-unciform contact, broad scapula with coracoid and metacromian process; see Thewissen, 1985) in combination with their unique dental configuration.

Position of Tubulidentata within Eutheria

Currently, there are three prevailing hypotheses for the position of aardvarks within Eutheria. All suggest a fairly ancient divergence (at least by the Paleocene) and the presence of a ghost lineage of significant duration. As for macroscelideans, there is both a Condylarthran and Afrotheria hypothesis; the third hypothesis may best be termed the "unresolved Epithere hypothesis."

AARDVARKS AS CONDYLARTHS. The idea that aardvarks may have arisen from some condylarth appears to have been first suggested by Smith (1898) and elaborated by Sonntag (1926), Colbert (1941), and Patterson (1975), among others. This line of thought has been well summarized in Patterson (1975) and Thewissen (1985). The argument is grounded in the similarities seen throughout the skeleton between condylarths (especially phenacodontids) and aardvarks. However, as many authors noted, most of these similarities are likely to be primitive morphologies for all ungulate mammals. The proposed relationship between condylarths and aardvarks has typically been put forth as one of possible unspecified ancestry from within the condylarthrans, although Wible (1987) did suggest that a unique origin from within Arctocyonidae may be indicated by certain petrosal and squamosal features.

AARDVARKS AS AFROTHERES. Tubulidentates have been closely associated with paenungulates in molecular phylogenetic analyses for over two decades (e.g., de Jong et al., 1981; Miyamoto and Goodman, 1986). In more recent analyses, tubulidentates occupy a position similar to macroscelidids nested within Afrotheria (Porter et al., 1996; Madsen et al., 1997; Springer et al., 1997; Stanhope et al., 1998a,b). *Orycteropus* is typically seen as the sister taxon to either Macroscelidea (Porter et al., 1996; Madsen et al., 1997; Springer et al., 1997; Stanhope et al., 1998a,b; van Dijk et al., 2001) or a larger clade containing Macroscelidea and Afrosoricida (Murphy et al., 2001b; Douady et al., 2002; Murata et al., 2003). However, aardvarks are also found at the base of the afrotherian clade (Malia et al., 2002) and as sister to Tenrecidae (Eizirik et al., 2001; Hudelot et al., 2003; Waddell and Shelley, 2003, for selected trees) in some analyses. According to Springer et al. (2003), the tubulidentate lineage originated approximately 76 million years ago, during the Late Cretaceous. This date is in sharp contrast to the first appearance of tubulidentates in the fossil record, but may be consistent with the long ghost lineage implied by the morphological evidence.

Not all molecular analyses confirm the afrotherian hypothesis for Tubulidentata. Árnason et al. (1999) examined the complete mitochondrial genome of *Orycteropus* and found that it resolves as the sister taxon to a clade containing Xenarthra and Cetferungulata. In that paper, Lagomorpha also nested within the *Orycteropus* + Xenarthra + Cetferungulata clade in certain analyses. The authors speculate that the reason for this grouping is related to the secondary structure of tRNA-Ser(UCN) in tubulidentates. *Orycteropus* has a unique structure, unlike any other mammals studied; its secondary structure represents a reversal to the structure of non-mammalian vertebrates. Árnason et al. chose *Orycteropus* to represent the "African clade," which includes all afrotheres. However, in a diagram of other mammalian tRNA-Ser(UCN) structures that portrayed the condition in macroscelidids, the elephant shrews did not share the aardvark condition. This evidence could lead investigators to suspect that Tubulidentata may not be the optimal representative for the African clade, and may, in fact, not nest within the clade at all. No analyses including the other members of the African clade were shown or described by Árnason et al. (1999), preventing any investigation of this hypothesis.

AARDVARKS AS UNRESOLVED EPITHERES. Attempts to resolve the position of tubulidentates through cladistic analysis of morphological characters within a broader eutherian context have resulted in varying placements, none of which are terribly robust or well resolved. Although tubulidentates consistently fall within Epitheria (Eutheria excluding Edentata and Pholidota), their position is unstable. Tubulidentates have fallen out as a distinct branch from a basal epitherian polytomy (Novacek and Wyss, 1986; Novacek et al., 1988; Novacek, 1989), as part of a polytomy at the base of ungulates (Novacek, 1982) or as the sister taxon to all other ungulates (Shoshani and McKenna, 1998). The characters used to unite tubulidentates with ungulates include spatulate distal phalanges, rounded patella, and the opening of the stylomastoid foramen above the level of cheek teeth.

However, Thewissen (1985) disputed the possibility of a relationship with ungulates. He examined cranial features across a variety of fossil and extant ungulates (including condylarthrans) and concluded that the mosaic of features seen in tubulidentates was not convincingly like that seen in ungulates (beyond shared primitive features). He further found that there were problems in establishing the polarities and the primitive condition of these cranial characters in both tubulidentates and many extant ungulate orders. In part, this problem in establishing the primitive condition for tubulidentates stems from Thewissen's placement of *Plesiorycteropus* (which demonstrates a number of primitive eutherian character states) within Tubulidentata. Nonetheless, the character variability he observed within ungulate orders is significant. Although Thewissen offered no alternative placement, his study is a cautionary one for the use of these cranial characters in attempting to establish a relationship with ungulates.

The instability of the position of tubulidentates within Eutheria is amply demonstrated in MacPhee's (1994) multiple analyses of a modified version of Novacek's (1989, 1992) character matrix. Using different assumption sets and selected deletions, MacPhee was able to produce Tubulidentata as sister taxon to Insectivora in a clade with Carnivora within Epitheria or as part of an edentate clade (with xenarthrans and pholidotans), whose sister taxon was Ungulata. MacPhee also examined the relationships that resulted using a character matrix based on Patterson's (1975) comparisons and obtained a tree with tubulidentates as a sister taxon to phenacodontids. The wide range of possible phylogenetic positions highlights the problems in assessing tubulidentate affinities.

DISCUSSION

There is no clear consensus on either elephant shrew or aardvark origins from the neontological, paleontological, and molecular evidence. However, several of the hypotheses based on this evidence are more consistent with one another than are others. As noted, paleontological hypotheses suggesting relationships with certain condylarthrans and/or tethytheres could be consistent with molecular studies implying the presence of an afrotherian clade. Also, close relationships among elephant shrews, aardvarks, and other African endemic groups are the most biogeographically parsimonious. However, hypotheses favoring condylarthran, tethytherian, or afrotherian affinities do not adequately explain the series of morphological resemblances observed in the crania and postcrania of elephant shrews, anagalids, and Glires. Nor do such hypotheses allow us to sort out the mélange of shared derived and homoplastic characters that suggest the "ungulate" affinities of aardvarks.

In part, we can point to the geographic and temporal gaps in our fossil record of African taxa and to the small sample sizes of most known taxa to explain this lack of consensus. However, the divergence among hypotheses equally stems from a number of methodological problems. Primary among these is taxon sampling: no study to date has examined the same set of taxa or a wide enough range of taxa to adequately test these hypotheses. Studies evaluating the Anagalida hypothesis have used extant macroscelideans and only incorporated fossils from within Glires and zalambdalestids, not fossil elephant shrews or condylarthrans. Studies explicitly testing the relationships among the fossil elephant shrews have only considered condylarthrans and tethytheres, but not representatives of Glires, anagalids, or zalambdalestids. MacPhee's (1994) cladistic analyses of selected eutherians using multiple data sets and alternate taxon lists illustrate a similar problem for understanding tubulidentate relationships.

As a demonstration of the possible effect that the inclusion of a wider range of taxa might have on understanding relationships of macroscelideans and other eutherian orders, let us reexamine the most comprehensive study of the morphological characters of extant and extinct Mammalia to date—that of Shoshani and McKenna (1998). Although they did not present the results graphically in that paper, the authors noted that they had run the analysis with fossil taxa included and provided these data as a supplement to the paper. Fig. 6.4 is a representation of their complete tree. Of interest is where the archaic groups fall out in a majority-rule consensus tree derived from that matrix. Anagalids are not sister taxon to Glires (although macroscelideans are), the condylarthran *Hyopsodus* appears as the sister taxon of aardvarks, the dinoceratan *Uintatherium* comes out in the Tethytheria, and other condylarthrans are the outgroup of a clade composed of Tethytheria, Cetartiodactyla, and Perissodactyla. Representatives of extinct orders fall out all over the tree as the basal branchings of many of the clades of extant mammals. The inclusion of these taxa alters the robustness of nodes, changes the sets of synapomorphies that define them, suggests other paleobiogeographic scenarios, and places different constraints on the minimum times of divergence. The implications of this are clear: we need to incorporate more fully representatives of extinct

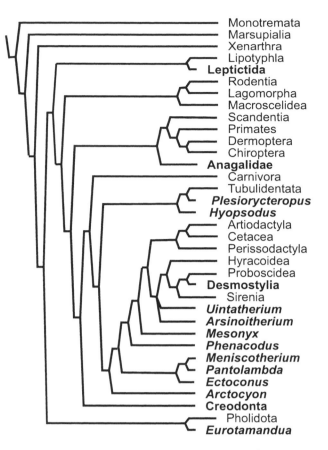

Fig. 6.4. Majority-rule consensus tree from the data matrix of Shoshani and McKenna (1998), incorporating both extant orders and selected fossil taxa. Representatives of extinct orders are shown in boldface.

groups in morphological and total evidence analyses of mammalian phylogeny.

A second issue of taxon sampling relevant to understanding basal splits within the mammalian tree is the composition of the taxa used in the available matrices. The matrices used in all the comprehensive eutherian phylogenies touched on in this chapter used composite taxa as OTUs. Characters were scored for orders as a whole, condensing all the variation inherent in each of these groups into a single score. As noted above, some of the matrices were based on studies of extant taxa only (e.g., Novacek and Wyss, 1986) or mixed characters observable in extant taxa (e.g., soft tissue anatomy) with osteological characters observed only in 50-million-year-old taxa (e.g., the scoring for Sirenia in the Shoshani and McKenna, 1998). For these reasons, it is important to recognize that the resultant phylogenetic hypotheses are fundamentally different from those that employed known taxa as OTUs (e.g., Meng and Wyss, 2001; Tabuce et al., 2001). Although such composite scoring does serve practical considerations (e.g., processing time, minimization of missing data), it is not a test of the most parsimonious evolution of characters as they appear in the fossil record. Rather, the composite scores serve as tests of the most parsimonious arrangement of our notions about distributions, inferred ancestral character states, and the other ad hoc hypotheses employed to construct these morphological amalgams: it would be more accurate to interpret them as readily falsifiable tests of our knowledge of character evolution. Bearing these caveats of taxon composition in mind, it is not surprising when other analyses based on fossils, molecules, or the morphology of specific taxa differ from these trees. If we want to take our understanding of eutherian diversification to a new level, we should stop testing the most parsimonious distribution of these encoded chimaeras and instead, work with the characters as presented in the fossil record, despite the gaps and inadequacies they may present.

SUMMARY

Macroscelidea (elephant shrews) and Tubulidentata (aardvarks) are both orders with morphologically derived extant representatives and limited fossil records that have obvious morphological and temporal gaps. Based on morphological analyses, macroscelideans have been suggested as closest to rodents and lagomorphs (Glires), as derived from condylarths, or related to tethytheres. Tubulidentates have also been suggested to derive from condylarths or simply regarded as a long-distinct lineage that branched off early in the history of eutherian mammals. Molecular analyses have now united these taxa with other groups with presumed African origins in the Afrotheria. Review of these varied hypotheses produces no clear resolution of the phylogenetic position of the two orders, although we find that most of the morphological hypotheses are not in conflict with the molecular analyses. The only incongruent morphological hypothesis is the suggested sister-group relationship of Glires with Macroscelidea, which is supported by neither the fossil record of elephant shrews nor molecular analyses. Future work on resolving the relationships of these orders must focus both on increased sampling of the fossil record of these groups and on more complete analyses of the archaic mammalian groups that have not yet been fully incorporated into higher-level mammalian systematic studies.

ACKNOWLEDGMENTS

We thank J. D. Archibald and K. D. Rose for their kind invitation to participate in this volume, T. A. Stidham for bringing important distributional references to our attention, D. Smith for his help with Fig. 6.4, and P. Butler, J.-L. Hartenberger, and E. Seiffert for their thoughtful reviews.

REFERENCES

Ameghino, F. 1905. Les Édentés fossils de France et d'Allemagne. Anales del Museo Nacionale de Buenos Aires 13: 175–250.

Amrine-Madsen, H., K.-P. Keopfli, R. K. Wayne, and M. S. Springer. 2003. A new phylogenetic marker, *apolipoprotein B*, provides compelling evidence for eutherian relationships. Molecular Phylogenetics and Evolution 28: 225–240.

Andrews, C. W. 1914. On the lower Miocene vertebrates from British East Africa collected by Dr. Felix Oswald. Quarterly Journal of the Geological Society of London 70: 163–186.

Archibald, J. D. 2003. Timing and biogeography of the eutherian radiation: Fossils and molecules compared. Molecular Phylogenetics and Evolution 28: 350–359.

Archibald, J. D., A. O. Averianov, and E. G. Ekdale. 2001. Late Cretaceous relatives of rabbits, rodents, and other extant eutherian mammals. Nature 414: 62–65.

Archibald, J. D., and D. Deutschman. 2001. Quantitative analysis of the timing of origin of extant placental orders. Journal of Mammal Evolution 8: 107–124.

Árnason, Ú., A. Gullberg, and A. Janke. 1999. The mitochondrial DNA molecule of the aardvark, *Orycteropus afer*, and the position of the Tubulidentata in the eutherian tree. Proceedings of the Royal Society of London B 266: 339–345.

Aymar, J. 1992. Essai de reconstitution de la "Faune de Perpignan"; Pliocène (Ruscinien). Naturalia Ruscinonensia 2: 33–47.

Boitani, L., F. Corsi, A. De Biase, I. D. Carranza, M. Ravagli, G. Reggiani, I. Sinibaldi, and P. Trapanese. 1999. African Mammals Databank: A Databank for the Conservation and Management of African Mammals. Instituto di Ecologia Applicata, Rome.

Bonaparte, C. L. 1838. Synopsis vertebratorum systematis. Nuovi Annali delle Scienze Naturali 1: 105–133.

Broom, R. 1948. Some South African Pliocene and Pleistocene mammals. Annals of the Transvaal Museum 21: 1–38.

Butler, P. M. 1956. The skull of *Ictops* and the classification of the Insectivora. Proceedings of the Zoological Society of London 126: 453–481.

———. 1984. Macroscelidea, Insectivora and Chiroptera from the Miocene of East Africa. Palaeovertebrata 14: 117–200.

———. 1987. Fossil insectivores from Laetoli; pp. 85–87 *in* M. D. Leakey and J. M. Harris (eds.), Laetoli: A Pliocene Site in Northern Tanzania. Oxford University Press, Oxford.

———. 1995. Fossil Macroscelidea. Mammal Review 25: 3–14.

Butler, P. M., and M. Greenwood. 1976. Elephant-shrews (Macroscelididae) from Olduvai and Makapansgat; pp. 1–56 *in* R.J.G. Savage and S. C. Coryndon (eds.), Fossil Vertebrates of Africa. Volume 4. Academic Press, London.

Butler, P. M., and A. T. Hopwood. 1957. Insectivora and Chiroptera from the Miocene rocks of Kenya Colony. Fossil Mammals of Africa 13: 1–35.

Colbert, E. H. 1941. A study of *Orycteropus gaudryi* from the island of Samos. Bulletin of the American Museum of Natural History 78: 305–351.

de Bonis, L., G. Bouvrain, D. Geraads, G. D. Koufos, and S. Sen. 1994. The first aardvarks (Mammalia) from the late Miocene of Macedonia, Greece. Neues Jahrbuch für Geologie und Paläontologie. Abhandlungen 194: 343–360.

de Jong, W. W. 1985. Supraordinal affinities of Rodentia studied by sequence analysis of eye lens protein; pp. 211–226 *in* W. P. Luckett and J.-L. Hartenberger (eds.), Evolutionary Relationships among Rodents: A Multidisciplinary Analysis. Plenum, New York.

———. 1993. Eye lens crystallins and the phylogeny of placental orders: Evidence for a macroscelid-paenungulate clade?; pp. 5–12 *in* F. S. Szalay, M. J. Novacek, and M. C. McKenna (eds.), Mammal Phylogeny. Volume 2. Placentals. Springer-Verlag, New York.

de Jong, W. W., A. Sweers, and M. Goodman. 1981. Relationships of aardvarks to elephants, hyraxes, and sea cows from alpha-crystallin sequences. Nature 292: 538–540.

Douady, C. J., F. Catzeflis, J. Raman, M. S. Springer, and M. J. Stanhope. 2003. The Sahara as a vicariant agent, and the role of Miocene climatic events, in the diversification of the mammalian order Macroscelidea (elephant shrews). Proceedings of the National Academy of Sciences USA 100: 8325–8330.

Douady, C. J., P. I. Chatelier, O. Madsen, W. W. de Jong, F. Catzeflis, M. S. Springer, and M. J. Stanhope. 2002. Molecular phylogenetic evidence confirming the Eulipotyphla concept and in support of hedgehogs as the sister group to shrews. Molecular Phylogenetics and Evolution 25: 200–209.

Douady, C. J., and E.J.P. Douzery. 2003. Molecular estimation of eulipotyphlan divergence times and the evolution of "Insectivora." Molecular Phylogenetics and Evolution 28: 285–296.

Eizirik, E., W. J. Murphy, and S. J. O'Brien. 2001. Molecular dating and biogeography of the early placental mammal radiation. Journal of Heredity 92: 212–219.

Emry, R. J. 1970. A North American Oligocene pangolin and other additions to the Pholidota. Bulletin of the American Museum of Natural History 142: 455–510.

Evans, F. G. 1942. The osteology and relationships of the elephant shrews (Macroscelididae). Bulletin of the American Museum of Natural History 80: 85–125.

Filhol, H. 1893. Observations concernant quelques mammifères fossils nouveaux du Quercy. Annales des Sciences Naturelles comprenant la zoologie, la botanique, l'anatomie et la physiologie comparée des deux règnes et l'histoire des corps organisés fossils, Série 7, 16: 129–150.

Fortelius, M. 1990. Less common ungulate species from Pasalar, middle Miocene ofAnatolia (Turkey). Journal of Human Evolution 19: 479–487.

Frechkop, S. 1931. Notes sure les mammiferes. V. Note preliminaire sur la dentition et la position systematique des Macroscelidae. Bulletin Musee Royal d'Histoire Naturelle de la Belgique 7(6): 1–11.

Geoffroy Saint-Hilaire, C. 1796. Extrait d'un mémoire sur le *Myrmecophaga* Capensis. Bulletin des Sciences par la Societe Philomatique de Paris 1: 102–103.

Gheerbrant, E., J. Sudre, and H. Cappetta. 1996. A Paleocene proboscidean from Morocco. Nature 383: 68–71.

Gheerbrant, E., J. Sudre, M. Iarochene, and A. Moumni. 2001. First ascertained African "condylarth" mammals (primitive ungulates: cf. Bulbulodentata and cf. Phenacodonta) from the earliest Ypresian of the Ouled Abdoun Basin, Morocco. Journal of Vertebrate Paleontology 21: 107–118.

Gill, T. 1872. Arrangement of the families of mammals. Smithsonian Miscellaneous Collections 230: 1–98.

Gray, J. E. 1821. On the natural arrangement of vertebrose animals. London Medical Repository, 15(1): 296–310.

Hartenberger, J.-L. 1986. Hypothèses paléontologiques sur l'origine des Macroscelidea (Mammalia). Comptes Rendus de l'Académie des Sciences, Sciences de la Terre et des Planètes B302, Série II, Paris 5: 247–249.

Hendey, Q. B. 1973. Fossil occurrences at Langebaanweg, Cape Province. Nature 244: 13–14.

Howell, F. C., and Y. Coppens. 1974. Les faunes de mammifères fossiles des formations plio-pleistocènes de l'Omo en Ethiopia (Tubulidentata, Hyracoidea, Lagomorpha, Roden-tia, Chiroptera, Insectivora, Carnivora, Primates). Comptes Rendus Hebdomadaires des Séances de l'Académie des Sciences, Série D, Sciences Naturelles 278: 2421–2424.

Hudelot, C., V. Gowri-Shankar, H. Jow, M. Rattray, and P. G. Higgs. 2003. RNA-based phylogenetic methods: Application to mammalian mitochondrial RNA sequences. Molecular Phylogenetics and Evolution 28: 241–252.

Huxley, T. H. 1872. A Manual of the Anatomy of Vertebrated Animals. Appleton and Company, New York.

Leakey, M. G. 1987. Fossil aardvarks from the Laetolil Beds; pp. 297–300 *in* M. D. Leakey and J. M. Harris (eds.), Laetoli: A Pliocene Site in Northern Tanzania. Clarendon Press, Oxford.

MacInnes, D. G. 1956. Fossil Tubulidentata from east Africa. Fossil Mammals of Africa 10: 1–38.

MacPhee, R.D.E. 1994. Morphology, adaptations and relationships of *Plesiorycteropus* and a diagnosis of a new order of Eutherian mammals. Bulletin of the American Museum of Natural History 220: 1–214.

Madsen, O., P.M.T. Deen, G. Pesole, C. Saccone, and W. W. de Jong. 1997. Molecular evolution of mammalian aquaporin-2: Further evidence that elephant shrew and aardvark join the paenungulate clade. Molecular Biology and Evolution 14: 363–371.

Malia, M. J., R. M. Adkins, and M. W. Allard. 2002. Molecular support for Afrotheria and the polyphyly of Lipotyphla based on analyses of the growth hormone receptor gene. Molecular Phylogenetics and Evolution 24: 91–101.

McDowell, S. B. 1958. The Greater Antillean insectivores. Bulletin of the American Museum of Natural History 115: 113–214.

McKenna, M. C. 1969. The origin and early differentiation of therian mammals. Annals of the New York Academy of Sciences 167: 217–240.

———. 1975. Toward a phylogenetic classification of the Mammalia; pp. 21–46 *in* W. P. Luckett and F. S. Szalay (eds.), Phylogeny of the Primates: A Multidisciplinary Approach. Plenum, New York.

———. 1987. Molecular and morphological analysis of high-level mammalian interrelationships; pp. 55–93 *in* C. Patterson (ed.), Molecules and Morphology in Evolution: Conflict or Compromise? Cambridge University Press, Cambridge.

McKenna, M. C., and S. K. Bell. 1997. Classification of Mammals above the Species Level. Columbia University Press, New York.

Meng, J., and A. R. Wyss. 2001. The morphology of *Tribosphenomys* (Rodentiaformes, Mammalia): Phylogenetic implications for basal Glires. Journal of Mammalian Evolution 8: 1–71.

Milledge, S.A.H. 2003. Fossil aardvarks from the Lothagam Beds; pp. 363–368 *in* M. G. Leakey and J. M. Harris (eds.), Lothagam: The Dawn of Humanity in Eastern Africa. Columbia University Press, New York.

Miyamoto, M. M., and M. Goodman. 1986. Biomolecular systematics of eutherian mammals: Phylogenetic patterns and classification. Systematic Zoology 35: 230–240.

Murata, Y., M. Nikaido, T. Sasaki, Y. Cao, Y. Fukumoto, M. Hasegawa, and N. Okada. 2003. Afrotherian phylogeny as inferred from complete mitochondrial genomes. Molecular Phylogenetics and Evolution 28: 253–260.

Murphy, W., E. Eizirik, W. E. Johnson, Y. P. Zhang, O. A. Ryder, and S. J. O'Brien. 2001a. Molecular phylogenetics and the origin of placental mammals. Nature 409: 614–618.

Murphy, W., E. Eizirik, S. J. O'Brien, O. Madsen, M. Scally, C. J. Douady, E. Teeling, O. A. Ryder, M. J. Stanhope, W. W. de Jong, and M. S. Springer. 2001b. Resolution of the early placental mammal radiation using Bayesian phylogenetics. Science 294: 2348–2351.

Novacek, M. J. 1980. Cranioskeletal features in tupaiids and selected eutheria as phylogenetic evidence; pp. 35–94 in W. P. Luckett (ed.), Advances in Primatology, Comparative Biology and Evolutionary Relationships of Tree Shrews. Plenum, New York.

———. 1982. Information for molecular studies from anatomical and fossil evidence on higher eutherian phylogeny; pp. 3–41 in M. Goodman (ed.), Macromolecular Sequences in Sytematic and Evolutionary Biology. Plenum, New York.

———. 1984. Evolutionary stasis in the elephant-shrew, Rhynchocyon; pp. 4–22 in N. Eldredge and S. M. Stanley (eds.), Living Fossils. Springer-Verlag, New York.

———. 1989. Higher mammal phylogeny: The morphological-molecular synthesis; pp. 421–435 in B. Fernholm and H. Jornvall (eds.), The Hierarchy of Life. Elsevier Science, New York.

———. 1990. Morphology, paleontology, and the higher clades of mammals; pp. 507–543 in H. H. Genoways (ed.), Current Mammalogy. Volume 2. Plenum, New York.

———. 1992. Fossils, topologies, missing data, and the higher level phylogeny of eutherian mammals. Systematic Biology 41: 58–73.

Novacek, M. J., and A. R. Wyss. 1986. Higher-level relationships of the recent eutherian orders: Morphological evidence. Cladistics 2: 257–287.

Novacek, M. J., A. R. Wyss, and M. C. McKenna. 1988. The major groups of eutherian mammals; pp. 31–71 in M. J. Benton (ed.), The Phylogeny and Classification of the Tetrapods. Volume 2. Mammals. Systematics Association Special Volume 35B. Clarendon Press, New York.

Patterson, B. 1965. The fossil elephant shrews (Family Macroscelididae). Bulletin of the Museum of Comparative Zoology at Harvard University 133: 297–336.

———. 1975. The fossil aardvarks (Mammalia: Tubulidentata). Bulletin of the Museum of Comparative Zoology at Harvard University 147: 185–237.

Peters, W.C.H. 1846. Bericht über die zur Bekanntmachung geeigneten Verhandlungen der Königl. Preuss. Akademie der Wissenschaften zu Berlin: 257–259.

———. 1847. Bericht über die zur Bekanntmachung geeigneten Verhandlungen der Königl. Preuss. Akademie der Wissenschaften zu Berlin: 36–37.

Pickford, M. 1975. New fossil Orycteropodidae (Mammalia, Tubulidentata) from East Africa. Orycteropus minutus sp. nov and Orycteropus chemeldoi sp. nov. Netherlands Journal of Zoology 25: 57–88.

———. 1978. New evidence concerning the fossil aardvarks (Mammalia, Tubulidentata) of Pakistan. Tertiary Research 2: 39–44.

———. 1994. Tubulidentata of the Albertine rift valley, Uganda; in B. Senut and M. Pickford (eds.), Geology and palaeobiology of the Albertine rift valley, Uganda-Zaire Volume II, Palaeobiology. Publication Occasionnelle—Centre international pour la formation et les échanges géologiques 29: 261–262.

———. 1996. Tubulidentata (Mammalia) from the middle and upper Miocene of southern Namibia. Comptes Rendus de l'Académie des Sciences, Série II, Sciences de la Terre et des Planètes 322: 805–810.

Porter, C., M. Goodman, and M. J. Stanhope. 1996. Evidence on mammalian phylogeny from sequences of Exon 28 of the von Willebrand Factor gene. Molecular Phylogenetics and Evolution 5: 89–101.

Rook, L., and F. Masini. 1994. Orycteropus cf. gaudryi (Mammalia, Tubulidentata) from the late Messinian of the Monticino Quarry (Faenza, Italy). Bollettino della Società Paleontologica Italiana 33: 369–374.

Schlosser, M. 1910. Über einige fossile Säugetiere aus dem Oligocan von Ägypten. Zoologischer Anzeiger 35: 500–508.

———. 1911. Beiträge zur Kenntnis der oligozanen Landsäugetiere aus dem Fayum, Ägypten. Beiträge zur Palaeontologie und Geologie Österreich-Ungarns und des Orients 24: 51–167.

Sen, S. 1994. Les gisements de mammifères du Miocène supérieur de Kemiklitepe, Turquie; 5, Rongeurs, Tubulidentes et Chalicotheres. Bulletin du Muséum National d'Histoire Naturelle, Section C, Sciences de la Terre, Paléontologie, Géologie, Minéralogie 16: 97–111.

Shoshani, J., and M. C. McKenna. 1998. Higher taxonomic relationships among extant mammals based on morphology, with selected camparisons of results from molecular data. Molecular Phylogenetics and Evolution 9: 572–584.

Simons, E. L., P. A. Holroyd, and T. M. Bown. 1991. Early Tertiary elephant-shrews from Egypt and the origin of the Macroscelidea. Proceedings of the National Academy of Sciences USA 88: 9734–9737.

Simpson, G. G. 1931. A new insectivore from the Oligocene, Ulan Gochu horizon, of Mongolia. American Museum Novitates 505: 1–22.

———. 1945. The principles of classification and a classification of mammals. Bulletin of the American Museum of Natural History 85: i–xvi, 1–350.

Smith, A. 1829. Contributions to the natural history of South Africa. Zoological Journal of London 4: 433–444.

Smith, G. E. 1898. The brain in the Edentata. Transactions of the Linnean Society of London 7: 277–394.

Sonntag, C. F. 1926. A monograph of Orycteropus afer. III. The skeleton of the trunk and limbs. General Summary. Proceedings of the Zoological Society of London 1926: 454–485.

Springer, M. S., A. Burk, J. R. Kavanagh, V. G. Waddell, and M. J. Stanhope. 1997. The interphotoreceptor retinoid binding gene in therian mammals: Implications for higher-level relationships and evidence for loss of function in the marsupial mole. Proceedings of the National Academy of Sciences USA 94: 13754–13759.

Springer, M. S., W. J. Murphy, E. Eizirik, and S. J. O'Brien. 2003. Placental mammal diversification and the Cretaceous-Tertiary boundary. Proceedings of the National Academy of Sciences USA 100: 1056–1061.

Stanhope, M., O. Madsen, V. G. Waddell, G. C. Cleven, W. W. de Jong, and M. S. Springer. 1998a. Highly congruent molecular support for a diverse superordinal clade of endemic African mammals. Molecular Phylogenetics and Evolution 9: 501–508.

Stanhope, M., V. G. Waddell, O. Madsen, W. W. de Jong, S. B. Hedges, G. C. Cleven, D. Kao, and M. S. Springer. 1998b. Molecular evidence for multiple origins of Insectivora and for a new order of endemic African insectivore mammals.

Proceedings of the National Academy of Sciences USA 95: 9967–9972.

Stromer, E. 1922. Erste Mitteilung über tertiäre Wirbeltier-Reste aus Deutsch-Südwestafrika. Sitzungsberichte Mathematik und Physik Kl Bayern Akademische Wissenschaft 1921: 331–340.

———. 1932. *Paleothentoides africanus* nov. gen., nov. spec., ein erstes Beuteltier aus Afrika. Sitzungsberichte der Mathematisch-Naturwissenschaftlichen Klasse der Bayerischen Akademie der Wissenschaften 1931: 177–190.

Szalay, F. S. 1977. Phylogenetic relationships and a classification of the eutherian Mammalia; pp. 315–374 *in* M. K. Hecht, P. C. Goody, and B. M. Hecht (eds.), Major Patterns in Vertebrate Evolution. Plenum, New York.

Tabuce, R., B. Coiffait, P.-E. Coiffait, M. Mahboubi, and J.-J. Jaeger. 2001. A new genus of Macroscelidea (Mammalia) from the Eocene of Algeria: A possible origin for elephant-shrews. Journal of Vertebrate Paleontology 21: 535–546.

Thewissen, J.G.M. 1985. Cephalic evidence for the affinities of the Tubulidentata. Mammalia 49: 257–284.

Thomas, O., and H. Schwann. 1906. The Rudd exploration of South Africa. V. List of mammals obtained by Mr. Grant in North East Transvaal. Proceedings of the Zoological Society of London 1906: 575–591.

van Dijk, M.A.M., O. Madsen, F. Catzeflis, M. J. Stanhope, W. W. de Jong, and M. Pagel. 2001. Protein sequence signatures support the African clade of mammals. Proceedings of the National Academy of Sciences USA 98: 188–193.

Waddell, P. J., and S. Shelley. 2003. Evaluating placental inter-ordinal phylogenies with novel sequences including RAG1, g-fibrinogen, ND6, and mt-tRNA, plus MCMC-driven nucleotide, amino acid, and codon models. Molecular Phylogenetics and Evolution 28: 197–224.

Wible, J. R. 1987. The eutherian stapedial artery: Character analysis and implications for superordinal relationships. Zoological Journal of the Linnean Society 91: 107–135.

EMMANUEL GHEERBRANT,
DARYL P. DOMNING,
AND PASCAL TASSY

7

PAENUNGULATA (SIRENIA, PROBOSCIDEA, HYRACOIDEA, AND RELATIVES)

CONCERNING THE PROBOSCIDEA, SIMPSON (1945: 244) wrote that "classification within the order is obviously a task for paleozoology rather than for neozoology." This is also true for higher relationships of the order, especially for the hypothesis of Paenungulata, a taxon Simpson himself named in 1945.

Simpson's Paenungulata ("near-ungulates") included the extant orders Proboscidea, Sirenia (including Desmostyliformes), and Hyracoidea, and such fossil taxa as Embrithopoda, Pantodonta, Pyrotheria, and Dinocerata. It was erected in place of the "Subungulata" *sensu* Schlosser, 1923—already used for a rodent taxon—with basically the same content. Recent studies show that this grouping is polyphyletic. Such primitive orders as Pantodonta, Pyrotheria, and Dinocerata were excluded from Paenungulata by McKenna and Manning (1977) and Novacek (1986). Besides the three extant orders Proboscidea, Sirenia, and Hyracoidea, the fossil taxa currently related to Paenungulata are Embrithopoda, "Anthracobunidae," some of "Phenacolophidae," and Desmostylia (Table 7.1). McKenna and Bell (1997: 490) erected the order Uranotheria with basically the same content: "for the most recent common ancestor of Hyracoidea, Embrithopoda, and Tethytheria, and all its descendants." Even in this modern concept, Paenungulata today is a controversial clade, and its content will be subject to changes, especially for fossil taxa. Paenungulata is supported by molecular systematics (based on extant Sirenia, Proboscidea, and Hyracoidea), whereas morphological studies are contradictory, some of them supporting alternatively a Perissodactyla-Hyracoidea clade. Most of the debate results from gaps in the fossil record, but some of it has other causes, such as molecular models.

Another higher grouping of ungulates related to the question of the origin and phylogeny of Paenungulata is Altungulata Prothero and Schoch, 1989, or Pantomesaxonia Franz, 1924, *sensu* Fischer (1986). Pantomesaxonia was named by Franz (1924) for a heterogeneous grouping of phenacodonts, meniscotheres, South American ungulates, hyracoids, proboscideans, and perissodactyls, but excluding sirenians. Fischer (1986) restricted Pantomesaxonia to the extant perissodactyls, hyracoids, proboscideans, and *added* sirenians, which corresponds to modern lophodont ungulates. Such a grouping of lophodont ungulates was initially advocated by Gregory (1910). It has been revived by McKenna

and Manning (1977: fig. 1, node 3), followed by Fischer (1986), Prothero et al. (1988), Tassy and Shoshani (1988), Court (1992b), and Thewissen and Domning (1992). More recently, Prothero and Schoch (1989) named the grandorder Altungulata with basically the same systematic content (also McKenna and Bell, 1997). Because Fischer's Pantomesaxonia is not synonymous with Franz's Pantomesaxonia—even in regard to the extant taxa—we use here instead the taxon Altungulata in the same sense as McKenna and Bell (1997). The classification of Altungulata used in this chapter, which includes modern and related fossil lophodont taxa, is given in Table 7.1.

Table 7.1 Synthetic phylogenetic classification of paenungulates and related major taxa used

Taxeopoda
  †Phenacodonta
  ?*Ocepeia*, early Eocene, Morocco
  †Family Phenacodontidae, early-late Paleocene, early-middle Eocene North America, early-middle Eocene, Europe
  Altungulata (=Pantomesaxonia)
    †Family ?, *Radinskya*, late Paleocene, Asia (China)
    Order Perissodactyla
    Paenungulata (=Uranotheria)
      Order Hyracoidea
†Family ?, *Seggeurius, Microhyrax,* early-middle Eocene, Morocco and Algeria
†Family "Pliohyracidae"
    Subfamily Geniohyinae, early Oligocene, Africa
    Subfamily Saghatheriinae, Eocene-Oligocene, Africa
    Subfamily Pliohyracinae, Miocene-Pliocene, Africa, Europe, Asia
Family Procaviidae, Miocene-Recent, Africa, Asia, Europe
      Tethytheria
        †Anthracobunia
Family ?, *Minchenella*, late Paleocene, Asia (China)
Family "Anthracobunidae," early-middle Eocene, Indo-Pakistan
        †Order Desmostylia
Family Paleoparadoxiidae, late Oligocene, Miocene, North Pacific
Family Desmostylidae, late Oligocene, Miocene, North Pacific
        †Order Embrithopoda
?Family "Phenacolophidae," *Phenacolophus;* late Paleocene, China, Mongolia
Family Palaeoamasiidae, middle Eocene, Anatolia; middle or late Eocene, Romania
Family Arsinoitheriidae, *Arsinoitherium,* late Eocene?, early Oligocene, Egypt, Libya, Angola, Oman
        Order Sirenia
†Family "Prorastomidae," middle-late Eocene, West Indies, North America
†Family "Protosirenidae," middle-late Eocene, western Atlantic Ocean, Mediterranean Sea, Indian Ocean
Family Trichechidae, late Oligocene-early Miocene, North Sea; middle Miocene-Recent, South America; ?Pliocene-Recent, western Atlantic Ocean-Caribbean-North America; Recent, West Africa
Family Dugongidae, middle Eocene-Pliocene, Mediterranean Sea, Europe, western Atlantic Ocean-Caribbean; Eocene-Recent, Indian-Pacific Oceans
        Order Proboscidea
†Family ?, *Phosphatherium, Daouitherium,* early Eocene, Morocco
†Family Numidotheriidae, early-middle Eocene Algeria, late Eocene, Egypt
†Family Barytheriidae, late Eocene, early Oligocene, North Africa
†Family Moeritheriidae, middle-late Eocene, early Oligocene, North Africa
†Family Deinotheriidae, Miocene-Pleistocene, Africa, Europe, Asia
†Family Palaeomastodontidae, early Oligocene, Africa
Elephantiformes, Miocene-Recent, Old World, North America, South America

Notes: Question mark indicates taxa of uncertain systematic and phyletic position. This classification differs from the original definition of Paenungulata by Simpson (1945). Classification of Paenungulata follows Novacek (1986) and Novacek and Wyss (1986), with the addition of related fossil taxa, and is broadly equivalent to the Uranotheria of McKenna and Bell (1997). Classification of Altungulata follows Prothero and Schoch (1989) and McKenna and Bell (1997); Altungulata is used here in place of the equivalent taxon Pantomesaxonia Franz 1924 *sensu* Fischer 1986. Classification of Paenungulata remains uncertain and controversial, especially for fossil taxa (†). Molecular studies restrict Paenungulata to endemic African orders of ungulates (see Table 7.2). Altungulata is also controversial. It corresponds mostly to lophodont ungulates and might not be monophyletic. For Perissodactyla, see Hooker (chapter 13, this volume). The position of *Phenacolophus* in Embrithopoda is uncertain: it is possibly a basal altungulate, as is *Radinskya. Minchenella* is a primitive bilophodont ungulate, considered by several authors as the ancestor of "Anthracobunidae." It was also referred to the paraphyletic family "Phenacolophidae." Anthracobunia was named as an infraorder within the Order Uranotheria and Suborder Tethytheria by Ginsburg et al. (1999). *Phosphatherium* is included in its own family Phosphatheriidae by Gheerbrant et al. (in press).

Table 7.2 Classification of Afrotheria

Afrotheria
  Afrosoricida
    Family Chrysochloridae
    Family Tenrecidae
  Africana
    Order Macroscelidea
    Order Tubulidentata
    Paenungulata
      Order Hyracoidea
      Order Sirenia
      Order Proboscidea

Note: Classification is according to Stanhope et al. (1996), Madsen et al. (1997), Springer et al. (1997a,b), and Stanhope et al. (1998).

Other high taxa relevant to the discussion of Paenungulata are Taxeopoda Cope, 1882, and Afrotheria Stanhope et al., 1998. Taxeopoda is used here in the sense of Archibald (1998): it includes phenacodontid condylarths and Altungulata as sister groups. Afrotheria is a controversial molecular supergroup that gathers together all endemic major lineages of African placentals, including Paenungulata and zalambdodont insectivores (Table 7.2). On the whole, the classification and relationships of Paenungulata are among the most vigorously debated and central questions in current phylogenetic studies of placental mammals.

EXTANT PAENUNGULATES

Order Sirenia

MAIN FEATURES AND SPECIALIZATIONS. Although formerly more diverse, sirenians comprise only three Recent genera in two families: *Trichechus,* with three species (Trichechidae), and the monospecific genera *Dugong* and the recently exterminated *Hydrodamalis* (Dugongidae). Sirenians are totally aquatic mammals, sharing the outward body form of derived cetaceans: short neck, flipperlike forelimbs, no hindlimbs, horizontal caudal fin providing propulsion by dorsoventral spinal undulation, nearly hairless skin, retracted nasal openings, and no external ear pinnae. The body is stiffened by a subdermal connective-tissue sheath of helically wound fibers. In contrast to cetaceans, a dorsal fin is never present, and the two nostrils are always separate. Body lengths ranged from about 2 m up to 9–10 m.

The thorax and lungs of sirenians are elongate, and the diaphragm is nearly horizontal. Skeletal pachyosteosclerosis (increased volume plus increased density of bones) is characteristic and extensive, providing ballast for shallow diving. A clavicle is absent; five digits are present in the forelimb. Seven cervical vertebrae are present, except in Recent *Trichechus,* which has only six.

Sirenians are herbivores, preferring a diet of marine angiosperms (seagrasses); in fresh water, a wide variety of plants is consumed. The dental formula of the most primi-

tive sirenians (3.1.5.3) may be a synapomorphy of the order. The large number of premolars (together with a primitively long, narrow rostrum and mandibular symphysis bearing parallel rows of incisors and canines) may have been an early mechanism for intraoral food transport. This mechanism, however, was quickly superseded by the evolution of broader rostral and symphyseal surfaces covered by horny pads; thereafter, the anterior dentition and permanent premolars were lost—except for I1, which is retained in some forms and enlarged for digging and/or social interactions. After the Eocene, the cheek teeth generally were reduced in number or totally lost (*Hydrodamalis*); but in derived trichechids, an unlimited number of horizontally replaced supernumerary molars has evolved. The molars are low crowned, primitively bunodont, and more or less bilophodont.

The 3.1.5.3 dental formula of Eocene sirenians raises questions about dental homologies with other mammals. The deciduous lower tooth in the fourth postcanine locus is trilobate, like the dp4 in Desmostylia and Artiodactyla, which might suggest their homology. The five premolar loci agree with McKenna's (1975) hypothesis that his superorder Tokotheria (sirenians, proboscideans, desmostylians, phenacodonts, and artiodactyls) retains dP5/dp5 and has lost the last molar except in sirenians. In most Eocene sirenians, the molariform dP5/dp5 is not replaced and is usually retained into adulthood as a fourth "molar." Cladistic analysis (e.g., Thewissen and Domning, 1992) suggests that the five premolars are a derived feature of Sirenia. However, if Sirenia is a more basal placental order, as the Afrotheria hypothesis suggests, the five-premolar condition could be a primitive retention.

PALEONTOLOGICAL DATA. Domning (2001a) briefly reviewed the fossil record of sirenians, and Domning (1996) provided an exhaustive bibliography, synonymy, and classification. Sirenian fossils are widespread and moderately abundant in many shallow marine deposits of formerly tropical and subtropical regions of the world. However, many of these fossils are rib fragments or other non-diagnostic elements; taxonomically useful specimens (skulls, mandibles) are much less common. Nonetheless, the broad outlines of sirenian diversity are apparent. The sirenian record extends from the middle Eocene to Recent times, and fossils are found on all continents except Antarctica.

Our knowledge of the earliest sirenians—all included in "Prorastomidae"—comes from the early middle Eocene of Jamaica, which has yielded two published taxa and a third, more derived, one that is yet to be described. *Prorastomus sirenoides* is both the earliest and most primitive of all, but is known only from a single skull with associated mandible and atlas vertebra (Savage et al., 1994). *Pezosiren portelli,* slightly younger and in some ways more derived, is known from abundant postcranial as well as cranial remains from a single quarry, which have made possible the reconstruction of a nearly complete composite skeleton, illustrated in Fig. 7.1 (Domning, 2001b).

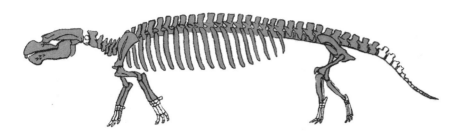

Fig. 7.1. Composite reconstruction of the skeleton of *Pezosiren portelli* from the early middle Eocene of Jamaica (Domning 2001b: fig. 1). This is the earliest and most primitive known sirenian except for *Prorastomus*. Length = 2.1 m.

Domning (1994) presented a preliminary cladistic analysis of the Sirenia, incorporating 36 of the best-known species and subspecies. Four families are traditionally recognized (Table 7.1). "Prorastomidae," Protosirenidae, and Dugongidae were already present in the middle Eocene, demonstrating a rapid early diversification. Domning's (1994) analysis indicated that manatees (Trichechidae) were derived from late Eocene or early Oligocene dugongids, but Sagne (2001) has adduced cladistic evidence to support the earlier supposition that trichechids arose instead from protosirenids.

BASAL PHYLOGENY, ORIGIN OF THE ORDER, AND ANCESTRAL MORPHOTYPE.

Preliminary cladistic analysis of cranial and dental characters indicates the relationship (*Prorastomus* (*Pezosiren* (Protosirenidae + Dugongidae + Trichechidae))) (Fig. 7.2). At least one major postcranial character, however, is incongruent with this tree: the sternum is derived (broad and platelike) in *Pezosiren*, dugongids, and trichechids, but still primitive (with numerous columnar sternebrae) in *Protosiren smithae* Domning and Gingerich, 1994. Possibly this derived state evolved independently in *Pezosiren*.

The basal sirenian morphotype, displayed by *Pezosiren*, comprises a long trunk (20 thoracic and 4 lumbar vertebrae) supported on relatively short legs, giving a somewhat dachshundlike appearance similar to that reconstructed for *Moeritherium* by Simons (1964). Four sacral vertebrae are present, with a firm sacroiliac articulation capable of supporting the body's weight on dry land. However, in *Pezosiren*, these sacrals are no longer ankylosed in most cases, pointing to an incipient increase in flexibility of the sacral region for swimming by spinal undulation—a convergence with early whales that were comparable in their stage of

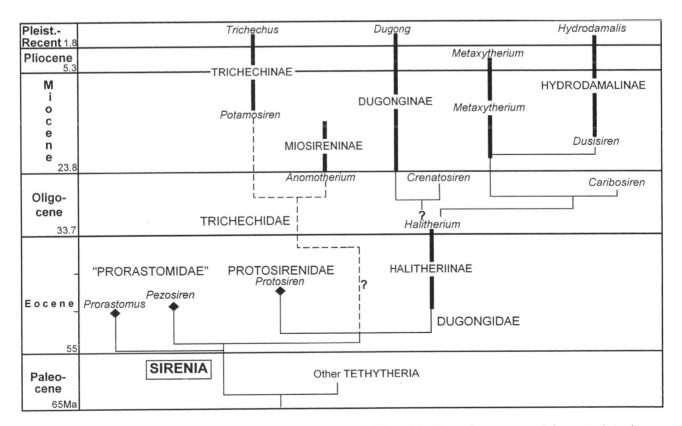

Fig. 7.2. Phylogeny of the Sirenia. Known stratigraphic distribution is indicated by bold lines; dashed lines indicate uncertain phylogenetic relationships.

evolution to *Rodhocetus*. The tail was long and strong compared to that of most terrestrial ungulates, but the caudal vertebrae still lacked the enlarged transverse processes for the attachment of powerful tail muscles seen in later sirenians and cetaceans. Instead, the transverse processes were very small, and pierced dorsoventrally by notches or foramina, as seen in several other aquatic and semiaquatic mammals. This tail was probably used for otterlike swimming, in which dorsoventral undulation of the spine transmitted propulsive force to the tail and (more importantly) to the hind feet. Few foot elements are preserved, but these suggest a relatively unmodified ungulate condition, probably with a digitigrade stance.

The seven cervical vertebrae are comparable in form and proportions to those of early condylarths, such as *Ectoconus,* and are not so compressed anteroposteriorly as in later sirenians. The anterior thoracic vertebrae have relatively long neural spines, suggesting support of the head and neck by a nuchal ligament. The ribs are massive and pachyosteosclerotic, as is typical for sirenians; this trait, together with the enlarged and retracted nasal opening and lack of paranasal air sinuses, indicates that middle Eocene prorastomids were already committed to the water and spent most of their time there, probably feeding as well as resting. This would make them more aquatic than modern hippopotami.

From these data, we can infer that the ancestral morphotype from which sirenians were derived was a herbivorous terrestrial ungulate, presumably digitigrade, with a long, well-muscled tail and possibly a long thorax, made up of about 20 pairs of ribs. Better knowledge of the forms presently grouped as "anthracobunids" may shed light on the source of such basal sirenian traits as premaxilla-frontal contact; enlarged premaxillary rostrum with slightly enlarged I1; long, deep, and narrow mandibular symphysis bearing parasagittally arranged incisor-canine toothrows; and the presence of five premolars.

PALEOBIOGEOGRAPHY.   Older literature, influenced by the discoveries a century ago of abundant sirenians and other paenungulates in the Fayum (as well as of Eocene sirenians in southern Europe), often identified the Mediterranean in general and North Africa in particular as the center of origin of these mammals. With new fossil finds of Eocene tethytheres in southern Asia, this view has been refined to include the latter continent as a candidate; thus we now place the origin of Sirenia in the Old World portion of the former Tethys Seaway. The best evidence for such a placement is that their near relatives (see Fig. 7.9) are all native to one or more parts of that broad region. Surprisingly, however, the earliest and most primitive sirenians, Prorastomidae, are so far known only from Jamaica and, perhaps, Florida (Savage et al., 1994)—although there is another possible specimen from Israel (Goodwin et al., 1998). By the beginning of the middle Eocene, sirenians had already dispersed to the Caribbean from the Eurasian-African shores of the Tethys. However, very few of their fossil remains have yet been found along the way.

Order Proboscidea

MAIN FEATURES AND SPECIALIZATIONS.   Proboscidea is one of the least diverse extant placental orders. Only two (or possibly three) surviving species of elephants are known in Africa and Asia. However, this group has a long history and was formerly very diversified and widely distributed, from the Old World to North and South America.

Proboscideans are the largest extant terrestrial mammals. The body masses of the two extant elephants range from 1 to 6 tons. Proboscideans are specialized herbivores, feeding on coarse plant material with the help of distinctive multilophed molars.

Aside from their gigantic size, modern proboscideans are characterized by a retracted nasal opening above the orbits and the correlated presence of a trunk (proboscis), which gave the order its name. This trunk results from the fusion of the muscles of the upper lip and the nostrils. The tusks of elephants are hypertrophied hypsodont second incisors and are the largest mammalian teeth known. Other important features are the absence of canines, horizontal serial tooth displacement with deciduous and permanent teeth lost anteriorly and erupting posteriorly until M3/m3 eruption, and graviportal stance (thick limb bones and pillarlike limbs) to sustain their huge body mass. Many of these features are, however, found only in the most derived members of the order, mostly in Neogene taxa. The sequence of their acquisition during evolution is documented by Shoshani and Tassy (1996).

PALEONTOLOGICAL DATA.   The proboscidean fossil record extends from the early Eocene to Recent times (Fig. 7.3). The Paleogene record is scanty and restricted to Africa, except for a striking discovery reported recently in the middle Eocene of Anatolia (Maas et al., 1998). The known diversity is modest, with 10 species, 7 genera, and 5 families described from the Eocene to the early Oligocene. Most data come from the late Eocene and early Oligocene of the Fayum (Egypt) and Dor El Talha (Libya) localities, which are by far the richest known from Africa. These localities have yielded *Barytherium* (Andrews, 1906; Savage, 1969), *Moeritherium* (Tassy, 1981), and the first elephantiforms *Phiomia* and *Palaeomastodon*. Another important locality is El Kohol, in the early/middle Eocene of Algeria, which has yielded rich material for *Numidotherium*. As is true for the whole African mammal fauna, the fossil record of proboscideans is especially poor in the middle Eocene (e.g., Bartonian) and between the early Oligocene and early Miocene, which corresponds to a gap of 10 million years. However, Sanders and Kappelman (2002) and Sanders et al. (2004) have recently reported the first remains of proboscideans from the late Oligocene of Africa. By contrast, the Miocene and subsequent fossil record is the best documented, with about 155 species described in the Old World and North and South America.

Minchenella and the "anthracobunids," which are related to Proboscidea by several authors, are in several respects much more primitive than the earliest African representatives of the order and may represent an offshoot of basal

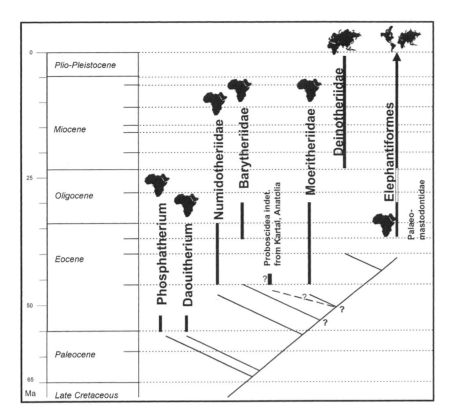

Fig. 7.3. Early proboscidean (*sensu stricto*) taxa. Stratigraphic and geographic distributions, and cladogram summarizing recent phylogenetic hypotheses on primitive proboscideans. The age of the divergence of taxa (position of nodes) is not given. The position of *Moeritherium* is according to Court (1995); position of *Phosphatherium* and *Daouitherium* according to Gheerbrant et al. (1998, 2002); that of the other taxa according to Shoshani et al. (1996). Dashed line indicates uncertain phylogenetic relationships; open line in Elephantiformes denotes gap in the fossil record.

tethytheres. They are excluded here from the true proboscideans (Proboscidea *sensu stricto*) and discussed separately (see the section on Anthracobunia).

The earliest proboscideans, *Phosphatherium* and *Daouitherium,* have been discovered in the Ypresian phosphates of the Ouled Abdoun Basin (Morocco; Gheerbrant et al., 1996, 1998, 2002). They are still poorly known, although new material of *Phosphatherium,* including the skull (Fig. 7.4), is under study (Gheerbrant et al., in press). *Phosphatherium* is very primitive with respect to *Moeritherium* and *Numidotherium.* It is small—the size of a fox—and plesiomorphic in its skull structure and anterior dentition. However, it displays unquestionable proboscidean synapomorphies, including true lophodonty and an orbit opening in the maxilla (Fig. 7.4). *Daouitherium,* known by the lower dentition, shares the striking dental pattern of *Phosphatherium* and is only slightly smaller than *Numidotherium.*

Numidotherium is known by two species from the middle Eocene (El Kohol) and the late Eocene (Dor El Talha). Most of its skeleton is described from El Kohol (Mahboubi et al., 1984, 1986; Court, 1994a, 1995). It was the size of a tapir. Its short, long-legged body resembles that of more advanced proboscideans, including its graviportal adaptations. Its skull is higher, has an elevated (retracted) nasal fossa, and is pneumatized. I2/i2 is enlarged, P1/p1 is lost, and a long diastema separates P2/p2 from the anterior teeth. The molars are lophodont. Striking primitive features are the presence of the os centrale and an entepicondylar foramen in the humerus, and its astragalar and otic construction (Court and Jaeger, 1991; Court, 1992b, 1994a,b).

Moeritherium, known by three or four species from the middle Eocene to the early Oligocene, was for decades considered the oldest and most primitive proboscidean. It is primitive in the retention of a canine, the anterior position of the external nares, the absence of contact between premaxilla and frontal, and its medium (piglike) size. However, *Moeritherium* is a peculiar proboscidean, specialized for a semiaquatic life. Court (1995) demonstrated that it is more advanced than *Numidotherium* in several respects, especially in the dentition, and that it may represent an early autapomorphic offshoot of the lineage of the modern graviportal proboscideans, rather than being morphotypic for the order. This is supported by recent discoveries of earlier proboscideans.

The diversified Eocene and early Oligocene true lophodont proboscideans exemplify an early radiation of the order. This radiation includes *Phosphatherium, Daouitherium, Numidotherium,* and *Barytherium,* which are successive stem fossil groups of Proboscidea (Fig. 7.3). It is uncertain if deinotheres are a surviving Neogene lineage representative of this early radiation. This conservative group was thought to be related to *Barytherium* on the basis of molar construction (Harris, 1978) and to *Barytherium* and *Numidotherium* on the basis of enamel microstructure (Bertrand, 1988). It is also associated by Shoshani et al. (2001) with other lophodont taxa in the suborder Plesielephantiformes. However, deinotheres were hypothesized to be the sister group of the Elephantiformes (Tassy, 1981; Tassy and Shoshani, 1988; Court, 1995; Shoshani et al., 1996), on the basis of synapomorphies such as the trilophodont M1/m1 and postcranial

Fig. 7.4. *Phosphatherium escuilliei*, the earliest and most primitive known proboscidean from the Ypresian of the Ouled Abdoun Basin, Morocco. (A) Skull (left lateral view; Gheerbrant et al. (in press). (B) Holotype, left maxilla with M2, M1, P4, P3 in occlusal view (Gheerbrant et al., 1998: fig. 2a). The molar morphology of *Phosphatherium* is identical to that of *Numidotherium* and *Barytherium*, and it supports a true lophodont ancestral morphotype of Proboscidea. *Phosphatherium* retains a vestigial mesostyle, which is probably a generalized feature of Paenungulata or Altungulata. The skull of *Phosphatherium* shows the opening of the orbit in the maxilla, which is an important synapomorphy of the Proboscidea. Abbreviation: i.o.f., infraorbital foramen. Scale bar = 5 mm.

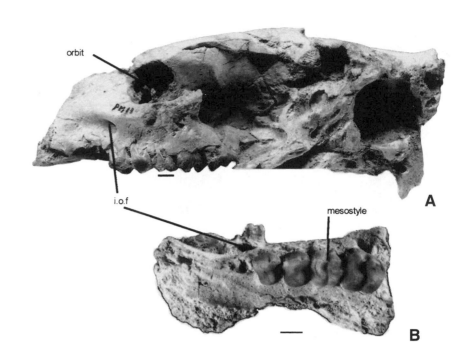

features. The postcranial skeleton of deinotheres is closer to that of elephantiforms than that of *Numidotherium*, *Barytherium*, and *Moeritherium*, although some postcranial similarities might be the result of parallelisms (Tassy, 1994: 75, 77).

The second and third major proboscidean radiations took their origin in the primitive bunolophodont taxa (e.g., *Moeritherium*) or at least in such primitive elephantiforms as *Phiomia* and *Palaeomastodon*. Elephantiformes Tassy 1988 are elephant-like proboscideans: there is a significant morphological gap between these first elephantiforms and other primitive (lophodont) proboscideans (Tassy, 1994). *Phiomia* and *Palaeomastodon* are large animals, decidedly more elephantine in general aspect (e.g., columnar limbs) than are other primitive proboscideans, according to the disarticulated known postcranial material, and they are at the origin of the modern lineages leading to the extant elephants.

BASAL PHYLOGENY, ORIGIN OF THE ORDER, AND ANCESTRAL MORPHOTYPE.

The monophyly of the Proboscidea is well established. Court (1995) demonstrated that *Numidotherium* is somewhat more representative of the ancestral condition of the order than is *Moeritherium*. The proboscidean synapomorphies that have been identified in *Moeritherium* are listed in Table 7.3. Most are also found in the oldest known proboscideans, *Phosphatherium* and *Daouitherium*, although several are unknown, absent, or in a more or less incipient stage (synapomorphy numbers 2?, 3, 5, 10, 16–21 in Table 7.3). These genera are characterized by primitive features previously unknown in the order, such as the condylarth-like general cranial morphology (reminiscent of arctocyonids and phenacodontids) and the anterior dentition, which retains C1/c1 and P1 and lacks a diastema. *Phosphatherium* and *Daouitherium* will help to refine the primitive ancestral morphotype of the order and its evolution. Several of the proboscidean features of

Moeritherium listed in Table 7.3 are also known in Embrithopoda, which is related to Proboscidea by Court (1992b).

The basal phylogeny of the proboscideans, illustrated in Fig. 7.2, shows a succession of lophodont stem lineages represented by *Phosphatherium*, *Daouitherium*, *Numidotherium*, and *Barytherium*, branching off before deinotheres, *Moeritherium*, and elephantiforms (Tassy, 1994; Gheerbrant et al., 2002). The phylogenetic position of *Moeritherium* and deinotheres is still debatable. *Phosphatherium* and *Daouitherium*, although of early Eocene age, are already quite specialized in some features (e.g., true lophodonty, large size of *Daouitherium*) known in other proboscideans. Moreover, they demonstrate an unexpected diversity of the earliest known proboscideans (Gheerbrant et al., 2002). This is very suggestive of at least a Paleocene origin of the order Proboscidea.

PALEOBIOGEOGRAPHY. Present paleontological data argue for a Paleocene African origin of the order Proboscidea *sensu stricto*. The alternative hypothesis of a north-Tethyan Asiatic origin is based on the supposed relationship with "anthracobunids" and *Minchenella*, which needs to be further supported and clarified. The early subsequent evolution of the order Proboscidea was endemic in Africa during most of the Paleogene. Africa was the theater of a first radiation of primitive lophodont lineages. This was possibly favored by the peculiar paleoecological conditions of Africa in the early Tertiary, and especially by the paleogeographic isolation, which prevented competition with successful Laurasiatic lophodont groups, such as perissodactyls. The African early endemic radiation also saw the emergence of primitive elephantiforms (*Palaeomastodon* and *Phiomia*) in the early Oligocene, and possibly of *Moeritherium* by the middle-late Eocene.

Recent discoveries in the middle Eocene of Anatolia (Maas et al., 1998) suggest a striking exception to this African en-

Table 7.3 Skeletal synapomorphies of Proboscidea identified in early African taxa, such as *Moeritherium*

Number	Description	Remarks
1	Jugal reduced anteriorly and orbit opened in the maxilla[a]	
2	Processus ascendens of the palatine absent from the orbitotemporal fossa and long suture between maxilla and frontal[a]	
3	External auditory meatus more or less closed ventrally by the posttympanic and postglenoid processes of the squamosal	Absent in *Phosphatherium* and *Numidotherium* (tethytherian convergence)
4	Jugal extended posteriorly to the posterior part of the glenoid fossa of the squamosal[a]	Convergence with hyracoideans
5	Condyloid foramen (= hypoglossal f.) absent: confluent with the foramen lacerum posterius[a]	
6	Cranial part of the squamosal expanded dorsally[a]	
7	Zygomatic process of the maxilla laterally enlarged and broadened[a]	
8	Presence of a submaxillary fossa	
9	Optic foramen (canalis opticus) and fissura sphenorbitale aligned under the crista orbitotemporalis of the orbitotemporal fossa[a]	
10	Pneumatized cranial bones (e.g., parietal)[a]	Known in *Phosphatherium*, but absent in *Moeritherium*, possibly secondarily
11	Loss of p1[a]	P1, C1, c1 retained in *Phosphatherium*
12	Metacone present on P4	Altungulate feature
13	Molars with a distocrista[a]	
14	Advanced bilophodonty: true lophodonty or bunolophodonty[a]	
15	One enlarged lower and upper pair of incisor present	Possibly not homologous in lophodont and bunolophodont taxa
16	Astragalus with enlarged calcaneal facet and protruding tuberculum mediale[a]	
17	Astragalus with a semicircular head and short neck[a]	
18	Crista capitalis lost from the astragalar head[a]	
19	Astragalar foramen lost[a]	
20	Radius positioned or fixed in a pronated position[a]	
21	Femur anteroposteriorly flattened	Feature of graviportal mammals

[a]Probable feature of the ancestral morphotype of the order.

demic early evolution of proboscideans, even if the material remains scarce. Antoine et al. (2003) also described a partial tusk with elephantoid dentine structure in the late Oligocene of Baluchistan. Several later dispersals occurred between Africa and Asia, in correlation with the closure of the Tethys.

Order Hyracoidea

MAIN FEATURES AND SPECIALIZATIONS. As in the case of sirenians and proboscideans, the poor diversity of extant hyraxes is relictual. They include one family, the Procaviidae, three genera, *Dendrohyrax, Heterohyrax* and *Procavia,* and seven to 12 species. All living hyraxes are small— the size of a rabbit (body mass, 1–5 kg). Tree hyraxes (*Dendrohyrax*) are solitary and nocturnal, whereas the terrestrial bush (*Heterohyrax*) and rock (*Procavia*) hyraxes live in colonies and are diurnal. As a whole, extant hyraxes are good climbers.

Hyraxes are specialized herbivores. *Procavia* has hypsodont teeth specialized for grazing; it feeds mostly on grasses. Other genera browse on less abrasive vegetation with brachyodont teeth. Living hyraxes have a unique, complex digestive system unknown in other ungulates (Gregory, 1910). A striking feature of the living hyraxes is their long period of gestation (8 months). It may reflect the large size of their Paleogene ancestors. Fischer (1989: 45) related this feature to "a former plains-dwelling life" (vs. biotopes requiring climbing adaptations).

Hyracoids show a striking mixture of features, which explains difficulties in clarifying their relationships. Extant species retain noticeable primitive features: an alisphenoid canal, an independent optic foramen, interparietal, a small third trochanter on the femur, and a free os centrale in the carpus between the trapezoid and the magnum, as in *Numidotherium*. Additional plesiomorphies are known in early hyraxes ("Pliohyracidae"): a complete dental formula, a double-rooted canine, orbits not shifted anteriorly, a lambdoid crest, the frontal and maxilla in broad contact (polarity? see Thewissen and Simons, 2001: 103), and the structure of the alisphenoid canal (Thewissen and Simons, 2001; De Blieux and Simons, 2002).

Many other features are specialized in hyracoids (Table 7.4). The claw-like nail of the second digit of the foot of extant procaviids (feature 20 in Table 7.4) is used for grooming. The third digit is the largest, as in cursorial perissodactyls. Foot and hand have a striking structure called "taxeopody" (features 2, 3): the carpal and tarsal bones are aligned and articulate within one column without lateral contact (serial arrangement). In the central axis of the carpus, articulation is restricted to the lunate and capitate (magnum). This construction characterizes paenungulates, but possibly as the result of a convergence of Hyracoidea and Tethytheria. It is presumed to be derived with respect to the generalized eutherian condition, in which the structure is alternate, with carpal and tarsal bones arranged in overlapping rows, but this is debatable (Thewissen, 1990). A hyracoid peculiarity is that supination of the hand is possible at the midcarpal joint; that is, between the proximal and distal carpal rows (Fischer, 1986). The morphology of the ankle of hyraxes is also peculiarly specialized (feature 4). The articulation between the tibial malleolus and the astragalus (cotylar fossa) forms a close hinge joint, at least in Paleogene and Miocene hyracoids (Fischer, 1986). The tusklike I1 (feature 1) has a long hypertrophied root and its enamel is restricted to a labial band. The molars have a basically lopho-selenodont pattern (feature 14). Several specialized features of extant hyraxes (features 21–23) are not representative of such Paleogene taxa as *Megalohyrax* and *Antilohyrax*. For instance, the canines, lost in extant taxa (feature 21), are present in Paleogene taxa. Most of these specializations of extant hyraxes are linked to the shortening of the rostrum, the anterior shift of the orbits, and the enlargement of the braincase (Janis, 1983). Some of the derived features of hyracoids cited by Andrews (1906) and Gregory (1910) are more generalized in altungulates (e.g., features 8, 9 in Table 7.4; see also Table 7.8).

The gait of extant hyracoids is plantigrade. However, Fischer (1986) has shown that in hyraxes, plantigrady is secondary. Study of the development of the limbs shows that hyraxes are basically unguligrade. The sole of the foot has a peculiar elastic pad, which is an adaptation for climbing and jumping on irregular substrata, such as rocks, trunks, or branches. These pads are kept moist by sudorific glands that facilitate their adherence like suction cups.

PALEONTOLOGICAL DATA. Hyracoids are known from Eocene to Recent times. Their Paleogene fossil record is poor and restricted to Arabo-Africa (see Table 7.1). However, hyracoids are among the most abundant and diversified mammals in African Paleogene sites. The earliest localities are in the early Eocene of the Ouled Abdoun and Ouarzazate basins in Morocco (Gheerbrant et al., 2003). However, the early-middle Eocene of Tunisia and Algeria (Chambi, El Kohol, Gour Lazib, and Glib Zegdou) has yielded more significant material, and the best localities are in the late Eocene–early Oligocene of the Fayum (Egypt). The Fayum locality L-41 is the richest; 90% of the fossils recovered there belong to Hyracoidea, mostly to *Saghatherium* and *Thyrohyrax* (Rasmussen and Simons, 1991). As for other mammals in Africa, a major fossil gap occurs between 20 and 35 million years ago. In the Miocene, hyracoids declined in Africa relative to immigrant ungulates from Eurasia, such as artiodactyls and perissodactyls.

The earliest and most primitive known hyracoid is *Seggeurius amourensis* Crochet, 1986, from the early-middle Eocene of El Kohol (Algeria; Fig. 7.5). This species, known only by its dentition, is noticeably primitive in its bunodont

Table 7.4 Derived skeletal features of Hyracoidea

Number	Description
1	The I1 enlarged and tusklike[a]
2	Serial tarsus[a]
3	Serial carpus[a]
4	Close hinge joint in the ankle: strong medial tibial malleolus articulated with a deep medial fossa of the astragalus (cotylar fossa)[a]
5	Flat navicular articular surface of the astragalus[a]
6	Postorbital bar made from a process of the frontal and from an original contribution of the parietal[a]
7	Posterior extension of the jugal into the glenoid fossa (where it bears an articular facet)[a]
8	V-shaped scapula, which lacks the acromion
9	Loss of the clavicle
10	Malleus suspended by a peculiar process of the hypohyal bone[a]
11	Lingual process (processus lingualis) of the hyoid apparatus issued from hypohyal bone (not from the basihyal)[a]
12	First and fifth digit lost in the foot (three digits), third digit enlarged[a]
13	Four digits in the hand or first digit reduced, third digit enlarged
14	Lopho-selenodont molar pattern
15	High number of vertebrae (dorsal, 20–21; lumbar, 7–9; sacral, 5–7)[a]
16	Humerus slender and without an entepicondylar foramen[a]
17	Ankylosis of the radius and the ulna
18	Third trochanter of the femur reduced[a]
19	Ankylosis of the tibia and proximal end of the fibula
20	Clawlike nail of second digit in the foot
21	Reduced dental formula
22	Reduced contact between frontal and maxilla in the face
23	Lambdoid crests absent

Note: Features 1–9 include "Pliohyracidae"; 10–13, 15–20 are unknown in "Pliohyracidae"; 14 includes "Pliohyracidae," but earliest forms are bunodont; 21–23 are not in "Pliohyracidae."

[a] Probable feature of the ancestral morphotype of the order.

teeth and simplified premolars. The genus *Seggeurius* may also occur in the earliest Eocene of Morocco (Fig. 7.5; Gheerbrant et al., 2003). The diversity of the order has been considerable since at least the early-middle Eocene, as seen at Algerian Saharan localities, which yielded one of the smallest known species, *Microhyrax lavocati* Sudre, 1979 (about 3 kg), and one of the largest known ones, *Titanohyrax mongereaui* Sudre, 1979 (about 800 kg).

Paleogene hyracoids are remarkably diverse, with about 30 species, all included in the paraphyletic family "Pliohyracidae" (Rasmussen, 1989). "Pliohyracid" features include large size, small brain, complete eutherian dentition, biradicular lower canine, and a mandibular chamber in the female of most taxa. However, the suprageneric systematics needs revision (Rasmussen, 1989; Rasmussen and Simons, 1991, 2000; Pickford et al., 1997; Tabuce et al., 2001b). The "Saghatheriinae" is the most diverse "pliohyracid" subfamily. Other Paleogene hyraxes, including Geniohyinae, are successive primitive stem lineages of Hyracoidea, which probably arose in the beginning of the Eocene.

Hyraxes were the dominant herbivores of the early Tertiary African mammal communities. They filled a variety of

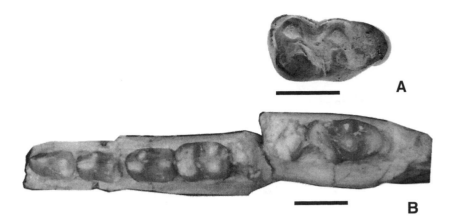

Fig. 7.5. *Seggeurius,* the most primitive and earliest known hyracoid. (A) *Seggeurius?* n. sp., right m3 from the Ypresian of the Ouled Abdoun Basin, Morocco (Gheerbrant et al. 2003). (B) *Seggeurius amourensis,* left lower jaw with m3, m1, p4–p2 from the early-middle Eocene of El Kohol, Algeria (Mahboubi et al., 1986). *Seggeurius* is characterized by primitive molars with a poorly advanced lopho-selenodont pattern and a high bunodonty, which is distinctive from the earliest known perissodactyls, such as *Hyracotherium.* Scale bar = 5 mm.

ecological niches—including large, rhino-sized species—that are occupied in Laurasiatic provinces by unrelated groups of artiodactyls and perissodactyls. Distinctive lineages are known, showing convergences with such Laurasiatic ungulates as suids, tapirs, equids, and even bovids. The order is a classic example of an endemic adaptive radiation.

As a whole, Paleogene hyraxes were generalized quadrupedal runners and leapers. However, their locomotor specializations remain poorly known in the absence of articulated skeletons.

BASAL PHYLOGENY, ORIGIN OF THE ORDER, AND ANCESTRAL MORPHOTYPE.

The order Hyracoidea is monophyletic. Its basal diversification is still poorly known. This is especially true for the relationships of early hyracoids, which are based mostly on teeth, and for their interordinal relationships, which are among the most debated questions of current mammalian phylogenetics. Our lack of knowledge about the roots of hyracoids is linked (1) to gaps in the Paleogene record, and (2) to the fact that several notable specializations have been acquired recently in the order and are not representative of the primitive hyracoid condition, as exemplified by the *Megalohyrax* skull (Thewissen and Simons, 2001; De Blieux and Simons, 2002).

The Hyracoidea are often described as a very primitive or conservative ungulate order (Gregory, 1910; Meyer, 1978). Kondrashov (1998) even wrote that they are direct descendants of condylarthrans. Extant hyraxes are indeed characterized by striking plesiomorphic features, and "pliohyracids" show additional primitive skull features. Even the bunodont, incipiently lopho-selenodont molar morphology of early hyracoids is primitive, as exemplified by *Seggeurius* (Fig. 7.5). It is especially distinct from that of perissodactyls (Rasmussen et al., 1990: 4691; Rasmussen, 1989: 69). Probable features of the ancestral morphotype are indicated in Table 7.4. Court and Mahboubi (1993) added possible morphotypic dental features, such as the compressed trigonid bearing a reduced paracristid and the absence of a postmetacrista.

Seggeurius, Microhyrax, Geniohyinae, and Saghatheriinae are successive primitive stem lineages of advanced hyracoids

(Pliohyracinae and Procaviidae; Fig. 7.6). The most primitive "pliohyracids" (*Seggeurius, Microhyrax, Bunohyrax,* and Geniohyinae) are characterized by a simple bunodont molar morphology and simplified premolars.

PALEOBIOGEOGRAPHY. Hyracoidea is one of the major endemic mammalian orders of African origin. This is supported both by paleontological data, which illustrate an autochthonous early Paleogene history for the "Pliohyracidae" since at least the early Eocene (*Seggeurius*), and by recent molecular studies. The initial radiation of the order (primitive "pliohyracids" and saghatheriines) was favored by Africa's isolation from Laurasiatic ungulate faunas. Hyracoidea survived the end of the isolation of Africa at the Oligocene-Miocene transition (closure of Tethys). They dispersed in the late Miocene (Fischer, 1992).

FOSSIL EARLY RELATIVES

Anthracobunia, Family "Anthracobunidae"

The family "Anthracobunidae" (Wells and Gingerich, 1983) includes several genera of large bunodont-lophodont mammals (body mass, 100–275 kg) from the early-middle Eocene of Indo-Pakistan. This family is known mostly by dental material, the scapula, and the astragalus. Thewissen et al. (2000) reported important new material from Pakistan, including an undescribed skull. Ginsburg et al. (1999) assigned the family to the new infraorder Anthracobunia.

Some features of the "Anthracobunidae" are strikingly primitive: dental formula complete (P1/p1 present), c1 and p1 two-rooted, molar cristid obliqua very lingual on the trigonid, and astragalar foramen present. Derived (altungulate and tethytherian) features include bilophodont (bunodont-bilophodont) molars (Fig. 7.7), developed postentoconulid, molarized premolars, and reduced centrocrista.

The family "Anthracobunidae" was thought to be close to the ancestry of *Moeritherium* (West, 1980, 1983). It is regarded as the sister group of proboscideans by Domning et al. (1986) and Thewissen et al. (2000), based on dental resemblances, including the occurrence of a postentoconulid.

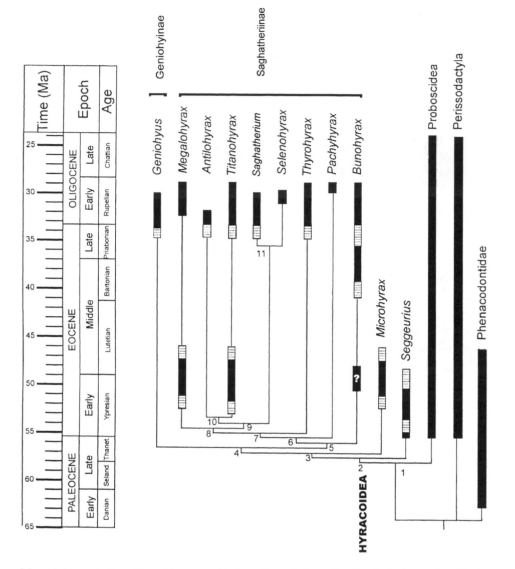

Fig. 7.6. Phylogeny of the early hyracoids. Extant Procaviidae are possible descendants of advanced small saghatheriines, such as *Thyrohyrax* (e.g., Rasmussen, 1989; Rasmussen and Simons, 1991). Modified from Tabuce et al. (2001b: fig. 4). Abbreviatons: Seland., Selandian; Thanet., Thanetian. Wide black lines indicate known stratigraphic distributions; wide patterned lines indicate uncertain fossil record or a gap in the fossil record; numbers denote nodes.

According to Fischer and Tassy (1993), the family "Anthracobunidae" is paraphyletic. In their analysis, *Jozaria, Pilgrimella, Anthracobune,* and *Minchenella* are considered to be successive stem taxa of Proboscidea, and *Lammidhania* is a basal tethythere with *Phenacolophus.* Consequently, Fischer and Tassy (1993) included anthracobunids (excluding *Lammidhania*) in the Proboscidea (see, e.g., Fig. 7.9A). Tassy and Shoshani (1988) also underlined the resemblances of *Anthracobune* to desmostylians. Wells and Gingerich (1983) relate "anthracobunids" to basal tethytheres, a stem position that is in accordance with their Eocene age. The same authors, followed by Domning et al. (1986), suggested that "Anthracobunidae" are descendants of phenacolophids, such as *Minchenella* from the late Paleocene, which seems to be currently the most acceptable hypothesis. *Minchenella* would be indeed better placed closer to anthracobunids than to "Phenacolophidae" (see Table 7.1).

Because the discovery of *Phosphatherium* and *Daouitherium* supports the hypothesis that true lophodonty is primitive for Proboscidea, some or all putative synapomorphies shared by anthracobunids and proboscideans (Table 7.5) could be questionable as being either possible synapomorphies of more inclusive taxa (e.g., bilophodonty), or possible convergences. The bunodont-bilophodont pattern of anthracobunids is much more primitive than the bunolophodonty of *Moeritherium* or the true lophodonty of *Phosphatherium* (see Gheerbrant et al., 1998). It may be representative of generalized tethytheres (or, perhaps, even more primitive). In anthracobunids, which are of similar or slightly later age than are *Phosphatherium* and *Daouitherium,* the association of such primitive features with derived features shared with proboscideans (Table 7.5) supports possible convergences. Although the relationship of anthracobunids and Proboscidea *sensu stricto* is not excluded (see, e.g., Fig.

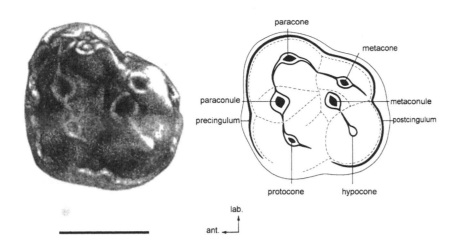

paracone

metacone

paraconule

metaconule

precingulum

postcingulum

protocone   hypocone

lab.

ant.

Fig. 7.7. Molar pattern of "anthracobunids." Upper molar of *Anthracobune pinfoldi*. It shows the typical bunodont-bilophodont pattern, which is primitive with respect to the true lophodonty of the earliest proboscideans, and which may be representative of the tethytherian ancestral morphotype. Anthracobunids show other striking primitive dental features, such as a birooted p1 and c1. Left image from West (1983: fig. 4); right image from Thenius (1989: fig. 740). Scale bar = 10 mm.

7.9A), it needs further support. New material from Pakistan might help to clarify anthracobunid relationships (Thewissen et al., 2000).

"Anthracobunidae" are generally regarded as amphibious (Wells and Gingerich, 1983), an idea based in part on the near-shore marine facies (coastal marshes) of their localities. This is supported by more complete material reported by Thewissen et al. (2000), who suggest that anthracobunids support an amphibious ancestry of tethytheres and that proboscideans are secondarily strictly terrestrial. Known distribution agrees with a likely origin of the "Anthracobunidae" on the Asiatic Tethyan shores.

Order Desmostylia

MAIN FEATURES AND SPECIALIZATIONS. The Desmostylia are the only extinct order of marine mammals. They were large amphibious quadrupeds resembling hippopotami, with massive limbs and a very short tail. The nasal opening is slightly to moderately retracted. The rostrum of the skull is more or less elongated and broadened; it usually bears procumbent incisor and canine tusks in both upper and lower jaws, and a long postcanine diastema resulting in part from a reduction in the number of premolars. The incisors are transversely aligned. The primitive dental formula is 3.1.4.3, with a trilobate dp4. The premolar and molar cusps tend to become closely appressed cylinders of thick enamel, giving the order its name (meaning "bundle of columns"). Primitively brachyodont and bunodont, the cheek teeth became hypsodont in advanced desmostylids, such as *Desmostylus,* which also acquired numerous supernumerary molar cusps. There is a prominent zygomatic arch, an elongated paroccipital process, and an epitympanic sinus that opens into the temporal fossa. There are indications of sexual size dimorphism in at least some taxa.

A clavicle is absent; the sternum consists of a series of heavy, platelike paired sternebrae; and the radius and ulna are articulated so as to preclude movement in adults. The metacarpals are longer than the metatarsals, and each foot has four well-developed digits (digit I is vestigial). Swimming

Table 7.5 Presumed synapomorphies of "Anthracobunidae" and Proboscidea

Number	Description
1	Medial tubercle (tuberculum mediale) of the astragalus prominent mediodistally
2	Flat tibial facet (trochlea tali) of the astragalus
3	Morphology of the scapula
4	?Dental features (e.g., paracone and metacone appressed on P⁴)
5	Loss of contact between astragalus and cuboid (taxeopody)

Note: Synapomorphy 1, Gingerich et al. (1990) and Fischer and Tassy (1993); 3, Shoshani et al. (1996); 2, 4–5, Fischer and Tassy (1993).

style seems to have resembled that of polar bears. Much controversy has attended the interpretation of desmostylian terrestrial posture, due to the peculiarities of the ankle joint and other features of the limbs (Inuzuka et al., 1995; Domning, 2002, and references therein).

PALEONTOLOGICAL DATA. Domning (2001a) reviewed the fossil record of desmostylians, and Domning (1996) provided exhaustive bibliography, synonymy, and classification. Desmostylians are known only from Oligocene and Miocene marine strata of the North Pacific (Japan to Baja California), and are most common in the Miocene of California, Oregon, and Japan. They are now represented by relatively complete skeletons of several taxa.

The earliest desmostylians known so far are early Oligocene in age, long postdating both the diversification of other paenungulates and the origin of the order Desmostylia itself. The oldest genera, *Behemotops* and *Ashoroa,* already seem to have the characteristic desmostylian body plan.

The cladistic relationships of the better-known genera and species are set forth by Domning et al. (1986), Ray et al. (1994), and Domning (2001a). Inuzuka (2000) published a cladogram recognizing only the families Paleoparadoxiidae (*Behemotops* and *Paleoparadoxia*) and Desmostylidae (*Ashoroa, Cornwallius, Desmostylus,* and *Kronokotherium;* see Table 7.1). Both families are traceable back into the late Oligocene of

Hokkaido, where they are represented by *Behemotops* and *Ashoroa,* respectively. *Cornwallius* also appears in the late Oligocene, and both *Desmostylus* and *Paleoparadoxia* are probably present at the base of the Miocene.

BASAL PHYLOGENY, ORIGIN OF THE ORDER, AND ANCESTRAL MORPHOTYPE.   What is known of the order's basal phylogeny is not really very close to its base. Probably no group of mammals has been more disparately assigned, but now it seems to have found a secure berth as an independent taxon within the Tethytheria (Tassy, 1981; Domning et al., 1986; Inuzuka et al., 1995). Its origins seem to lie closest to those of proboscideans and anthracobunids; Domning et al. (1986; Ray et al., 1994) proposed the late Paleocene Asiatic form *Minchenella* as a potential ancestor of all these groups. They also argued for uniting the Desmostylia more closely to the Proboscidea than to the Sirenia, based on the presence of a postentoconulid in m3 and on the high position of the external auditory meatus and its being nearly enclosed ventrally by contact of the squamosal posttympanic and postglenoid processes in both desmostylians and proboscideans.

The ancestral morphotype of the order is poorly known. The upper molars would have been bunodont and trapezoidal, with two rows of prominent cusps slightly oblique to the midline (three in the anterior row and two in the posterior). The lower molars were rectangular, with two rows of two cusps each, and a well-developed hypoconulid on m3. Both *Minchenella* and "anthracobunids" fit this description.

PALEOBIOGEOGRAPHY.   As stated earlier in the chapter, desmostylians are known only from Oligocene and Miocene strata of the littoral North Pacific. Their relationships suggest that they originated in Asia, but they quickly spread to North America. By their first appearance in the Oligocene, at least, they were already present on both continents. Their distribution on both sides of the Pacific indicates considerable vagility, resulting from their swimming ability.

Order Embrithopoda

The order Embrithopoda is based on the genus *Arsinoitherium* Beadnell, 1901, from the early Oligocene of Egypt. *Arsinoitherium* was studied by Andrews (1906) and more recently by Court (1992b). It is a very large and singular ungulate, the size of a large rhino. Most of its skeleton was reconstructed by Andrews (1906). It is characterized by a pair of huge (but hollow) nasal horn cores (with a pair of smaller horns behind them) and by peculiar hypsodont and bilophodont cheek teeth, which led Andrews (1906) to distinguish it in its own order (Embrithopoda). The bilophodont-like molar pattern illustrated in Fig. 7.8 is secondary, derived from an exaggeration of the dilambdodont pattern (Court 1992a). The dental formula is complete, and there is no diastema. The teeth are nearly homodont.

Arsinoitherium zitteli and *A. andrewsi* are described from the Fayum. Scarce remains of *Arsinoitherium* have also been reported from other early Oligocene localities in Oman, Libya, and Angola. Three other genera from southeastern Europe have recently been referred to the primitive family Palaeoamasiidae. McKenna and Manning (1977), followed by Prothero et al. (1988), also related the "phenacolophid" *Phenacolophus* from the late Paleocene of Asia to the ancestry of the embrithopods. They stated that *Phenacolophus, Crivadiatherium,* and *Arsinoitherium* belong to a single "morphocline," based on their dental resemblances. The postulated synapomorphies are not supported by *Palaeoamasia* and *Crivadiatherium,* and the embrithopod relationships of *Phenacolophus* remain poorly substantiated (Radulescu and Sudre, 1985; Maas et al., 1998).

Initially, Embrithopoda were thought to be related to the Hyracoidea (Andrews, 1906; Gregory, 1910). Simpson (1945) included Embrithopoda in his superorder Paenungulata. McKenna and Manning (1977), followed by Prothero et al. (1988), considered embrithopods the sister group of other altungulates, on the basis of the development of the entolophid and the distal shift of the metaconule (i.e., basically, the bilophodont pattern). However, the bilophodonty of embrithopods has been interpreted by Court (1992a) as a *secondary* feature, derived from a hyperspecialized dilambdodonty (Fig. 7.8). This interpretation raises the question of whether the bilophodonty of embrithopods and other altungulates is homologous. Features shared with proboscideans and sirenians, such as amastoidy and the serial carpal bone arrangement, suggested close relationships with Tethytheria to Tassy and Shoshani (1988). Other tethythere features are lacking in *Arsinoitherium;* for example, orbits in primitive posterior position (Court 1992b), although the infraorbital canal is short (Novacek and Wyss, 1986). Embrithopods and *Phenacolophus* are also distinguished from proboscideans and sirenians by the alternate pattern of the tarsals (astragalus articulating with the cuboid). Nevertheless, Court (1992b) reported seven synapomorphies of *Arsinoitherium* and Proboscidea. Of these, Fischer and Tassy (1993) retain only the loss of the hypoglossal foramen and the reduced processus ascendens of the palatine, and they add the presence of a prominent tuberculum mediale on the astragalus (which is also known in "anthracobunids"; Gingerich et al., 1990), and a ventrally protruding coracoid process on the scapula (which is also present in the sirenian *Pezosiren*). They support a sister-group relationship of the Embrithopoda and "Anthracobunidae" + Proboscidea, in agreement with Court (1992b). The different patterns of the bilophodonty and the enamel microstructure (Koenigswald et al., 1993), however, argue against close relationship of embrithopods and proboscideans.

The diversity and antiquity of the order Embrithopoda suggest its origin in southeastern Europe, which might have formed a distinct province in the Eocene (Radulescu and Sudre, 1985). This is supported by the recent discovery of *Hypsamasia* (Maas et al., 1998), and it is better substantiated

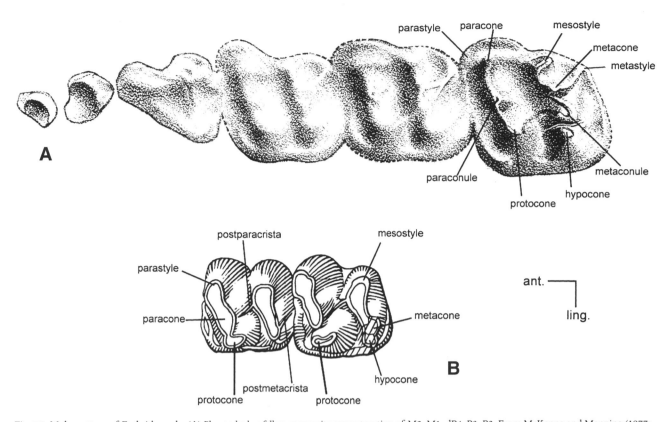

Fig. 7.8. Molar pattern of Embrithopoda. (A) *Phenacolophus fallax,* composite reconstruction of M2, M1, dP4, P3, P2. From McKenna and Manning (1977: Plate 3, fig. 1, reversed). (B) *Palaeoamasia kansui,* left M1–M2. Modified from Radulescu and Sudre (1985: fig. 7), according to the interpretation of Court (1992a: fig. 3). *Phenacolophus* shows a primitive classical bilophodont molar pattern, whereas embrithopods show a secondary bilophodont-like specialized molar pattern corresponding to a hyperdilambdodonty (lingual cusps reduced and lophs made mostly by lingually placed labial cusps), as demonstrated by Court (1992a).

than an earlier Asiatic origin (relationship with *Phenacolophus?*). This implies that the lineage of *Arsinoitherium* is derived from an Eocene trans-Tethyan dispersal into Africa.

Family "Phenacolophidae"

Several bilophodont genera from the late Paleocene of Asia, initially referred to "condylarths," have been discussed as primitive tethytheres or altungulates: *Minchenella* and *Radinskya* from China and *Phenacolophus* from China and Mongolia, all included in the family "Phenacolophidae."

The family "Phenacolophidae" is poorly known and clearly paraphyletic. It is described by McKenna and Manning (1977) as an intermediate step between the primitive phenacodontids and the derived tethytheres. *Phenacolophus* Matthew and Granger, 1925, is known from several dental, cranial, and postcranial remains. It shows a classical bilophodont pattern, in contrast to embrithopods (Court 1992b), and it lacks a postentoconulid, in contrast to tethytheres. It was considered a primitive embrithopod by McKenna and Manning (1977), and is included in Embrithopoda by McKenna and Bell (1997). But this relationship is poorly supported, and it may be more reasonably identified as a primitive altungulate or Altungulata *incertae sedis* (see Table

7.1). This is indeed the position adopted by McKenna and Bell (1997) regarding *Radinskya*.

Minchenella Zhang, 1980, known only by its lower teeth, is included in "Phenacolophidae" (e.g., McKenna and Bell, 1997). However, it was also discussed as a possible ancestor of the "Anthracobunidae" (Wells and Gingerich, 1983), in which it is included here (see Table 7.1). Domning et al. (1986) considered it a possible ancestor of "Anthracobunidae," Desmostylia, and Proboscidea. The pasimony analysis of altungulates (=pantomesaxonians) by Fischer and Tassy (1993) yielded two possible hypotheses for the relationships of *Minchenella* : (1) it is the sister group of Anthracobunidae" + Proboscidea; or (2) it is the sister group of Desmostylia (Fig. 7.9A). A key feature of *Minchenella* shared with the tethytheres is the presence of a postentoconulid, an additional accessory cusp of the molar talonid behind the entoconid, and usually more developed on m3.

Radinskya McKenna, Chow, Ting, and Luo, 1989, known from one skull preserving the cheek teeth, was described as the most primitive "phenacolophid," and as more closely related to perissodactyls (as their sister group) than are phenacodontids. Its supposed relationship with perissodactyls is based especially on its bilophodonty and large M3. *Radinskya* was subsequently identified as a primitive

Fig. 7.9. (A) Cladogram of Altungulata (= Pantomesaxonia), according to Fischer and Tassy (1993: fig. 16.1, modified), who reject the Paenungulata. (B) Alternative hypothesis of Paenungulata, summarizing the opinions of various authors (see also Table 7.2). Fischer and Tassy's (1993) original cladogram does not include Desmostylia in Tethytheria, contrary to the systematics used here (see Table 7.1). The dashed line indicates that these researchers do not exclude a relationship of *Minchenella* and Desmostylia.

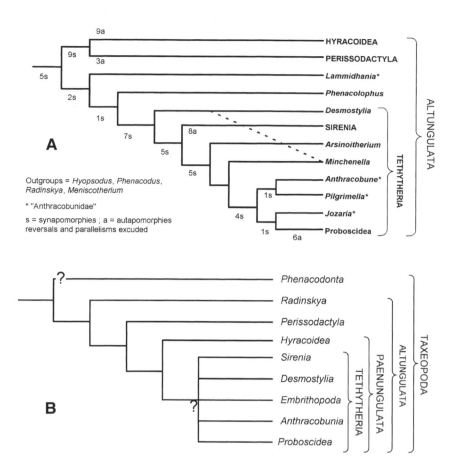

Outgroups = *Hyopsodus, Phenacodus, Radinskya, Meniscotherium*

* "Anthracobunidae"

s = synapomorphies ; a = autapomorphies reversals and parallelisms excluded

member of Altungulata, because of the reduced exposure of the mastoid (Fischer and Tassy, 1993; McKenna and Bell, 1997).

The "phenacolophids" support an Asiatic origin of the Altungulata. However, it should be stressed that (1) the relationships of the "phenacolophids" within Altungulata are poorly resolved, and (2) convergent early evolution of bilophodonty cannot be excluded.

TETHYTHERIA

Tethytheria McKenna, 1975, which means literally "beasts of Tethys," is the best supported supraordinal monophyletic grouping among the ungulates. This clade includes the extant Proboscidea and Sirenia, and the extinct "Anthracobunidae," Desmostylia, and Embrithopoda. The inclusion of "Anthracobunidae" in Tethytheria is supported by Wells and Gingerich (1983), Domning et al. (1986), Tassy and Shoshani (1988), and Fischer and Tassy (1993). The inclusion of Embrithopoda in Tethytheria is supported by Tassy and Shoshani (1988), Court (1992b), and Fischer and Tassy (1993).

Tethytheria is supported by recent molecular (Lavergne et al., 1996; Murphy et al., 2001a) and anatomical (Tassy and Shoshani, 1988) studies. However, it is noticeable that molecular studies provide stronger support for Paenungu-

lata than for Tethytheria. Some molecular analyses support the Hyracoidea-Tethytheria sister-group relationships (e.g., Stanhope et al., 1998); but others also infer a close relationship of Hyracoidea with Proboscidea (Springer et al., 1997a; Madsen et al., 2001) or with Sirenia (Murphy et al., 2001b). Moreover, Court (1994b) demonstrated the unexpected homoplastic nature of the auditory region in tethytheres, previously considered as strikingly synapomorphic of this taxon (Fischer, 1990).

Note that the relationships within Tethytheria are poorly resolved, especially for the fossil taxa (see Table 7.1), such as Embrithopoda (sister group of Proboscidea in Court, 1992b; sister group of Proboscidea and Sirenia for Novacek and Wyss, 1986, and Novacek et al., 1988), "Anthracobunidae" (Proboscidea or stem Tethytheria), and Desmostylia (sister group of Proboscidea for Domning et al., 1986; sister group of Sirenia for Shoshani, 1986; or sister-group of Sirenia + Proboscidea for Tassy, 1981; Court, 1992b; and Fischer and Tassy, 1993).

PAENUNGULATA AND PERISSODACTYLA-HYRACOIDEA

The interordinal relationships of hyracoids are among the most debated questions in current phylogenetic studies of ungulates. The present controversy concerns the relative position of Hyracoidea with respect to Tethytheria (Sirenia, Proboscidea, Desmostylia, "Anthracobunidae," and Embrithopoda) and Perissodactyla. There are two competing hypotheses: the Paenungulata clade versus the Hyracoidea + Perissodactyla clade. The debate results in particular because the early hyracoids remain poorly known.

Paenungulata

The monophyly of the Paenungulata (Fig. 7.9B; see Table 7.2) is disputed by both Fischer and Prothero (e.g., Fischer, 1986, 1989; Prothero et al., 1988; Fischer and Tassy, 1993), who argue for an alternative relationship of Hyracoidea with Perissodactyla (Fig. 7.9). Most allegedly shared morphological features of paenungulates are questioned or refuted by these authors.

However, molecular data provide the best support for the clade Paenungulata (de Jong et al., 1981; Hedges et al., 1996; Lavergne et al., 1996; Stanhope et al., 1996, 1998; Madsen et al., 1997, 2001; Springer et al., 1997a,b; Murphy et al., 2001a,b; Van Dijk et al., 2001). Paenungulata is one of the best-supported superordinal clades in mammalian molecular phylogeny, together with the more inclusive Afrotheria (see Table 7.2).

Recent morphological works supporting Paenungulata are those of Novacek (1986), Novacek and Wyss (1986), Shoshani (1986, 1993), Novacek et al. (1988), and Thewissen and Madar (1999). Shoshani (1993) records six to ten important myological features supporting the Paenungulata. Fischer (1986) and Tassy and Shoshani (1988) mention the distal doubling of the styloglossus muscle as "the best paenungulate feature," although this cannot be checked in fossils. Recent studies of the enamel microstructure of early Eocene hyracoids by Tabuce (2002) support the Paenungulata hypothesis. Thewissen and Madar (1999: 28) describe similarities of the tarsal anatomy (calcaneo-cuboid joint narrow and extended on the lateral side of the foot) in early Perissodactyla, Cetacea, Artiodactyla and Mesonychia, "to the exclusion of the paenungulates." Some molar features quite distinct from those of primitive perissodactyls also support the Paenungulata: the bunodont, incipiently lophoselenodont morphology of the most primitive hyraxes (e.g., *Seggeurius*), and the absence of conules in early hyracoids (*Seggeurius*) and proboscideans (*Phosphatherium* and *Numidotherium*).

Other anatomical (mostly skeletal) features shared by paenungulates are interpreted as primitive or convergent in Hyracoidea and Tethytheria (Table 7.6). Shared features, such as the developmental pattern of the fetal membranes and testicondy (i.e., permanent retention of the testes within the body cavity) are primitive and more inclusive, most being altungulate or taxeopode features (see Table 7.8). A classical feature of Paenungulata (e.g., Fischer, 1986; Novacek and Wyss, 1986) is the serial arrangement of the carpal and tarsal bones, which is called "taxeopody" (feature 1 in Table 7.6). The tarsals are unknown in Sirenia, including

Table 7.6 Features shared by Hyracoidea and Tethytheria interpreted as possible convergences rather than paenungulate synapomorphies

Number	Description
1	Serial carpus
2	Amastoidy
3	?Distal extension of the jugal to the posterior part of the glenoid fossa[a]
4	Fenestra rotunda with vertical orientation and partly covered ventrally by an expansion of the promontorium cochleae[b]
5	Enlargement of one pair of incisors
6	Reduction of the lambdoid crest[c]
7	Short face[c]
8	Reduction of the acromion process of the scapula (present in *Arsinoitherium*)

Note: Features 1, 2, 3, 5, Fischer and Tassy (1993); 3, Tassy and Shoshani (1988), Fischer (1989), Court (1992b); 6, 7, Thewissen and Simons (2001); 8, Court (1992b).

[a] Absent in sirenians or desmostylians, present in *Arsinoitherium*.

[b] Homology questionable in Fischer (1990).

[c] Not in "pliohyracids."

Pezosiren. The feet of Desmostylia and Embrithopoda (*Arsinoitherium*) have an alternating arrangement, whereas the hand is serial in Embrithopoda: hand and foot may have evolved independently. Fischer and Tassy (1993) suggested (1) convergent evolution of the serial arrangement in Paenungulata, especially for the carpus of Hyracoidea and Tethytheria (only Proboscidea in Court, 1992b); and (2) that the serial tarsus shared by Hyracoidea and Tethytheria might be a more inclusive feature (Altungulata or Taxeopoda), as it is known in *Tetraclaenodon,* and as *Hyracotherium* has only an incipient alternate structure. Fischer (1989) related this feature in Hyracoidea only to peculiar lineages with specialized carpal rotation for climbing and grasping (i.e., to peculiar arboreal lineages). However, Rasmussen et al. (1990) have described taxeopod carpals and tarsals in large Paleogene "pliohyracids," which were not arboreal. Taxeopody is a generalized condition in Hyracoidea.

Within Paenungulata, a sister-group relationship of Hyracoidea and Tethytheria is generally accepted. However, molecular studies are not consensual on the relationships of Hyracoidea within Paenungulata (see the section on Tethytheria earlier in the chapter).

Molecular data support a very early and African origin of Paenungulata. Yet fossil taxa, such as *Minchenella, Phenacolophus,* embrithopods, anthracobunids, and desmostylians, indicate a more complicated, still unresolved paleobiogeographic history of paenungulates, and one not confined to Africa.

Hyracoidea-Perissodactyla

Relationships between Perissodactyla and Hyracoidea were initially suggested by Cuvier (1800), Blainville (1816), and Owen (1848). Recent works supporting this controversial

(especially for *Seggeurius* and *Khamsaconus*) and the absence of conules, and bunodonty are distinctive with respect to early perissodactyls, in agreement with the Paenungulata hypothesis.

SUMMARY

Paenungulata includes the two extant orders of African ungulates Hyracoidea and Proboscidea and the aquatic order Sirenia, as well as several fossil groups. The relationships of Hyracoidea are still debated, and the supraordinal clade Paenungulata is controversial. Paenungulata is supported by molecular systematics, which suggests a Late Cretaceous (67–80 million years ago) African origin. Morphological data are much less explicit, or even reject Paenungulata: such shared paenungulate features as taxeopody might be interpreted as homoplastic and/or more inclusive, and several striking shared features of Hyracoidea and Perissodactyla (e.g., eustachian sac, cursorial features) are possible synapomorphies. In this respect, Tethytheria (Proboscidea + Sirenia) is morphologically better supported. The confusion may result from considerable fossil gaps, especially for African taxa, but also for some key Laurasiatic extinct taxa, such as "Anthracobunidae." Recently discovered fossil data support an African origin of Proboscidea and Hyracoidea, at least by the Paleocene, but the origins of such higher clades as Tethytheria or Paenungulata remain uncertain. Sirenia originated at least by the early Eocene from a quadrupedal digitigrade paenungulate or "condylarth" ancestor, presumably on eastern Tethyan continental shores. With respect to the hypothesis of Perissodactyla-Hyracoidea relationships, some features distinguishing the earliest hyraxes and proboscideans from Perissodactyla, such as dental morphology and enamel microstructure, are reported. Recent studies of early macroscelidids and louisinines also support Paenungulata, and part of the molecular Afrotheria. Recent fossil discoveries of importance for the understanding of paenungulate relationships are early taxa, such as *Ocepeia* (Taxeopoda *incertae sedis*), *Phosphatherium* (Proboscidea), *Seggeurius* (Hyracoidea), *Pezosiren* (Sirenia), and "Anthracobunidae" (Tethytheria).

ACKNOWLEDGMENTS

We thank the editors for their invitation to contribute to this volume, and for all their support. Photographs for Figures 7.4A, 7.5A, and 7.8B were taken by D. Serrette, C. Chancogne, and R. Tabuce, respectively. We thank the two reviewers, P. D. Gingerich and M. C. McKenna, for their constructive comments.

REFERENCES

Andrews, C. W. 1906. A Descriptive Catalogue of the Tertiary Vertebrata of the Fayum, Egypt, Based on the Collection of the Egyptian Government in the Geological Museum, Cairo, and on the Collection of the British Museum (Natural History). British Museum (Natural History), London.

Antoine, P.-O., J. L. Welcomme, L. Marivaux, I. M. Baloch, B. Benammi, and P. Tassy. 2003. First record of Paleogene Elephantoidea (Proboscidea, Mammalia) from the Bugti Hills of Pakistan. Journal of Vertebrate Paleontology 23: 977–980.

Archibald, J. D. 1998. Archaic ungulates ("Condylarthra"); pp. 292–331 in C. M. Janis, K. M. Scott, and L. L. Jacobs (eds.), Evolution of Tertiary Mammals of North America. Volume 1. Terrestrial Carnivores, Ungulates, and Ungulate-like Mammals. Cambridge University Press, Cambridge.

Bertrand, P. 1988. Evolution de la structure de l'émail chez les proboscidiens primitifs. Aspects phylogénétiques et fonctionnels; pp. 109–124 in D. Russell, J. P. Santora, and D. Sigogneau-Russell (eds.), Proceedings of the VIIth International Symposium on Dental Morphology, Paris 1986, Muséum National d'Histoire Naturelle, Paris, Série C, 53.

Blainville, H. de 1816. Prodrome d'une nouvelle distribution systématique du règne animal. Bulletin des Sciences, Société Philomathique de Paris, Série 3, 3: 105–124.

Cope, E. D. 1882. The Taxeopoda, a new order of Mammalia. American Naturalist 16: 522–523.

Court, N. 1990. The periotic of *Arsinoitherium* and its phylogenetic implications. Journal of Vertebrate Paleontology 10: 170–182.

———. 1992a. A unique form of dental bilophodonty and a functional interpretation of peculiarities in the masticatory system of *Arsinoitherium* (Mammalia, Embrithopoda). Historical Biology 6: 91–111.

———. 1992b. The skull of *Arsinoitherium* (Mammalia, Embrithopoda) and the higher order interrelationships of ungulates. Palaeovertebrata 22: 1–43.

———. 1994a. Limb posture and gait in *Numidotherium koholense*, a primitive proboscidean from the Eocene of Algeria. Zoological Journal of the Linnean Society 111: 297–338.

———. 1994b. The periotic of *Moeritherium* (Mammalia, Proboscidea): Homology or homoplasy in the ear region of Tethytheria McKenna, 1975? Zoological Journal of the Linnean Society 112: 13–28.

———. 1995. A new species of *Numidotherium* (Mammalia: Proboscidea) from the Eocene of Libya and the early phylogeny of the Proboscidea. Journal of Vertebrate Paleontology 15: 650–671.

Court, N., and J.-J. Jaeger. 1991. Anatomy of the periotic bone in the Eocene proboscidean *Numidotherium koholense*: An example of parallel evolution in the inner ear of tethytheres. Comptes Rendus de l'Académie des Sciences, Série II, Paris 312: 559–565.

Court, N., and M. Mahboubi. 1993. Reassessment of lower Eocene *Seggeurius amourensis*: Aspect of primitive dental morphology in the mammalian order Hyracoidea. Journal of Paleontology 67: 889–893.

Cuvier, G. 1800. Leçons d'anatomie comparée. Baudoin, Paris.

De Blieux, D. D., and E. L. Simons. 2002. Cranial and dental anatomy of *Antilohyrax pectidens*: A late Eocene hyracoid (Mammalia) from the Fayum, Egypt. Journal of Vertebrate Paleontology 22: 122–136.

De Jong W. W., A. Zweers, and M. Goodman 1981. Relationships of aardvark to elephants, hyraxes and sea cows from a-crystallin sequence. Nature 292: 538–540.

Domning, D. P. 1982. Evolution of manatees: A speculative history. Journal of Paleontology 56: 599–619.

———. 1994. A phylogenetic analysis of the Sirenia; pp. 177–189 *in* A. Berta and T. A. Deméré (eds.), Contributions in Marine Mammal Paleontology Honoring Frank C. Whitmore, Jr. Volume 29. Proceedings of the San Diego Society of Natural History. San Diego Society of Natural History, San Diego.

———. 1996. Bibliography and Index of the Sirenia and Desmostylia. Volume 80. Smithsonian Contributions to Paleobiology. Smithsonian Institution Press, Washington, D.C.

———. 2001a. Evolution of the Sirenia and Desmostylia; pp. 151–168 *in* J.-M. Mazin and V. de Buffrénil (eds.), Secondary Adaptation of Tetrapods to Life in Water. Proceedings of the International Meeting, Poitiers, 1996. Verlag Dr. Friedrich Pfeil, Munich.

———. 2001b. The earliest known fully quadrupedal sirenian. Nature 413: 625–627.

———. 2002. The terrestrial posture of desmostylians; pp. 99–111 *in* R. J. Emry (ed.), Later Cenozoic Mammals of Land and Sea: Tributes to the Career of Clayton E. Ray. Volume 93. Smithsonian Contributions to Paleobiology. Smithsonian Institution Press, Washington, D.C.

Domning, D. P., C. E. Ray, and M. C. McKenna. 1986. Two New Oligocene Desmostylians and a Discussion of Tethytherian Systematics. Volume 59. Smithsonian Contributions to Paleobiology. Smithsonian Institution Press, Washington, D.C.

Fischer, M. S. 1986. Die Stellung der Schliefer (Hyracoidea) im phylogenetischen System der Eutheria. Courier Forschungs-Institut Senckenberg, Frankfurt, 84: 1–132.

———. 1989. Hyracoids, the sister-group of perissodactyls; pp. 36–56 *in* D. R. Prothero and R. M. Schoch (eds.), The Evolution of Perissodactyls. Oxford University Press, New York.

———. 1990. Un trait unique de l'oreille des éléphants et des siréniens (Mammalia): Un paradoxe phylogénétique. Comptes Rendus de l'Académie des Sciences, Série III, Paris 311: 157–162.

———. 1992. Hyracoidea. Handbuch der Zoologie, Bd. 8 Mammalia, Td. 58. Walter de Gruyter, Berlin.

Fischer, M. S., and P. Tassy. 1993. The interrelation between Proboscidea, Sirenia, Hyracoidea, and Mesaxonia: The morphological evidence; pp. 217–234 *in* F. S. Szalay, M. J. Novacek, and M. C. McKenna (eds.), Mammal Phylogeny: Placentals. Springer-Verlag, New York.

Franz, V. 1924. Geschichte der Organismen. Gustav Fischer, Jena.

Gheerbrant, E., J. Sudre, and H. Cappetta. 1996. A Palaeocene proboscidean from Morocco. Nature 383: 68–71.

Gheerbrant, E., J. Sudre, H. Cappetta, and G. Bignot. 1998. *Phosphatherium escuilliei* du Thanétien du bassin des Ouled Abdoun (Maroc), plus ancien proboscidien (Mammalia) d'Afrique. Geobios 30: 247–269.

Gheerbrant, E., J. Sudre, H. Cappetta, M. Iarochene, M. Amaghzaz, and B. Bouya. 2002. A new large mammal from the Ypresian of Morocco: Evidence of a surprising diversity of early proboscideans. Acta Palaeontologica Polonica 47: 493–506.

Gheerbrant, E., J. Sudre, H. Cappetta, M. Iarochene, and A. Moumni. 2001. First ascertained African "condylarth" mammals (primitive ungulates: cf. Bulbulodentata and cf. Phenacodonta) from the earliest Ypresian of the Ouled Abdoun Basin, Morocco. Journal of Vertebrate Paleontology 21: 107–117.

Gheerbrant, E., J. Sudre, H. Cappetta, C. Mourer-Chauvire, E. Bourdon, M. Iarochene, M. Amaghzaz, and B. Bouya. 2003. Les localités à mammifères des carrières de Grand Daoui, Bassin des Ouled Abdoun, Maroc, Yprésien: Premier état des lieux. Bulletin de la Société Géologique de France 174: 279–293.

Gheerbrant E., J. Sudre, P. Tassy, M. Amaghzaz, B. Bouya, M. Iarochène. In press. Nouvelles données sur *Phosphatherium escuilliei* de l'Eocène inférieur du Maroc, apports à la phylogénie des Proboscidea et des ongulés lophodontes. Geodiversitas.

Gingerich, P. D., D. E. Russell, and N. A. Wells. 1990. Astragalus of *Anthracobune* (Mammalia, Proboscidea) from the early-middle Eocene of Kashmir. Contributions from the Museum of Paleontology, University of Michigan 28: 71–77.

Ginsburg, L., K. H. Durrani, A. M. Kassi, and J. L. Welcomme. 1999. Discovery of a new Anthracobunidae (Tethytheria, Mammalia) from the Lower Eocene lignite of the Kach-Harnai Area in Baluchistan (Pakistan). Comptes Rendus de l'Académie des Sciences de Paris, Sciences de la Terre et des Planètes 328: 209–213.

Godinot, M., T. Smith, and R. Smith. 1996. Mode de vie et affinités de *Paschatherium* (Condylarthra, Hyopsodontidae) d'après ses os du tarse. Palaeovertebrata, Vol. jubil. D. E. Russell, 25: 225–242.

Goodwin, M. B., D. P. Domning, J. H. Lipps, and C. Benjamini. 1998. The first record of an Eocene (Lutetian) marine mammal from Israel. Journal of Vertebrate Paleontology 18: 813–815.

Gregory, W. K. 1910. The orders of mammals. Bulletin of the American Museum of Natural History 27: 1–524.

Harris, J. M. 1978. Deinotherioidea and Barytherioidea; pp. 315–332 *in* V. J. Maglio and H.B.S. Cooke (eds.), Evolution of African Mammals. Harvard University Press, Cambridge, Massachusetts, and London.

Hedges, S. B, P. H. Parker, C. G. Sibley, and S. Kumar. 1996. Continental breakup and the ordinal diversification of birds and mammals. Nature 381: 226–229.

Inuzuka, N. 2000. Primitive Late Oligocene desmostylians from Japan and phylogeny of the Desmostylia. Bulletin of the Ashoro Museum of Paleontology 1: 91–123.

Inuzuka, N., D. P. Domning, and C. E. Ray. 1995. Summary of taxa and morphological adaptations of the Desmostylia; *in* L. G. Barnes, N. Inuzuka, and Y. Hasegawa (eds.), Evolution and Biogeography of Fossil Marine Vertebrates in the Pacific Realm. The Island Arc 3(4): 522–537.

Janis, C. M. 1983. Muscles of the masticatory apparatus in two genera of hyraces (*Procavia* and *Heterohyrax*). Journal of Morphology 176: 61–87.

Koenigswald, W. von, T. Martin, and H. V. Pfretzschner. 1993. Phylogenetic interpretation of enamel structures in mammalian teeth: Possibilities and problems; pp. 303–314 *in* F. S. Szalay, M. J. Novacek, and M. C. McKenna (eds.), Mammal Phylogeny: Placentals. Springer-Verlag, New York.

Kondrashov, P. E. 1998. The taxonomic position and relationships of the order Hyracoidea (Mammalia, Eutheria) within the Ungulata *sensu lato*. Paleontological Journal 32: 418–428.

Lavergne, A., Y. T. Douzer, T. Stichler, F. M. Catzeflis, and M. S. Springer. 1996. Interordinal mammalian relationship: Evidence for paenungulate monophyly is provided by complete mitochondrial 12s rRNA sequences. Molecular Phylogeny Evolution 6: 245–258.

Maas, M. C., J. G. M. Thewissen, and J. Kappelman. 1998. *Hypsamasia seni* (Mammalia: Embrithopoda) and other mammals from the Eocene Kartal formation of Turkey; pp. 286–297 *in* K. C. Beard and M. R. Dawson (eds.), Dawn of the Age of Mammals in Asia. Volume 43. Bulletin of the Carnegie Museum of Natural History.

Madsen, O., P.M. Deen, G. Pesole, C. Saccone, and W. W. de Jong. 1997. Molecular Evolution of mammalian aquaporin-2: Further evidence that elephant shrew and aardvark join the paenungulate clade. Molecular Biology and Evolution 14: 363–371.

Madsen, O., M. Scally, C. J. Douady, D. J. Kao, R. W. Debry, R. Adkins, H. M. Amrine, M. J. Stanhope, W. W. de Jong, and M. S. Springer. 2001 Parallel adaptive radiations in two major clades of placental mammals. Nature 409: 610–614.

Mahboubi, M., R. Ameur, J.-Y. Crochet, and J.-J. Jaeger. 1984. Earliest known proboscidean from early Eocene of northwest Africa. Nature 308: 543–544.

———. 1986. El Kohol (Saharan Atlas, Algeria): A new Eocene mammal locality in Northwestern Africa. Palaeontographica 192: 15–49.

Matthew, W. D., and W. Granger. 1925. Fauna and correlation of the Gashato Formation of Mongolia. American Museum Novitates 189: 1–12.

McKenna, M. C. 1975. Toward a phylogenetic classification of the Mammalia; pp. 21–46 *in* W. P. Luckett and F. S. Szalay (eds.), Phylogeny of the Primates. Plenum, New York and London.

McKenna, M. C., and S. K. Bell 1997. Classification of Mammals above the Species Level. Columbia University Press, New York.

McKenna, M. C., M. Chow, S. Ting, and Z. Luo. 1989. *Radinskya yupingae,* a perissodactyl-like mammal from the late Paleocene of China; pp. 24–36 *in* D. R. Prothero and R. M. Schoch (eds.), The Evolution of Perissodactyls. Oxford University Press, New York.

McKenna, M. C., and E. Manning. 1977. Affinities and paleobiogeographic significance of the Mongolian Paleogene genus *Phenacolophus.* Mémoire Special 1, Colloque International du CNRS, Montpellier 1976. Geobios 1: 61–85.

Meyer, G. E. 1978. Hyracoidea; pp. 284–314 *in* V. J. Maglio and H.B.S. Cooke (eds.), Evolution of African Mammals. Harvard University Press, Cambridge, Massachusetts, and London.

Murphy, W. J., E. Eizirik, W. E. Johnson, P. Y. Zhang, O. A. Ryder, and S. J. O'Brien. 2001a. Molecular phylogenetics and the origins of placental mammals. Nature 409: 614–618.

Murphy, W. J., E. Eizirik, S. J. O'Brien, O. Madsen, M. Scally, C. J. Douady, E. Teeling, O. A. Ryder, M. J. Stanhope, W. W. de Jong, and M. S. Springer. 2001b. Resolution of the early placental mammal radiation using Bayesian phylogenetics. Science 294: 2348–2351.

Novacek, M. J. 1986. The skull of leptictid insectivorans and the higher-level classification of eutherian mammals. Bulletin of the American Museum of Natural History 183: 1–111.

———. 1992. Mammalian phylogeny: Shaking the tree. Nature 356: 121–125.

Novacek, M. J., and A. R. Wyss 1986. Higher-level relationships of the recent eutherian orders: Morphological evidence. Cladistics 2: 257–287.

Novacek, M. J., A. R. Wyss, and M. C. McKenna. 1988. The major groups of eutherian mammals; pp. 31–71 *in* M. J. Benton (ed.), The Phylogeny and Classification of the Tetrapods. Volume 2: Mammals. Systematics Association Special Volume 35B. Clarendon Press, Oxford.

Owen, R. 1848. Description of teeth and portions of jaws in two extinct anthracotheroid quadrupeds (*Hyopotamus vectianus* and *Hyop. bovinus*) discovered by the Marchioness of Hastings in the Eocene deposits on the N.W. coast of the Isle of Wight: With an attempt to develop Cuvier's idea of the classification of Pachyderms by the number of their toes. Quarterly Journal of the Geological Society of London 4: 103–141.

Pickford, M., S. Moyà Solà, and P. Mein. 1997. A revised phylogeny of Hyracoidea (Mammalia) based on new specimens of Pliohyracidae from Africa and Europe. Neues Jahrbuch für Geologie und Paläontologie Abhandlungen 205: 265–288.

Prothero, D. R., E. M. Manning, and M. Fischer. 1988. The phylogeny of the ungulates; pp. 201– 234 *in* M. J. Benton (ed.), The Phylogeny and Classification of the Tetrapods. Volume 2: Mammals. Systematics Association Special Volume 35B. Clarendon Press, Oxford.

Prothero, D. R., and R. M. Schoch. 1989. Origin and evolution of the Perissodactyla: summary and synthesis; pp. 504–529 *in* D. R. Prothero and R. M. Schoch (eds.), The Evolution of Perissodactyls. Oxford University Press, New York.

Radinsky, L. B. 1966. The adaptive radiation of phenacodontid condylarths and the origin of the Perissodactyla. Evolution 20: 408–417.

Radulescu, C., and J. Sudre. 1985. *Crivadiatherium iliescui* n. sp., un nouvel Embrithopode (Mammalia) dans le Paléogène de la dépression de Hateg (Roumanie). Palaeovertebrata 15: 139–157.

Rasmussen, D. T. 1989. The evolution of the Hyracoidea: A review of the fossil evidence; pp. 57–78 *in* D. R. Prothero and R. M. Schoch (eds.), The Evolution of Perissodactyls. Oxford University Press, New York.

Rasmussen, D. T., M. Gagnon, and E. L. Simons. 1990. Taxeopody in the carpus and tarsus of Oligocene Pliohyracidae (Mammalia: Hyracoidea) and the phylogenetic position of hyraxes. Proceedings of the National Academy of Sciences USA 87: 4688–4691.

Rasmussen, D. T., and E. L. Simons. 1991. The oldest hyracoids (Mammalia: Pliohyracidae): New species of *Saghatherium* and *Thyrohyrax* from the Fayum. Neues Jahrbuch für Geologie und Paläontologie Abhandlungen 182: 187–209.

———. 2000. Ecomorphological diversity among Paleogene hyracoids (Mammalia): A new cursorial browser from the Fayum, Egypt. Journal of Vertebrate Paleontology 20: 167–176.

Ray, C. E., D. P. Domning, and M. C. McKenna. 1994. A new specimen of *Behemotops proteus* (Mammalia: Desmostylia) from the marine Oligocene of Washington; pp. 205–222 *in* A. Berta and T. A. Deméré (eds.), Contributions in Marine Mammal Paleontology Honoring Frank C. Whitmore, Jr. Volume 29. Proceedings of the San Diego Society of Natural History. San Diego Society of Natural History, San Diego.

Sagne, C. 2001. La diversification des siréniens à l'Eocène (Sirenia, Mammalia): Etude morphologique et analyse phylogénétique du sirénien de Taulanne, *Halitherium taulannense.* Ph.D. thesis, Muséum National d'Histoire Naturelle, Paris, 2 volumes.

Sanders, W., and J. Kappelman. 2002. A new transitional Late Oligocene proboscidean assemblage from Chilga, Ethiopia. Journal of Vertebrate Paleontology 22(supplement to no. 3): 102A.

Sanders, W. J., J. Kappelman, and D. T. Rasmussen. 2004. New large-bodied mammals from the late Oligocene site of Chilga, Ethiopia. Acta Palaeontologica Polonica 49: 365–392.

Savage, R.J.G. 1969. Early Tertiary mammal locality in southern Libya. Proceedings of the Geological Society of London 1657: 167–171.

Savage, R.J.G., D. P. Domning, and J.G.M. Thewissen. 1994. Fossil Sirenia of the West Atlantic and Caribbean region. V. The most primitive known sirenian, *Prorastomus sirenoides* Owen, 1855. Journal of Vertebrate Paleontology 14: 427–449.

Schlosser, M. 1923. Mammalia; pp. 402–689 *in* K. A. Von Zittel, F. Broili, and M. Schlosser (eds.), Grundzüge der Paläontologie (Paläozoologie), Vierte Auflage. R. Oldenbourg, Munich.

Shoshani, J. 1986. Mammalian phylogeny: Comparison of morphological and molecular results. Molecular Biology and Evolution 3: 222–242.

———. 1993. Hyracoidea-Tethytheria affinity based on myological data; pp. 235–256 *in* F. S. Szalay, M. J. Novacek, and M. C. McKenna (eds.), Mammal Phylogeny: Placentals. Springer-Verlag, New York.

Shoshani, J., P. M. Sanders, and P. Tassy. 2001. Elephants and other proboscideans: A summary of recent findings and new taxonomic suggestions; pp. 676–679 *in* G. Cavarretta, P. Giola, M. Mussi, and M. R. Palombo (eds.), The World of Elephants. Proceedings of the First International Congress. Consiglio Nazionale delle Ricerche, Rome, 1–20 October 2001.

Shoshani, J., and P. Tassy. 1996. Summary, conclusions, and a glimpse into the future; pp. 335–348 *in* J. Shoshani and P. Tassy (eds.), The Proboscidea. Evolution and Palaeoecology of Elephants and Their Relatives. Oxford University Press, Oxford.

Shoshani J., R. M. West, N. C. Court, R.J.G. Savage, and J. M. Harris. 1996. The earliest proboscideans: General plan, taxonomy and palaeoecology; pp. 57–75 *in* J. Shoshani and P. Tassy (eds.), The Proboscidea. Evolution and Palaeoecology of Elephants and Their Relatives. Oxford University Press, Oxford.

Simons, E. L. 1964. Yale Peabody Museum. Society of Vertebrate Paleontology News Bulletin 70: 14–15.

Simpson, G. G. 1945. The principles of classification and a classification of mammals. Bulletin of the American Museum of Natural History 85: 1–350.

Springer, M. S., A. Burk, J. R. Kavanagh, V. G. Waddell, and M. J. Stanhope. 1997a. The interphotoreceptor retinoid binding protein gene in therian mammals: Implications for higher level relationships and evidence for loss of function in the marsupial mole. Proceedings of the National Academy of Sciences USA 94: 13754–13759.

Springer, M. S., G. C. Cleven, O. Madsen, W. W. de Jong, V. G. Waddell, H. M. Amrine, and M. J. Stanhope. 1997b. Endemic African mammals shake the phylogenetic tree. Nature 388: 61–64.

Stanhope, M. J., M. R. Smith, V. G. Waddell, C. A. Porter, M. S. Shiviji, and M. Goodman. 1996. Mammalian evolution and the IRBP gene: Convincing evidence for several superordinal clades. Journal of Molecular Evolution 43: 83–92.

Stanhope, M. J., V. G. Waddell, O. Madsen, W. W. de Jong, B. S. Hedges, G. C. Cleven, D. Kao, and M. S. Springer. 1998. Molecular evidence for multiple origins of Insectivora and for a new order of endemic African insectivore mammals. Proceedings of the National Academy of Sciences USA 95: 9967–9972.

Tabuce, R. 2002. Mammifères du Paléogène ancien d'Afrique du Nord occidentale: systématique, paléobiogéographie et apport à la phylogénie des ordres endémiques africains. Ph.D. thesis, Université Montpellier II, Institut des Sciences de l'Evolution, Montpellier.

Tabuce, R., B. Coiffait, P. E. Coiffait, M. Mahboubi, and J.-J. Jaeger 2001a. A new genus of Macroscelidea (Mammalia) from the Eocene of Algeria: A possible origin for elephant-shrews. Journal of Vertebrate Paleontology, 21: 535–546.

Tabuce, R., M. Mahboubi, and J. Sudre. 2001b. Reassessment of the Algerian Eocene hyracoid *Microhyrax*. Consequences on the early diversity and basal phylogeny of the order Hyracoidea (Mammalia). Eclogae Geologicae Helvetiae 94: 537–545.

Tassy, P. 1981. Le crâne de *Moeritherium* (Proboscidea, Mammalia) de l'Eocène de Dor El Talha (Libye) et le problème de la classification phylogénétique du genre dans les Tethytheria McKenna 1975. Bulletin du Muséum National d'Histoire Naturelle, 4 Série, Paris 3 (sec. C): 87–147.

———. 1994. Origin and differentiation of the Elephantiformes (Mammalia, Proboscidea). Verhandlungen Naturwissenschaftlicher Verein Hamburg, N.F. 34: 73–94.

Tassy, P., and J. Shoshani. 1988. The Tethytheria: Elephants and their relatives; pp. 283–315 *in* M. J. Benton (ed.), The Phylogeny and Classification of the Tetrapods. Volume 2: Mammals. Systematics Association Special Volume 35B. Clarendon Press, Oxford.

Thenius, E. 1989. Zähne und Gebiss der Säugetiere. Handbuch der Zoologie, Volume VIII. Mammalia, Part 56. Walter de Gruyter, Berlin and New York.

Thewissen, J.G.M. 1990. Evolution of Paleocene and Eocene Phenacodontidae (Mammalia, Condylarthra). University of Michigan, Papers on Paleontology 29: 1–107.

Thewissen, J.G.M., and D. P. Domning. 1992. The role of phenacodontids in the origin of the modern orders of ungulate mammals. Journal of Vertebrate Paleontology 12: 494–504.

Thewissen, J.G.M., and S. I. Madar. 1999. Ankle morphology of the earliest cetaceans and its implications for the phylogenetic relations among ungulates. Systematic Biology 48: 21–30.

Thewissen, J.G.M., and E. L. Simons. 2001. Skull of *Megalohyrax eocaenus* (Hyracoidea, Mammalia) from the Oligocene of Egypt. Journal of Vertebrate Paleontology 21: 98–106.

Thewissen, J.G.M., E. M. Williams, and S. T. Hussain. 2000. Anthracobunidae and the relationships among Desmostylia, Sirenia and Proboscidea. Journal of Vertebrate Paleontology 20: 73A.

Van Dijk, M.A.M., O. Madsen, F. Catzeflis, M. J. Stanhope, W. W. de Jong, and M. Pagel. 2001. Protein sequence signatures support the African clade of mammals. Proceedings of the National Academy of Sciences USA 98: 188–193.

Wells, N. A., and P. D. Gingerich. 1983. Review of the Eocene Anthracobunidae (Mammalia, Proboscidea) with a new genus and species, *Jozaria palustris,* from the Kuldana Formation of Kohat (Pakistan). Contributions from the Museum of Paleontology of the University of Michigan 26: 117–139.

West, R. M. 1980. Middle Eocene large mammal assemblage with Tethyan affinities, Ganda Kas Region, Pakistan. Journal of Paleontology 54: 508–533.

———. 1983. South Asian Middle Eocene moeritheres (Mammalia: Tethytheria). Annals of the Carnegie Museum 52: 359–373.

Wislocki, G. B., and O. P. van der Westhuysen. 1940. The placentation of *Procavia capensis,* with a discussion of the placental affinities of the Hyracoidea. Contributions to Embryology 28: 65–88.

KENNETH D. ROSE,
ROBERT J. EMRY,
TIMOTHY J. GAUDIN,
AND GERHARD STORCH

XENARTHRA AND PHOLIDOTA

X ENARTHRA (ARMADILLOS, SLOTHS, ANTEATERS, AND their
extinct relatives) and Pholidota (*sensu stricto,* i.e., living and fossil pangolins)
are widely accepted as monophyletic groups, most of whose members are
characterized by fossorially adapted skeletons and a tendency toward reduced dentition, sometimes in association with a myrmecophagous diet. Because of these similarities, the two orders have sometimes been united in a higher taxon, Edentata, or
grouped informally as "edentates." However, the composition of the Edentata has
been unstable, variously being restricted to Xenarthra alone, or broadened to include
such additional groups as the aardvarks (Tubulidentata) and the extinct taeniodonts,
palaeanodonts, and gondwanatheres. As detailed in this chapter, there is little compelling morphological or molecular support for a cohort Edentata consisting of the
orders Xenarthra and Pholidota *sensu stricto,* although consideration of fossil evidence
makes this conclusion less certain.

   In this chapter, we begin by reviewing the various concepts of the Edentata, and
then proceed to explore the following questions:

- What evidence, if any, supports a monophyletic Edentata consisting of
  Xenarthra + Pholidota?
- What is the relationship of the extinct Palaeanodonta to Xenarthra and
  Pholidota, and what does that imply about the origin and relationships of
  the extant orders?

- How do various other early Tertiary fossil taxa relate to Xenarthra or Pholidota, and what do they imply about the origin and relationships of the extant orders?

Besides Palaeanodonta, fossil taxa of particular interest in this regard include *Eomanis* and *Eurotamandua* from the middle Eocene of Germany, and *Ernanodon* and *Asiabradypus* from the late Paleocene of Asia. Considerable diversity of opinion exists concerning the relationships of these and other fossil "edentates." Although this disagreement results in part from new fossil data that have altered previous views, it is also due to differing interpretations of character polarity and of whether similar characters are synapomorphic or homoplastic. Such dissension can best be resolved through an improved fossil record and by more detailed and comprehensive phylogenetic analyses.

COMPOSITION AND CHARACTERIZATION OF HIGHER TAXA OF "EDENTATES"

Edentata Cuvier, 1798

Two centuries ago, Vicq d'Azyr and Cuvier were the first to apply the name Edentata (Edentati Vicq d'Azyr, 1792) to the armadillos, sloths, xenarthran anteaters or vermilinguas, pangolins, and aardvarks. The name implies the absence of teeth, but only vermilinguas and pangolins are toothless, and each presumably achieved this state independently. The others retain simplified, mostly homodont teeth, usually rootless and often reduced (but sometimes increased) in number compared to the primitive eutherian condition. It was not until late in the nineteenth century that it became evident that this was not a natural group, and Huxley (1872) removed the pangolins and aardvarks to separate orders, now known as Pholidota and Tubulidentata, respectively. (Huxley's name, Squamata, for pangolins was preoccupied, so Weber's [1904] name, Pholidota, proposed some three decades later, is the one recognized today.) Soon thereafter, Cope (1889) proposed the name Xenarthra for the three groups of New World "edentates." Since then, various fossil groups have been added to the "edentates," including the early Tertiary Palaeanodonta (Matthew, 1918) and Taeniodonta (=Ganodonta of Wortman, 1896), the Cretaceous-Paleocene gondwanatheres (Mones, 1987), and Paleogene genera, such as *Chungchienia, Eurotamandua,* and *Ernanodon* (Chow, 1963; Storch, 1981; Ding, 1987). Of these, only palaeanodonts, *Eurotamandua,* and *Ernanodon* are still considered to have possible links to one or more of the original members of Edentata. Aardvark is now generally considered to be related to ungulates, tethytheres, or afrotheres (see Holroyd and Mussell, chapter 6, this volume), and in any case, is no longer believed to have any close relationship to other "edentates;" but the nature of the relationship between Pholidota and Xenarthra remains a contentious issue.

Gregory (1910) recognized a superorder Edentata encompassing the order Xenarthra and, questionably, the orders Pholidota, Tubulidentata, and Taeniodonta; but he was clearly not convinced of their close relationship. Matthew (1918: 653–655) used the Edentata to comprise Xenarthra, Pholidota, his new suborder Palaeanodonta, and possibly Taeniodonta (which he later removed from Edentata). Szalay (1977) adopted Matthew's concept of Edentata (excluding Taeniodonta), but most authors subsequent to Matthew have used Edentata as a synonym of Xenarthra (with or without Palaeanodonta) and considered it to have no special relationship to Pholidota. Simpson (1931, 1945) restricted Edentata to Xenarthra + Palaeanodonta, separating Pholidota to a distinct order probably not closely related to xenarthrans or palaeanodonts. Emry (1970), however, rejected palaeanodont-xenarthran ties and transferred the two families of palaeanodonts to Pholidota. Ding (1987) accepted Emry's composition of Pholidota and added the suborder Ernanodonta to the Edentata, which she apparently otherwise limited to Xenarthra. McKenna (1975) used Edentata as a synonym of Xenarthra and separated it from all other eutherians, which he called Epitheria. He considered Pholidota and Tubulidentata to be unrelated to one another or to Edentata. The same arrangement was adopted by McKenna and Bell (1997).

A prominent exception to this prevailing view is that of Novacek (e.g., 1986, 1992; Novacek and Wyss, 1986), who revived the notion that Xenarthra + Pholidota compose a clade (cohort Edentata in Novacek, 1986) separate from Epitheria (see also Stucky and McKenna, 1993), although he confusingly continued to use the term "edentates" in reference to xenarthrans only. A century earlier, Oldfield Thomas (1887) had proposed the name Paratheria for the same clade, which he considered of equivalent rank to Eutheria and Metatheria. Paratheria has been resurrected by some recent workers to accommodate gondwanatheres, as well as Xenarthra and Pholidota (e.g., Scillato-Yané and Pascual, 1985; Bonaparte, 1990). The affinities of gondwanatheres remain ambiguous, but there is little evidence that they are related to xenarthrans or pholidotans.

Glass (1985: 2) suggested that the name Edentata could "more properly . . . be considered as a synonym of the Pholidota and Tubulidentata" to the exclusion of Xenarthra. This conflicts with both the original meaning and current usage, however, and has not been widely adopted. In our opinion, unless convincing evidence can be marshaled to corroborate Novacek's concept of Edentata, this taxonomic name should be avoided and the term "edentate" should only be used informally (if at all) to apply to a non-taxonomic eco-morphological grouping of mammals (e.g., Szalay and Schrenk, 1998).

Characterization: "Edentates" have been grouped together because of their typically robust skeletons, which usually show fossorial adaptations, and their tendency toward reduced dentition, associated in some forms with a myrmecophagous diet. In particular, the forelimbs of most "edentates"

Table 8.1 Synoptic classification and distribution of Xenarthra and Pholidota

Order Xenarthra Cope, 1889
  Suborder Cingulata Illiger, 1811
      Superfamily Dasypodoidea Gray, 1821
        Family Dasypodidae Gray, 1821 (mPal–R, SA; lPlio–R, NA)
        Family Peltephilidae Ameghino, 1894 (mEoc–lMio, SA)
      Superfamily Glyptodontoidea Gray, 1869
        Family Pampatheriidae Paula Couto, 1954 (mMio–Pleist, SA; lPlio–Pleist, NA; m-lEoc, SA if *Machlydotherium* included)
        Family Glyptodontidae Gray, 1869 (lEoc–Pleist, SA; lPlio–Pleist, NA)
  Suborder Pilosa Flower, 1883
    Infraorder Vermilingua Illiger, 1811
        Family Myrmecophagidae Gray, 1825 (eMio–R, SA; Pleist–R, NA; mEoc, Eu if *Eurotamandua* included)[a]
        Family Cyclopedidae Pocock, 1924 (lMio–R, SA)
    Infraorder Phyllophaga Owen, 1842
        Pseudoglyptodon Engelmann, 1987 (Eoc/Olig boundary–late Olig, SA)
        Family Entelopidae Ameghino, 1889 (eMio, SA)
        Family Bradypodidae Gray, 1821 (R, SA)
      Parvorder Mylodonta McKenna and Bell, 1997
        Family Scelidotheriidae Ameghino, 1889 (mMio–Pleist, SA)
        Family Mylodontidae Gill, 1872 (lOlig–Pleist, SA; lMio–Pleist, NA
      Parvorder Megatheria McKenna and Bell, 1997
        Family Megatheriidae Gray, 1821 (eMio–Pleist, SA; Pleist, NA)
        Family Nothrotheriidae Ameghino, 1920 (lMio–Pleist, SA; Pleist–R, NA)
        Family Megalonychidae P. Gervais, 1855 (eMioc–R, SA, W Indies; lMio–Pleist, NA)
Order Pholidota Weber, 1904
        Family Eomanidae Storch, 2003 (mEoc, Eu)
        Family Patriomanidae Szalay and Schrenk, 1998 (lEoc, NA, Asia; eOlig–lMio, Eu)
        Family Manidae Gray, 1821 (Pleist, Eur; lMio–R, Asia; ?eOlig, ePlio–R, Afr)
?Order Pholidota
  Suborder Palaeanodonta Matthew, 1918
        Family Escavadodontidae Rose and Lucas, 2000 (ePal, NA)
        Family Epoicotheriidae Simpson, 1927 (lPal–lEoc, NA; eEoc, Asia; eOlig, Eu)
        Family Metacheiromyidae Wortman, 1903 (lPal–mEoc, NA)
  Suborder Ernanodonta Ding, 1987
        Family Ernanodontidae Ding, 1979 (lPal, Asia)

Source: Modified after McKenna and Bell (1997).

Notes: Itaboraian is considered middle Paleocene, Casamayoran is middle Eocene, Mustersan is late Eocene, and Tinguiririran is near the Eocene/Oligocene boundary. Epoch abbreviations: Pal, Paleocene; Eoc, Eocene; Olig, Oligocene; Mio, Miocene; Plio, Pliocene; Pleist, Pleistocene; R, Recent; e, m, or l preceding epochs designates early, middle, or late, respectively.

[a] See text for further discussion.

elevated spine, elongated acromion, and a secondary spine associated with an expanded fossa for the origin of the teres major muscle (Hoffstetter, 1958; Jenkins, 1970; Gaudin, 1999). Many of these features also are not restricted to xenarthrans but are seen in various other fossorial mammals (Rose and Emry, 1993). The presence of dermal ossicles has been suggested to be a derived feature unique to Xenarthra, although it is known only in cingulates and some mylodont and megatheriid sloths (Engelmann, 1985; Cartelle and Bohórquez, 1986).

Cingulates, the most diverse extant xenarthrans, are characterized by their bony dermal armor, or carapace. It is composed of a mosaic of bony plates covered by epidermal scales, forming a shield with a variable number of mobile bands that covers the back and sides of the trunk, as well as a plate covering the top of the braincase and a bony tube encircling the tail. Individual plates readily fossilize and provide the oldest known evidence of Xenarthra (middle or late Paleocene). Besides the carapace, cingulates are united by the presence of a large, distinct postglenoid fossa containing the postglenoid foramen; fused cervical vertebrae behind the atlas (except in the earliest armadillos); an enlarged greater trochanter of the femur extending well proximal to the femoral head; and fusion of the tibia and fibula (Engelmann, 1985). The primitive members of the group were almost certainly fossorial, with an omnivorous diet that perhaps tended toward myrmecophagy. Armadillos have simple, peglike teeth, cylindrical or elliptical in shape, and often reduced in size. However, there are a number of herbivorous cingulates, including glyptodonts, pampatheres, and perhaps some extinct eutatine armadillos (Vizcaíno and Bargo, 1998). These taxa have enlarged, lobate cheek teeth that resemble those of some extinct sloths, although differing in histological details. Cingulates are known from the Paleocene to the Recent, but evidently were not particularly common or diverse until the Miocene.

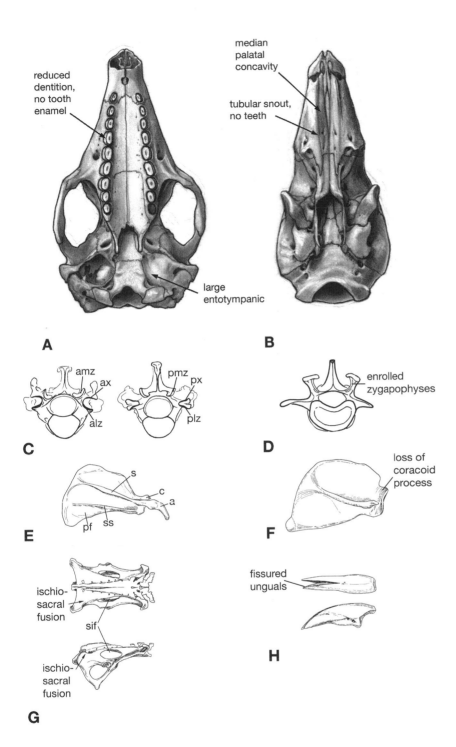

Fig. 8.3. Osteological traits of xenarthrans and pholidotans. (A) Skull of *Euphractus sexcinctus* (ventral view), showing xenarthran cranial characteristics. (B) Skull of *Manis tricuspis*, showing pholidotan cranial characteristics. (C) Posterior thoracic vertebrae of *Tamandua mexicana*, illustrating xenarthrous articulations (left: T14 in anterior view; right: T13 in posterior view). (D) Lumbar vertebra of *Patriomanis americanus* (USNM P299960) in anterior view, showing enrolled zygapophyses. (E) Right scapula of *Euphractus sexcinctus*. (F) Right scapula of *Manis javanica*. (G) Sacrum and pelvis of *Euphractus sexcinctus* (dorsal and right lateral views), showing synsacrum and ischiosacral fusion. (H) Ungual phalanx of digit III from *Manis javanica* (dorsal and lateral views). Abbreviations: a, acromion process; alz, anterior lateral zygapophyseal facet; amz, anterior medial zygapophyseal facet; ax, anterior xenarthrous facet; c, coracoid process; pf, postscapular (teres) fossa; plz, posterior lateral zygapophyseal facet; pmz, posterior medial zygapophyseal facet; px, posterior xenarthrous facet; s, scapular spine; sif, sacroischial foramen; ss, secondary scapular spine; USNM, National Museum of Natural History, Smithsonian Institution, Washington, D.C. (C) is from Gaudin (1999); (E)–(H) from Rose and Emry (1993).

Vermilinguans are the only truly edentulous xenarthrans. They are characterized by elongate, tubular skulls with a curved basicranial/basifacial axis (Gaudin and Branham, 1998). The hard palate is prolonged posteriorly, extending back to the ear region. The ossified bulla in the living genera *Tamandua* and *Myrmecophaga* is connected to a large hypotympanic sinus in the pterygoid bone. The tail in all but *Myrmecophaga* is prehensile, suggesting that the group may have been primitively arboreal. However, all have enlarged unguals on the central (third) digit of the manus, which probably evolved as an adaptation for digging into ant and termite nests. Extant xenarthran anteaters have an extremely long tongue that originates on the sternum and a specialized keratinized stomach, in which ingested ants and termites are ground (Grassé, 1955). Vermilinguans range from the Miocene to the Recent in the New World, but could extend much earlier in time if Eocene specimens from Europe (*Eurotamandua*) and Antarctica prove to be vermilinguan (see below).

Phyllophagans are generally characterized by an herbivorous diet. They typically have five upper and four lower teeth. The anteriormost upper and lower teeth are usually caniniform or incisiform, but the uppers occlude anterior to the lowers, in contrast to true canines. The posterior teeth are composed of layers of cementum, orthodentine, and vascularized orthodentine, which vary in hardness and form lophs or crests with wear. These posterior "molariforms" are lobate in outline in extinct mylodontid sloths, quadrangular or triangular in other extinct sloths, and cylindrical in tree sloths. Skull modifications—including large ascending and descending processes on the jugal and a large, platelike descending lamina of the pterygoid—provide more attachment area for jaw muscles. Postcranially, sloths are characterized by reduction of the astragalar trochlea to a single medial convexity, a feature correlated with the tendency to support their weight on the outer edge rather than on the plantar surface of the feet. Many extinct sloths attained large size and had robust limbs. The proximal radius is circular (to promote supination), the ilium tends to flare laterally, and the thorax is elongated and relatively rigid (Hoffstetter, 1958; Gaudin, 1999). Most extinct sloths probably bore their anterior body weight on the ulnar side of the manus (Coombs, 1983) or on the dorsal surface of the large and hooded claws, the tips of which were turned inward and upward toward the palm (e.g., Stock, 1925: fig. 4, plate 16). Unlike most of their extinct relatives, living tree sloths are suspensory arboreal creatures, with slender, elongate limbs. The forelimbs are longer than the hind in tree sloths (especially in the three-toed sloth *Bradypus*) and in extinct sloths except mylodontids. Both manus and pes in tree sloths have two or three syndactylous functional digits, with claws modified into hooks for suspensory posture and locomotion. The skull tends to be short, with a reduced snout (but complex nasoturbinals). Sloths range from the Eocene/Oligocene boundary (Tinguiririran) to the present, but could be even older if Antarctic specimens (discussed below) are correctly referred to this group.

Pholidota Weber, 1904

The modern ordinal name for pangolins or scaly anteaters has been in use for a century, although earlier terms, including Huxley's Squamata and Cope's Nomarthra, refer to a similar ordinal-level concept (Nomarthra included aardvarks as well as pangolins). Emry (1970) assigned the two families of palaeanodonts to the Pholidota but did not recognize a formal clade Palaeanodonta. This view was also adopted by McKenna and Bell (1997). In this chapter, we restrict usage of the name Pholidota to the extant pangolins and their closest fossil relatives; palaeanodonts are considered separately, in order to assess their relationships objectively.

Unlike Xenarthra, Pholidota *sensu stricto* have evidently never been diverse or particularly common (Fig. 8.4, Table 8.1). Excluding palaeanodonts, all living and some fossil pholidotans are assigned to the family Manidae. The three extinct genera (Eocene *Eomanis* and *Patriomanis,* and Oligo-Miocene *Necromanis*) have been assigned to Patriomanidae and Eomanidae. McKenna and Bell (1997) recognize four

extant genera for the seven modern species, but most current authorities recognize only one (*Manis;* e.g., Wilson and Reeder, 1993) or two genera (*Manis* and *Phataginus;* Patterson, 1978; Corbet and Hill, 1991). Recent species are restricted to Africa and southeast Asia, but Tertiary forms are also known from Europe and North America, as well as Asia and Africa. The oldest fossil pholidotans (excluding palaeanodonts) date from the middle Eocene of Germany (Lutetian *Eomanis*) and the late Eocene of western North America (Chadronian *Patriomanis*) and Asia (unnamed form).

Characterization: The most obvious synapomorphy of extant pholidotans is the presence of imbricating, keratinous, epidermal scales covering the top of the head, body, limbs, and tail. Keratinous structures rarely fossilize, however, and only in Eocene *Eomanis* among fossil forms have scales been verified (Koenigswald et al., 1981). Nonetheless, these fossils seem to confirm the early acquisition of this uniquely derived character. The snout is tubular and edentulous, a highly derived feature shared with vermilinguans, and the mandible bears a pair of bony, anterolaterally directed prongs on the outside of the symphysis on each side (incipient in *Eomanis;* better developed in all other pholidotans). Gaudin and Wible (1999) listed several other cranial synapomorphies of Pholidota, including the presence of a deep median palatal groove. Additional skeletal synapomorphies of Pholidota include the loss of the coracoid process of the scapula, presence of a distinctive unciform process on the hamate, and reduction in size of the obturator foramen (Gaudin, unpublished data). A large epitympanic sinus is present in the squamosal in *Manis* and apparently in *Eomanis,* as in palaeanodonts and xenarthrans. This sinus is not found in *Patriomanis,* however (Gaudin and Emry, 2002). Other characteristics include deeply fissured ungual phalanges (except in *Eomanis,* where they are unfissured), scapholunar fusion (probably not in *Eomanis*), and an osseous tentorium cerebelli separating the cerebrum from the cerebellum.

Palaeanodonta Matthew, 1918

Palaeanodonts (Fig. 8.4) are a group of small, early Tertiary mammals that show trends toward increasing fossorial habits and reduction and loss of teeth, presumably associated with a diet rich in ants and termites. Consequently, they were initially classified as a suborder of Edentata, together with Xenarthra and Pholidota (Matthew, 1918) or with Xenarthra alone (Simpson, 1931, 1945). They have been considered to be closely related to Xenarthra (e.g., Simpson, 1931; Szalay, 1977; Patterson et al., 1992), Pholidota (Emry, 1970; Rose and Emry, 1993; McKenna and Bell, 1997), or both (e.g., Matthew, 1918; Patterson, 1978; Rose, 1978). In fact, palaeanodonts seem to show a mix of xenarthran and pholidotan anatomical features. Three families are currently recognized: Escavadodontidae, Metacheiromyidae, and Epoicotheriidae (Table 8.1). Although most common and diverse in North American faunas (early Paleocene–late Eocene, 12 genera in three families), they are also known from Europe

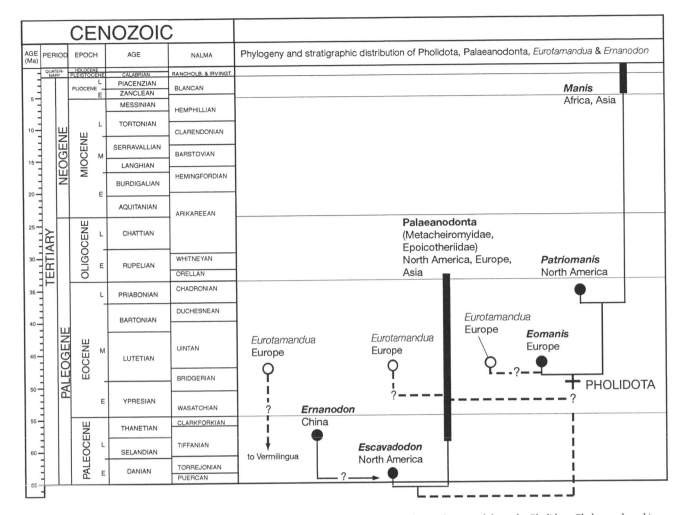

Fig. 8.4. Temporal distribution and phylogenetic relationships of *Eurotamandua, Ernanodon,* Palaeanodonta, and the order Pholidota. Phylogeny based in part on Rose and Lucas (2000) and Gaudin and Emry (2002). Time scale, as in Fig. 8.2, from Geological Society of America (1999). North American Land Mammal Ages (NALMA) modified after Prothero (1998). Broad bars represent known distributions; solid circles are isolated occurrences. The relationship between Pholidota and Palaeanodonta is represented by dashed lines, to indicate uncertainty as to whether the two are sister taxa, or Pholidota evolved from within Palaeanodonta sometime in the early Eocene. Because the affinities of *Eurotamandua* are ambiguous, it is shown in lighter font, with open circles reflecting its restricted occurrence. Three separate possibilities are indicated: (1) *Eurotamandua* is the oldest vermilinguan and the only xenarthran known from outside the Western Hemisphere; (2) *Eurotamandua* is related to palaeanodonts; or (3) *Eurotamandua* is the sister taxon of *Eomanis,* the oldest pangolin, and is related, along with that taxon, to pholidotans and palaeanodonts.

(early Eocene and early Oligocene, two genera in two families) and Asia (early Eocene, one genus) (Tong and Wang, 1997; Nel et al., 1999; Storch and Rummel, 1999).

Characterization: The skull of palaeanodonts is typically short-snouted, with a broad occipital region for attachment of the neck muscles. The mastoid is inflated, and the epitympanic sinus extends dorsally into the squamosal (Patterson et al., 1992). The mandible is thickened posteromedially, forming a "medial buttress." The dentition (except canines) shows increasing simplification and reduction in the size and number of teeth, with short diastemata separating postcanines. The canines (the upper occluding behind the lower) are large and distinctive, triangular in horizontal section, and form large honing facets. The postcranial skeleton is robust and fossorially adapted. The scapula has an elevated spine and long, bifid acromion. The most distinctive and probably the most diagnostic postcranial element is the humerus, with its very long deltopectoral crest; primitively elevated, shelflike, and medially directed large entepicondyle; and very prominent lateral supracondylar ridge (=supinator crest). The ulnar olecranon process is long and medially inflected. The manual digits of palaeanodonts are short and robust, the second and third metacarpals have extensor tubercles and a characteristic distal articulation, and the third digit is the largest.

Afredentata Szalay and Schrenk, 1994

Szalay and Schrenk (1994, 1998) proposed this order to accommodate Eocene *Eurotamandua* from Germany, which was described as the earliest myrmecophagid xenarthran (vermilinguan) by Storch (1981). Szalay and Schrenk (1998:

98) concluded that *Eurotamandua* represents an ancient clade of "non-xenarthrous ... 'edentates' probably of African origin, distinct from Xenarthra, Pholidota, Palaeanodonta, Tubulidentata, and from *Ernanodon* and *Plesiorycteropus*."

Characterization: Eurotamandua (Fig. 8.5) is similar in general form to the extant vermilinguan *Tamandua*. The rostrum is tubular and edentulous, and the postcranial skeleton is robust and fossorially adapted. Although initially reported as xenarthrous (Storch, 1981), reanalysis indicates that *Eurotamandua* lacks xenarthrous processes (Szalay and Schrenk, 1998; Storch, 2003), the key synapomorphy of Xenarthra. Storch (1981, 2003) and Storch and Habersetzer (1991) cite the following additional xenarthran and/or vermilinguan features: an ossified auditory bulla with a large hypotympanic sinus (=anterior accessory bulla), short skull region behind the tympanic, thin and slightly downturned mandibular symphysis, large subarcuate fossa, small and loosely attached C-shaped premaxilla, well-developed facial part of the lacrimal that contains the lacrimal foramen, loosely attached jugal, no maxillary exposure in the orbit, a large orbital wing of the parietal, prominent coracoid process on the scapula, and ischiocaudal fusion. To this list, Gaudin and Branham (1998) added two additional features shared with vermilinguans: loss of the postorbital process of the frontal and presence of a lateral tuberosity on the fifth metatarsal. Both features are also shared with Pholidota, including *Eomanis*, and the metatarsal tubercle is also present in palaeanodonts (Rose et al., 1992). However, *Eurotamandua* seems to lack several vermilinguan features, including a curved basicranial/basifacial axis, a greatly reduced temporal fossa, a reduced number of lumbar vertebrae (true vermilinguans have no more than three; *Eurotamandua* has seven), expanded ribs, a prehensile tail, and a narrow, erect dorsal process of the premaxilla (Gaudin and Branham, 1998). The first two features are controversial, because it is difficult to substantiate the form of the basicranial/basifacial axis in the somewhat crushed skull, and the large temporal fossa depicted by Szalay and Schrenk (1998: fig. 9) lacks well-defined muscle scars and may be exaggerated by distortion of the nuchal crest. Nonetheless, the temporalis of *Eurotamandua* appears to have been larger than in extant vermilinguans. Gaudin (1999) also observed that the anterior thoracic vertebrae are more similar to those of armadillos than to those of vermilinguans.

The holotype skeleton of *Eurotamandua joresi* comes from Messel, Germany, renowned for its abundance of well-preserved, articulated skeletons. Like other skeletons from the Messel oil shale, however, *Eurotamandua* is somewhat crushed and is now partly encased in epoxy resin for preservation, obscuring many critical details and rendering others ambiguous. Radiography has clarified some features but, for the most part, has not led to consensus about some of the most important characters, many of which have been questioned by other researchers. Thus Szalay and Schrenk (1998) suggest that *Eurotamandua* lacks an accessory auditory bulla and ischiocaudal fusion. We disagree with some

A

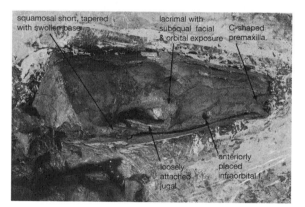

B

Fig. 8.5. Holotype of *Eurotamandua joresi* (HLMD-Me 17.000), from the middle Eocene of Messel, Germany. (A) Skeleton in right lateral view. (B) Detail of skull, showing loosely attached jugal shared with vermilinguans; anteriorly situated infraorbital foramen and subequal orbital and facial exposures of the lacrimal shared with palaeanodonts; C-shaped premaxilla shared with *Eomanis* and palaeanodonts; and short, tapering zygomatic process of the squamosal shared only with *Eomanis*. Abbreviation: HLMD, Hessisches Landesmuseum Darmstadt, Darmstadt, Germany. Skull image courtesy of Dr. Norbert Micklich, Hessisches Landesmuseum Darmstadt (HLMD).

of Szalay and Schrenk's observations on the ear region and note that new radiographs confirm the presence of a large accessory sinus connected to the tympanic cavity. Whether it is a hypotympanic sinus (as in *Myrmecophaga* and *Tamandua*) or an epitympanic sinus (as in xenarthrans, pholidotans, and palaeanodonts) is difficult to establish.

ANATOMICAL EVIDENCE FOR PROPOSED RELATIONSHIPS

Xenarthra + Pholidota

A clade composed of Xenarthra + Pholidota (Edentata) was considered the most primitive placental clade by Novacek (1986, 1992) and Novacek and Wyss (1986). In these and other papers, Novacek proposed that Xenarthra + Pholidota (i.e., Edentata) form a clade separate from Epitheria, based on features including a reduced subarcuate fossa, the arrangement of orbital bones, reduction or loss of teeth, and extensive sacro-innominate fusion, whereas epitheres were said to share certain features not found in Edentata (a bicrurate stapes and an alisphenoid-enclosed foramen ovale). Subsequent comparisons suggested that the utility or accuracy of most supposed synapomorphies of such a clade is suspect (Rose and Emry, 1993, among others). Patterson et al. (1992), however, proposed that presence of a large epitympanic sinus in the squamosal and a large circular posterior lacerate foramen are potential synapomorphies of an edentate clade. But aside from these basicranial traits, morphological support for this clade is weak at best, and there is virtually no molecular support for this concept of Edentata.

Engelmann (1985; see also Novacek and Wyss, 1986) proposed a close relationship between Pilosa and Pholidota, based on the position of the posterior lacerate foramen and the presence of an epitympanic sinus, an incomplete zygoma, a concave navicular facet on the astragalus, and pelvic kidneys. The significance of these characters is open to question. For example, the absence of a concave navicular facet in primitive pholidotans and pilosans indicates that this character evolved independently in the two groups (Rose and Emry, 1993). Moreover, pholidotans lack a number of key xenarthran synapomorphies (xenarthry, ischiosacral ossification), as well as other features characteristic of xenarthrans (septomaxilla, large facial exposure of the lacrimal, ossified larynx, ossified sternal ribs, scapular fenestration [pilosans only], secondary scapular spine and fossa for teres major, and presence of a large entotympanic), while retaining a postglenoid foramen and pterygoid hamulus, both lost in pilosans (Patterson et al., 1992; Gaudin, 1993, 1995; Gaudin and Wible, 1999). The large number of conflicting characters indicates that the resemblances between vermilinguans and pangolins arose through convergence—hence, the combination Pilosa + Pholidota does not constitute a natural group.

Xenarthra

Many authorities regard Xenarthra alone as the most primitive placental group. Gregory (1910: 468) depicted American Edentata (i.e., Xenarthra) as the basal group of placental mammals, and McKenna and Bell (1997; see also McKenna, 1975; Shoshani and McKenna, 1998) separated Xenarthra from all other placentals (Epitheria), considering each of them magnorders. McKenna (1975: 28) cited such plesiomorphic features as the presence of a septomaxillary bone, ossified sternal ribs, low body temperature, and poor thermoregulatory ability as evidence of their early separation from other mammals. However, the purported primitiveness of all these features has been questioned (Gaudin et al., 1996). As McKenna noted, there are no clear intermediates linking xenarthrans to any other group, with the possible exception of palaeanodonts.

Pholidota

Apart from Xenarthra, Pholidota have most often been linked with Carnivora among extant mammals. Morphological evidence supporting this seems to be limited to a single character, however: the presence of a bony tentorium cerebelli. Other features sometimes cited—fusion of the scaphoid and lunar bones and fissured unguals (the latter present in some creodonts, but not in carnivorans, and therefore only weakly suggestive of relationship to a Carnivora-Creodonta clade)—are now known to be absent in the most primitive members of one or both orders. Accordingly, the case for a close pholidotan-carnivoran relationship based on anatomical similarity is extremely tenuous.

However, morphological comparisons between palaeanodonts and the oldest pholidotans, especially *Eomanis* (as discussed in the next section), reveal considerable evidence that the two groups are either sister taxa or ancestor-descendant.

Palaeanodonta

Emry (1970) showed that many of the supposed resemblances between palaeanodonts and xenarthrans—particularly those enumerated by Simpson (1931)—were less precise than originally believed and probably not indicative of a special link between the two. Nonetheless, palaeanodonts do share a number of basicranial features with xenarthrans, including the size of the entotympanic and its relationship to the basicranium (the entotympanic forms the entire medial wall of the auditory bulla, as well as part of the posterior and ventral walls; it is present in the wall of the jugular foramen and contains the internal carotid foramen); the presence of a groove for the occipital artery on the occipital bone (not present in pholidotans); and the position of the tympanohyal relative to the facial nerve (the nerve exits the skull anterior to the tympanohyal) (Patterson et al., 1992; Gaudin, 1995; Gaudin and Wible, 1999). The derived entotympanic traits are also found in extant Carnivora, but early

Tertiary miacids apparently lack an entotympanic, suggesting an independent origin of these traits in carnivorans.

In the postcranial skeleton of palaeanodonts, the large scapular acromion process that overhangs the humeral head resembles that of armadillos and vermilinguans and is lacking in manids, but a similar acromion also occurs in unrelated digging mammals. Szalay (1977) suggested that the astragalus of palaeanodonts is similar to that of xenarthrans (e.g., dasypodids), but the resemblance does not appear to be any closer than that to primitive pholidotans such as *Patriomanis*. There is, however, a much closer resemblance between the astragali of *Palaeanodon* and middle Eocene *Eomanis krebsi* (see Szalay and Schrenk, 1998: fig. 7). Although palaeanodonts resemble xenarthrans more than pholidotans in having anapophyses and large metapophyses on the trunk vertebrae (presumably Simpson's incipient xenarthry), the anapophyses are non-articulating, and many other non-xenarthran mammals have such non-articulating processes (notably, some Carnivora). Palaeanodonts lack the diagnostic xenarthran traits of xenarthrous vertebral articulations and ischiosacral ossification. In summary, except for some basicranial features in common, there appear to be few compelling morphological characters shared by palaeanodonts and Xenarthra.

The anatomy of palaeanodonts generally favors a close relationship with Pholidota (Emry, 1970), based on particular similarities between *Eomanis* and palaeanodonts, including the reduction and loss of teeth (e.g., in metacheiromyids); presence of a mandibular medial buttress and a loosely attached, C-shaped premaxilla; and detailed resemblances in the postcranial skeleton, especially the forelimb (Rose and Emry, 1993; Storch, 2003). The latter similarities include a scapula of similar shape, with elevated spine, long acromion, and thickened caudal border; humerus with a raised, broad deltopectoral shelf, expanded supinator crest, and prominent entepicondyle; ulna with long, medially inflected olecranon (*Eomanis, Necromanis,* and *Manis,* but not *Patriomanis;* also present in *Eurotamandua*); radius expanded distally; short and broad metapodials and proximal phalanges; and similar enlarged, curved, unfissured manual ungual phalanges. Although some of these conditions have evolved in some xenarthrans (and other groups), the premaxillary features, medial buttress, and detailed resemblances in humeral morphology are specific to *Eomanis* and palaeanodonts. Notably, *Eurotamandua* resembles palaeanodonts in many of the same features, including the otherwise unique premaxillary and humeral characteristics. Palaeanodonts apparently lacked dermal ossicles, but whether scales were present is unknown. Although it cannot be entirely ruled out that the similarities between palaeanodonts and pholidotans arose independently in response to similar habits, the weight of currently known anatomical evidence suggests that palaeanodonts are more closely related to Pholidota than to Xenarthra.

Palaeanodonta has usually been thought to be related to, if not derived from, early Tertiary Pantolestidae, based on resemblances in the dentition and tarsus (Szalay, 1977; Rose,

1978). A more distant relationship to Leptictidae has also been suggested, but the evidence is relatively weak (Szalay, 1977). The oldest known palaeanodont, early Paleocene *Escavadodon,* exhibits derived traits distinctive of Palaeanodonta in its forelimb and hindlimb skeleton. Its dentition, however, is very primitive, showing no reduction in the number or size of the teeth, which closely resemble in crown morphology those of primitive leptictids and pantolestids (Rose and Lucas, 2000). Indeed, the resemblance to leptictids is so great that the specimens of *Escavadodon* were initially believed to represent a new leptictid (Williamson et al., 1994; Williamson, 1996).

MOLECULAR EVIDENCE FOR PROPOSED RELATIONSHIPS

Little molecular evidence has been put forth to support Edentata (Xenarthra + Pholidota). Only one molecular study (eye-lens proteins; McKenna, 1992) supported a Xenarthra + Pholidota clade, but other eye-lens protein studies (de Jong et al., 1985, 1993) rejected such a relationship. Most molecular studies strongly support a monophyletic Xenarthra, either as a distinct clade or as the sister taxon of Afrotheria, but with no special relationship to Pholidota (e.g., Czelusniak et al., 1990; Honeycutt and Adkins, 1993; Springer et al., 1997; Van Dijk et al., 1999; Eizirik et al., 2001; Madsen et al., 2001; Murphy et al., 2001; Delsuc et al., 2002). Most of these studies place Xenarthra (or Xenarthra and Afrotheria) at or very near the base of Eutheria and suggest a Gondwanan origin for both groups. In contrast, studies of mitochondrial DNA led Árnason et al. (1997) to propose that Xenarthra is closely related to Ferungulata, a broad grouping of carnivorans and ungulates.

Most molecular research places Pholidota as the sister taxon of Carnivora (de Jong et al., 1985; Czelusniak et al., 1990; Honeycutt and Adkins, 1993; Madsen et al., 2001; Murphy et al., 2001; Delsuc et al., 2002; Amrine-Madsen et al., 2003). However, Madsen et al. (2001) also found support for a relationship between Pholidota and Chiroptera + Eulipotyphla.

Supertree analysis also supports Xenarthra as the most primitive eutherians and a clade composed of Pholidota + Carnivora (Liu et al. 2001).

TIME AND PLACE OF ORIGIN OF PHOLIDOTA, XENARTHRA, AND POTENTIALLY RELATED TAXA

Smith and Peterson (2002) noted the difficulty of establishing the time of origin of major clades, because of preservational biases in the fossil record (especially the relatively incomplete Cretaceous record) and problems with molecular clocks (determining reliable calibration points, accurate rooting of phylogenies, and estimates of branch length). With these caveats in mind, in the following section we assess the morphological (including paleontological) evidence for the time and place of origin of Pholidota and Xenarthra.

Pholidota

Non-palaeanodont pholidotans are first known from the middle Eocene of Europe and the late Eocene of Asia and North America.

EOMANIS. *Eomanis* (Figs. 8.4, 8.6), from the middle Eocene of Messel, Germany, is the oldest known pholidotan (excluding palaeanodonts). Storch (2003) places it in the new family Eomanidae, whereas Szalay and Schrenk (1998) assigned it to Patriomanidae. Pholidotan synapomorphies present in *Eomanis* include the absence of teeth, presence of keratinous scales, anteriorly-directed prong-like projections at the front of the jaw (incipient in *Eomanis*), a large epitympanic sinus in the squamosal, loss of the coracoid process of the scapula, a reduced lesser trochanter, reduction in the size of the obturator foramen, and double transverse processes on some caudals (Storch, 1978, 2003). Like other pholidotans, *Eomanis* has a distally deep radial shaft and an enlarged ungual phalanx on the middle digit of the manus. *Eomanis* is more primitive than other pholidotans *sensu stricto,* but not more so than palaeanodonts, in retaining a

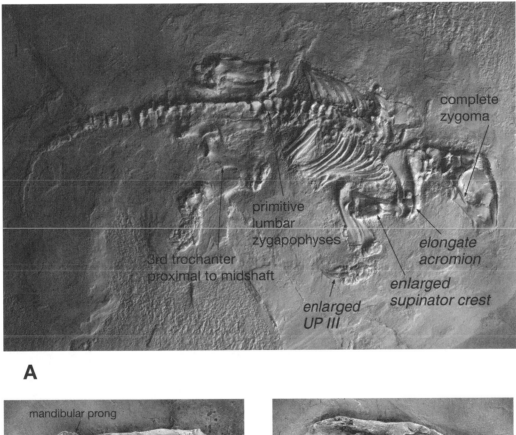

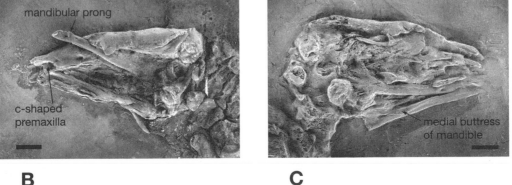

Fig. 8.6. (A) Skeleton of *Eomanis waldi* (SMF MEA 263), from Messel, in right lateral view. Features labeled in normal type are primitive characteristics of *Eomanis;* those labeled in italics are derived features shared with palaeanodonts. UP indicates ungual phalanx. (B) Skull of *Eomanis waldi* (Pohl specimen) in left ventrolateral view, showing the derived C-shaped premaxilla shared with palaeanodonts and *Eurotamandua,* and the incipient mandibular prong shared with pangolins. (C) Skull of *Eomanis waldi* (LNK ME 718) in right ventrolateral view, showing the derived medial buttress of the lower jaw shared with palaeanodonts. Scale bars = 1 cm for B and C. Institutional abbreviations: LNK, Landessammlungen für Naturkunde, Karlsruhe, Germany; SMF, Senckenberg Museum, Frankfurt, Germany.

clavicle, a long scapular acromion, separate scaphoid and lunar bones in the carpus, a higher third trochanter (proximal to midshaft), and unfissured ungual phalanges, and in lacking embracing (tightly interlocking) lumbar zygapophyses.

Resemblances between palaeanodonts and the most primitive pholidotans (*Eomanis* and *Patriomanis*) suggest that Palaeanodonta is the sister taxon, or possibly the direct ancestor, of Pholidota *sensu stricto*. No more plausible sister taxon is known. The closest anatomical similarities exist between the middle Eocene palaeanodont *Metacheiromys* and the middle Eocene pholidotan *Eomanis*. Thus, Pholidota probably either diverged from a common ancestor with Palaeanodonta very early in the Paleocene, or evolved directly from a palaeanodont as late as the early Eocene. Like early pholidotans, palaeanodonts are known from the three northern continents, so the place of origin of Pholidota is uncertain. Synapomorphies of *Eomanis* and palaeanodonts include a medial buttress on the dentary (previously considered an autapomorphy of palaeanodonts); a loosely attached, C-shaped premaxilla; a raised, shelflike deltopectoral crest and expanded supinator crest on the humerus; and an elevated spine and long acromion process of the scapula (humeral and scapular traits also in *Ernanodon*; scapular traits also in xenarthrans; Fig. 8.6).

PATRIOMANIS. *Patriomanis* from the late Eocene (Chadronian) of North America is the only North American pangolin. It is more primitive than other non-palaeanodont pholidotans (except *Eomanis*) in having a coronoid process on the mandible, a well-developed lambdoid crest, a globose petrosal promontorium, a large lateral iliac crest and a gluteal fossa on the ilium, a fovea capitis on the femoral head, a more proximal third trochanter, a well-developed trochanteric fossa, a tibial crest, a more convex astragalar head (navicular facet), metapodials lacking a dorsal keel, less sharply keeled articular facets of manual phalanges, and in retaining a dorsal astragalar foramen. Like other pholidotans, *Patriomanis* is completely edentulous and has well developed anterior mandibular prongs and a spoutlike symphysis. *Patriomanis* has a number of pangolin synapomorphies not present in *Eomanis,* including a posterodorsally inclined dorsal premaxillary process, embracing lumbar zygapophyses, probable loss of the clavicle, a distal pulley for the biceps at the end of the deltopectoral crest, a fused scapholunar, and fissured unguals (Emry, 1970; Gaudin and Emry, 2002).

EOCENE PHOLIDOTAN FROM ASIA. The existence of an unnamed pangolin in the late Eocene of Inner Mongolia confirms the presence of pholidotans *sensu stricto* in the early Cenozoic of Asia. Although generally similar to *Patriomanis,* it is larger and clearly represents a distinct genus and species (Gaudin and Emry, 2002; see also Rose and Emry, 1993: figs. 7.9, 7.10). The new Asian genus is represented by a single skeleton that includes complete hind limbs, partial forelimbs, and the vertebral column, but no skull. It has most of the pholidotan synapomorphies found in the postcranial skeleton of *Patriomanis,* such as fissured unguals, fused scapholunar, and embracing lumbar zygapophyses. At the same time, it retains several primitive features retained in *Patriomanis,* including a proximally situated third trochanter (even more proximal than in *Patriomanis*), a fovea capitis on the femur, a well-developed tibial crest, and a convex astragalar head (navicular facet).

Xenarthra

The unusual autapomorphies of xenarthrans, in combination with multiple supposed plesiomorphic features, suggest that Xenarthra could have diverged from other eutherians "well back in the Cretaceous" (McKenna, 1975: 28), but the fossil record documents xenarthrans (dasypodids) only as far back as the late Paleocene (Itaboraian). The phylogenetic source and precise date of origin of Xenarthra remain virtually unknown.

Evolution of Xenarthra seems to have been largely restricted to South America, but several controversial forms complicate the issue: *Eurotamandua, Asiabradypus,* Antarctic supposed xenarthrans, *Ernanodon,* Bibymalagasia, and *Chungchienia.*

DASYPODIDAE. The oldest armadillos (*Riostegotherium*) are based on scutes from Itaboraí, Brazil (Scillato-Yané, 1976; Oliveira and Bergqvist, 1998, 1999), and astragali and other postcrania (Cifelli, 1983; Bergqvist and Oliveira, 1995). Itaboraí is considered middle or late Paleocene in age, preceding the late Paleocene Riochican South American Land-Mammal Age (SALMA; Bond et al., 1995). Judging from osteoderms, dasypodids appear to have been moderately diverse during both the Itaboraian and the Riochican SALMAs (Oliveira and Bergqvist, 1999), suggesting that they must have existed by the early Paleocene, if not before. Vizcaíno et al. (1998), however, note that no dasypodids have yet been found in older beds (middle Paleocene) of Patagonia, although other mammals have been found; they also note that dasypodids are relatively rare during the Paleogene, suggesting that southern South America may not have been their center of origin. The oldest dasypodid known from craniodental material, Casamayoran *Utaetus,* had simple cylindrical or elliptical teeth similar to those of the extant armadillo *Euphractus* (Simpson, 1948). The Casamayoran SALMA has conventionally been considered early Eocene (e.g., Flynn and Swisher, 1995), but recent isotopic dates suggest an age as young as late Eocene (Kay et al., 1999). No dasypodids have been found in Antarctica, but the fossil record from there remains very poor. The oldest relatively complete dasypodid skeletons come from the early Miocene Santacrucian beds of Patagonia (Fig. 8.7A).

Bergqvist et al. (2004) report tantalizing new forelimb elements from Itaboraí (partial humeri and ulna), which they attribute to Xenarthra *incertae sedis.* The proximal ulna resembles that of dasypodids in having a long, straight olecranon, with only the tip medially inflected. The humeri, however, are more like those of palaeanodonts than those of xenarthrans, in having a prominent teres tubercle; an

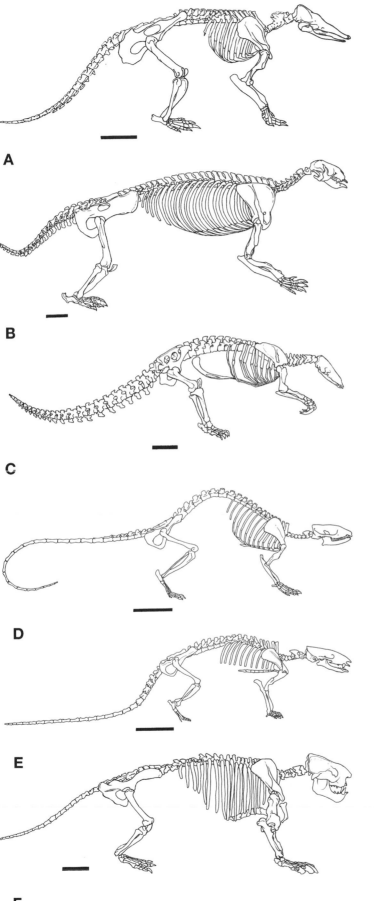

Fig. 8.7. Skeletons of "edentates" in right lateral view. (A) Miocene (Santacrucian) armadillo *Stegotherium tesselatum;* modified from Scott (1903–1904; carapace excluded). (B) Miocene (Santacrucian) ground sloth *Hapalops longiceps;* modified from Scott (1903–1904). (C) Extant pangolin *Manis gigantea;* modified from Kingdon (1974). (D) Early Paleocene palaeanodont *Escavadodon zygus;* modified from Rose and Lucas (2000). (E) Middle Eocene palaeanodont *Metacheiromys tatusia;* modified from Simpson (1931). (F) Late Paleocene *Ernanodon antelios;* modified from Ding (1987). Scale bars = 5 cm for (A), (D), (E), and (F); scale bars = 10 cm for (B) and (C).

A

B

C

D

E

F

expanded supinator crest; and a long, broad, shelflike delto-pectoral crest that is directed anteromedially. That these are both the oldest and the most palaeanodont-like postcrania from South America lends additional support to a possible link between palaeanodonts and Xenarthra.

GLYPTODONTOIDEA.   Based on scutes described by Ameghino, glyptodonts (e.g., *Glyptatelus;* possibly others) were present by the Mustersan (Simpson, 1948). The Mustersan SALMA is generally regarded as middle Eocene in age, but new evidence suggests that it is late Eocene (Kay et al., 2002). In addition, *Machlydotherium,* a possible pampathere, is known from scutes and a tooth, the latter distinctly pampathere-like, from the Mustersan and possibly the underlying Casamayoran (Simpson, 1948). An astragalus from Itoboraí—questionably referred to Glyptodontidae by Cifelli (1983)—is probably dasypodoid (Bergqvist and Oliveira, 1998). Simpson (1948: fig. 23) attributed a jaw fragment from the Mustersan with a single trilobed cheek tooth and part of another to an indeterminate glyptodont, but this specimen is now believed to belong to the phyllophagan *Pseudoglyptodon* (Wyss et al., 1994); however, later undoubted glyptodonts have trilobed cheek teeth. Relatively complete glyptodonts are not known until the early Miocene (Santacrucian SALMA).

PHYLLOPHAGA.   Assuming *Asiabradypus* (see below) is not a xenarthran, the oldest known sloth is *Pseudoglyptodon,* known from a partial cranium and mandible with complete dentition from the Tinguiririrican SALMA (late Eocene?) of Chile, and a mandible from the Deseadan (late Oligocene) of Bolivia (Engelmann, 1987; Wyss et al., 1990, 1994). *Pseudoglyptodon* has a reduced dental formula (5/4) with caniniform anterior teeth, but the other teeth have trilobate crowns, similar to those of glyptodonts. These trilobate teeth are, however, histologically more slothlike. The early Miocene pilosan *Entelops* had a heterodont dentition (3.1.3.3), including bilobed molars, all lacking enamel. These fossils suggest that relatively unreduced cheek teeth with lobate crowns could be primitive for Xenarthra, or at least for phyllophagans (Engelmann, 1987). The relationships of *Pseudoglyptodon* and *Entelops* to other sloths, however, are uncertain.

Octodontotherium and *Orophodon,* also from the Deseadan, are the oldest genera with clear relationships to later sloths. Although sometimes placed in their own family (e.g., McKenna and Bell, 1997), they share a number of derived cranial features with Mylodontidae that support allocation to that family, including loss of the epitympanic sinus, similar glenoid morphology, an inflated pterygoid, laterally-directed tympanohyal, mastoid and stylohyal fossa exposed laterally, a reduced entoglenoid process, and a more anterior eustachian tube (Patterson et al., 1992; Gaudin, 1995). The cheek teeth of *Octodontotherium* are lobate, like those of mylodontids. As for dasypodids, the oldest good skeletons of sloths date from the early Miocene Santacrucian deposits (Fig. 8.7B).

VERMILINGUA.   The oldest South American anteater is from the early Miocene (Colhuehuapian) of Patagonia (Carlini et al., 1992). From the succeeding Santacrucian-Friasian (early-middle Miocene) come *Protamandua* and *Neotamandua* (Hirschfeld, 1976), which are similar to modern myrmecophagids. Because of their relatively advanced morphology, these fossils provide little indication of the source of vermilinguans. If xenarthran anteaters and sloths are sister taxa (Pilosa), and anteaters did not evolve from sloths, they must have a ghost lineage extending back at least to the late Eocene.

ANTARCTIC XENARTHRANS.   Very fragmentary fossils identified as xenarthran are known from the La Meseta Formation of Seymour Island, of late Eocene age. Carlini et al. (1990) reported an ungual phalanx which they attributed to a megatherioid sloth. It appears to belong to a pilosan, but it may lack diagnostic traits to determine whether it represents a sloth or a vermilinguan (Marenssi et al., 1994). Vizcaíno and Scillato-Yané (1995) reported a fragmentary caniniform tooth which they assigned to an indeterminate tardigrade (=phyllophagan), seemingly confirming the presence of Xenarthra in the late Eocene of Antarctica. If the identity of these fossils is confirmed, the origin of Pilosa must extend back at least to the late Eocene, and could have occurred outside of South America.

Problematic Taxa

EUROTAMANDUA.   *Eurotamandua* (Fig. 8.5), from the middle Eocene of Messel, Germany, has been considered to be the oldest known myrmecophagid xenarthran (vermilinguan) by Storch (1981 and later), but its relationships have been highly controversial. It has alternatively been regarded as a close relative of *Eomanis* (and by extension, palaeanodonts; Rose and Emry, 1993; Rose, 1999), the sister taxon of all other vermilinguans (if constrained to be a vermilinguan) or of Pilosa (Gaudin and Branham, 1998), a pholidotan (McKenna and Bell, 1997; Shoshani et al., 1997), or a possible member of the pholidotan-palaeanodont assemblage (Cifelli, 1983). Szalay and Schrenk (1998) rejected both vermilinguan and xenarthran affinities and instead assigned it to its own clade, Afredentata. However, they considered it likely that Xenarthra, Palaeanodonta, and Afredentata form a monophyletic clade that originated in Africa during the early Cenozoic. If confirmed as a xenarthran, *Eurotamandua* would be the only one known from the Old World. Rose and Emry (1993) suggested that *Eurotamandua* is, in many respects, anatomically similar to *Eomanis*. Furthermore, the recent accounts by Szalay and Schrenk (1998), Shoshani et al. (1997), and Rose (1999) show that *Eurotamandua* shares many derived skeletal traits with palaeanodonts. Szalay and Schrenk (1998: 177) allowed the possibility that *Eurotamandua* might eventually prove to be the sister group of palaeanodonts.

According to Storch (1981, 2003), *Eurotamandua* shares a number of cranial features with xenarthrans and/or vermilinguans (listed above in the Afredentata section). An additional feature shared with pilosans is the absence of a postglenoid foramen. *Eurotamandua* also shares several derived features with palaeanodonts, particularly *Metacheiromys*, including a loosely attached, C-shaped premaxilla (also present in *Eomanis*); a lacrimal with subequal orbital and facial portions (in manids the lacrimal is restricted to the orbit, whereas xenarthrans have a large facial and small orbital exposure); a more anteriorly situated infraorbital foramen; and short, robust digits (also found in *Eomanis*). *Eurotamandua* further resembles palaeanodonts in having enlarged metapophyses and anapophyses of the posterior thoracic and lumbar vertebrae (cf. Rose et al., 1992), and in having an ossified auditory bulla (also present in vermilinguans and most other xenarthrans; Storch, 1981). Both *Eurotamandua* and *Eomanis* are derived in having small coronoid and angular processes of the mandible. In both genera the zygomatic process of the squamosal is short, tapered, and swollen at the base. Storch (1981) compared the forelimb anatomy of *Eurotamandua* to that of myrmecophagids, whereas Rose (1999) contended that derived details of the forelimb (especially the humerus and third metacarpal, including its unique distal joint morphology) are shared with palaeanodonts and *Eomanis* (as far as can be compared) and contrast with those of myrmecophagids. *Eurotamandua* resembles living pangolins in having enclosed spinal nerve foramina in some caudal vertebrae, a ventral projection on the transverse process of the third sacral vertebra for the sacrospinous ligament, and a vertically oriented distal articular surface on the radius (Gaudin, unpublished observations). These features are either not present or not preserved in *Eomanis*.

The interpretation of *Eurotamandua* continues to be controversial, largely due to its state of preservation, as noted earlier. One of us (GS) maintains that the most critical features link *Eurotamandua* with myrmecophagid xenarthrans. The other authors (RJE, TJG, and KDR) believe that *Eurotamandua* and *Eomanis* bear close phenetic similarity to palaeanodonts—for *Eurotamandua*, closer than to Xenarthra—which strengthens the probability that the Messel taxa are closely related to Palaeanodonta, although not members of that group. If *Eomanis* is a pholidotan *sensu stricto*, as we believe, then this suggests that *Eurotamandua* is also a pholidotan, or closely allied to Pholidota, rather than a xenarthran or an afredentate, and that Pholidota is the sister taxon or direct descendant of Palaeanodonta.

ASIABRADYPUS. *Asiabradypus,* from the late Paleocene of Kazakhstan, was proposed as a pilosan, questionably a mylodontid sloth, by Nessov (1987). McKenna and Bell (1997) assigned it to Ernanodonta, which they included in the eclectic order Cimolesta. *Asiabradypus* is based on a jaw fragment with part of two teeth. It superficially resembles a sloth in the reduced number and complexity of teeth, apparent lack of enamel, gabled wear, and large alveoli. Closer examination of the single known specimen (which is much smaller than any known sloth), suggests that what have been

interpreted as two single-rooted, peglike teeth are actually the broken roots of a single posterior molar. Accordingly, we consider *Asiabradypus* to be an indeterminate non-xenarthran mammal that is irrelevant to xenarthran origins. A jaw fragment from the earliest Eocene of Wyoming referred to cf. *Asiabradypus* (Gingerich, 1989) differs in jaw shape and size and configuration of the alveoli, and is probably not referable to *Asiabradypus* (whatever its true affinities).

ERNANODON. *Ernanodon* (Fig. 8.7F), from the late Paleocene of China, was assigned to a new suborder of Edentata (*sensu* Xenarthra) by Ding (1987). The suborder Ernanodonta was included in the order Cimolesta by McKenna and Bell (1997), close to Pholidota (including palaeanodonts) but unrelated to Xenarthra. *Ernanodon* is characterized by a very robust skull and skeleton, giving it a superficial resemblance to ground sloths. It has a xenarthran-like, spout-shaped mandibular symphysis, which also occurs in palaeanodonts and pangolins in association with loss of the incisors. The dental formula is 0.1.3.3 / 1.1.4.3, and the teeth are single-rooted, pointed pegs (except m2, which is two-rooted) with thin enamel. The caniniform teeth occlude like true canines, unlike the caniniforms of sloths. The posterior thoracic vertebrae of *Ernanodon,* like those of xenarthrans, have enlarged metapophyses and anapophyses, but the anatomy differs in detail, and the processes of *Ernanodon* are non-articulating (Gaudin, 1999). The lumbar zygapophyses are embracing, like those of pangolins. Other potential xenarthran similarities include ossified sternal ribs that articulate with the sternum; a scapula with an elevated spine, a bifid acromion, and a caudal border thickened into a "secondary scapular spine"; a humerus with an elevated, very wide deltopectoral shelf and large entepicondyle and supinator crest; and an ulna with a prominent, medially inflected olecranon. The forelimb specializations, however, are more like those of palaeanodonts than those of xenarthrans. There is widespread agreement that *Ernanodon* belongs to a distinct clade, but its broader relationships are in dispute. It is possible that it represents an independent lineage convergent to other "edentates," but its overall anatomy and particular specializations of the forelimb suggest that it may be closer to palaeanodonts than to any other group.

CHUNGCHIENIA. *Chungchienia,* from the middle Eocene of China, was based on a jaw fragment with one hypselodont cheektooth with limited enamel. It was initially considered to be a megalonychoid sloth (Chow, 1963) and subsequently was attributed to Taeniodonta or Tillodontia. More complete specimens clearly establish it as a highly derived tillodont (Chow et al., 1996).

BIBYMALAGASIA. MacPhee (1994) named the new order Bibymalagasia to accommodate the extinct Holocene genus *Plesiorycteropus* from Madagascar, which was long considered a tubulidentate (Patterson, 1975). *Plesiorycteropus* is relevant to the present discussion because of its striking resemblance throughout the skeleton to digging mammals other than aardvarks—in particular, armadillos, pangolins, and, to a lesser extent, vermilinguans (MacPhee, 1994). Which skeletal features are phylogenetically significant and which ones evolved convergently have confounded researchers for a century. MacPhee's phylogenetic analyses, based on his painstakingly thorough anatomical study of the skeleton, found *Plesiorycteropus* to be broadly related to ungulates but not specifically to aardvarks or, as one alternative (with a more limited taxon set), the sister taxon to Edentata (Xenarthra + Pholidota). He assigned it to a new order to highlight its distinctiveness. It is likely that *Plesiorycteropus* had a long pedigree, as yet unknown, in Madagascar or Africa. It may prove to be part of a clade of African ungulates, afrotheres, or "edentates," but determining which, if any, of these alternatives is correct seems impossible at present.

Palaeanodonta

The oldest palaeanodont is the late early Paleocene (Torrejonian) *Escavadodon* from western North America (Fig. 8.7D). Its anatomy is more primitive than any other palaeanodont, and its dentition is particularly close to that of leptictids and pantolestids (both of which are first known from the Puercan). This combination of features indicates that *Escavadodon* is not far removed from the base of the palaeanodont clade. General derived resemblances in the skeleton and dentition of primitive palaeanodonts and pantolestids suggest that these two groups may share a close common ancestor. As the earliest members of all these groups are from the early Paleocene, Palaeanodonta must have originated no later than the very early Paleocene, and, most likely, before.

SUMMARY

Skeletal attributes related to digging and ant-eating have clearly evolved multiple times in unrelated mammals and have sometimes attained astonishing similarity. But to dismiss such features as phylogenetically uninformative because of their association with function may result in the loss of significant data on relationships. The challenge is to discriminate meaningful similarities and differences in characters adapted for the same purpose. Pholidota and most Xenarthra share general anatomical similarities associated with fossorial habits and a tendency toward dental reduction sometimes associated with myrmecophagy, but many of the resemblances do not stand up to careful scrutiny and appear to be homoplasious. The morphological evidence to support Edentata, encompassing Xenarthra + Pholidota (excluding palaeanodonts), therefore is weak at best. If palaeanodonts are included within Pholidota, it becomes more difficult to rule out a possible close relationship to Xenarthra, but at present, there seems to be no compelling evidence, either morphological or molecular, to support such a link. Likewise, morphological evidence upholding a

special relationship between Pholidota and Carnivora is very meager, although molecular support for such an arrangement is somewhat stronger. Xenarthra possess a number of features that have been claimed to be primitive, which, if true, is consistent with a position near the base of Placentalia, although not necessarily the most basal group. This is generally supported by molecular data as well. Otherwise, the position of both Xenarthra and Pholidota relative to other extant orders is poorly understood.

Surely contributing to the ambiguity of the relationships of Xenarthra and Pholidota to one another and to other mammals is their very meager early fossil record. Although a few excellent non-palaeanodont pholidotan skeletons are known from the Eocene, only five genera are known from the entire Tertiary. Fragmentary fossils record the presence of xenarthrans in the Paleocene, but the first relatively complete skulls and skeletons of undoubted xenarthrans are not known until the Miocene.

Early Tertiary palaeanodonts show similar adaptations to Pholidota *sensu stricto* and many Xenarthra, and the oldest members predate known fossils of both extant orders, which has led many authors to suggest that they are pertinent to the origins and relationships of one or both orders. The earliest known palaeanodont, from the late early Paleocene (Torrejonian), was dentally unspecialized, but the postcranial skeleton was already relatively derived, suggesting that the group probably originated early in the Paleocene (Puercan), if not before. They may have shared a common ancestor with pantolestoids. Palaeanodonts show a mix of xenarthran and pholidotan (pangolin) anatomical features, but the resemblances to primitive pangolins, such as *Eomanis* and *Patriomanis,* seem to be both more numerous and more detailed, suggesting that Palaeanodonta is the sister taxon or includes the direct ancestor of Pholidota *sensu stricto*. This does not necessarily rule out a relationship between palaeanodonts and Xenarthra as well, for which there is limited basicranial and postcranial evidence. Otherwise, however, the origin of Xenarthra remains nebulous. All that is known for certain is that xenarthrans existed in South America by the late Paleocene (Itaboraian). Paleocene *Ernanodon* seems to represent an independent Asian clade, perhaps more closely related to Palaeanodonta than to any other group.

Middle Eocene *Eomanis* is clearly a primitive pangolin, but in many ways it is structurally intermediate between palaeanodonts and other pangolins, strongly suggesting a palaeanodont source for Pholidota *sensu stricto*, perhaps in the early Eocene. However, if palaeanodonts are included within Pholidota, as evidence currently supports, the order dates back to at least the early Paleocene.

There is no consensus on the phylogenetic position of *Eurotamandua*. It may be a xenarthran and the oldest vermilinguan (GS) or more closely related to palaeanodonts and *Eomanis* (RJE, TG, KDR). Others have argued that *Eurotamandua* represents a new order, Afredentata, independent from all of these groups but possibly related to Palaeanodonta and/or Xenarthra (Szalay and Schrenk, 1998).

ACKNOWLEDGMENTS

We are grateful to Lílian Bergqvist for providing her pre-publication manuscript and images of new Itaboraian xenarthrans. We thank Alexander Averianov for permission to examine *Asiabradypus* and David Archibald for images of this fossil. Norbert Micklich (Hessisches Landesmuseum, Darmstadt, Germany) kindly supplied the skull image of *Eomanis* in Fig. 8.5. Julia Morgan Scott prepared Fig. 8.7. We are indebted to Drs. Greg McDonald, Fred Szalay, and Sergio Vizcaíno, who suggested numerous improvements to the manuscript, although they do not necessarily endorse all the views presented here. Support from National Science Foundation RUI Grant DEB 0107922 to TJG is gratefully acknowledged.

REFERENCES

Ameghino, F. 1889. Contribucion al conocimiento de los mamíferos fósiles de la Republica Argentina. Actas de la Academia Nacional de Ciencias Córdoba, Buenos Aires 6, atlas, 98 pls.

———. 1894. Enumération synoptique des espèces de mammifères fossiles des formations éocènes de Patagonie. Boletin de la Academia Nacional de Ciencias Córdoba, Buenos Aires 13: 259–445.

———. 1920. Sur les édentés fossiles de l'Argentine. Examen critique, révision et correction de l'ouvrage de la M. R. Lydekker. Obras Completas y Correspondencia Cientifica 11: 447–909.

Amrine-Madsen, H., K.-P. Koepfli, R. K. Wayne, and M. S. Springer. 2003. A new phylogenetic marker, apolipoprotein B, provides compelling evidence for eutherian relationships. Molecular Phylogenetics and Evolution 28: 225–240.

Árnason, Ú., A. Gullberg, and A. Janke. 1997. Phylogenetic analyses of mitochondrial DNA suggest a sister group relationship between Xenarthra (Edentata) and ferungulates. Molecular Biology and Evolution 14: 762–768.

Bergqvist, L. P., E.A.L. Abrantes, and L. Avilla. 2004. The Xenarthra of São José de Itaboraí Basin (upper Paleocene, Itaboraian), Rio de Janeiro, Brazil. Geodiversitas 26: 323–337.

Bergqvist, L. P., and E. V. Oliveira. 1995. Novo material póscraniano de Cingulata (Mammalia-Xenarthra) do Paleoceno médio (Itaboraiense) do Brasil. 14° Congresso Brasileiro de Paleontologia (Sociedade Brasiliera de Paleontologia, Uberaba): 16–17.

———. 1998. Comments on the xenarthran astragali from the Itaboraí basin (middle Paleocene) of Rio de Janeiro, Brazil. Acta Geológica Lilloana, Tucumán 18 (1): 153–154.

Bonaparte, J. 1990. New Late Cretaceous mammals from the Los Alamitos Formation, northern Patagonia. National Geographic Research 6: 63–93.

Bond, M., A. A. Carlini, F. J. Goin, L. Legarreta, E. O. Ortiz-Jaureguizar, R. Pascual, and M. A. Uliana. 1995. Episodes in South American land mammal evolution and sedimentation: Testing their apparent concurrence in a Paleocene succession from central Patagonia. 6° Congreso Argentino de Paleontología y Bioestratigrafia (Trelew): 47–58.

Carlini, A. A., R. Pascual, M. A. Reguero, G. J. Scillato-Yané, E. P. Tonni, and S. F. Vizcaíno. 1990. The first Paleogene land placental mammal from Antarctica: Its paleoclimatic and paleogeographical bearings; p. 325 *in* Abstracts, IV International Congress of Systematic and Evolutionary Biology. University of Maryland, College Park, Maryland.

Carlini, A. A., G. J. Scillato-Yané, S. F. Vizcaíno, and M. T. Dozo. 1992. Un singular Myrmecophagidae (Xenarthra, Vermilingua) de Edad Colhuehuapense (Oligoceno tardío-Mioceno temprano) de Patagonia, Argentina. Ameghiniana 29: 176.

Cartelle, C., and G. A. Bohórquez. 1986. Presença de ossículos dérmicos em *Eremotherium laurillardi* (Lund) Cartelle and Bohórquez, 1982 (Edentata, Megatheriidae). Iheringia, Série Geologia, Porto Alegre 11: 3–8.

Chow, M. 1963. A xenarthran-like mammal from the Eocene of Honan. Scientia Sinica 12: 1889–1893.

Chow, M., J. Wang, and J. Meng. 1996. A new species of *Chungchienia* (Tillodontia, Mammalia) from the Eocene of Lushi, China. American Museum Novitates 3171: 1–10.

Cifelli, R. L. 1983. Eutherian tarsals from the late Paleocene of Brazil. American Museum Novitates 2761: 1–31.

Coombs, M. C. 1983. Large mammalian clawed herbivores: A comparative study. Transactions of the American Philosophical Society 73(7): 1–96.

Cope, E. D. 1889. The Edentata of North America. American Naturalist 23: 657–664.

Corbet, G. B., and J. E. Hill. 1991. A World List of Mammalian Species. Third Edition. Oxford University Press, Oxford.

Cuvier, G.L.C.F.D. 1798. Tableau élémentaire de l'histoire naturelle des animaux. J. B. Baillière, Paris.

Czelusniak, J., M. Goodman, B. F. Koop, D. A. Tagle, J. Shoshani, G. Braunitzer, T. K. Kleinschmidt, W. W. de Jong, and G. Matsuda. 1990. Perspectives from amino acid and nucleotide sequences on cladistic relationships among higher taxa of Eutheria. Current Mammalogy 2: 545–572.

de Jong, W. W., J.A.M. Leunissen, and G. J. Wistow. 1993. Eye lens crystallins and the phylogeny of placental orders: Evidence for a macroscelidid-paenungulate clade; pp. 5–12 *in* F. S. Szalay, M. J. Novacek, and M. C. McKenna (eds.), Mammal Phylogeny: Placentals. Springer-Verlag, New York.

de Jong, W. W., A. Zweers, K. A. Joysey, J. T. Gleaves, and D. Boulter. 1985. Protein sequence analysis applied to xenarthran and pholidote phylogeny; pp. 65–76 *in* G. G. Montgomery (ed.), The Evolution and Ecology of Armadillos, Sloths, and Vermilinguas. Smithsonian Institution Press, Washington, D.C.

Delsuc, F., M. Scally, O. Madsen, M. J. Stanhope, W. W. de Jong, F. M. Catzeflis, M. S. Springer, and E.J.P. Douzery. 2002. Molecular phylogeny of living xenarthrans and the impact of character and taxon sampling on the placental tree rooting. Molecular Biology and Evolution 19: 1656–1671.

Ding, S.-Y. 1979. A new edentate from the Paleocene of Guangdong. Vertebrata PalAsiatica 17: 57–64.

———. 1987. A Paleocene edentate from Nanxiong Basin, Guangdong. Palaeontologia Sinica, new series C 173: 1–118.

Eizirik, E., W. J. Murphy, and S. J. O'Brien. 2001. Molecular dating and biogeography of the early placental mammal radiation. Journal of Heredity 92: 212–219.

Emry, R. J. 1970. A North American Oligocene pangolin and other additions to the Pholidota. Bulletin of the American Museum of Natural History 142: 455–510.

Engelmann, G. F. 1985. The phylogeny of the Xenarthra; pp. 51–64 *in* G. G. Montgomery (ed.), The Evolution and

Ecology of Armadillos, Sloths, and Vermilinguas. Smithsonian Institution Press, Washington, D.C.

———. 1987. A new Deseadan sloth (Mammalia: Xenarthra) from Salla, Bolivia, and its implications for the primitive condition of the dentition in edentates. Journal of Vertebrate Paleontology 7: 217–223.

Flower, W. H. 1883. On the arrangement of the orders and families of existing Mammalia. Proceedings of the Zoological Society of London 1883: 178–186.

Flynn, J. J., and C. C. Swisher, III. 1995. Cenozoic South American land mammal ages: Correlation to global geochronologies; pp. 317–333 in W. A. Berggren, D. V. Kent, M.-P. Aubry, and J. Hardenbol (eds.), Geochronology, Time Scales and Global Stratigraphic Correlation. SEPM Special Publication 54. Society for Sedimentary Geology, Tulsa.

Gaudin, T. J. 1993. Phylogeny of the Tardigrada (Mammalia, Xenarthra) and the evolution of locomotor function in the Xenarthra. Ph.D. Thesis, University of Chicago, Chicago.

———. 1995. The ear region of edentates and the phylogeny of the Tardigrada (Mammalia, Xenarthra). Journal of Vertebrate Paleontology 15: 672–705.

———. 1999. The morphology of xenarthrous vertebrae (Mammalia: Xenarthra). Fieldiana Geology, new series 41: 1–38.

Gaudin, T. J., and A. A. Biewener. 1992. The functional morphology of xenarthrous vertebrae in the armadillo *Dasypus novemcinctus* (Mammalia, Xenarthra). Journal of Morphology 214: 63–81.

Gaudin, T. J., and D. G. Branham. 1998. The phylogeny of the Myrmecophagidae (Mammalia, Xenarthra, Vermilingua) and the relationship of *Eurotamandua* to the Vermilingua. Journal of Mammalian Evolution 5: 237–265.

Gaudin, T. J., and R. J. Emry. 2002. The late Eocene pangolin *Patriomanis* from North America, and a new genus of pangolin from the late Eocene of Nei Mongol, China (Mammalia, Pholidota). Journal of Vertebrate Paleontology 22 (supplement to no. 3): 57A.

Gaudin, T. J., and J. R. Wible. 1999. The entotympanic of pangolins and the phylogeny of the Pholidota (Mammalia). Journal of Mammalian Evolution 6: 39–65.

Gaudin, T. J., J. R. Wible, J. A. Hopson, and W. D. Turnbull. 1996. Reexamination of the morphological evidence for the cohort Epitheria (Mammalia, Eutheria). Journal of Mammalian Evolution 3: 31–79.

Geological Society of America. 1999. *www.geosociety.org/science/timescale/timescl.htm.*

Gervais, F.L.P. 1855. Mammifères. Animaux nouveau, ou rares, recueillis pendant l'expédition dans les parties centrales de l'Amérique du Sud. P. Bertrand, Paris.

Gill, T. 1872. Arrangement of the families of mammals with analytical tables. Smithsonian Miscellaneous Collections 11(1): 1–98.

Gingerich, P. D. 1989. New earliest Wasatchian mammalian fauna from the Eocene of northwestern Wyoming: Composition and diversity in a rarely sampled high-floodplain assemblage. University of Michigan Papers on Paleontology 28: 1–97.

Glass, B. 1985. History of classification and nomenclature in Xenarthra (Edentata); pp. 1–3 in G. G. Montgomery (ed.), The Evolution and Ecology of Armadillos, Sloths, and Vermilinguas. Smithsonian Institution Press, Washington, D.C.

Grassé, P.-P. 1955. Ordre des Édentés; pp. 1182–1266 in P.-P. Grassé (ed.), Traité de Zoologie. Volume XVII (2). Mammifères. Masson et Cie, Paris.

Gray, J. E. 1821. On the natural arrangement of vertebrose animals. London Medical Repository 15(1): 296–310.

———. 1825. Outline of an attempt at the disposition of the Mammalia into tribes and families with a list of the genera apparently appertaining to each tribe. Annals of Philosophy, new series 10: 337–344.

———. 1869. Catalogue of Carnivorous, Pachydermatous, and Edentate Mammalia in the British Museum, London. British Museum (Natural History), London.

Gregory, W. K. 1910. The orders of mammals. Bulletin of the American Museum of Natural History 27: 3–524.

Hirschfeld, S. E. 1976. A new fossil anteater (Edentata, Mammalia) from Colombia, S.A. and evolution of the Vermilingua. Journal of Paleontology 50: 419–432.

Hoffstetter, R. 1958. Édentés Xénarthres; pp. 535–636 in J. Piveteau (ed.), Traité de Paléontologie. Volume VI (2). Mammifères, évolution. Masson et Cie, Paris.

Honeycutt, R. L., and R. M. Adkins. 1993. Higher level systematics of eutherian mammals: An assessment of molecular characters and phylogenetic hypotheses. Annual Review of Ecology and Systematics 24: 279–305.

Huxley, T. H. 1872. A Manual of the Anatomy of Vertebrated Animals. D. Appleton and Co., New York.

Illiger, C. 1811. Prodromus systematis mammalium et avium additis terminis zoographicis utriudque classis. C. Salfeld, Berlin.

Jenkins, F. A., Jr. 1970. Anatomy and function of expanded ribs in certain edentates and primates. Journal of Mammalogy 51: 288–301.

Kay, R. F., R. H. Madden, M. G. Vucetich, A. A. Carlini, M. M. Mazzoni, G. H. Re, M. Heizler, and H. Sandeman. 1999. Revised geochronology of the Casamayoran South American Land Mammal Age: Climatic and biotic implications. Proceedings of the National Academy of Sciences USA 96: 13235–13240.

Kay, R. F., R. H. Madden, A. A. Carlini, M. G. Vucetich, M. Bond, E. Bellosi, M. Heizler, G. H. Re, and J. F. Vilas. 2002. The Mustersan interval at Gran Barranca. Journal of Vertebrate Paleontology 22 (supplement to no. 3): 73A.

Kingdon, J. 1974. East African Mammals. An Atlas of Evolution in Africa. Volume 1. University of Chicago Press, Chicago.

Koenigswald, W. V., G. Richter, and G. Storch. 1981. Nachweis von Hornschuppen bei *Eomanis waldi* aus der "Grube Messel" bei Darmstadt (Mammalia, Pholidota). Senckenbergiana Lethaea 61: 291–298.

Liu, F.-G.R., M. M. Miyamoto, N. P. Freire, P. Q. Ong, M. R. Tennant, T. S. Young, and K. F. Gugel. 2001. Molecular and morphological supertrees for eutherian (placental) mammals. Science 291: 1786–1789.

MacPhee, R.D.E. 1994. Morphology, adaptations, and relationships of *Plesiorycteropus,* and a diagnosis of a new order of eutherian mammals. Bulletin of the American Museum of Natural History 220: 1–214.

Madsen, O., M. Scally, C. J. Douady, D. J. Kao, R. W. DeBry, R. Adkins, H. M. Amrine, M. J. Stanhope, W. W. de Jong, and M. S. Springer. 2001. Parallel adaptive radiations in two major clades of placental mammals. Nature 409: 610–614.

Marenssi, S. A., M. A. Reguero, S. N. Santillana, and S. F. Vizcaíno. 1994. Eocene land mammals from Seymour Island,

Antarctica: Palaeobiogeographical implications. Antarctic Science 6: 3–15.

Matthew, W. D. 1918. A revision of the lower Eocene Wasatch and Wind River faunas. Part V. Insectivora (continued), Glires, Edentata. Bulletin of the American Museum of Natural History 38: 565–657.

McKenna, M. C. 1975. Toward a phylogenetic classification of the Mammalia; pp. 21–46 *in* W. P. Luckett and F. S. Szalay (eds.), Phylogeny of the Primates. Plenum, New York.

———. 1992. The alpha crystallin A chain of the eye lens and mammalian phylogeny. Annales Zoologici Fennici 28: 349–360.

McKenna, M. C., and S. K. Bell. 1997. Classification of Mammals above the Species Level. Columbia University Press, New York.

Mones, A. 1987. Gondwanatheria, un nuevo orden de mamíferos sudamericanos (Mammalia: Edentata: ?Xenarthra). Comunicaciones Paleontológicas del Museo de Historia Natural de Montevideo 1(18): 237–240.

Murphy, W. J., E. Eizirik, W. E. Johnson, Y. P. Zhang, O. A. Ryder, and S. J. O'Brien. 2001. Molecular phylogenetics and the origins of placental mammals. Nature 409: 614–618.

Nel, A., G. de Plöeg, J. Dejax, D. Dutheil, D. de Franceschi, E. Gheerbrant, M. Godinot, S. Hervet, J.-J. Menier, M. Augé, G. Bignot, C. Cavagnetto, S. Duffaud, J. Gaudant, S. Hua, A. Jossang, F. de Lapparent de Broin, J.-P. Pozzi, J.-C. Paicheler, F. Beuchet, and J.-C. Rage. 1999. Un gisement sparnacien exceptionnel à plantes, arthropodes et vertébrés (Éocène basal, MP7): Le Quesnoy (Oise, France). Comptes Rendus de l'Académie des Sciences, Série II, Sciences de la Terre et des Planètes, Paris 329: 65–72.

Nessov, L. 1987. Rezultaty poiskov i issledovanija melovych i rannepaleogenovych mlekopitajushchi ich na territorii SSSR [Research on Cretaceous and Paleocene mammals of the territory of the USSR]. Ezhegodnik Vsesoyuznogo Paleontologicheskogo Obshchestva (Akademiya Nauk SSSR) [Annual of the All-Union Paleontological Society] 30: 199–218. (In Russian.)

Novacek, M. J. 1986. The skull of leptictid insectivorans and the higher-level classification of eutherian mammals. Bulletin of the American Museum of Natural History 183: 1–112.

———. 1992. Mammalian phylogeny: Shaking the tree. Nature 356: 121–125.

Novacek, M. J., and A. Wyss. 1986. Higher-level relationships of the recent eutherian orders: Morphological evidence. Cladistics 2: 257–287.

Oliveira, E. V., and L. P. Bergqvist. 1998. A new Paleocene armadillo (Mammalia, Dasypodoidea) from the Itaboraí Basin, Brazil. Asociación Paleontológica Argentina (Buenos Aires), Publicación Especial 5, Paleógeno de América del Sur y de la Península Antártica: 35–40.

———. 1999. A new Paleocene armadillo (Mammalia, Xenarthra, Astegotheriini) from Itaboraí, Brazil, and phylogeny of early Tertiary astegotheriines. Anais da Academia Brasileira de Ciências 71: 814–815.

Owen, R. 1842. Description of the Skeleton of an Extinct Gigantic Sloth, *Mylodon robustus,* Owen, with Observations on the Osteology, Natural Affinities, and Probable Habits of the Megatherioid Quadrupeds in General. Royal College of Surgeons of England, London.

Patterson, B. 1975. The fossil aardvarks (Mammalia: Tubulidentata). Bulletin of the Museum of Comparative Zoology, Harvard University 147: 185–237.

———. 1978. Pholidota and Tubulidentata; pp. 268–278 *in* V. J. Maglio and H.B.S. Cooke (eds.), Evolution of African Mammals. Harvard University Press, Cambridge.

Patterson, B., W. Segall, W. Turnbull, and T. Gaudin. 1992. The ear region in xenarthrans (=Edentata: Mammalia). Part II. Pilosa (sloths, anteaters), palaeanodonts, and a miscellany. Fieldiana Geology, new series 24: 1–79.

Paula Couto, C. de. 1954. Sôbre um gliptodonte do Uruguai e um tatú fóssil do Brasil. Notas Preliminares e Estudos, Divisão de Geologia e Mineralogia 80: 1–10.

Pocock, R. I. 1924. The external characters of the South American edentates. Proceedings of the Zoological Society of London 65: 983–1031.

Prothero, D. R. 1998. The chronological, climatic, and paleogeographic background to North American mammalian evolution; pp. 9–36 *in* C. M. Janis, K. M. Scott, and L. L. Jacobs (eds.), Evolution of Tertiary Mammals of North America. Volume 1. Terrestrial Carnivores, Ungulates, and Ungulatelike Mammals. Cambridge University Press, Cambridge.

Rose, K. D. 1978. A new Paleocene epoicotheriid (Mammalia), with comments on the Palaeanodonta. Journal of Paleontology 52: 658–674.

———. 1999. *Eurotamandua* and Palaeanodonta: Convergent or related? Paläontologische Zeitschrift 73: 395–401.

Rose, K. D., and R. J. Emry. 1993. Relationships of Xenarthra, Pholidota, and fossil "edentates": The morphological evidence; pp. 81–102 *in* F. S. Szalay, M. J. Novacek, and M. C. McKenna (eds.), Mammal Phylogeny: Placentals. Springer-Verlag, New York.

Rose, K. D., R. J. Emry, and P. D. Gingerich. 1992. Skeleton of *Alocodontulum atopum,* an early Eocene epoicotheriid (Mammalia, Palaeanodonta) from the Bighorn Basin, Wyoming. Contributions from the Museum of Paleontology, University of Michigan 28: 221–245.

Rose, K. D., and S. G. Lucas. 2000. An early Paleocene palaeanodont (Mammalia, ?Pholidota) from New Mexico, and the origin of Palaeanodonta. Journal of Vertebrate Paleontology 20: 139–156.

Scillato-Yané, G. 1976. Sobre un Dasypodidae (Mammalia, Xenarthra) de Edad Riochiquense (Paleoceno Superior) de Itaboraí, Brasil. Anais da Academia Brasileira de Ciências 48: 527–530.

Scillato-Yané, G., and R. Pascual. 1985. Un peculiar Xenarthra del Paleoceneo Medio de Patagonia (Argentina). Su importancia en la sistemática de los Paratheria. Ameghiniana 21: 173–176.

Scott, W. B. 1903–1904. Mammalia of the Santa Cruz Beds. Part 1: Edentata. Reports of the Princeton University Expeditions to Patagonia, 1896–1899. Volume 5. Princeton University, Princeton, New Jersey.

Shoshani, J., and M. C. McKenna. 1998. Higher taxonomic relationships among extant mammals based on morphology, with selected comparisons of results from molecular data. Molecular Phylogenetics and Evolution 9: 572–584.

Shoshani, J., M. C. McKenna, K. D. Rose, and R. J. Emry. 1997. *Eurotamandua* is a pholidotan not a xenarthran. Journal of Vertebrate Paleontology 17 (supplement to no. 3): 76A.

Simpson, G. G. 1927. A North American Oligocene edentate. Annals of Carnegie Museum 17: 283–298.

———. 1931. *Metacheiromys* and the Edentata. Bulletin of the American Museum of Natural History 59: 295–381.

———. 1945. The principles of classification and a classification of mammals. Bulletin of the American Museum of Natural History 85: 1–350.

———. 1948. The beginning of the Age of Mammals in South America. Bulletin of the American Museum of Natural History 91: 1–232.

Smith, A. B., and K. J. Peterson. 2002. Dating the time of origin of major clades: Molecular clocks and the fossil record. Annual Review of Earth and Planetary Sciences 30: 65–88.

Springer, M. S., A. Burk, J. R. Kavanagh, V. G. Waddell, and M. J. Stanhope. 1997. The interphotoreceptor retinoid binding protein gene in therian mammals: Implications for higher level relationships and evidence for loss of function in the marsupial mole. Proceedings of the National Academy of Sciences USA 94: 13754–13759.

Stock, C. 1925. Cenozoic gravigrade edentates of western North America, with special reference to the Pleistocene Megalonychinae and Mylodontidae of Rancho La Brea. Publication 331. Carnegie Institution of Washington, Washington, D.C.

Storch, G. 1978. *Eomanis waldi,* ein Schuppentier aus dem Mittel-Eozän der "Grube Messel" bei Darmstadt (Mammalia: Pholidota). Senckenbergiana Lethaea 59: 503–529.

———. 1981. *Eurotamandua joresi,* ein Myrmecophagide aus dem Eozän der "Grube Messel" bei Darmstadt (Mammalia, Xenarthra). Senckenbergiana Lethaea 61: 247–289.

———. 2003. Fossil Old World "edentates." Senckenbergiana Biologica 83: 51–60.

Storch, G., and J. Habersetzer. 1991. Rückverlagerte Choanen und akzessorische Bulla tympanica bei rezenten Vermilingua und *Eurotamandua* aus dem Eozän von Messel (Mammalia: Xenarthra). Zeitschrift Säugetierkunde 56: 257–271.

Storch, G., and M. Rummel. 1999. *Molaetherium heissigi* n. gen., n. sp., an unusual mammal from the early Oligocene of Germany (Mammalia: Palaeanodonta). Paläontologische Zeitschrift 73: 179–185.

Stucky, R. K., and M. C. McKenna. 1993. Mammalia; pp. 739–771 *in* M. J. Benton (ed.), The Fossil Record. Volume 2. Chapman and Hall, London.

Szalay, F. S. 1977. Phylogenetic relationships and a classification of the eutherian Mammalia; pp. 315–374 *in* M. K. Hecht, P. C. Goody, and B. M. Hecht (eds.), Major Patterns in Vertebrate Evolution. Plenum, New York.

Szalay, F. S., and F. Schrenk. 1994. Middle Eocene *Eurotamandua* and the early differentiation of the Edentata. Journal of Vertebrate Paleontology 14 (supplement to no. 3): 48A.

———. 1998. The middle Eocene *Eurotamandua* and a Darwinian phylogenetic analysis of "edentates." Kaupia-Darmstädter Beiträge zur Naturgeschichte 7: 97–186.

Thomas, O. 1887. On the homologies and succession of the teeth in the Dasyuridae, with an attempt to trace the history of the evolution of mammalian teeth in general. Philosophical Transactions of the Royal Society of London 1887B: 443–462.

Tong, Y.-S., and J. Wang. 1997. A new palaeanodont (Mammalia) from the early Eocene of Wutu Basin, Shandong Province. Vertebrata PalAsiatica 35: 110–120.

Van Dijk, M.A.M., E. Paradis, F. M. Catzeflis, and W. W. de Jong. 1999. The virtues of gaps: Xenarthran (edentate) monophyly supported by a unique deletion in *A-Crystallin*. Systematic Biology 48: 94–106.

Vicq d'Azyr, M. F. 1792. Système anatomique des Quadrupèdes. Encyclopédie méthodique. Volume 2. Vve. Agasse, Paris.

Vizcaíno, S. F., and M. S. Bargo. 1998. The masticatory apparatus of the armadillo *Eutatus* (Mammalia, Cingulata) and some allied genera: Paleobiology and evolution. Paleobiology 24: 371–383.

Vizcaíno, S. F., R. Pascual, M. A. Reguero, and F. J. Goin. 1998. Antarctica as background for mammalian evolution. Asociación Paleontológica Argentina (Buenos Aires), Publicación Especial 5, Paleógeno de América del Sur y de la Península Antártica: 199–209.

Vizcaíno, S. F., and G. J. Scillato-Yané. 1995. An Eocene tardigrade (Mammalia, Xenarthra) from Seymour Island, West Antarctica. Antarctic Science 7: 407–408.

Weber, M. 1904. Die Säugetiere. Einfürung in die Anatomie und Systematik der recenten und fossilen Mammalia. Gustav Fischer, Jena.

Williamson, T. E. 1996. The beginning of the age of mammals in the San Juan Basin, New Mexico: Biostratigraphy and evolution of Paleocene mammals of the Nacimiento Formation. New Mexico Museum of Natural History and Science Bulletin 8: 1–141.

Williamson, T. E., S. G. Lucas, and J. W. Froehlich. 1994. New genus and species of early Paleocene (Torrejonian) leptictid? from the Nacimiento Formation, San Juan Basin, New Mexico. Journal of Vertebrate Paleontology 14 (supplement to no. 3): 52A.

Wilson, D. E., and D. M. Reeder (eds.). 1993. Mammal Species of the World. Second Edition. Smithsonian Institution Press, Washington, D.C.

Wortman, J. L. 1896. *Psittacotherium,* a member of a new and primitive suborder of Edentata. Bulletin of the American Museum of Natural History 8: 259–262.

———. 1903. Studies of Eocene Mammalia in the Marsh collection, Peabody Museum. Part II. Primates. Suborder Cheiromyoidea. American Journal of Science 16: 345–368.

Wyss, A. R., J. J. Flynn, M. A. Norell, C. C. Swisher III, M. J. Novacek, M. C. McKenna, and R. Charrier. 1994. Paleogene mammals from the Andes of Central Chile: A preliminary taxonomic, biostratigraphic, and geochronologic assessment. American Museum Novitates 3098: 1–31.

Wyss, A. R., M. A. Norell, J. J. Flynn, M. J. Novacek, R. Charrier, M. C. McKenna, C. C. Swisher III, D. Frassinetti, P. Salinas, and J. Meng. 1990. A new early Tertiary mammal fauna from central Chile: Implications for Andean stratigraphy and tectonics. Journal of Vertebrate Paleontology 10: 518–522.

MARY T. SILCOX,
JONATHAN I. BLOCH,
ERIC J. SARGIS, AND
DOUGLAS M. BOYER

EUARCHONTA (DERMOPTERA, SCANDENTIA, PRIMATES)

GREGORY (1910) NAMED ARCHONTA (DERIVED from the Greek for chief: Αρχον) for a supraordinal group composed of Menotyphla (Tupaiidae + Macroscelididae), Dermoptera, Chiroptera, and Primates. Gregory suggested an origin for these taxa from a common ancestor in the Upper Cretaceous that may have resembled a tree shrew in form. A monophyletic Archonta was not immediately widely accepted. For example, Simpson (1945) considered Archonta to be an unnatural group and did not include it in his landmark classification of mammals. Later, however, support began to grow for a modified version of Archonta, excluding macroscelidids (e.g., McKenna, 1975; Szalay, 1977; Novacek and Wyss, 1986). A key component of archontan monophyly was the proposed relationship between dermopterans and chiropterans, referred to as Volitantia Illiger, 1811. Morphological support suggested for the monophyly of Volitantia (e.g., Novacek and Wyss, 1986; Wible and Novacek, 1988; Szalay and Lucas, 1993, 1996; Simmons and Geisler, 1998; Shoshani and McKenna, 1998; Stafford and Thorington, 1998) includes a substantial number of features such as the elongation of the forelimbs; the fusion of the distal part of the ulna to the radius; and the presence of the humeropatagialis muscles, a scaphocentralunate, and a tendon-locking mechanism on the digits of the feet (see Simmons and Geisler, 1998: table 4, for a more complete list).

Starting with Carlsson (1922) and continuing into the 1960s, tupaiids were often included in the order Primates—a classification that received particular support from the work of Le Gros Clark (1925, 1926). Van Valen (1965) and McKenna (1966) advocated removing tupaiids from the order, based on the premise that many of the

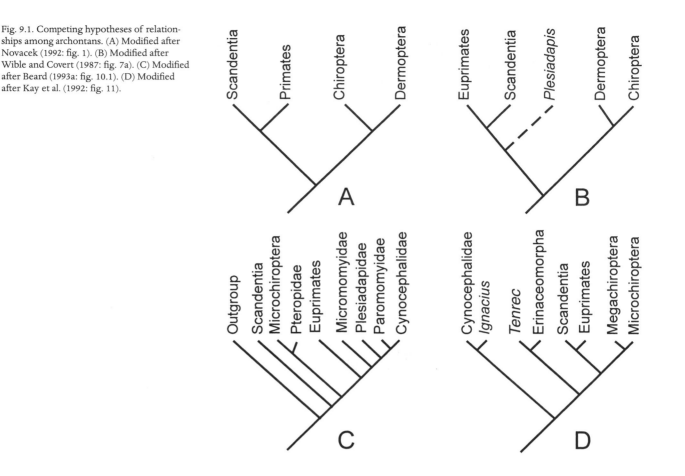

Fig. 9.1. Competing hypotheses of relationships among archontans. (A) Modified after Novacek (1992: fig. 1). (B) Modified after Wible and Covert (1987: fig. 7a). (C) Modified after Beard (1993a: fig. 10.1). (D) Modified after Kay et al. (1992: fig. 11).

features that they were supposed to share with Euprimates (=primates of "modern aspect"; i.e., Primates not including plesiadapiforms; see Hoffstetter, 1977) were mis-characterizations, convergences, or symplesiomorphies. Butler (1972) grouped tupaiids in their own order, Scandentia Wagner, 1855, and this classification has since been followed by most mammal systematists. In spite of their removal from Primates, a euprimate-scandentian relationship was still supported by some authors (e.g., Wible and Covert, 1987; Kay et al., 1992). Novacek (1992; Fig. 9.1A) supported a monophyletic Archonta with sister taxon relationships between Primates and Scandentia, and between Dermoptera and Chiroptera. Subsequent morphological debates about the intraordinal relationships of living archontans have largely focused on conflicts between these phylogenetic hypotheses and apparently inconsistent evidence from the fossil record (see reviews in Szalay and Lucas, 1996; Sargis, 2002c).

Even though phylogenetic analyses of morphological data often result in hypotheses with a monophyletic Archonta and Volitantia, these particular relationships have consistently not been supported by molecular analyses. In fact, no phylogenetic hypothesis based strictly on molecular data has ever supported a close relationship between chiropterans and other archontans—instead, chiropterans have generally fallen out with carnivores and ungulates (e.g., Miyamoto et al., 2000; Liu et al., 2001; Murphy et al., 2001a,b; Springer et al., 2003, 2004). Although results of early mo-

lecular studies provided conflicting hypotheses regarding the monophyly of a version of Archonta that excludes bats (Allard et al., 1996), a consensus now appears to be developing in support of a clade composed of Scandentia, Dermoptera, and Primates (Adkins and Honeycutt, 1991; Waddell et al., 1999; Liu et al., 2001; Murphy et al., 2001a,b; Springer et al., 2003, 2004). Waddell et al. (1999) proposed the name Euarchonta for this group.

Within Euarchonta, there have been various hypothesized interordinal relationships based on molecular studies, although a clade including Dermoptera and Scandentia has appeared in several recent results (Liu et al., 2001; Murphy et al., 2001a,b; Springer et al., 2003, 2004). Euarchonta has recently been grouped in several molecular analyses with Glires (rodents + lagomorphs; e.g., Waddell et al., 1999; Murphy et al., 2001a,b; Springer et al., 2003, 2004) in a clade that Murphy et al. (2001b) named Euarchontoglires.

Our purpose in this chapter is to review the morphological evidence, particularly the fossil material, that is central to debates about the origins and relationships of Euarchonta and to discuss recent viewpoints about these issues.

INSTITUTIONAL ABBREVIATIONS

UM, University of Michigan, Museum of Paleontology (Ann Arbor, Michigan)

TF, Collections of the Palaeontological Section, Department of Mineral Resources, Bangkok (Ducrocq et al., 1992)

DISTINCTIVE MORPHOLOGICAL FEATURES AND FOSSIL REPRESENTATIVES OF EXTANT EUARCHONTAN GROUPS

Euarchonta

A list of hypothesized synapomorphies for Archonta or Euarchonta is given in Table 9.1 and illustrated in Figs. 9.2–9.3 (see also Figs. 9.4, 9.5). A recent reconstruction of the euarchontan morphotype (Bloch et al., 2003) suggests that the common ancestor of this group was a small-bodied arboreal form similar to the living scandentian *Ptilocercus* (Sargis, 2002c). This common ancestor may have frequently adopted vertical postures on large-diameter supports—this is suggested, for example, by the presence of an elliptical acetabulum (see Fig. 9.5A) and deep, narrow unguals (see Fig. 9.5D,E). Of the other features listed in Table 9.1, only a few (characters 1 and 6) are thought to be present in all living archontans. If bats are excluded, then characters 5 and 2 can also be added to this list. It is worth noting that of these four characters, only two features of the ankle (characters 5 and 6) are potentially identifiable in fossils. Although the variability in these features in modern archontans does create some methodological difficulties, if they were all present in the ancestral archontan or euarchontan, then more re-

cent modifications to these traits do not eliminate them as support for the monophyly of Archonta or Euarchonta.

Dermoptera

Living dermopterans are nocturnal gliders that are found only on a group of islands in Southeast Asia (Stafford and Szalay, 2000). Stafford and Szalay (2000) argued convincingly for a separation of living members into two monospecific genera, *Cynocephalus volans* and *Galeopterus variegatus,* although most other recent authors have included both extant species in the genus *Cynocephalus.* Stafford and Szalay's (2000) argument was based on the presence of more robust masticatory features in *C. volans* than are found in *G. variegatus.*

Overall, modern dermopterans are profoundly odd animals, with many very unusual morphological characteristics. One of the most distinctive features of living dermopterans is their patagium, which is more extensive than in other extant gliding mammals, extending between the digits of the hand (interdigital patagium; the reason for the common name "mitten gliders"), between the limbs, and even between the hindlimb and the tail (uropatagium). Dermopterans have numerous unusual postcranial features that are possibly associated with this structure, including a distinctive lengthening of the intermediate phalanges of the hand (Beard, 1989, 1990). Dentally, dermopterans are unusual in having multi-tined, pectinate lower incisors that may be used in certain grooming behaviors (Aimi and Inagaki, 1988). The first upper incisor is lost, and there is a central gap in the upper dental arcade, so that i1 and i2 articulate

Table 9.1 Hypothesized synapomorphies of Archonta and Euarchonta

Number	Character	Number	Character
1	Pendulous penis suspended by a reduced sheath between the genital pouch and abdomen (Smith and Madkour, 1980)	9	Robust, medially protruding lesser tuberosity (missing in tupaiines; Sargis, 2002c)
2	Tectopetal connections to the superior or anterior colliculus of one side with progressively more projections from the ipsilateral retina (absent in microchiropterans; Johnson and Kirsch, 1993)	10	Spheroidal capitulum of the humerus (missing in chiropterans and tupaiines), circular and excavated radial central fossa (missing in chiropterans and tupaiines; Szalay and Lucas, 1996; Sargis, 2002c)
3	Presence of at least one entotympanic element, with a close proximity to the tubal cartilage, tegmen tympani, internal carotid artery and nerve, and the greater petrosal nerve (missing in euprimates; Wible and Martin, 1993)	11	Acetabulum elliptical in shape with buttressing cranially (missing in tupaiines; Sargis, 2002c)
4	Epitympanic wing of the petrosal absent (missing in euprimates and tupaiines; Wible and Martin, 1993)	12	Enlarged, flattened triangular areas for attachment of quadratus femoris (missing in tupaiines, plesiadapids, and chiropterans; Sargis, 2002c)
5	Sustentacular facet of the astragalus in distinct medial contact with the navicular facet of the astragalus (missing in microchiropterans; Szalay and Drawhorn, 1980; Novacek and Wyss, 1986)	13	Short, shallow patellar groove (missing in euprimates; Szalay and Lucas, 1996; Sargis, 2002c)
6	Large distal sustentacular facet of the calcaneum articulating with a ventral extension of the navicular facet of the astragalus (Hooker, 2001)	14	Synovial distal tibiofibular joint (missing in some euprimates, tupaiines, and chiropterans; Sargis, 2002c)
7	Craniocaudal expansion of the ribs (missing in primates and tupaiines; Sargis, 2002c)	15	Concave cuboid facet on the calcaneus (missing in tupaiines and chiropterans; Szalay and Lucas, 1996; Sargis, 2002c)
8	"Weak" or absent spinous processes on C3–C7 (missing in euprimates and tupaiines; Sargis, 2002c)	16	Wide distal facet on the entocuneiform (missing in tupaiines and chiropterans; Szalay and Lucas, 1996; Sargis, 2002c)
		17	Deep, narrow unguals (missing in euprimates; Sargis, 2002c)

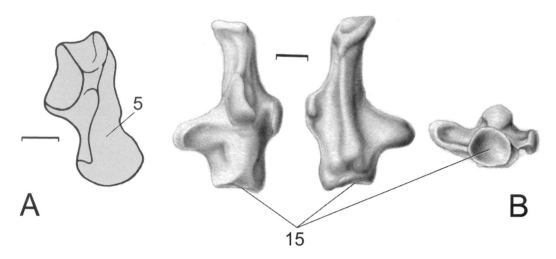

Fig. 9.2. Tarsal bones of *Ptilocercus lowii,* demonstrating characteristic archontan features. Numbers refer to characters in Table 9.1. (A) Left astragalus in plantar view. The sustentacular facet is in distinct medial contact with the navicular facets. Redrawn after Szalay and Drawhorn (1980: fig. 2). (B) Left calcaneus in dorsal (left), plantar (middle), and distal (right) views. This specimen has a round and concave cuboid facet. Redrawn after Szalay and Drawhorn (1980: fig. 9). Scale bar = 1 mm.

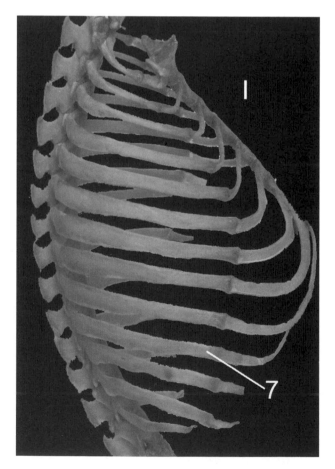

Fig. 9.3. Thorax of *Cynocephalus volans* in lateral view, demonstrating the characteristic euarchontan feature of craniocaudally expanded ribs (Table 9.1, character 7). Scale bar = 5 mm.

with an edentulous premaxillary pad (Stafford and Szalay, 2000). *Galeopterus variegatus* also has multi-tined canines, although these teeth are more typically caniniform in *C. volans* (Stafford and Szalay, 2000). The upper and lower premolars are enlarged and extensively molarized in both species. The upper molars exhibit very large conules, no hypocones, and a transverse valley running buccolingually between the paracone/paraconule and metacone/metaconule (Mac-Phee et al., 1989). The lower molars (and p4) have a distinctly crestiform paracristid that is well separated from the protoconid and metaconid, and strong shearing crests. In the cranium, the majority of the auditory bulla is formed by the ectotympanic, plus one or two small entotympanic elements associated with the auditory tube (Wible and Martin, 1993). The ear region is extensively pneumatized, and the orientation of the eardrum is unusual in remaining nearly horizontal into adulthood (Hunt and Korth, 1980). The internal carotid artery involutes in ontogeny, and the brain is supplied with blood by the vertebral arteries (Hunt and Korth, 1980).

The fossil record of dermopterans that are clearly related to modern Cynocephalidae (not including the more questionably assigned dermopterans, e.g., Plagiomenidae, Paromomyidae) consists of only one specimen, a poorly preserved mandibular fragment from late Eocene sediments from the Krabi Basin, southern Thailand. This specimen (TF 2580) is the holotype of *Dermotherium major* (Ducrocq et al., 1992) and is classified in the same family as modern dermopterans (Cynocephalidae; often referred to as Galeopithecidae, but Stafford and Szalay, 2000, argue that because *Cynocephalus* is the type genus, the valid family name is Cynocephalidae).

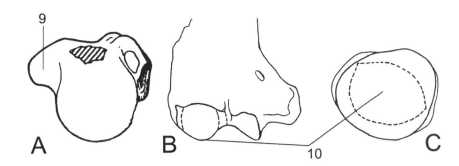

Fig. 9.4. Characteristic archontan features. Numbers refer to characters in Table 9.1. (A) Right humerus of *Phenacolemur simonsi* in proximal view, demonstrating a robust, medially protruding lesser tuberosity. Redrawn after Beard (1989: fig. 67). (B) Right distal humerus (anterior view) and (C) proximal radius (proximal view) of a characteristic euarchontan. Redrawn after Beard (1989: fig. 69). Panels (B) and (C) illustrate a spheroidal capitulum and corresponding circular and deeply excavated central fossa of the radius.

Stafford and Szalay (2000) disputed the hypothesized relationship between *Dermotherium major* and Cynocephalidae based on inconsistencies in the description of the specimen, an inadequate accounting for the differences between the two genera of modern dermopterans, and the widespread distribution in Eutheria of the traits supposedly linking *Dermotherium* to Dermoptera. To deal with these criticisms, new comparisons of a cast of TF 2580 were made to both *Galeopterus variegatus* and *Cynocephalus volans*. In these comparisons, the differences between the two modern species were taken into account, and any inconsistencies in the description were rendered irrelevant by fresh observations (Silcox, 2001). These comparisons show that *Dermotherium major* is remarkably similar in most comparable aspects of morphology to the modern taxa, in ways that are more distinctive and unique than Stafford and Szalay (2000) imply. The morphology of the molar talonid basin, for example, is identical. This similarity includes not only the (admittedly widespread) close approach of the hypoconulid to the entoconid, but the manner in which this occurs (i.e., both cusps are distinct, pointed, and are not developed along a clear postcristid). Like modern dermopterans, the talonid in *Dermotherium* is extremely broad relative to the trigonid. Although these breadth relationships are found in some other mammals (e.g., primates), the wide gap between the hypoconulid and hypoconid and the deepening of the basin to a "pit" along the midline just behind the postvallid is more distinctive and less commonly found (e.g., these features are generally missing in primates). Modern cynocephalids share with *Dermotherium* an indistinct, lingually shifted paraconid. Again, this feature is found in other mammals (e.g., paromomyids), but the similarities are not as detailed. In modern cynocephalids and *Dermotherium*, this trigonid morphology is associated with a paracristid that extends mesiolingually as a relatively straight crest. The mesial orientation of the paracristid makes the trigonid basin relatively long and the front of the tooth triangular (rather than rectangular) in shape. This configuration differs, for example, from paromomyids, in which the lingual shift of the paraconid is associated with a shortening of the trigonid and a less mesially inclined paracristid. Overall, then, the similarities between *Dermotherium* and modern dermopterans do not consist of vague similarities in broadly distributed features, but rather to very detailed and narrowly distributed character complexes. This impression is not contradicted by including both modern genera in the assessment.

Based on these comparisons, it seems that the fossil record of definitive Dermoptera does extend back to the late Eocene. As a result of the damage to TF 2580 and the similarity of its preserved morphology to that in modern cynocephalids, the specimen does not add much new information regarding the primitive morphology of Cynocephalidae or its close relatives. It does, however, demonstrate that the divergence of Dermoptera from other mammals must have predated the late Eocene. The location of this specimen, near the modern distribution of dermopterans, could suggest that the group has never existed outside of Southeast Asia, although this view is invalidated if plagiomenids were dermopterans (see below).

Scandentia

Although extant scandentians are more speciose and extensively distributed than dermopterans, they are also found only in Southern Asia. Extant members of this order are typically included in one family, Tupaiidae, with two subfamilies, Tupaiinae and Ptilocercinae. The latter of the two subfamilies is represented by only a single species, *Ptilocercus lowii*, which is the most arboreal and likely the most primitive member of the group (Sargis, 2002a–c). Distinctive features of Scandentia include the presence of a postorbital bar and divergent orbits. A list of possible scandentian synapomorphies is given in Table 9.2.

The fossil record of Scandentia is very limited. The only records of the order before the Miocene are fragments of teeth from the Middle Eocene of China that, although placed in a new species and genus (*Eodendrogale parvum*), differ only in relatively minor details from the modern genus *Dendrogale* (Tong, 1988). The Miocene material is also quite limited, including a few skull fragments and various isolated teeth, all of which are closely comparable to modern scandentians (Dutta, 1975; Chopra and Vasishat, 1979; Chopra et al., 1979; Jacobs, 1980; Qiu et al., 1985; Qiu, 1986; Mein and Ginsburg, 1997; Ni and Qiu, 2002). The only possible postcranial specimen known for a fossil scandentian is a rib cage from Pliocene deposits of the Upper Siwaliks of India, attributed to *Tupaia* (Dutta, 1975). As such, the fossil record of scandentians is of limited value in providing information on the early events in the evolution of the group or the likely

Table 9.2 Hypothesized synapomorphies of Scandentia

Number	Character
1	Enclosure of the intratympanic portion of the internal carotid artery in a bony canal floored proximally and distally by the entotympanic, and by the petrosal in between (Wible and Zeller, 1994)
2	Enclosure of the intratympanic portion of the stapedial artery by the petrosal in a canal on the promontorium, and within the epitympanic crest beneath the tympanic roof (Wible and Zeller, 1994)
3	Absence of an exit for arteria diploëtica magna (Wible and Zeller, 1994)
4	Alisphenoid canal present (Wible and Zeller, 1994)
5	Maxillary artery passes medial to the mandibular nerve, beneath the foramen ovale (Wible and Zeller, 1994)
6	Laryngeopharyngeal artery present (Wible and Zeller, 1994)
7	Scaphoid and lunate fused into scapholunate (Sargis, 2002a)
8	Articular facet for the medial malleolus on the posterior side of the sustentaculum (Szalay and Lucas, 1996; Sargis, 2002c)

form of ancestors to the order, beyond indicating that such ancestors must have predated the middle Eocene. As with definitive dermopterans, the fossil record of scandentians is entirely Asian in distribution.

Euprimates

Cartmill (1992: 105) recognized five adaptive complexes that he identified as the "chief primate peculiarities" found in most modern members of the order: grasping extremities, claw loss, optical convergence and orbital approximation (including the presence of a postorbital bar), enhanced vision, and brain enlargement. To this list, we can add the presence of a petrosal bulla (Cartmill, 1972) and dental features that are associated with increased omnivory (e.g., low-crowned molars with broad talonid basins; Szalay, 1968b).

Compared to the other euarchontan groups, the fossil record of Euprimates is better sampled. Several reviews of the fossil record of primates exist (e.g., Szalay and Delson, 1979; Hartwig, 2002) and our discussion focuses only on those forms most relevant to debates about the origins and supraordinal relationships of the group.

Many fossils traditionally grouped in two families, Adapidae and Omomyidae (sometimes raised to superfamilial or infraordinal rank), have been recovered from early Eocene sediments in North America and Europe. The earliest and most primitive representatives of these two families are the adapids *Cantius* and *Donrussellia* (Godinot, 1978, 1992; Godinot et al., 1987; Gingerich et al., 1991) and the omomyids *Teilhardina* and *Steinius* (Simpson, 1940; Bown, 1976; Rose and Bown, 1991; Rose, 1995). These basal taxa are dentally very similar to one another, with a recent study (Rose and Bown, 1991) identifying only a single known, derived trait of omomyids that separates them from primitive adapids (a single-rooted, rather than double-rooted, p2). A

few other subtle dental features have been found that distinguish the most primitive adapids from omomyids, including the loss of the postmetaconule crista and the buccal position of the hypoconulid (Godinot, 1992). The implication of these dental similarities may be that we are approaching, in the early Eocene, the common root for euprimates, or at least for Adapidae + Omomyidae.

The fossil record of primitive euprimates in Africa and Asia is much more limited. Although some of this material can be confidently attributed to Adapidae or Omomyidae, other specimens remain more controversial. Of particular importance to issues of euprimate, primate, and archontan (or euarchontan) origins are *Petrolemur brevirostre* (late Paleocene), *Decoredon anhuiensis* (early Paleocene), and *Altanius orlovi* (early Eocene) from Asia, and *Altiatlasius koulchii* (late Paleocene) from Africa. It has been suggested that *Petrolemur brevirostre* and *Decoredon anhuiensis* are not primates (Rose, 1995; Wang et al., 1998). *Petrolemur* was classified by McKenna and Bell (1997) as an oxyclaenid ungulate, and this seems a more likely attribution than Primates (see also Szalay, 1982). Although McKenna and Bell (1997) included *Decoredon* in Primates, it lacks distinguishing primate features (Rose, 1995) and is more likely to be a condylarth than a primate.

That leaves *Altiatlasius* from the Late Paleocene of Morocco as the oldest known possible euprimate. The systematic position of *Altiatlasius* has also been a subject of debate. Although *Altiatlasius* was originally described as an omomyid, Sigé et al. (1990) also suggested a possible relationship to basal "simiiforms" (i.e., anthropoids), a view supported by Godinot (1994). Gingerich (1990) suggested a plesiadapiform attribution, and Hooker et al. (1999) argued for assignment to the family Toliapinidae in the Plesiadapiformes. Silcox (2001) disagreed—her cladistic analysis supported a basal euprimate position for *Altiatlasius*. Neither Silcox (2001) nor Hooker et al. (1999) analyzed any anthropoids, however, and thus the anthropoid hypothesis has not been tested. Although the affinities of this taxon remain somewhat uncertain, it appears to be a euprimate, making it the oldest known for that group.

Altanius is likely not as old as *Altiatlasius* (coming from the Bumbanian of Mongolia; although Bowen et al., 2002, suggest that the Bumbanian may straddle the Paleocene-Eocene boundary), but may be equally important to an understanding of the earliest part of the euprimate fossil record. Dashzeveg and McKenna (1977), in describing the initial material of *Altanius*, likened it to anaptomorphine omomyids. These authors also noted similarities to carpolestid plesiadapiforms, and Rose and Krause (1984) expanded on these comparisons, arguing that there were at least as many similarities to carpolestids as to omomyids. Gingerich et al. (1991) retained *Altanius* in the Omomyidae, based on overall similarity, including small size. They also documented some important new anatomical findings, including a 2.1.4.3 lower dental formula, rendering this taxon more primitive than the derived anaptomorphine omomyids with which it shares the most postcanine simi-

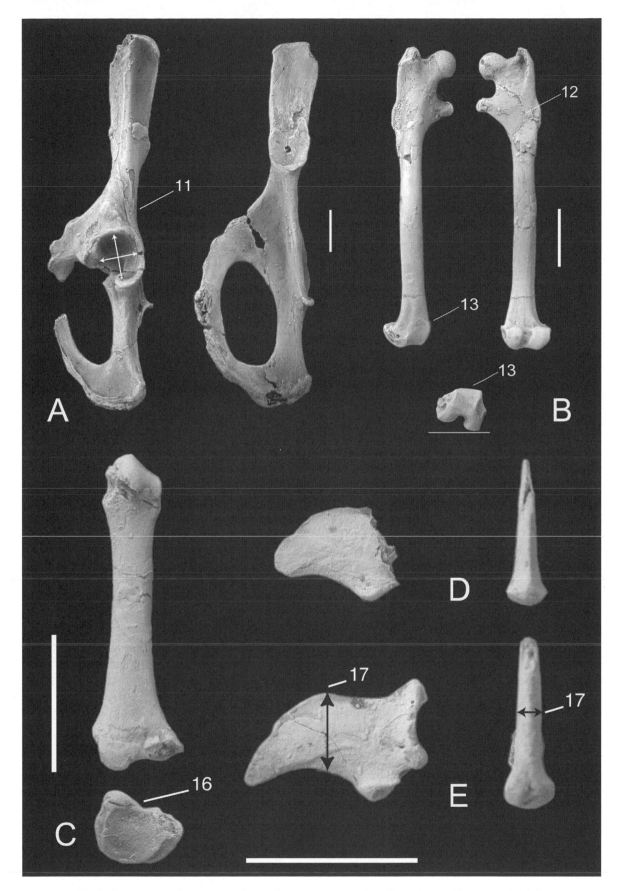

Fig. 9.5. Paromomyid fossils, demonstrating characteristic archontan features. Numbers refer to characters in Table 9.1. (A) Clarkforkian *Ignacius* sp. (UM 82606), left innominate. This specimen illustrates cranial buttressing above an elliptically shaped acetabulum. Scale bar = 5 mm. (B) Clarkforkian *Ignacius* sp. (UM 82606), right femur. Lines point to the short, shallow patellar groove, and the enlarged attachment area for quadratus femoris. Scale bar = 10 mm. (C) Clarkforkian *Ignacius* sp. (UM 82606), right metatarsal I. The broad proximal facet on the first metatarsal reflects the wide distal facet on the entocuneiform. Scale bar = 5 mm. (D) *Acidomomys hebeticus* (UM 08209), distal phalanx is both deep and narrow. (E) Clarkforkian *Ignacius* sp. (UM 82606), distal phalanx is both deep and narrow. Scale bar = 5 mm.

larities. This observation suggests that such similarities are likely to be homoplasies. Beard (1998) hypothesized that *Altanius* was a basal euprimate, a position also supported by Silcox (2001).

Both *Altiatlasius koulchii* and *Altanius orlovi* can be plausibly interpreted as basal euprimates outside Adapidae and Omomyidae. Interestingly, these two taxa are very different morphologically (Rose, 1995; Silcox, 2001). *Altanius orlovi* has a relatively sharp-cusped, high-crowned dentition, with a well-demarcated postprotocingulum and hypocone. *Altiatlasius koulchii,* however, lacks a postprotocingulum and a hypocone and has a wide stylar shelf, which are likely to be primitive characteristics. This taxon also has very bunodont cusps and is quite low-crowned, which might not be expected in a basal euprimate based on the taller, more acute cusps of the most primitive adapids and omomyids (*Donrussellia* and *Teilhardina*). In sum, these two taxa give a rather different picture of dental morphology at the base of the euprimate clade, and suggest that we may be missing a significant portion of the Old World diversity at the base of the euprimate radiation. The age of *Altiatlasius* indicates that euprimates, and by extension Euarchonta, must have already diverged by the late Paleocene.

PALEOCENE POSSIBLE ARCHONTANS

Apart from *Altiatlasius* and possibly *Altanius,* there are currently no known Paleocene members of Euprimates, Dermoptera, or Scandentia. There are, however, a number of Paleocene groups that have been considered possible archontan, euarchontan, or euprimate relatives that may provide additional information about the origin and evolution of these groups.

Plesiadapiformes

Plesiadapiforms are Paleocene-Eocene fossil mammals known from North America, Europe, and Asia. Currently, 11 distinct families are recognized (Purgatoriidae, Micromomyidae, Microsyopidae, Palaechthonidae, Paromomyidae, Plesiadapidae, Carpolestidae, Saxonellidae, Picrodontidae, Picromomyidae, and Toliapinidae; Silcox, 2001). The term "Plesiadapiformes" is not included as a formal taxonomic label in Table 9.3 because it seems likely that it is not a monophyletic group (see Fig. 9.6). Nonetheless, these 11 families do generally share a suite of characteristics that include enlarged, procumbent upper and lower central incisors, low-crowned molars with wide talonid basins, an enlarged m3 hypoconulid, and a P4 with a postprotocingulum. Many plesiadapiforms also exhibit enlargement and elaboration of the premolars (or m1, in the case of picrodontids) and substantial dental reduction. Postcranially, the group shares features for non-leaping arboreality (Szalay et al., 1975, 1987; Szalay and Drawhorn, 1980; Bloch and Boyer, 2002, 2003, in press).

Table 9.3   Classification of Euarchonta

Cohort Euarchonta
    Family Mixodectidae
  Order Scandentia
    Family Tupaiidae
      Subfamily Ptilocercinae
      Subfamily Tupaiinae
  Order Dermoptera
    Family Cynocephalidae
    Family Plagiomenidae
  Order Primates
    Family Microsyopidae
    Family Purgatoriidae
    Family Picromomyidae
    Family Micromomyidae
    Family Toliapinidae
    Superfamily Paromomyoidea
    Family Palaechthonidae
    Family Paromomyidae
    Family Picrodontidae
    Superfamily Plesiadapoidea
      Chronolestes simul
    Family Plesiadapidae
    Family Carpolestidae
    Family Saxonellidae
  Semiorder Euprimates
    Altiatlasius koulchii
    Altanius orlovi
    Family Adapidae
    Family Omomyidae
    Many other families

Note: The classification given here is that used in this chapter.

As plesiadapiform fossils from North America began to be discovered and described in the early part of the twentieth century, they were generally included in Primates, often specifically in Tarsiidae (e.g., Matthew and Granger, 1921; Gidley, 1923), largely based on dental similarities. Martin (1968) and Cartmill (1972) were among the first to seriously call this ordinal placement into question. Both Martin and Cartmill emphasized only the formation of a clearer definition of the order Primates based on the presence of characters relating to the unique ecological adaptations of modern members of the group. Neither offered any substantively new views on the broader relationships of plesiadapiforms to other archontans. More compelling arguments for the exclusion of plesiadapiforms from primates were made by authors who argued that they were not the sister group of euprimates and/or may be more closely related to some other group. Wible and Covert (1987) argued that Scandentia was a more likely sister taxon to Euprimates than Plesiadapiformes, based on characters derived entirely from the basicranium. They published a "preferred" cladogram, not backed up by a data matrix, which positioned *Plesiadapis* as the sister taxon to Scandentia + Euprimates (their "Primates"), with Dermoptera + Chiroptera being the sister group to that clade (see Fig. 9.1B). They argued that the dental evidence linking plesiadapiforms to eu-

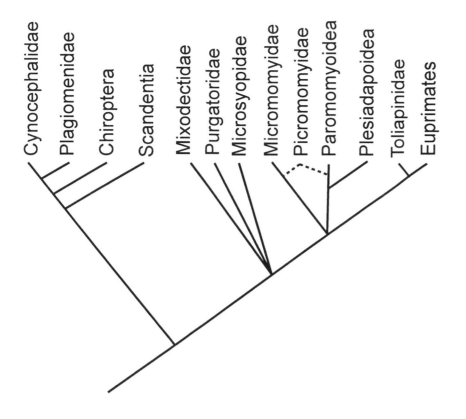

Fig. 9.6. Hypothesis of phylogenetic relationships among euarchontans. This tree is based on a parsimony analysis run in PAUP 4.0*β (Swofford, 2003) using 181 dental, cranial, and postcranial characters (Silcox, 2001). This is a simplified version of the Adams consensus tree, calculated from 20 equally parsimonious trees of length = 788 steps, CI = 0.490, RI = 0.521, and RC = 0.255. On this tree, Plesiadapoidea includes (*Chronolestes simul* (Saxonellidae (Plesiadapidae + Carpolestidae))). Paromomyoidea includes (Picrodontidae (Paromomyidae + "Palaechthonidae")). Toliapinidae on this tree includes only *Toliapina* and *Avenius*. These results support the inclusion of plesiadapiforms in Primates and plagiomenids in Dermoptera. Also supported are the monophyly of Volitantia (Dermoptera + Chiroptera) and a clade including Volitantia and Scandentia. Removal of Chiroptera from the analysis does not substantively change the other relationships on the tree, but a Dermoptera + Scandentia clade is supported. A more comprehensive analysis, including new data for extant taxa (e.g., *Ptilocercus lowii*) and undescribed fossil specimens is currently being undertaken. Preliminary results (Bloch et al., 2002, 2003; Bloch and Boyer, 2003) are basically in agreement with the findings of Silcox (2001), although a more basal position within primates is supported for Micromomyidae, as is a sister-group relationship between Plesiadapoidea and Euprimates.

primates consisted only of ill-defined "trends" (1987: 9) rather than well-documented synapomorphies. This conclusion was not based on a cladistic analysis that included dental data, however, and amounts to an unfounded dismissal of an entire partition of morphological data. This is particularly problematic, as the euprimate-plesiadapiform relationship had always been based largely on dental evidence. Furthermore, one of the critical basicranial features that they cited in support of a Scandentia-Euprimate clade has since been found in a paromomyid plesiadapiform (a bony tube for the internal carotid nerves and/or artery; Silcox, 2003).

Two papers published in the same issue of Nature in 1990 suggested different possible relationships for plesiadapiforms. These two papers documented new cranial (Kay et al., 1990) and postcranial (Beard, 1990) material of paromomyid plesiadapiforms, which was interpreted to support a closer relationship between some plesiadapiforms and Dermoptera. Subsequent, more detailed cladistic analyses (Kay et al., 1992; Beard, 1993a) supported this interpretation, and in both cases the results did not support a monophyletic Volitantia. Instead, at least some plesiadapiforms were found to be sister taxa to Dermoptera, to the exclusion of Chiroptera.

Although Beard (1990, 1993a,b) and Kay and coauthors (1990, 1992) did form some similar conclusions about paromomyids and dermopterans, their results are actually incongruent regarding broader relationships within Archonta.

Kay et al. (1992) failed to support archontan or euarchontan monophyly, with paromomyids, plesiadapids, and dermopterans falling outside of erinaceomorph and tenrecoid insectivores, and presumably far distant from bats, primates, and tree shrews (see Fig. 9.1D). Like Wible and Covert (1987), these authors did not include dental data in their analysis and obtained results that supported a sister-group relationship between scandentians and euprimates. Kay et al. (1992) also presented evidence that paromomyids and plesiadapids form a monophyletic clade to the exclusion of dermopterans. Beard (1993a; see Fig. 9.1C), however, argued for a paraphyletic Plesiadapiformes, with paromomyids being more closely related to modern dermopterans than to any other plesiadapiforms, including plesiadapids. Beard recognized this relationship by including plesiadapiforms with cynocephalids (=his galeopithecids; see above) in Dermoptera and proposing a new group name, Eudermoptera, for modern dermopterans + paromomyids. In a subsequent paper, Beard (1993b) expanded Eudermoptera to include Micromomyidae, based on the hypothesis that gliding had already evolved in this group. A key element of this argument was the presence of relatively long, thin intermediate manual phalanges in Paromomyidae and Micromomyidae. This morphology was interpreted to reflect the presence of the uniquely dermopteran interdigital patagium, based on the same functional argument that alleged a relationship between this feature and improved control of the patagium in modern cynocephalids. In contrast to Kay et al. (1992), Beard (1993a,b) argued that it was this reconstituted Dermoptera (*sensu lato*) that was the sister taxon to euprimates, proposing the name Primatomorpha for Dermoptera (including plesiadapiforms) + Primates. The sister group to this clade in Beard's (1993a) analysis was Chiroptera, with Scandentia falling out as the basalmost group of archontans. By excluding nonarchontan mammals, Beard's (1993a) analysis did not test the monophyly of Archonta as a whole.

Both the results of Beard (1990, 1993a,b) and Kay et al. (1990, 1992) suggested alternatives not only to the systematic position of plesiadapiforms, but also to the relationships of archontans in general. An implication of these papers is that an understanding of the systematic position of plesiadapiforms is central to an understanding of archontan relationships. In particular, these hypotheses represent competing viewpoints as to the monophyly of Volitantia. Some authors have tried to reconcile these views by simply classifying some plesiadapiforms in Dermoptera (e.g., Thewissen and Babcock, 1992; Simmons and Quinn, 1994). These implied phylogenetic hypotheses are not based on cladistic analysis, however, and are directly at odds with the results of Beard (1990, 1993a,b) and Kay et al. (1990, 1992).

The hypotheses of Kay et al. (1990, 1992) and Beard (1990, 1993a,b) have not gone unchallenged, with many papers questioning elements of both the systematic and functional conclusions reached by these authors (e.g., Krause, 1991; Szalay and Lucas, 1993, 1996; Wible, 1993; Wible and Martin, 1993; Van Valen, 1994; Runestad and Ruff, 1995; Stafford and Thorington, 1998; Hamrick et al., 1999; Stafford and Szalay, 2000; Bloch and Silcox, 2001; Boyer et al., 2001; Bloch and Boyer, 2002, 2003; Bloch et al., 2002; Sargis, 2002c; Silcox, 2003). The issues raised by these authors include problems with the identification of some of the phalanges attributed to paromomyids and micromomyids, demonstrating that these taxa do not have the unusual hand proportions found in dermopterans (Krause, 1991; Boyer et al., 2001; Hamrick, 2001; Bloch and Boyer, 2002). Moreover, the discovery of "primatomorphan" features in the most plesiomorphic extant scandentian greatly weakens the postcranial support for the monophyly of Primatomorpha (Sargis, 2002c). New discoveries, and new interpretations of the cranial material have also called into question the degree of similarity between dermopterans and paromomyids in the ear region (Wible, 1993; Wible and Martin, 1993; Bloch and Silcox, 2001; Silcox, 2003).

New phylogenetic analyses and new fossil finds have suggested a return to earlier views (e.g., Szalay and Delson, 1979; Szalay et al., 1987) that Primates should be defined in such a way that plesiadapiforms are included. For example, the first associated skeletal material ever recovered for a carpolestid has revealed the presence of a divergent, opposable hallux with a nail, a feature previously thought to be limited to euprimates (Bloch and Boyer, 2002). A cladistic analysis of plesiadapiforms and other archontans, incorporating cranial, dental, and postcranial features, also failed to support the Eudermoptera or Primatomorpha hypotheses. Instead,

it supports a sister-group relationship between Euprimates and a paraphyletic Plesiadapiformes (Fig. 9.6; Silcox, 2001; see also Bloch and Boyer, 2002, 2003; Bloch et al., 2002). Although the results of this analysis support Volitantia, contrary to the molecular results mentioned above, the inclusion of bats is not critical to the overall topology, as no substantive changes in the other relationships on the tree occur when the analysis is run without Chiroptera. In particular, a relationship between Dermoptera and Scandentia was maintained, which matches the findings of some recent molecular work (e.g., Liu et al., 2001; Murphy et al., 2001a,b; Springer et al., 2003, 2004). In light of the exclusion of ungulates and carnivorans from this dataset (taxa that are supposed to be closer to bats than are euarchontans, according to molecular results), this analysis cannot yet be seen as a test of the monophyly of either Voliantia or Archonta, leaving the systematic position of Chiroptera equivocal.

Plesiadapiforms include the oldest known archontans (or euarchontans), in the family Purgatoriidae. The earliest possible member of this family, *Purgatorius ceratops,* dates from the latest Cretaceous (Van Valen and Sloan, 1965; Van Valen, 1994) or earliest Paleocene (Lofgren, 1995; Clemens, in press) and comes from North America.

Plagiomenidae

Plagiomenidae is known from the Paleocene and Eocene of North America. The dental similarities between Plagiomenidae and modern dermopterans have been recognized ever since *Plagiomene* was first described in 1918 (Matthew, 1918), and until recently, most authors treated members of this group as the earliest representatives of that order (e.g., Rose, 1973). MacPhee et al. (1989) provided the first description of cranial material of a plagiomenid. The specimen they described was heavily damaged, but their reconstruction hypothesized some profound differences from modern dermopterans. In particular, they reconstructed the auditory bulla as a highly complex, composite structure quite unlike the bulla of extant dermopterans.

Although not providing any additional evidence for a plagiomenid-dermopteran relationship, this basicranial evidence also did not convincingly support any other set of relationships—it only demonstrated the morphology to be oddly autapomorphous. As such, the ear morphology does not necessarily detract from the phylogenetic signal of the dentition. New results from an analysis including both dental and cranial traits (Silcox, 2001; Fig. 9.6) support the traditional alliance of plagiomenids with modern dermopterans, suggesting again that this family may be the earliest representation of Dermoptera known in the fossil record.

Mixodectidae

Mixodectidae, from the Torrejonian (early Paleocene) of North America, has often been linked to Plesiadapiformes, based on dental similarities between this group and Microsyopidae (see the historical review in Szalay, 1969). Szalay (1969) argued against such an association, and also rejected a mixodectid-plagiomenid link. Gunnell (1989) performed a cladistic analysis of plagiomenids and mixodectids. Although this analysis might seem to support a sister-group relationship between these families, failure to include other taxa means that other possibilities were not tested. Gunnell (1989: 67) concluded that mixodectids were "difficult to evaluate systematically or functionally." Szalay and Lucas (1996) published the first postcranial material attributed to the group, and documented a number of similarities to plesiadapiforms, *Ptilocercus,* and their reconstruction of the "protoeuprimate," suggesting that Mixodectidae is referable to Archonta (or Euarchonta). Where preserved, this postcranial material suggests that mixodectids are an aberrant group of basal archontans, but does not provide evidence of supposed volitantian features (Silcox, 2001). This argues weakly against a special relationship with plagiomenids if Volitantia is monophyletic, although plagiomenids are unknown from the postcranium.

Nyctitheriidae

Nyctitheriidae is a family of insectivores from the Paleocene and Eocene of North America and Asia, and the Paleocene-Oligocene of Europe (McKenna and Bell, 1997). Dental similarities between nyctitheriids and chiropterans have long been recognized. For example, *Wyonycteris,* a taxon that was initially described as the earliest bat (Gingerich, 1987) is now generally included in Nyctitheriidae (e.g., McKenna and Bell, 1997), following challenges to its chiropteran status (Hand et al., 1994; Hooker, 1996).

Recently, postcranial material has been described for the family (Bloch and Boyer, 2001; Hooker, 2001; Bloch et al., 2003). The tarsal elements studied by Hooker (2001) revealed some surprising similarities to archontans. When Hooker included these features—along with cranial and dental traits—in a cladistic analysis, he found a sister-group relationship between nyctitheriids and a Euarchonta clade that included a plesiadapiform, a euprimate, a dermopteran, a scandentian, and *Mixodectes.* Hooker also supported a close relationship between Nyctitheriidae + Euarchonta and a late Cretaceous primitive mammal from Asia, *Deccanolestes,* with microchiropterans lying outside of this clade. *Deccanolestes* was included in Hooker's analysis, based on a hypothesized relationship with Archonta from tarsal evidence (Prasad and Godinot, 1994), which, if upheld, would make it the earliest known archontan. Although Hooker's analysis included relatively little taxonomic diversity, his findings suggest that nyctitheriids warrant additional consideration as possible primitive archontans or euarchontans.

Apatemyidae

Apatemyidae is a broadly dispersed and long-lived family of odd fossil mammals from the Paleocene and Eocene of Europe and the Paleocene-Oligocene of North America (McKenna and Bell, 1997). Apatemyids exhibit some general dental similarities to plesiadapiforms, including a set of enlarged upper and lower central incisors. Gingerich (1989) suggested that the family could belong in the same taxonomic assemblage as plesiadapiforms ("Proprimates"), and MacPhee et al. (1983) placed apatemyids as one of several groups possibly related to the origin of Primates. Other authors, however, have questioned a close association of this family with early primates (e.g., McKenna, 1963, 1966; Szalay, 1968a; Silcox, 2001). In their recent classification of mammals, McKenna and Bell (1997) classified Apatemyidae in the order Cimolesta with several other primitive mammalian groups (e.g., palaeoryctids), under the grandorder Ferae, indicating a closer relationship to carnivorans than to primates.

A cladistic analysis capable of assessing a possible apatemyid-primate relationship has never been performed. The recent discovery of relatively complete material for apatemyids, including well-preserved crania and postcranials, (e.g., Koenigswald, 1990; Bloch and Boyer, 2001) suggests that a reassessment of possible archontan or euarchontan affinities for this group may be warranted (Godinot, pers. comm.). The material from Messel that has been described indicates arboreality, and presumably convergent adaptive similarities to the euprimate *Daubentonia.* The very long fingers of apatemyids are also seen in many archontans (part of the gliding/flying complex in dermopterans and chiropterans, and associated with grasping in euprimates and non-plesiadapid plesiadapiforms; Simmons and Geisler, 1998; Hamrick, 2001; Bloch and Boyer, 2002, 2003). Although a tie with Archonta remains tentative at the current time, further analysis of undescribed specimens is warranted to more fully assess this possibility.

TIMES OF ORIGIN

Various studies have attempted to estimate the time of origin of Primates using molecular data. These estimates have generally produced dates for the divergence of primates that extend well into the Cretaceous (e.g., Hedges et al., 1996; Árnason et al., 1998; Eizirik et al., 2001), averaging around 90 million years ago (Tavaré et al., 2002) and implying an even earlier divergence date for Archonta or Euarchonta. The most recent molecular estimates, from an analysis that controls for many of the problems that have plagued previous studies, indicate somewhat later divergences. Springer et al. (2003) support late Cretaceous dates for the major events in euarchontan diversification (including origination of the group from a common ancestor with Glires), with all such branching events likely postdating the 90 million year mark, but predating the K/T boundary. These estimates are closer to the age of the oldest euarchontan fossils.

The very limited paleontological record for scandentians and crown-clade dermopterans makes it impossible to formulate a plausible fossil-based hypothesis for a time of origin for these groups, as it seems clear that they have very long ghost lineages. The oldest possible dermopteran, however, is the early Paleocene (Torrejonian) plagiomenid *Elpidophorus,* indicating that Dermoptera had likely arisen by that time. The oldest known clear archontans (or euarchontans; i.e., not including the poorly known *Deccanolestes*) are plesiadapiform primates. The earliest proposed date for a plesiadapiform is the latest Cretaceous age suggested for *Purgatorius ceratops* (Van Valen and Sloan, 1965; Van Valen, 1994), based on a single tooth of questionable identity. This date has been the source of considerable debate. The only known specimen of the species comes from a time-averaged Cretaceous-Paleocene (Lancian/Puercan-1) deposit, making its proposed Cretaceous age questionable (Lofgren, 1995; Clemens, in press). The next oldest euarchontan material consists of isolated teeth of *Purgatorius* from Rav-W1 (Pu-1; earliest Paleocene; Johnston and Fox, 1984). This material is better constrained geologically, and the one well-preserved upper molar exhibits features that support its interpretation as the most primitive known member of the genus (Silcox, 2001). The fossil record thus indicates that primates diverged by at least the earliest Paleocene. The earliest euprimate is *Altiatlasius,* which is late Paleocene in age (Sigé et al., 1990; Gheerbrant et al., 1998), implying that euprimates had arisen by that time.

The date of the earliest recognized fossils indicates only a minimum age for the divergence of a given clade. Various approaches have been attempted to estimate the length of the gap between the earliest fossils and the actual point of divergence for Euprimates. Gingerich and Uhen (1994) used a probability model to evaluate Cretaceous ages for the origin of euprimates (their "Primates"), concluding that it was improbable that this group arose before 63 million years ago. This analysis was based, however, on an earliest date for euprimates of 55 million years ago. Such an estimate would need to be recalculated with a late Paleocene date (based on *Altiatlasius*) for Euprimates, and an earliest Paleocene date for Primates that includes *Purgatorius,* which would likely push both of these divergences (as well as the origin of Archonta or Euarchonta) into the late Cretaceous.

Tavaré et al. (2002) reached a very different conclusion from that of Gingerich and Uhen (1994). They estimated the last common ancestor of living primates at 81.5 million years ago, which they suggested is in better agreement with molecular estimates than are other fossil-based approaches. There are several fundamental problems with the techniques applied by Tavaré et al., however, which reduce confidence in their results and in their assertion of a close agreement with molecular dates. Molecular estimates are generally based on a divergence point—that is, the date at which the stem lineage leading to some modern group branched off. Tavaré et al. dealt only with the origination time of the last common ancestor of living primates (i.e., euprimates), not taking into account the length of the stem separating divergence from origination. This is particularly problematic if plesiadapiforms are stem primates (Fig. 9.6),

as they include a broad radiation of forms, classified into more than 120 different species. These stem taxa sit in the gap between the point of divergence of the earliest primate and the point of origination of the group including all living primates. To be truly comparable with molecular estimates, their analysis would have to take these forms into consideration. In light of the speciose nature of the plesiadapiform radiation, doing so would likely push their estimated date back in time significantly, to a point even earlier than that given by molecular estimates.

Furthermore, there are some fundamental problems with the model used by Tavaré et al. (2002). By assuming a logistic diversification model, these authors do not account for two key issues in estimating early primate evolutionary rates. First, it seems likely from the relatively sudden appearance of so many plesiadapiforms and euprimates that a very rapid diversification accompanied each of these adaptive radiations. These diversifications likely resulted from the opening up of niches following the extinction of the dinosaurs, and the creation of new niches by a major radiation of angiosperms (Sussman, 1991). Second, primates, as tropical, forest-living animals, are heavily influenced by climate and the distribution of tropical forests. Consequently, rates of diversification have likely varied considerably over the history of the group. This provides an additional reason why very rapid diversification during the late Paleocene and early Eocene—the warmest period in the Cenozoic—is plausible. Tavaré et al. do not allow for such large rate variations, and particularly for unusually rapid rates at the base of the tree. When these issues are taken into consideration, it seems quite plausible that the speciose Eocene euprimate fossil record could have arisen very quickly, and did not require some 25 million years to accrue.

What can be said with some confidence is that the fossil record demonstrates that primates (and by extension Archonta or Euarchonta) had diverged by the earliest Paleocene or possibly the latest Cretaceous. There is currently no concrete paleontological support for an earlier origin for Archonta, particularly in light of the primitive nature of *Purgatorius.* Thus, the discrepancy between even the more recent molecular divergence estimates of Springer et al. (2003, 2004) and the evidence from the fossil record is still significant and must be reconciled. Future collecting may fill this gap, or it may indicate that the earliest archontans were not much separated from *Purgatorius* in either time or morphology.

PLACES OF ORIGIN

The place of origin of Archonta (or Euarchonta) remains unclear in the absence of definitive fossil stem taxa or a consensus on the group's sister taxon. If recent molecular results supporting Euarchontoglires (e.g., Waddell et al., 1999; Murphy et al., 2001a,b; Springer et al., 2003, 2004) are correct, they may be supportive of an Asian origin for Euarchonta, based on the Asian distribution of most proposed

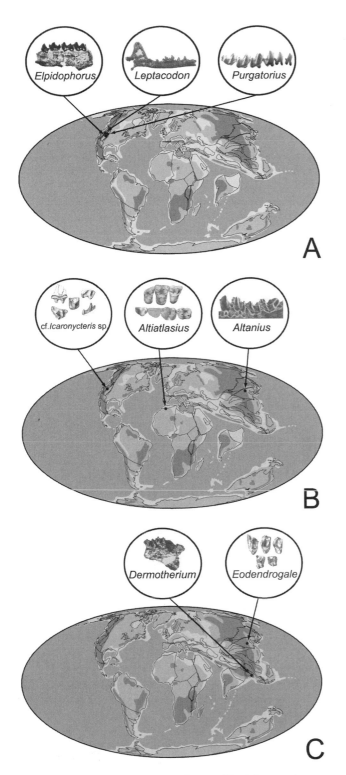

relatives or ancestors to Glires, including zalambdalestids (Archibald et al., 2001), anagalids, and pseudictopids (McKenna and Bell, 1997). The results of Silcox's analysis (2001; see Fig. 9.6) suggest a North American origin for Archonta, however, based on the North American location (Fig. 9.7) of the earliest plagiomenids (*Elpidophorus*), plesiadapiforms (*Purgatorius*), and chiropterans (*Icaronycteris*). A North American origin would also apply to Euarchonta. If mixodectids, apatemyids, and nyctitheriids are primitive archontans, then the presence of the earliest representatives of these groups in North America adds support to this idea. This pattern may be a result of preservation bias, because the fossil record from North America is more complete than that from other possible source areas, including particularly Africa, which has a very poor Cretaceous and Paleocene mammalian fossil record. If *Deccanolestes* were an archontan, this would strengthen the case for an Asian origin, although this specimen comes from the Indian subcontinent, which was isolated from the rest of Asia until approximately 55 million years ago (Clyde et al., 2003).

As noted above, the fossil record of definitive dermopterans and scandentians is extremely limited and entirely Asian in distribution. Therefore, the existing evidence supports an Asian origin for these clades, although the earliest known fossils likely significantly postdate the origins of these groups (i.e., they have ghost lineages that stretch at least the length of the Paleocene and early Eocene). If plagiomenids were dermopterans, this would substantially reduce the ghost lineage of Dermoptera. The exclusively North American distribution of plagiomenids would then suggest a North American origin for Dermoptera. The earliest record of a fossil bat, cf. *Icaronycteris* sp. (Gingerich, 1987), also comes from North America, implying that if Volitantia is valid, it may have a North American origin.

Primates can also be reconstructed as having a North American origin, based on the exclusively North American distribution of the most primitive members of this group (i.e., Purgatoriidae, Palaechthonidae, and Micromomyidae; Silcox, 2001; Bloch et al., 2003; see Fig. 9.6). Unlike that of Beard (1998), Silcox's (2001) analysis does not support an unequivocal resolution to the place of origin of Euprimates (his "Primates"). Although Asia is one possible place of origin (based largely on the primitive features of *Altanius;* Beard, 1998), the African location of the oldest fossil taxon

Fig. 9.7. Geographic and temporal distribution of fossils key to archontan or euarchontan origins and relationships. All maps are from Scotese (2001). (A) Early Paleocene. The only euarchontans known from the early Paleocene are the plagiomenid dermopteran *Elpidophorus* and the plesiadapiform primate *Purgatorius,* both from North America. If nyctitheriids are archontans (as suggested by Hooker, 2001), Paleocene *Leptacodon* from North America may prove key to issues of archontan (or euarchontan) origins. *Elpidophorus:* left mandible of *E. minor,* modified from Szalay (1969: plate 24 [4]); *Leptacodon:* right mandible of *L. tener,* modified from McKenna (1968: fig. 4); *Purgatorius:* right mandible of *P. janisae,* modified from Clemens (1974: fig. 2). (B) Late Paleocene. The earliest euprimate (*Altiatlasius*) and earliest chiropteran (cf. *Icaronycteris* sp.; Gingerich, 1987) appear in the late Paleocene of Africa and North America, respectively. *Altanius,* from Asia, may be latest Paleocene or earliest Eocene in age (Bowen et al., 2002). cf. *Icaronycteris:* isolated teeth of cf. *Icaronycteris* sp., modified from Gingerich (1987: fig. 23); *Altiatlasius:* isolated teeth of *A. koulchii,* modified from Sigé et al. (1990: plate 4); *Altanius:* left mandible of *A. orlovi,* modified from Rose and Krause (1984: fig. 2c). (C) Eocene. The earliest definitive dermopteran (*Dermotherium*) and scandentian (*Eodendrogale*) do not appear until the Eocene, with both specimens coming from Asia. By this time, Chiroptera and Primates were already broadly distributed across the Northern Hemisphere and, in the case of Chiroptera, Australia. *Dermotherium:* left mandible of *D. major,* modified from Ducrocq et al. (1992: fig. 1A); *Eodendrogale:* isolated teeth of *E. parvum,* modified from Tong (1988: fig. 1).

(*Altiatlasius*) assigned to Euprimates in Silcox's (2001) analysis makes the latter continent slightly more likely. The predominantly North American distribution of plesiadapiforms might support North America as a place of origin for Euprimates. However, because plesiadapiforms have only been known from Asia since 1995 (Beard and Wang, 1995) and are currently unknown in Africa, it is possible that a more complete record for the group would change that perception. Recent findings (Bowen et al., 2002) indicate that, contrary to previous assertions, euprimates do not occur in the North American record earlier than they do in Asia. The relatively sudden appearance of euprimates in Europe without clear precursors at the beginning of the Eocene might argue against an origin of Euprimates on this continent, in spite of the presence of the most primitive adapid (*Donrussellia*) and omomyid (*Teilhardina;* also known from North America). Silcox (2001; see Fig. 9.6), however, found support for a sister-group relationship between the plesiadapiform family Toliapinidae *sensu stricto* (i.e., including only the European genera *Toliapina* and *Avenius*) and Euprimates. Even though the support for this grouping is extremely weak (largely because Toliapinidae is known only from a handful of isolated teeth), this finding does suggest the possibility that Europe needs to be considered as a possible source area for Euprimates. In sum, the place of origin of Euprimates remains equivocal, and could conceivably be Asia, Africa, Europe, or North America.

CONCLUSIONS

These are exciting times for the study of archontan relationships. The controversy sparked by the bold interpretations by Beard (1990, 1993a) and Kay et al. (1990, 1992) of new fossil specimens has inspired new, detailed morphological descriptions of previously poorly known archontan groups (e.g., Stafford and Szalay, 2000; Sargis, 2002a–c), the introduction of new methods for analyzing and interpreting form and phylogeny (e.g., Runestad and Ruff, 1995; Hamrick et al., 1999), projects to find and study additional fossils (e.g., Bloch and Boyer, 2002, Bloch et al., 2003), and comprehensive cladistic analyses (e.g., Silcox, 2001; Bloch et al., 2002). There are two major issues that remain unresolved about this section of the mammalian evolutionary tree, however. First, stem archontan (or euarchontan) fossil taxa have yet to be convincingly identified. Second, the immediate sister taxon to Archonta (or Euarchonta) also needs to be recognized. Resolving these issues is central to a better understanding of this group, as these identifications are needed to accurately polarize characters.

Recent molecular studies have forced morphologists to query their basic assumptions and to re-evaluate interpretations of paleontological data. There are two major areas of conflict that remain between morphological and molecular results. The first relates to the timing of the origin of the group. Molecular studies conclude that the divergence point is much earlier than the fossil record would suggest. The second relates to the phylogenetic position of bats, and by extension, to the validity of Archonta vs. Euarchonta. Molecular studies suggest that bats are members of Laurasiatheria (Murphy et al., 2001a) whereas morphology has consistently placed them in Archonta, although the molecular hypothesis has not been explicitly tested with morphological data using cladistics. Understanding and accounting for the reasons behind these divergent results remains a major challenge, which could fundamentally shake up our understanding of the early history of Archonta.

SUMMARY

Morphological evidence is supportive of a supraordinal clade including Scandentia, Dermoptera, Chiroptera, and Primates, which can be referred to as Archonta. Recent molecular studies have supported a modified version of this clade that excludes Chiroptera, called Euarchonta. Molecular estimates of origination times place Euarchonta's origin well into the Cretaceous. Fossil evidence is supportive of a more recent origin—by the latest Cretaceous or earliest Paleocene. The place of origin of Euarchonta is reconstructed as North America, although this may be a product of uneven geographic sampling in the Cretaceous and Paleocene. The fossil record of definitive members of two euarchontan groups, Dermoptera and Scandentia, is very limited. The specimens available are weakly supportive of an Asian origin for both groups and a divergence by the late Eocene. If Plagiomenidae are dermopterans, as we suggest, then this might alter the place of origin for Dermoptera to North America and the age of origin to the early Paleocene. Primates are much better known from the fossil record, and include the earliest euarchontan (*Purgatorius*) from the latest Cretaceous or early Paleocene of North America. The presence of the most primitive primates in North America supports an origin for the group on that continent. The place of origin of Euprimates remains equivocal, and may be Africa, Asia, North America, or even Europe. In light of the age of *Altiatlasius*, Euprimates had already diverged by the late Paleocene.

ACKNOWLEDGMENTS

We thank K. D. Rose and J. D. Archibald for inviting us to contribute to this volume. F. S. Szalay and G. F. Gunnell made many helpful comments in their reviews. We also thank K. D. Rose for access to the cast of TF 2580, kindly provided to him by S. Ducrocq. Thanks are due to P. D. Gingerich, K. D. Rose, F. S. Szalay, and A. C. Walker for valuable conversations and ongoing support over the course of this project. Research was funded, in part, by grants from the Wenner-Gren Foundation for Anthropological Research, the Paleobiological Fund, Sigma Xi, and the National Science Foundation (Doctoral Dissertation Improvement grant 9815884) to MTS, and by a grant from the National Science

Foundation (BCS-0129601) to G. F. Gunnell, P. D. Gingerich, and JIB. EJS received funding from the Yale University Social Science Faculty Research Fund.

REFERENCES

Adkins, R. M., and R. L. Honeycutt. 1991. Molecular phylogeny of the superorder Archonta. Proceedings of the National Academy of Sciences USA 88: 10317–10321.

Aimi, M., and H. Inagaki. 1988. Grooved lower incisors in flying lemurs. Journal of Mammalogy 69: 138–140.

Allard, M. W., B. E McNiff, and M. M. Miyamoto. 1996. Support for interordinal eutherian relationships with an emphasis on primates and their archontan relatives. Molecular Phylogenetics and Evolution 5: 78–88.

Archibald J. D., A. O. Averianov, and E. G. Ekdale. 2001. Late Cretaceous relatives of rabbits, rodents, and other extant eutherian mammals. Nature 414: 62–65.

Árnason, Ú., A. Gullberg, and A. Janke. 1998. Molecular timing of primate divergences as estimated by two non-primate calibration points. Journal of Molecular Evolution 47: 718–727.

Beard, K. C. 1989. Postcranial anatomy, locomotor adaptations, and paleoecology of Early Cenozoic Plesiadapidae, Paromomyidae, and Micromomyidae (Eutheria, Dermoptera). Ph.D. Dissertation, The Johns Hopkins University School of Medicine, Baltimore.

———. 1990. Gliding behavior and palaeoecology of the alleged primate family Paromomyidae (Mammalia, Dermoptera). Nature 345: 340–341.

———. 1993a. Phylogenetic systematics of the Primatomorpha, with special reference to Dermoptera; pp. 129–150 in F. S. Szalay, M. J. Novacek, and M. C. McKenna (eds.), Mammal Phylogeny: Placentals. Springer-Verlag, New York.

———. 1993b. Origin and evolution of gliding in Early Cenozoic Dermoptera (Mammalia, Primatomorpha); pp. 63–90 in R.D.E. MacPhee (ed.), Primates and their Relatives in Phylogenetic Perspective. Plenum, New York.

———. 1998. East of Eden: Asia as an important center of taxonomic origination in mammalian evolution; pp. 5–39 in K. C. Beard and M. R. Dawson (eds.), Dawn of the Age of Mammals in Asia. Bulletin of the Carnegie Museum of Natural History 34.

Beard, K. C., and J. Wang. 1995. The first Asian plesiadapoids (Mammalia: Primatomorpha). Annals of the Carnegie Museum 64: 1–33.

Bloch, J. I., and D. M. Boyer. 2001. Taphonomy of small mammals in freshwater limestones from the Paleocene of the Clarks Fork Basin; pp. 185–198 in P. D. Gingerich (ed.), Paleocene-Eocene Stratigraphy and Biotic Change in the Bighorn and Clarks Fork Basins, Wyoming. University of Michigan Papers on Paleontology 33.

———. 2002. Grasping primate origins. Science 298: 1606–1610.

———. 2003. Response to comment on "Grasping primate origins." Science 300: 741c.

———. In press. New skeletons of Paleocene-Eocene Plesiadapiformes: A diversity of arboreal positional behaviors in early primates; in M. J. Ravosa and M. Dagosto (eds.), Primate Origins and Adaptations: A Multidisciplinary Perspective. Plenum, New York.

Bloch, J. I., D. M. Boyer, and P. Houde. 2003. Skeletons of Paleocene-Eocene micromomyids (Mammalia, Primates):

Functional morphology and implications for euarchontan relationships. Journal of Vertebrate Paleontology 23(supplement to no. 3): 35A.

Bloch, J. I., and M. T. Silcox. 2001. New basicrania of Paleocene-Eocene *Ignacius:* Re-evaluation of the plesiadapiform-dermopteran link. American Journal of Physical Anthropology 116: 184–198.

Bloch, J. I., M. T. Silcox, and E. J. Sargis. 2002. Origin and relationships of Archonta (Mammalia, Eutheria): Re-evaluation of Eudermoptera and Primatomorpha. Journal of Vertebrate Paleontology 22(supplement to no. 3): 37A.

Bowen, G. J., W. C. Clyde, P. L. Koch, S. Ting, J. Alroy, T. Tsubamoto, Y. Wang, and Y. Wang. 2002. Mammalian dispersal at the Paleocene/Eocene boundary. Science 295: 2062–2065.

Bown, T. M. 1976. The affinities of *Teilhardina* (Primates, Omomyidae) with description of a new species from North America. Folia Primatologica 25: 62–72.

Boyer, D. M., J. I. Bloch, and P. D. Gingerich. 2001. New skeletons of Paleocene paromomyids (Mammalia, ?Primates): Were they mitten gliders? Journal of Vertebrate Paleontology 21(supplement to no. 3): 35A.

Butler, P. M. 1972. The problem of insectivore classification; pp. 253–265 in K. A. Joysey and T. S. Kemp (eds.), Studies in Vertebrate Evolution. Oliver and Boyd, Edinburgh.

Carlsson, A. 1922. Über die Tupaiidae und ihre Beziehungen zu den Insectivora und den Prosimiae. Acta Zoologica 3: 227–270.

Cartmill, M. 1972. Arboreal adaptations and the origin of the order Primates; pp. 97–122 in R. Tuttle (ed.), The Functional and Evolutionary Biology of Primates. Aldine-Atherton, Chicago.

———. 1992. New views on primate origins. Evolutionary Anthropology 1: 105–111.

Chopra, S.R.K., S. Kaul, and R. N. Vasishat. 1979. Miocene tree shrews from the India Sivaliks. Nature 218: 213–214.

Chopra, S.R.K., and R. N. Vasishat. 1979. Sivalik fossil tree shrew from Haritalyangar, India. Nature 281: 214–215.

Clemens, W. A. 1974. *Purgatorius,* an early paromomyid primate (Mammalia). Science 184: 903–905.

———. in press. *Purgatorius* (Plesiadapiforms, Primates?, Mammalia), a Paleocene immigrant into northeastern Montana: Stratigraphic occurrences and incisor proportions; in M. R. Dawson and J. A. Lillegraven (eds.), Fanfare for an Uncommon Paleontologist: Papers in Honor of Malcolm C. McKenna. Bulletin of Carnegie Museum of Natural History.

Clyde, W. C., I. H. Khan, and P. D. Gingerich. 2003. Stratigraphic response and mammalian dispersal during initial India-Asia collision: Evidence from the Ghazij Formation, Baluchistan, Pakistan. Geology 31: 1097–1100.

Dashzeveg, D., and M. C. McKenna. 1977. Tarsioid primate from the early Tertiary of the Mongolian People's Republic. Acta Palaeontologica Polonica 22: 119–137.

Ducrocq, S., E. Buffetaut, H. Buffetaut-Tong, J.-J. Jaeger, Y. Jongkanjanasoontorn, and Y. Suteethorn. 1992. First fossil flying lemur: A dermopteran from the Late Eocene of Thailand. Palaeontology 35: 373–380.

Dutta, A. K. 1975. Micromammals from Siwaliks. Indian Minerals 29: 76–77.

Eizirik, E., W. J. Murphy, and S. J. O'Brien. 2001. Molecular dating and biogeography of the early placental mammal radiation. Journal of Heredity 92: 212–219.

Gheerbrant, E., J. Sudre, S. Sen, C. Abrial, B. Marandat, B. Sigé, and M. Vianey-Liaud. 1998. Nouvelles données sur les mammifères du Thanetien et de l'Ypresien du Bassin d'Ouarzazate (Maroc) et leur contexte stratigraphique. Palaeovertebrata 27: 155–202.

Gidley, J. W. 1923. Paleocene primates of the Fort Union, with discussion of relationships of Eocene primates. Proceedings of the U.S. National Museum 63: 1–38.

Gingerich, P. D. 1987. Early Eocene bats (Mammalia, Chiroptera) and other vertebrates in freshwater limestones of the Willwood Formation, Clarks Fork Basin, Wyoming. Contributions from the Museum of Paleontology, University of Michigan 27: 275–320.

———. 1989. New earliest Wasatchian mammalian fauna from the Eocene of northwestern Wyoming: Composition and diversity in a rarely sampled high-floodplain assemblage. University of Michigan Papers on Paleontology 28: 1–97.

———. 1990. African dawn for primates. Nature 346: 411.

Gingerich, P. D., D. Dashzeveg, and D. E. Russell. 1991. Dentition and systematic relationships of Altanius orlovi (Mammalia, Primates) from the early Eocene of Mongolia. Géobios 24: 637–646.

Gingerich, P. D., and M. D. Uhen. 1994. Time of origin of primates. Journal of Human Evolution 27: 443–445.

Godinot, M. 1978. Un nouvel Adapidé (primate) de l'Éocène inférieur de Provence. Comptes Rendus de l'Académie des Sciences, Série D, Paris 286: 1869–1872.

———. 1992. Apport à la systématique de quatre genres d'Adapiformes (Primates, Éocène). Comptes Rendus de l'Académie des Sciences, Série II: Mécanique-Physique-Chimie, Sciences de la Terre et de l'Univers, Paris 314: 237–242.

———. 1994. Early North African primates and their significance for the origin of Simiiformes (=Anthropoidea); pp. 235–295 in J. G. Fleagle and R. F. Kay (eds.), Anthropoid Origins. Plenum, New York.

Godinot, M., J.-Y. Crochet, J.-L. Hartenberger, B. Lange-Badré, D. E. Russell, and B. Sigé. 1987. Nouvelles données sur les mammifères de Palette (Eocène inférieur, Provence). Münchner Geowissenschaftliche Abhandlungen, Reihe A: Geologie und Paläontologie 10: 273–288.

Gregory, W. K. 1910. The orders of mammals. Bulletin of the American Museum of Natural History 27: 1–524.

Gunnell, G. F. 1989. Evolutionary history of Microsyopoidea (Mammalia, ?Primates) and the relationship between Plesiadapiformes and Primates. University of Michigan Papers on Paleontology 27: 1–157.

Hamrick, M. W. 2001. Primate origins: Evolutionary change in digital ray patterning and segmentation. Journal of Human Evolution 40: 339–351.

Hamrick, M. W., B. A. Rosenman, and J. A. Brush. 1999. Phalangeal morphology of the Paromomyidae (?Primates, Plesiadapiformes): The evidence for gliding behavior reconsidered. American Journal of Physical Anthropology 109: 397–413.

Hand, S., M. Novacek, H. Godthelp, and M. Archer. 1994. First Eocene bat from Australia. Journal of Vertebrate Paleontology 14: 375–381.

Hartwig, W. C. 2002. The Primate Fossil Record. Cambridge University Press, Cambridge.

Hedges, S. B., P. H. Parker, C. G. Sibley, and S. Kumar. 1996. Continental break up and the ordinal diversification of birds and mammals. Nature 381: 226–229.

Hoffstetter, R. 1977. Phylogénie des primates. Bulletins et Mémoires de la Société d'Anthropologie de Paris, Série XIII 4: 327–346.

Hooker, J. J. 1996. A primitive emballonurid bat (Chiroptera, Mammalia) from the earliest Eocene of England. Palaeovertebrata 25: 287–300.

———. 2001. Tarsals of the extinct insectivoran family Nyctitheriidae (Mammalia): Evidence for archontan relationships. Zoological Journal of the Linnean Society 132: 501–529

Hooker, J. J., D. E. Russell, and A. Phélizon. 1999. A new family of Plesiadapiformes (Mammalia) from the Old World lower Paleogene. Palaeontology 42: 377–407.

Hunt R. M., Jr., and W. K. Korth. 1980. The auditory region of Dermoptera: Morphology and function relative to other living mammals. Journal of Morphology 164: 167–211.

Illiger, C. 1811. Prodromus systematis mammalium et avium additus terminis zoographis utriudque classis. C. Salfeld, Berlin.

Jacobs, L. L. 1980. Siwalik fossil tree shrews; pp. 205–216 in W. P. Luckett (ed.), Comparative Biology and Evolutionary Relationships of Tree Shrews. Plenum, New York.

Johnson, J. I., and J.A.W. Kirsch. 1993. Phylogeny through brain traits: Interordinal relationships among mammals including Primates and Chiroptera; pp. 293–331 in R.D.E. MacPhee (ed.), Primates and Their Relatives in Phylogenetic Perspective. Plenum, New York.

Johnston, P. A., and R. C. Fox. 1984. Paleocene and late Cretaceous mammals from Saskatchewan, Canada. Palaeontographica Abteilung A 186: 163–222.

Kay, R. F., J.G.M. Thewissen, and A. D. Yoder. 1992. Cranial anatomy of Ignacius graybullianus and the affinities of the Plesiadapiformes. American Journal of Physical Anthropology 89: 477–498.

Kay, R. F., R. W. Thorington, Jr., and P. Houde. 1990. Eocene plesiadapiform shows affinities with flying lemurs not primates. Nature 345: 342–344.

Koenigswald, W. von. 1990. Die Paläobiologie der Apatemyiden (Insectivora s.l.) und die Ausdeutung der Skelettfunde von Heterohyus nanus aus dem Mitteleozän von Messel bei Darmstadt. Palaeontographica Abteilung A 210: 41–77.

Krause, D. W. 1991. Were paromomyids gliders? Maybe, maybe not. Journal of Human Evolution 21: 177–188.

Le Gros Clark, W. E. 1925. On the skull of Tupaia. Proceedings of the Zoological Society of London 1925: 559–567.

———. 1926. On the anatomy of the pen-tailed tree shrew (Ptilocercus lowii). Proceedings of the Zoological Society of London 1926: 1179–1309.

Liu, F.-G.R., M. M. Miyamoto, N. P. Freire, P. Q. Ong, M. R. Tennant, T. S. Young, and K. F. Gugel. 2001. Molecular and morphological supertrees for eutherian (placental) mammals. Science 291: 1786–1789.

Lofgren, D. L. 1995. The Bug Creek problem and the Cretaceous-Tertiary boundary at McGuire Creek, Montana. University of California Publications in Geological Science 140: 1–185.

MacPhee, R.D.E, M. Cartmill, and P. D. Gingerich. 1983. New Paleogene primate basicrania and the definition of the order Primates. Nature 301: 509–511.

MacPhee, R.D.E., M. Cartmill, and K. D. Rose. 1989. Craniodental morphology and relationships of the supposed Eocene dermopteran Plagiomene (Mammalia). Journal of Vertebrate Paleontology 9: 329–349.

Martin, R. D. 1968. Towards a new definition of Primates. Man 3: 377–401.

Matthew, W. D. 1918. Part V—Insectivora (continued), Glires, Edentata; pp. 565–567 *in* W. D. Matthew and W. Granger (eds.), A Revision of the Lower Eocene Wasatch and Wind River Faunas. Bulletin of the American Museum of Natural History 38.

Matthew, W. D., and W. Granger. 1921. New genera of Paleocene mammals. American Museum Novitates 13: 1–7.

McKenna, M. C. 1963. Primitive Paleocene and Eocene Apatemyidae (Mammalia, Insectivora) and the Primate-Insectivore boundary. American Museum Novitates 2160: 1–39.

———. 1966. Paleontology and the origin of the Primates. Folia Primatologica 4: 1–25.

———. 1968. *Leptacodon,* an American Paleocene nyctithere (Mammalia, Insectivora). American Museum Novitates 2317: 1–12.

———. 1975. Toward a phylogenetic classification of the Mammalia; pp. 21–46 *in* W. P. Luckett and F. S. Szalay (eds.), Phylogeny of the Primates. Plenum, New York.

McKenna, M. C., and S. K. Bell. 1997. Classification of Mammals Above the Species Level. Columbia University Press, New York.

Mein P., and L. Ginsburg. 1997. Les mammifères du gisement Miocène inférieur de Li Mae Long, Thaïlande: Systématique, biostratigraphie et paléoenvironnement. Geodiversitas 19: 783–844.

Miyamoto, M. M., C. A. Porter, and M. Goodman. 2000. *c-Myc* gene sequences and the phylogeny of bats and other eutherian mammals. Systematic Biology 49: 501–514.

Murphy, W. J., E. Eizirik, W. E. Johnson, Y. P. Zhang, O. A. Ryder, and S. J. O'Brien. 2001a. Molecular phylogenetics and the origins of placental mammals. Nature 409: 614–618.

Murphy, W. J., E. Eizirik, S. J. O'Brien, O. Madsen, M. Scally, C. J. Douady, E. C. Teeling, O. A. Ryder, M. J. Stanhope, W. W. de Jong, and M. S. Springer. 2001b. Resolution of the early placental mammal radiation using Bayesian phylogenetics. Science 294: 2348–2351.

Ni, X., and Z. Qiu. 2002. The micromammalian fauna from the Leilao, Yuanmou hominoid locality: Implications for biochronology and paleoecology. Journal of Human Evolution 42: 535–546

Novacek, M. J. 1992. Mammalian phylogeny: Shaking the tree. Nature 356: 121–125.

Novacek, M. J., and A. R. Wyss. 1986. Higher-level relationships of the recent eutherian orders: Morphological evidence. Cladistics 2: 257–287.

Prasad, G.V.R., and M. Godinot. 1994. Eutherian tarsal bones from the late Cretaceous of India. Journal of Paleontology 68: 892–902.

Qiu, Z., 1986. Fossil tupaiid from the hominoid locality of Lufeng, Yunnan. Vertebrata PalAsiatica 24: 308–319.

Qiu, Z., D. Han, Q. Qi, and Y. Lin. 1985. A micromammalian assemblage from the hominoid locality of Lufeng, Yunnan. Acta Anthropologica Sinica 4: 13–32. (Translated by Will Downs.)

Rose, K. D. 1973. The mandibular dentition of *Plagiomene* (Dermoptera, Plagiomenidae). Breviora 411: 1–17.

———. 1995. The earliest primates. Evolutionary Anthropology 3: 159–173.

Rose, K. D., and T. M. Bown. 1991. Additional fossil evidence on the differentiation of the earliest euprimates. Proceedings of the National Academy of Sciences USA 88: 98–101.

Rose, K. D., and D. W. Krause. 1984. Affinities of the primate *Altanius* from the early Tertiary of Mongolia. Journal of Mammalogy 65: 721–726.

Runestad, J. A., and C. B. Ruff. 1995. Structural adaptations for gliding in mammals with implications for locomotor behavior in paromomyids. American Journal of Physical Anthropology 98: 101–119.

Sargis, E. J. 2002a. Functional morphology of the forelimb of tupaiids (Mammalia, Scandentia) and its phylogenetic implications. Journal of Morphology 253: 10–42.

———. 2002b. Functional morphology of the hindlimb of tupaiids (Mammalia, Scandentia) and its phylogenetic implications. Journal of Morphology 254: 149–185.

———. 2002c. The postcranial morphology of *Ptilocercus lowii* (Scandentia, Tupaiidae): An analysis of primatomorphan and volitantian characters. Journal of Mammalian Evolution 9: 137–160.

Scotese, C. R. 2001. Digital Paleogeographic Map Archive on CD-ROM. PALEOMAP Project, Arlington, Texas.

Shoshani, J., and M. C. McKenna. 1998. Higher taxonomic relationships among extant mammals based on morphology, with selected comparisons of results from molecular data. Molecular Phylogenetics and Evolution 9: 572–584.

Sigé, B., J.-J. Jaeger, J. Sudre, and M. Vianey-Liaud. 1990. *Altiatlasius koulchii* n. gen et sp., primate omomyidé du Paléocène supérieur du Maroc, et les origines des euprimates. Palaeontographica Abteilung A 212: 1–24.

Silcox, M. T. 2001. A phylogenetic analysis of Plesiadapiformes and their relationship to Euprimates and other archontans. Ph.D. Dissertation, The Johns Hopkins School of Medicine, Baltimore.

———. 2003. New discoveries on the middle ear anatomy of *Ignacius graybullianus* (Paromomyidae, Primates) from ultra high resolution X-ray computed tomography. Journal of Human Evolution 44: 73–86.

Simmons, N. B., and G. H. Geisler 1998. Phylogenetic relationships of *Icaronycteris, Archaeonycteris, Hassianycteris,* and *Palaeochiropteryx* to extant bat lineages, with comments on the evolution of echolocation and foraging strategies in Microchiroptera. Bulletin of the American Museum of Natural History 235: 1–182.

Simmons, N. B., and T. H. Quinn. 1994. Evolution of the digital tendon locking mechanism in bats and dermopterans:a phylogenetic perspective. Journal of Mammalian Evolution 2: 231–254.

Simpson, G. G. 1940. Studies on the earliest Primates. Bulletin of the American Museum of Natural History 77: 185–212.

———. 1945. The principles of classification and a classification of mammals. Bulletin of the American Museum of Natural History 85: 1–350.

Smith, J. D., and G. Madkour. 1980. Penial morphology and the question of chiropteran phylogeny; pp. 347–365 *in* D. E. Wilson and A. L. Gardner (eds.), Proceedings: Fifth International Bat Research Conference. Texas Tech Press, Lubbock.

Springer, M. S., W. J. Murphy, E. Eizirik, and S. J. O'Brien. 2003. Placental mammal diversification and the Cretaceous-Tertiary boundary. Proceedings of the National Academy of Sciences USA 100: 1056–1061.

Springer, M. S., M. J. Stanhope, O. Madsen, and W. W. de Jong. 2004. Molecules consolidate the placental mammal tree. Trends in Ecology and Evolution 19: 430–438.

Stafford, B. J., and F. S. Szalay. 2000. Craniodental functional mor-phology and taxonomy of dermopterans. Journal of Mam-malogy 81: 360–385.

Stafford, B. J., and R. W. Thorington, Jr. 1998. Carpal develop-ment and morphology in archontan mammals. Journal of Morphology 235: 135–155.

Sussman, R. W. 1991. Primate origins and the evolution of angiosperms. American Journal of Primatology 23: 209–223.

Swofford, D. A. 2003. PAUP* 4.0. Sinauer Associates, Sunderland, Massachusetts.

Szalay, F. S. 1968a. Origins of the Apatemyidae (Mammalia, Insectivora). American Museum Novitates 2352: 1–11.

———. 1968b. The beginnings of primates. Evolution 22: 19–36.

———. 1969. Mixodectidae, Microsyopidae, and the insectivore-primate transition. Bulletin of the American Museum of Natural History 140: 195–330.

———. 1977. Phylogenetic relationships and a classification of the eutherian Mammalia; pp. 315–374 in M. K. Hecht, P. C. Goody, and B. M. Hecht (eds.), Major Patterns in Vertebrate Evolution, Plenum, New York.

———. 1982. A critique of some recently proposed Paleogene primate taxa and suggested relationships. Folia Primatologica 37: 152–182

Szalay, F. S., and E. Delson. 1979. Evolutionary History of the Primates. Academic Press, New York.

Szalay, F. S., and G. Drawhorn. 1980. Evolution and diversifi-cation of the Archonta in an arboreal milieu; pp. 133–169 in W. P. Luckett (ed.), Comparative Biology and Evolutionary Relationships of Tree Shrews. Plenum, New York.

Szalay, F. S., and S. G. Lucas. 1993. Cranioskeletal morphology of Archontans, and diagnoses of Chiroptera, Volitantia, and Archonta; pp. 187–226 in R.D.E. MacPhee (ed.), Primates and Their Relatives in Phylogenetic Perspective. Plenum, New York

———. 1996. The postcranial morphology of Paleocene *Chriacus* and *Mixodectes* and the phylogenetic relationships of archon-tan mammals. New Mexico Museum of Natural History and Science Bulletin 7: 1–47.

Szalay, F. S., A. L. Rosenberger, and M. Dagosto. 1987. Diagnosis and differentiation of the order Primates. Yearbook of Physi-cal Anthropology 30: 75–105.

Szalay, F. S., I. Tattersall, and R. L. Decker. 1975. Phylogenetic relationships of *Plesiadapis*—postcranial evidence; pp. 136–166 in F. S. Szalay (ed.), Approaches to Primate Paleobiology. Karger, Basel.

Tavaré, S., C. R. Marshall, O. Will, C. Soligo, and R. D. Martin. 2002. Using the fossil record to estimate the age of the last common ancestor of extant primates. Nature 416: 726–729.

Thewissen, J.G.M., and S. K. Babcock. 1992. The origin of flight in bats. BioScience 42: 340–345.

Tong, Y. 1988. Fossil tree shrews from the Eocene Hetaoyuan formation of Xichuan, Henan. Vertebrata PalAsiatica 26: 214–220.

Van Valen, L. M. 1965. Tree shrews, primates, and fossils. Evolu-tion 19: 137–151.

———. 1994. The origin of the plesiadapid primates and the nature of *Purgatorius*. Evolutionary Monographs 15: 1–79.

Van Valen, L. M., and R. E. Sloan. 1965. The earliest primates. Science 150: 743–745.

Waddell, P. J., N. Okada, and M. Hasegawa. 1999. Towards re-solving the interordinal relationships of placental mammals. Systematic Biology 48: 1–5.

Wagner, J. A. 1855. Die Säugethiere in Abbildungen nach der Natur. Weiger, Leipzig.

Wang, Y., Y. Hu, M. Chow, and C. Li. 1998. Chinese Paleocene mammal faunas and their correlation; pp. 89–123 in K. C. Beard and M. R. Dawson (eds.), Dawn of the Age of Mam-mals in Asia. Bulletin of the Carnegie Museum of Natural History 34.

Wible, J. R. 1993. Cranial circulation and relationships of the colugo *Cynocephalus* (Dermoptera, Mammalia). American Museum Novitates 3072: 1–27.

Wible, J. R., and H. H. Covert. 1987. Primates: Cladistic diagnosis and relationships. Journal of Human Evolution 16: 1–22.

Wible, J. R., and J. R. Martin. 1993. Ontogeny of the tympanic floor and roof in archontans; pp. 111–146 in R.D.E. MacPhee (ed.), Primates and Their Relatives in Phylogenetic Perspec-tive. Plenum, New York.

Wible, J. R., and M. J. Novacek. 1988. Cranial evidence for the monophyletic origin of bats. American Museum Novitates 2911: 1–19.

Wible, J. R., and U. Zeller. 1994. Cranial circulation of the pen-tailed tree shrew *Ptilocercus lowii* and relationships of Scan-dentia. Journal of Mammalian Evolution 2: 209–230.

JIN MENG AND ANDRÉ R. WYSS

10

GLIRES (LAGOMORPHA, RODENTIA)

G LIRES, WHICH INCLUDES TWO EXTANT mammalian clades, Lagomorpha and Rodentia, is characterized by (among many other diagnostic attributes) a pair of enlarged, evergrowing incisors, in which enamel is restricted to the buccal surfaces, in both upper and lower jaws. Nearly half the species of extant mammals are members of Glires. Rodentia (rats, squirrels, guinea pigs, and kin) includes 29 Recent major clades—traditionally carrying the rank of family (about 468 genera, 2,052 species; Nowak, 1999) and 743 extinct genera (McKenna and Bell, 1997) that date to the late Paleocene of North America (Wood, 1962; Dawson et al., 1984; Korth, 1984; Dawson and Beard, 1996). Lagomorpha (pikas, rabbits, and hares) consists of two living major clades (13 genera, 81 species; Nowak, 1999) and 56 fossil genera (McKenna and Bell, 1997), the nearest fossil allies of which first appeared in the Paleogene of Asia (Dawson, 1977; McKenna, 1982).

Morphologists and paleontologists long emphasized the numerous distinctions between rodents and lagomorphs (Gidley, 1912; Simpson, 1945; Hartenberger, 1985; Wilson, 1989), but the phylogenetic implication—if any—of these distinctions (Glires monophyly versus polyphyly) has proven one of the most enduring controversies in higher-level mammalian phylogenetics of the twentieth century (Wilson, 1949; Wood, 1957; Dawson, 1967; Landry, 1999; Luckett and Hartenberger, 1985; Li and Ting, 1985; Jaeger, 1988; Novacek, 1990; Luckett and Hartenberger, 1993). In addition to proposals of a sister-group relationship between rodents and lagomorphs, each group has for a time been allied with other mammalian taxa, based on morphological data. Rodents have been considered related to multituberculates (Major,

1893; Hinton, 1926; Friant, 1932), mixodectids (Osborn, 1902), tillodonts and taeniodonts (Cope, 1888), primates (McKenna, 1961; Van Valen, 1966, 1971; Wood, 1962, 1977; Lillegraven, 1969; Patterson and Wood, 1982), leptictids (Szalay, 1977, 1985), and eurymylids (Li, 1977; Gingerich and Gunnell, 1979; Li and Yan, 1979; Gingerich and McKenna, 1980; Hartenberger, 1980; Dawson et al., 1984; Korth, 1984; Wilson, 1989; Meng et al., 1994; McKenna and Meng, 2001; Meng and Wyss, 2001; Meng et al., 2003). Lagomorphs, however, have been regarded as related to triconodonts (Gidley, 1906; Ehik, 1926), artiodactyls (Gidley, 1912; Hürzeler, 1936; Moody et al., 1949), condylarths (Wood, 1957), zalambdodont insectivorans (Russell, 1959), pseudictopids (Van Valen, 1964), anagalids (Szalay and McKenna, 1971), macroscelidids (McKenna, 1975), eurymylids (Wood, 1942), zalambdalestids (Szalay and McKenna, 1971; McKenna, 1975, 1994), and mimotonids (Li and Yan, 1979; Li and Ting, 1985, 1993; Meng et al., 1994; Meng and Wyss, 2001).

The monophyly of Glires has been disputed by many molecular studies (Easteal, 1990; Penny et al., 1991; Honeycutt and Adkins, 1993; Ma et al., 1993; Martignetti and Brosius, 1993; Porter et al., 1996; Stanhope et al., 1996; Huchon et al., 1999; Misawa and Janke, 2003). In addition to casting doubt on a close relationship between rodents and lagomorphs, some molecular analyses have even called into question the monophyly of Rodentia itself (Graur et al., 1991, 1996; Li et al., 1992; D'Erchia et al., 1996; Huchon et al., 1999; Mouchaty et al., 2001; Árnason et al., 2002; Janke et al., 2002). The divergence time between clades within Glires (at a variety of taxonomic levels) is another area of ongoing debate. For instance, one molecular-clock-based estimate places the divergence time of rodents from other eutherians at about 110 million years ago (Kumar and Hedges, 1998), suggesting that rodents are the oldest placental group after Xenarthra. Similarly, Lagomorpha are argued to have diverged from other placental mammals at about 90 million years ago (Kumar and Hedges, 1998). These molecular-based divergence estimates contradict conventional interpretations of a post Cretaceous/Tertiary boundary radiation of most modern placental groups—as suggested by a literal reading of their known fossil records (Gingerich, 1977; Novacek, 1992; Archibald and Deutschman, 2001).

Herein we provide a brief synopsis of the key morphological features that are unique to Glires, distinguish rodents from lagomorphs, and ally stem taxa to rodents and lagomorphs. Phylogenetic results deriving from morphological and molecular data sets are contrasted, as are the discrepant estimates of divergence times between major clades.

TAXONOMIC TERMINOLOGY

We employ the taxonomic terminology proposed in earlier studies (Wyss and Meng, 1996; Meng and Wyss, 2001), noting that contrasting schemes are also currently used (Mc-

Kenna and Bell, 1997; Landry, 1999). Definitions of the taxonomic names we use in this chapter are:

Glires is defined as the clade stemming from the most recent common ancestor of Lagomorpha and Rodentia.

Simplicidentata is defined as all Glires sharing a more recent common ancestor with Rodentia than with Lagomorpha. It is the stem-based counterpart to Rodentia.

Rodentia is defined as the clade stemming from the most recent common ancestor of *Mus* and all Recent mammals more closely related to it than to Lagomorpha or any other eutherian "order," *sensu* Simpson (1945).

Duplicidentata is defined as all members of Glires sharing a more recent common ancestor with Lagomorpha than with Rodentia.

Lagomorpha is defined as the clade stemming from the most recent common ancestor of *Ochotona* and Leporidae.

MORPHOLOGY

Evolutionary Novelties Diagnostic of Glires

Given our paleontological emphasis, the morphological comparisons made here are primarily osteological and dental. Members of Glires are characterized by many highly distinctive morphological features (Figs. 10.1–10.4). Perhaps the most obvious derived resemblance shared by members of Glires is their enlarged pair of evergrowing upper and lower incisors, which extend deep into the maxilla and dentary. These teeth are the retained deciduous second incisors —dI2/di2 of the typical eutherian dentition, which go unreplaced during the life cycle (Luckett, 1985). The first incisors (lower and upper) do not develop. The third lower incisor also fails to develop in all other members of Glires except "mimotonids." The enlarged incisors are transversely compressed, and their convex anterior surfaces are covered with thickened enamel. In addition to a reduced number of incisors, members of Glires have also lost the upper and lower canines, the first upper and lower premolars, and the second lower premolar. The loss of these teeth creates a lengthy gap (diastema) between the incisors and cheek teeth in both the upper and lower toothrows. Members of Glires primitively share many detailed resemblances of the cheek-tooth crowns, including the lack of a centrocrista on the upper molars and the reduction or loss of the paraconid on the lower molars (Fig. 10.2). The horizontal ramus of the mandible is relatively deep and short, bearing a reduced coronoid process and a somewhat expanded angular process. The mandibular condyle is oriented anteroposteriorly, articulating with a longitudinally elongate and anterodorsally shifted glenoid fossa. The postglenoid process is greatly reduced. In combination, these morphologies permit broad anteroposterior, or propalinal, movement of the mandible, and thus, gnawing (incisor biting) and incisor

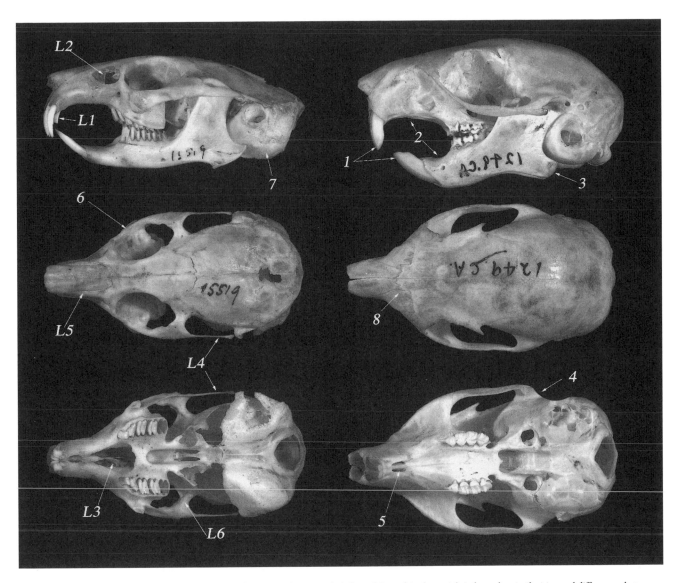

Fig. 10.1. Lateral, dorsal, and ventral views of skulls of *Ochotona* (lagomorph, left) and *Sciurus* (rodent, right) show the similarities and differences between lagomorphs and rodents. Diagnostic features shown in the figure for rodents: 1, upper and lower incisors (dI2/di2); 2, canines, P1/p1, and p2 are lost so that a diastema is present between the incisors and cheek teeth in both the upper and lower toothrows; 3, horizontal ramus of the mandible; 4, glenoid fossa; 5, incisive foramina; 6, anterior zygomatic root and orbit; 7, bulla; 8, premaxilla. Some diagnostic features for lagomorphs: L1, diminutive I3 immediately behind dI2; L2, fenestrae in the maxilla on the facial region; L3, incisive foramina and the hard palate; L4, posterior projection of the jugal; L5, anterior projection of the frontal inserts between the premaxilla and maxilla; L6, shortened glenoid fossa longitudinally oriented. L3–L6 are also present in "mimotonids."

sharpening (particularly of the lowers against the uppers). Numerous additional cranial similarities are shared by members of Glires, such as elongate incisive foramina, short infraorbital canals, a narrow premaxilla-frontal contact on the facial region of the skull, an anteriorly shifted posterior edge of anterior zygomatic root, anteriorly shifted orbits, lack of a jugal-lacrimal contact, and an ectotympanic bulla.

Many of these features are related to the unique masticatory apparatus of Glires. As pointed out by Meng et al. (2003), transformation of these features during the origin of Glires and its subsequent diversification involved various divergent and convergent trends, creating a mosaic of derived character combinations. This complex and protracted transformation of the masticatory apparatus during the evolution of Glires is a key source of phylogenetic information at various levels within the group.

Distinctions between Rodents and Lagomorphs

Lagomorphs differ from rodents in many respects (see Fig. 10.1). In lagomorphs, dI2/di2 are shorter than in rodents, dI2 extends usually only to beneath p3 or m1, and dI2 (which is anteroposteriorly compressed) is restricted to the premaxilla. Even more striking, two pairs of upper incisors occur in lagomorphs, the more diminutive set (I3— permanent teeth) occurring immediately behind the major

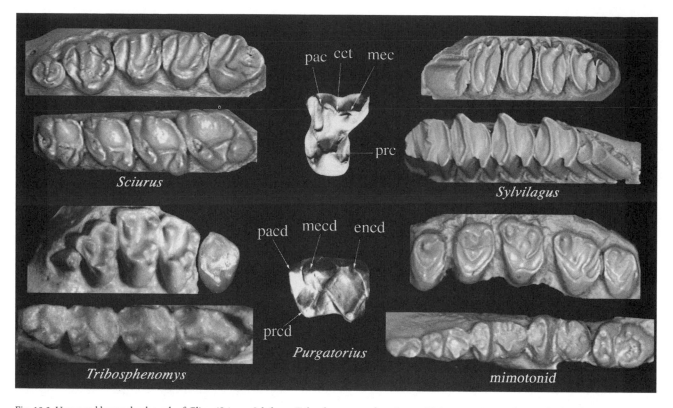

Fig. 10.2. Upper and lower cheek teeth of Glires (*Sciurus*, *Sylvilagus*, *Tribosphenomys*, and a mimotonid) in comparison with typical tribosphenic molars of *Purgatorius*. Abbreviations: cct, centrocrista; encd, entoconid; mec, metacone; mecd, metaconid; pac, paracone; pacd, paraconid; prc, protocone; prcd, protoconid. Images are from Meng and Wyss (2001) for *Tribosphenomys*, Kielan-Jaworowska et al. (1979) for *Purgatorius*, and an unpublished Bumban specimen of Mongolia for mimotonid. Images are not to the same scale.

set (dI2). The cheek teeth of the lagomorphs differ significantly from those of rodents (see Fig. 10.2). The third upper and lower premolars are large, high-crowned, molariform or submolariform, and bear enamel folds. The molars are similarly hypsodont. Transverse lophs are well developed, obscuring the usual mammalian dental cusps. The lower cheek teeth are basically bilobed. The M3/m3 are reduced in size or lost. The nasals are generally broad throughout their length, bearing a V-shaped posterior margin. Abundant fenestrae occur in the maxilla on the facial region. The alveoli of the upper cheek teeth dorsally intrude the orbit floor. Compared to most rodents, many aspects of the lagomorph postcranial skeleton are specialized. The distal tibia and fibula are fused at midshaft, and the fibula is slim. The astragalus is relatively narrow and elongated (see Fig. 10.3). The fibular facet on the calcaneus is broad and faces dorsally, and there is a distinct calcaneal canal (Bleefeld and Bock, 2002).

Another important distinction between rodents and lagomorphs is the double-layer incisor enamel with Hunter-Schreger bands in rodents. Rodent upper cheek teeth are quadrate and usually increase in width posteriorly. The precingulum is distinctive along the anterior edge of the teeth. A mesostyle and mure (an enamel ridge connecting the protocone and hypocone) usually occur on the upper molars. The entoconid of the lower molars is shifted anteriorly. The angular process is usually strongly curved medially.

Morphology of Stem Taxa

Fossil outgroups to lagomorphs, such as *Mimotona* and *Gomphos*, are commonly referred to as "mimotonids," which collectively are probably paraphyletic. The clade stemming from the last common ancestor of "mimotonids" and lagomorphs (Duplicidentata) is characterized by (among other things) the distinctive arrangement of two pairs of upper incisors. "Mimotonids" differ from lagomorphs mainly in possessing two pairs of lower incisors; that is, in retaining i3, which is greatly reduced and probably non-functional (see Figs. 10.1, 10.4).

In relation to other eutherians, "mimotonids" share many derived features with lagomorphs. The anterior face of the enlarged incisors is covered in single-layered enamel formed of Hunter-Schreger bands. The mandibular condyle is dorsally situated relative to the cheek-tooth row, and the coronoid process is greatly reduced. The incisive foramina are elongate and expanded posteriorly. The hard palate is short. The postglenoid foramen is laterally positioned, occurring between the squamosal and ectotympanic or petrosal. The optic foramina are confluent anteriorly at the sagittal plane. The nasals are broad, whereas the premaxillae bear a long, needle-shaped posterior process. A posterior projection of the jugal is present. An anterior projection of the frontal inserts between the premaxilla and maxilla. The glenoid fossa is not only longitudinally oriented, but is also

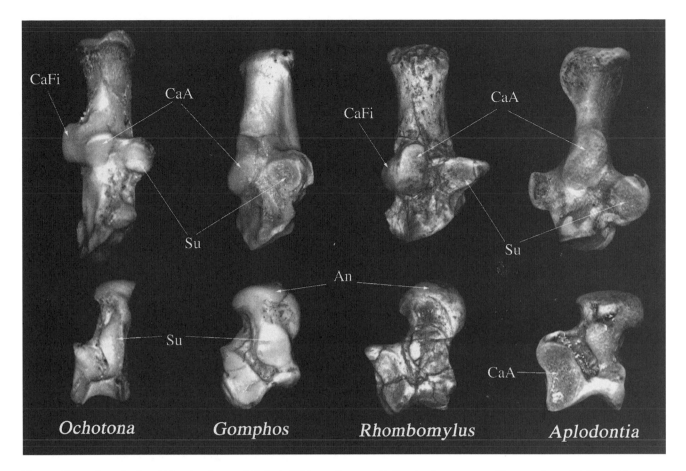

Fig. 10.3. Comparison of calcanei (above) and astragali (below) of *Ochotona* (lagomorph), *Gomphos* (a stem taxon of lagomorph; modified from Meng et al., 2004), *Rhombomylus* (a stem taxon of rodent; modified from Meng et al., 2003), and *Aplodontia* (rodent). The calcanei are in dorsal view, and the astragali are in ventral view. Abbreviations: An, astragalonavicular facet; CaA, calcaneoastragalar facet; CaFi, calcaneofibular facet; and Su, sustentacular facet. Images are not to the same scale.

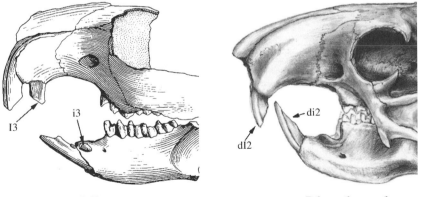

Mimotona   **Rhombomylus**

Fig. 10.4. Lateral views of rostrum of *Mimotona* (modified from Li and Ting, 1993) and *Rhombomylus* (modified from Meng et al., 2003). Note the presence of two pairs of incisors in both upper and lower jaws of *Mimotona*.

shortened and anteriorly shifted. The alisphenoid-parietal contact in the orbit is absent, owing to a broad squamosal-frontal contact. The navicular facet on the astragalus is dorsoventrally expanded. The astragalar facet on the calcaneus is narrow and roughly parallels the long axis, and the sustentaculum of the calcaneus is situated immediately medial to astragalar facet (see Fig. 10.3).

Proximal fossil outgroups to rodents, such as *Heomys, Rhombomylus,* and *Matutinia,* are commonly referred to collectively as "eurymylids" (Fig. 10.4). Eurymylids have frequently been considered a paraphyletic group, but new evidence points to the likely monophyly of some of these forms (Meng and Wyss, 2001; Meng et al., 2003; Fig. 10.5). The least inclusive clade encompassing rodents and eurymylids, termed the "Simplicidentata," differs from duplicidentates mainly in being characterized by a single pair of upper incisors, in addition to the other features listed above. Eurymylids share several additional derived features with

Fig. 10.5. Phylogeny and distribution of selected Glires and related taxa (modified from Meng et al., 2003). Solid lines are geological distributions of taxa based on McKenna and Bell (1997); dashed lines are inferred presence. Shaded area embraces Glires.

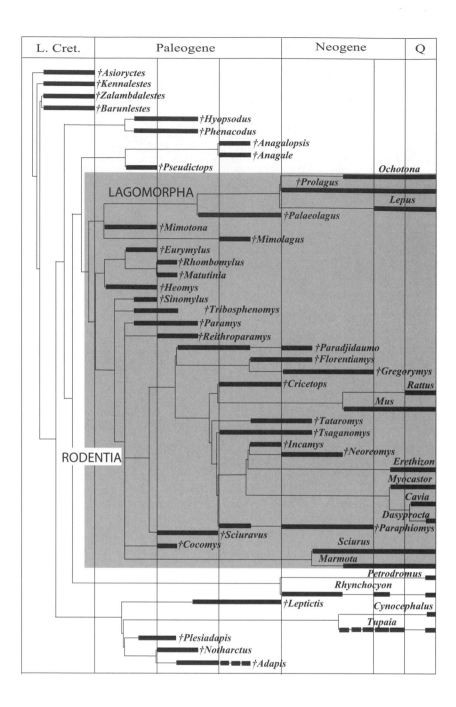

rodents: the upper diastema is significantly larger than the lower diastema, the second upper premolar is absent, and a posterior process occurs on the distal end of the tibia.

Some stem taxa to rodents, such as *Rhombomylus* (Meng et al., 2003), also exhibit similarities to lagomorphs. The upper cheek teeth are considerably wider than the lower ones and bear transverse shearing ridges, allowing transverse mastication. The main shearing facets occur on the anterior surface of the upper cheek teeth and on the posterior wall of the trigonid of their lower counterparts, where the enamel is thickest. The fibular facet on the calcaneus is present, indicating contact with the fibula. This contact is absent in rodents, *Tribosphenomys,* and such stem taxa to lagomorphs as *Mimolagus* (Bohlin, 1951; Szalay, 1985) and *Gomphos* (Meng

et al., 2004), but is present in lagomorphs, as well as in macroscelideans and *Pseudictops* (Li and Ting, 1985; Szalay, 1985; see Fig. 10.3). In lagomorphs, this contact faces more dorsally and is broader than in *Rhombomylus.*

GLIRES PHYLOGENY

Based on Morphological Data

Although rodents and lagomorphs have a checkered history of phylogenetic placement in relation to various mammalian groups, the following proposed affiliations are most commonly cited in the literature and are of particular interest. Based on cheek-tooth similarities between *Acritoparamys* on the one hand and *Plesiadapis* and *Phenacolemur* on the other, McKenna (1961) proposed a possible relationship between rodents and primates, a notion endorsed by others (Van Valen, 1966; Wood, 1962, 1977; Lillegraven, 1969; Patterson and Wood, 1982). A rodent-leptictid relationship was proposed on the basis of shared similarities in postcranial osteology (Szalay, 1977, 1985), and the postcranial dissimilarity of rodents and lagomorphs (Szalay, 1977, 1985; Bleefeld and McKenna, 1985; Bleefeld and Bock, 2002). A lagomorph-zalambdalestid relationship, to the exclusion of rodents, was proposed mainly on the basis of the enlarged lower incisors of zalambdalestids (Szalay and McKenna, 1971; McKenna, 1975). Material recovered from Mongolia relatively recently led to the proposal that the Late Cretaceous *Barunlestes* (a zalambdalestid) is a possible distant lagomorph relative (McKenna, 1994). A thorough analysis of those *Barunlestes* specimens, however, shows that zalambdalestids are only distantly related to Glires (Wible et al., 2004). Those hypotheses rejecting sister-group pairing between rodents and lagomorphs tacitly assume convergence as the mechanism accounting for the numerous derived similarities between rodents and lagomorphs.

The view that rodents and lagomorphs share an exclusive common ancestry with such early diverging fossil taxa as eurymylids and "mimotonids"—in essence, a more inclusive conception of Glires—has been espoused by many morphological workers (Li, 1977; Gingerich and Gunnell, 1979; Li and Yan, 1979; Gingerich and McKenna, 1980; Hartenberger, 1980, 1985; Dawson et al., 1984; Korth, 1984; Flynn et al., 1986; Li et al., 1987; Wilson, 1989; Novacek, 1992; Luckett and Hartenberger, 1993; Meng et al., 1994, 2003; Shoshani and McKenna, 1998; Meng and Wyss, 2001). Although details about the relationships within Glires are inconsistent among these various studies, most suggest the affiliation of rodents and eurymylids within the Simplicidentata, and lagomorphs plus mimotonids within the Duplicidentata (Li and Ting, 1985, 1993; Flynn et al., 1986, 1987; Dashzeveg et al., 1987; Dashzeveg and Russell, 1988; Luckett and Hartenberger, 1993; Meng et al., 1994, 2003; McKenna and Meng, 2001; Meng and Wyss, 2001).

The monophyly of Glires may be tested by considering whether either Rodentia or Lagomorpha can be shown to be more closely related to some third extant mammalian taxon (e.g., Macroscelidea, Dermoptera, Scandentia, or Primates in particular) than the two are to one another. For instance, if Rodentia could be linked to Primates to the exclusion of Lagomorpha, then the monophyly of Glires (in the traditional sense of that name) would have been falsified. If extinct taxa, such as zalambdalestids, prove to be more closely linked to Lagomorpha than to Rodentia, the monophyly of Glires is not challenged, but the name Glires would now refer to a more inclusive clade than it had previously. In this case, some of the features currently regarded as diagnosing Glires would have to be reinterpreted as having arisen convergently in Lagomorpha and Rodentia (or to have reversed in zalambdalestids), our conception of the features diagnosing the group ancestrally would change substantially, and the divergence of Glires would have to be extended back in time to at least the first record of zalambdalestids.

Two recent studies (Meng and Wyss, 2001; Meng et al., 2003) specifically addressed this question, the latter including a much more comprehensive suite of potential relatives. Rather than a full-scale phylogenetic analysis of higher-level relationships within Placentalia, these two studies emphasized taxa that have previously been posited as close rodent or lagomorph allies in either morphological or molecular studies. In Meng et al. (2003) the non-gliroid mammals sampled include *Asioryctes, Kennalestes, Zalambdalestes, Barunlestes, Hyopsodus, Phenacodus, Anagalopsis, Anagale, Leptictis, Cynocephalus, Tupaia, Petrodromus, Rhynchocyon, Pseudictops, Plesiadapis, Notharctus,* and *Adapis.* Among these taxa, *Asioryctes* and *Kennalestes* were used as outgroups (Fig. 10.5). Taxa generally considered to be only distantly related to Glires, such as ungulates and carnivores, were not included. The earlier analysis (Meng and Wyss, 2001) was limited to fossil taxa, whereas the later (Meng et al., 2003) also included extant forms. Both studies strongly corroborated the monophyly of Glires, but yielded little evidence for a close relationship with zalambdalestids. These relationships are particularly strongly supported in the study based on the most comprehensive morphological character matrix assembled to date (Meng et al., 2003). A recent study of Cretaceous and Cenozoic eutherians (Archibald et al., 2001) argued for the existence of a superordinal clade including zalambdalestids and Glires. As pointed out by Meng et al. (2003), however, Archibald and coworkers' study does not adequately test the monophyly of Glires and its relationship to other eutherian orders, because few relevant taxa were included in the analysis (see below).

Based on Molecular Data

Although some morphologists questioned the monophyly of Glires, another challenge to the concept has emerged from molecular studies over the past decade. Graur et al. (1991) first questioned whether the guinea pig (*Cavia porcellus*) is, in fact, a rodent, based on an analysis of amino-acid sequence data. Their results indicated a closer relationship of *Cavia* to primates than to other rodents, thus indicating the non-monophyly of the conventional Rodentia, and therefore, Glires. The hypothesis of non-monophyly of Glires was corroborated by some later molecular studies (Li et al., 1992; D'Erchia et al., 1996; Graur et al., 1996), whereas other studies left the relationship of Rodentia and Glires more ambiguous (Honeycutt and Adkins, 1993; Ma et al., 1993; Martignetti and Brosius, 1993; Porter et al., 1996; Stanhope et al., 1996; Huchon et al., 1999). Other recent molecularly based proposals include a primate-lagomorph linkage (Easteal, 1990; Penny et al., 1991), a primate-rodent-lagomorph grouping that is in turn linked with tree shrews (Miyamoto and Goodman, 1986; Czelusniak et al., 1990), and a clade consisting of Primates, Dermoptera, Scandentia, Lagomorpha, and Rodentia (Stanhope et al., 1993).

The early molecular studies stimulated a wave of debate not only about the monophyly of Rodentia and Glires, but also about methodological issues, including data quality in phylogenetic analysis (Allard et al., 1991; Hasegawa et al., 1992; Catzeflis, 1993; Graur, 1993; Honeycutt and Adkins, 1993; Luckett and Hartenberger, 1993; Novacek, 1993; Sullivan and Swofford, 1997). More recent molecular studies are generally based on larger data sets, including a broader diversity of both the taxa and gene sequences sampled. Although some of these more comprehensive studies have found the relationship of rodents and lagomorphs to be poorly resolved (Huchon et al., 1999; Mouchaty et al., 2001; Árnason et al., 2002; Janke et al., 2002; Misawa and Janke, 2003), others have argued in favor of a monophyletic Glires (DeBry and Sagel, 2001; Eizirik et al., 2001; Madsen et al., 2001; Murphy et al., 2001a,b; Scally et al., 2001; Adkins et al., 2003; Delsuc et al., 2002; Huchon et al., 2002; Springer et al., 2003), consistent with recent morphological studies (Meng and Wyss, 2001; Meng et al., 2003). Note also that some recent molecular studies recognize a clade encompassing Glires, Dermoptera, Scandentia, and Primates (Stanhope et al., 1993, 1996; Madsen et al., 2001; Murphy et al., 2001a,b; Springer et al., 2003), termed "Euarchontoglires," for which morphological evidence is lacking.

GLIRES DIVERGENCE

Fossil Record

The timing of the divergence of Glires and its subgroups is controversial; estimates derived from paleontological and molecular evidence are widely discrepant. The paleontological view of the eutherian diversification is one of a post-K/T boundary radiation of most modern major clades (Gingerich, 1977; Novacek, 1992; Alroy, 1999; Archibald and Deutschman, 2001), Glires included. The phylogeny and stratigraphic distributions of basal members of Glires show that the area of early divergence (if not the center of origin) of the group is Asia (Fig. 10.5). Among basal taxa of Glires, only one, *Alagomys*, is known outside Asia (Dawson and Beard, 1996). *Tribosphenomys*, the nearest relative of *Alagomys*, is known from the late Paleocene (Meng et al., 1994; Meng and Wyss, 2001) and early Eocene (Dashzeveg, 2003) of Asia. The earliest known members of Glires, such as *Heomys* and *Mimotona*, are restricted to the early or middle Paleocene of Asia (Li and Ting, 1993; Fig. 10.5). The earliest fossils conventionally regarded as lagomorphs are known from the early or middle Eocene of Asia and North America (McKenna and Bell, 1997); the earliest forms conventionally termed rodents, *Acritoparamys* and *Paramys*, occurr in the late Paleocene Clarkforkian of North America, about 57 million years ago (Woodburne and Swisher, 1995; Dawson and Beard, 1996). These early forms, however, likely fall outside the crown clade (to which the name "Rodentia" is tied in this chapter). Nonetheless, as stem rodents, they establish a minimum divergence time for these clades. Hystricognath

rodents first appear in the fossil record during the late Eocene or early Oligocene of Africa, Asia, and South America (Wood, 1968; Gagnon, 1992; Gingerich, 1993; Wyss et al., 1993, 1994; Bryant and McKenna, 1995; McKenna and Bell, 1997). The paleontologically derived age of divergence between *Mus* and *Rattus*, common subjects in molecular analyses, is about 8–14 million years ago (Jacobs and Pilbeam, 1980; Jaeger et al., 1986). The first record of *Mus* may only date back to 5.7 million years ago (Jacobs and Downs, 1994). The oldest fossil occurrence of a given group indicates only the minimum age of divergence for the group, because there always remains the potential that earlier fossils may exist, but have yet to be sampled.

Molecular Dating

Despite varied potential pitfalls of molecular-clock–derived estimates of divergence times, including rate heterogeneity of gene evolution over time, across lineages, and at various loci (Ayala, 1997; Bromham et al., 1999; Huchon et al., 2000, 2002; Eizirik et al., 2001; Murphy et al., 2001b; Rodriguez-Trelles et al., 2001; Adkins et al., 2003; Smith and Peterson, 2002), numerous divergence ages for various mammalian groups have been postulated. Molecularly based estimates usually place the divergence times of Glires far earlier than is indicated by the fossil record. Some molecular studies (Easteal, 1990; Kumar and Hedges, 1998) identify rodents as having made their first appearance about 112 million years ago—roughly twice as old as indicated by fossils. Divergence ages within Rodentia estimated from one set of molecular data are equally surprising from a paleontological perspective, with sciurognath rodents reported to originate about 112 ± 3.5 million years ago, followed by the divergences of hystricognaths (109 ± 3.2 Ma), Gerbillidae/Muridae (66.2 ± 7.6 Ma), Muridae/Cricetidae (65.8 ± 2.2 Ma), and *Mus/Rattus* (40.7 ± 0.9 Ma; Kumar and Hedges, 1998). Kumar and Hedges (1998) also place the time of divergence of Lagomorpha at about 90.8 ± 2.0 million years ago—that is, younger than some divergences within Rodentia, and hence, incompatible with a monophyletic Glires. Other molecular studies give younger estimated divergence times for Glires, but in general, these estimates are still far older than supported by the fossil record, placing the origin of rodents (and Glires) well into the Cretaceous (Janke et al., 1994; Huchon et al., 2000; Eizirik et al., 2001; Murphy et al., 2001b; Adkins et al., 2003 Springer et al., 2003). The inconsistencies within these sequence-based estimates undoubtedly stem from the different suites of genes employed and the variation in their rates of change, choice of calibration points, phylogenetic relationships, and statistical methodologies. At least one recent molecular study (Huchon et al., 2002), however, advocates Glires divergence dates compatible with the fossil record. According to this study, the "rodent diversification" took place near the Paleocene/Eocene transition, 55.8 million years ago (median of the 95% confidence intervals: 49.4–63.7 Ma), and the split between rodents and lagomorphs is estimated to be 64.5 million years ago (57.3–

73.3 Ma). But the meaning of the phrase "rodent diversification" (Huchon et al., 2002) is unclear. If it signifies the time at which rodents diverged from their nearest known living relatives, there would be a nine-million-year gap between "rodent diversification" and the "split between rodents and lagomorphs," given the sister-group relationship of rodents and lagomorphs accepted in that study. It is probable, however, that the age of divergence between extant rodents and lagomorphs estimated by Huchon et al. (2002) is actually the split between simplicidentates and duplicidentates. As mentioned, the earliest record and diversity of rodents and stem taxa is across the Paleocene/Eocene boundary, and the oldest stem taxa to rodents and lagomorphs (e.g., *Heomys* and *Mimotona,* respectively) are early-middle Paleocene. Thus, the discrepancy between paleontological and molecular data on divergence times of Glires is minimal at least from the perspective of the analysis of currently available molecular data by Huchon et al. (2002). Given this scenario, simplicidentates and duplicidentates must have diverged at the same time, but the first appearances of the crown rodents and lagomorphs could be different.

Recent Controversy Surrounding the Glires Divergence

If two taxa are nearest relatives, the documented presence of one implies the coeval existence of the other. An unsampled temporal interval, or ghost lineage (Norell, 1992), can be inferred for the group with the younger first-known occurrence, based on the assumed phylogeny. To demonstrate the divergence of a modern clade back into the Cretaceous, one has to find outgroups to that crown clade in the Cretaceous. For instance, acceptance of a lagomorph-zalambdalestid relationship (McKenna, 1994) and the monophyly of Glires would imply that lagomorphs extend at least as far back in time (the Late Cretaceous) as does the earliest zalambdalestid. A robust phylogeny is crucial to such inferences. Unambiguous evidence of close outgroups of Rodentia or Lagomorpha in the Late Cretaceous is lacking, let alone that for early simplicidentates or duplicidentates, as indicators of ancient divergences of placental orders would predict. Paleontological and most interpretations of molecular data part ways primarily over the question of how old the extant clades themselves are.

Glires is perhaps the only superordinal eutherian grouping in which morphologists place much confidence. In that context, the recognition (Archibald et al., 2001) of a clade consisting of zalambdalestids (first known from rocks about 85 million years old) and Glires merits scrutiny. Together, the assumed phylogenetic relationship and geochronological information led these authors to infer the pre-Cenozoic existence of the least-inclusive clade of which Glires is a member. This, Archibald et al. contended, is congruent with molecularly based estimates of a Late Cretaceous superordinal diversification of placentals. In turn, this conclusion was cited in molecular studies as paleontological evidence for the early divergence of Glires, in which zalambdalestids

were considered possible members of Euarchontoglires (Murphy et al., 2001b). More recently, it has been postulated that, because zalambdalestids appear to be more closely related to Glires than to other groups within Euarchontoglires, this would imply an even older split (more than 90 Ma) between Euarchontoglires and Laurasiatheria (Archibald, 2003). Several problems persist, however, in relation to the postulated Cretaceous divergence of Glires as advocated primarily by some molecular workers.

First, assuming zalambdalestids to represent the proximal outgroup to Glires, as argued by Archibald et al. (2001), the logic of ghost lineages would predict only that the divergences of the lineage ultimately giving rise to Glires and the one containing *Zalambdalestes* are equally ancient, but we do not know a priori how old the crown clade (Glires) might be. Given that relationship, Glires may still be no older than we currently have paleontological evidence for (Paleocene), even though its stem may extend all the way to the divergence of zalambdalestids. Therefore, a zalambdalestid-Glires pairing does not necessarily support the molecular dating of divergences of the crown Glires, Rodentia, and Lagomorpha.

Second, among the groups of Cretaceous eutherians surveyed by Archibald et al. (2001), zalambdalestids were argued to show the closest relationship to Glires (represented by *Tribosphenomys* and *Mimotona* in their analysis), but because of limited taxon sampling, Glires may share a still closer relationship with some other placental clade. As noted previously, various morphological and molecular studies have entertained the possibility that rodents, lagomorphs, and Glires are related to a wide variety of Cenozoic mammal clades, such as primates, leptictids, tree shrews, pseudictopids, anagalids, and macroscelideans. A phylogenetic analysis rigorously testing Glires relationships should involve all these taxa. When relevant taxa are included in such an analysis (Meng and Wyss, 2001; Meng et al., 2003; Meng, 2004; Fig. 10.5), zalambdalestids and Glires are not identified as nearest relatives. Hence, the divergence between Glires and other superordinal placental clades some 85–90 million years ago is not supported by the morphological data.

Finally, whether zalambdalestids are stem taxa of Euarchontoglires is a matter quite apart from the question of a close zalambdalestid-Glires relationship, because Euarchontoglires is a more inclusive grouping. These hypotheses are mutually exclusive. There is no convincing published evidence of a close relationship between zalambdalestids and Euarchontoglires. None of the classical archontan taxa was included in the analysis of Archibald et al. (2001). Naturally, molecular studies cannot consider fossil taxa. Although other works (Meng and Wyss, 2001; Meng et al., 2003) include archontan taxa, they do not sufficiently address interordinal relationships among placentals. Establishing whether zalambdalestids and Euarchontoglires are closely related requires a phylogenetic analysis at the level of Eutheria. Future studies may demonstrate such a relationship, but even so, the presence of the crown Euarchontoglires in the Cretaceous would not necessarily be demonstrated. At present, phylogenies based on morphological data and the stratigraphic occurrence of Glires and other eutherian mammals indicate that divergences of Glires from other placental clades—Lagomorpha from Rodentia, and sciurognath rodents from hystricognath rodents—do not predate the K/T boundary.

SUMMARY

We briefly review the morphological, mainly dental and osteological, features that characterize Glires, a mammalian clade including extant Lagomorpha and Rodentia and all descendants from their most recent common ancestor. The most obvious diagnostic feature of Glires is their enlarged pair of evergrowing upper and lower incisors, which extend deep into the maxilla and dentary. Rodentia and Lagomorpha are distinguished from one another by numerous characters, and each group is further distinct from its own stem taxa by many derived features. Although the higher-level phylogeny of Glires has been controversial, many recent morphological and molecular studies concur on the monophyly of Glires. Estimates of the timing of the origin of Glires and its major subgroups—as derived from molecular and paleontological data—are generally widely discrepant, with the molecularly based dates being substantially older than those indicated by the fossil record. Investigations intended to narrow this discrepancy, including phylogenetic analyses based on newly discovered fossils, have met with little success.

ACKNOWLEDGMENTS

We thank Kenneth D. Rose and J. David Archibald for the invitation to participate in the symposium and to contribute to this volume. We are grateful to J. David Archibald, Mary R. Dawson, Lawrence J. Flynn, and Kenneth D. Rose for comments on the manuscript. JM's research has been supported by National Science Foundation grants (DEB-9796038, EAR-0120727) and the National Natural Science Foundation of China (special funds for major state basic research projects of China [G200007707] and NNSFC grant 49928202).

REFERENCES

Adkins, R. M., A. H. Walton, and R. L. Honeycutt. 2003. Higher-level systematics of rodents and divergence time estimates based on two congruent nuclear genes. *Molecular Phylogenetics and Evolution* 26: 409–420.

Allard, M. W., M. M. Miyamoto, and R. L. Honeycutt. 1991. Tests for rodent polyphyly. Nature 353: 610–611.

Alroy, J. 1999. The fossil record of North American mammals: Evidence for a Paleocene evolutionary radiation. Systematic Biology 48: 107–118.

Archibald, J. D. 2003. Timing and biogeography of the eutherian radiation: Fossils and molecules compared. Molecular Phylogenetics and Evolution 28: 350–359.

Archibald, J. D., A. O. Averianov, and E. G. Ekdale. 2001. Late Cretaceous relatives of rabbits, rodents, and other extant eutherian mammals. Nature 414: 62–65.

Archibald, J. D., and D. H. Deutschman. 2001. Quantitative analysis of the timing of the origin and diversification of extant placental orders. Journal of Mammalian Evolution 8: 107–124.

Árnason, Ú., J. A. Adegoke, K. Bodin, E. W. Born, Y. B. Esa, A. Gullberg, M. Nilsson, R. V. Short, X. Xu, and A. Janke. 2002. Mammalian mitogenomic relationships and the root of the eutherian tree. Proceedings of the National Academy of Sciences USA 99: 8151–8156.

Ayala, F. J. 1997. Vagaries of the molecular clock. Proceedings of the National Academy of Sciences USA 94: 7776–7783.

Bleefeld, A. R., and W. J. Bock. 2002. Unique anatomy of lagomorph calcaneus. Acta Palaeontologica Polonica 47: 181–183.

Bleefeld, A. R., and M. C. McKenna. 1985. Skeletal integrity of *Mimolagus rodens* (Lagomorpha, Mammalia). American Museum Novitates 2806: 1–5.

Bohlin, B. 1951. Some mammalian remains from Shi-her-macheng, Hui-hui-pu area, western Kansu. Report from the scientific expedition to the northwestern provinces of China under leadership of Dr. Sven Hedin. Vertebrate Palaeontology 5: 1–47.

Bromham, L., M. J. Phillips, and D. Penny. 1999. Growing up with dinosaurs: Molecular dates and the mammalian radiation. Trends in Ecology and Evolution 14: 113–118.

Bryant, J. D., and M. C. McKenna. 1995. Cranial anatomy and phylogenetic position of *Tsaganomys altaicus* (Mammalia: Rodentia) from the Hsanda Gol Formation (Oligocene), Mongolia. American Museum Novitates 3156: 1–42.

Catzeflis, F. M. 1993. Mammalian phylogeny: Morphology and molecules. Trends in Ecology and Evolution 8: 340–341.

Cope, E. D. 1888. The mechanical causes of the origin of the dentition of the Rodentia. American Naturalist 22: 3–11.

Czelusniak, J., M. Goodman, B. F. Koop, W. W. de Jong, and G. Matsuda. 1990. Perspective from amino acid and nucleotide sequences on cladistic relationships among higher taxa of Eutheria; pp. 545–572 in H. H. Genoways (ed.), Current Mammalogy. Volume 2. Plenum, New York.

Dashzeveg, D. 2003. A new species of *Tribosphenomys* (Glires, Rodentia) from the early Eocene of Nemegt Basin, Mongolia and its implication for alagomyid phylogeny. Proceedings of the Mongolian Academy of Sciences 2003 (1): 49–62.

Dashzeveg, D., and D. E. Russell. 1988. Palaeocene and Eocene Mixodontia (Mammalia, Glires) of Mongolia and China. Palaeontology 31: 129–164.

Dashzeveg, D., D. E. Russell, and L. J. Flynn. 1987. New Glires (Mammalia) from the early Eocene of the People's Republic of Mongolia. 1. Systematics and description. Proceedings of the Koninklijke Nederlandse Akademie Van Wetenschappen, Series B, Physical Sciences 90: 133–154.

Dawson, M. R. 1967. Lagomorph history and the stratigraphic record; pp. 287–316 in C. Teichert and E. L. Yochelson (eds.), Essays in Paleontology and Stratigraphy. Special Publication 2. University of Kansas, Department of Geology, Lawrence, Kansas.

———. 1977. Late Eocene rodent radiations: North America, Europe and Asia. Géobios Mémoire Spécial 1: 195–209.

Dawson, M. R., and C. K. Beard. 1996. New Late Paleocene rodents (Mammalia) from Big Multi Quarry, Washakie Basin, Wyoming. Palaeovertebrata 25: 301–321.

Dawson, M. R., C.-K. Li, and T. Qi. 1984. Eocene ctenodactyloid rodents (Mammalia) of eastern central Asia. Special Publication, Carnegie Museum of Natural History 9: 138–150.

DeBry, R. W., and R. M. Sagel. 2001. Phylogeny of Rodentia (Mammalia) inferred from the nuclear-encoded gene *IRBP*. Molecular Phylogenetics and Evolution 19: 290–301.

Delsuc, F., M. Scally, O. Madsen, M. J. Stanhope, W. W. de Jong, F. M. Catzeflis, M. S. Springer, and E.J.P. Douzery. 2002. Molecular phylogeny of living xenarthrans and the impact of character and taxon sampling on the placental tree rooting. Molecular Biology and Evolution 19: 1656–1671.

D'Erchia, A. M., C. Gissi, G. Pesole, C. Saccone, and Ú. Árnason. 1996. The guinea-pig is not a rodent. Nature 381: 597–600.

Easteal, S. 1990. The pattern of mammalian evolution and the relative rate of molecular evolution. Genetics 124: 165–173.

Ehik, J., 1926. The right interpretation of the cheekteeth tubercles of *Titanomys*. Annales Historico-Naturales Musei Nationalis Hungarici 23: 178–186.

Eizirik, E., W. J. Murphy, and S. J. O'Brien. 2001. Molecular dating and biogeography of the early placental mammal radiation. Journal of Heredity 92: 212–219.

Flynn, L. J., L. L. Jacobs, and I. U. Cheema. 1986. Baluchimyinae, a new ctenodactyloid rodent subfamily from the Miocene of Baluchistan. American Museum Novitates 2841: 1–58.

Flynn, L. J., D. E. Russell, and D. Dashzeveg. 1987. New Glires (Mammalia) from the early Eocene of the People's Republic of Mongolia. 2. Incisor morphology and enamel microstructure. Proceedings of the Koninklijke Nederlandse Akademie Van Wetenschappen, Series B, Physical Sciences 90: 143–154.

Friant, M. 1932. Contributions à l'étude de la différenciation des dents jugales chez les mammifères. Publications Diverses du Muséum National d'Histoire Naturelle 1: 1–132.

Gagnon, M. 1992. Succession of mammalian communities in the Fayum of Egypt in relation to the age of the Jebel Qatrani Formation. Journal of Vertebrate Paleontology 12: 29A.

Gidley, J. W. 1906. Evidence bearing on tooth-cusp development. Proceedings of the Washington Academy of Sciences 8: 91–110.

———. 1912. The lagomorphs as an independent order. Science 36: 285–286.

Gingerich, P. D. 1977. Patterns of evolution in the mammalian fossil record; pp. 469–500 *in* A. Hallam (ed.), Patterns of Evolution as Illustrated by the Fossil Record. Elsevier, Amsterdam.

———. 1993. Oligocene age of the Gebel Qatrani Formation, Fayum, Egypt. Journal of Human Evolution 24: 207–218.

Gingerich, P. D., and G. F. Gunnell. 1979. Systematics and evolution of the genus *Esthonyx* (Mammalia, Tillodontia) in the early Eocene of North America. University of Michigan Contributions to Paleontology 25: 125–153.

Gingerich, P. D., and M. C. McKenna. 1980. Mammalian paleontology in China. Society of Vertebrate Paleontology News Bulletin 118: 42–44.

Graur, D. 1993. Molecular phylogeny and the higher classification of eutherian mammals. Trends in Ecology and Evolution 8: 141–147.

Graur, D., L. Duret, and M. Gouy. 1996. Phylogenetic position of the order Lagomorpha (rabbits, hares and allies). Nature 379: 333–335.

Graur, D., W. A. Hide, and W.-H. Li. 1991. Is the guinea-pig a rodent? Nature 351: 649–652.

Hartenberger, J.-L. 1980. Données et hypothèses sur la radiation initiale des rongeurs. Palaeovertebrata, Mémoire Jubilaire en Hommage à René Lavocat: 285–301.

———. 1985. The order Rodentia: Major questions on their evolutionary origin, relationships and suprafamilial systematics; pp. 1–33 *in* W. P. Luckett and J.-L. Hartenberger (eds.), Evolutionary Relationships among Rodents: A Multidisciplinary Analysis. Plenum, New York.

Hasegawa, M., Y. Cao, J. Adachi, and T.-A. Yano. 1992. Rodent polyphyly? Nature 355: 595.

Hinton, M.A.C. 1926. Monograph of the Voles and Lemmings (Microtinae) Living and Extinct. Volume 1. British Museum (Natural History), London.

Honeycutt, R. L., and R. M. Adkins. 1993. Higher level systematics of eutherian mammals: An assessment of molecular characters and phylogenetic hypotheses. Annual Review of Ecology and Systematics 24: 279–305.

Huchon, D., F. M. Catzeflis, and E.J.P. Douzery. 1999. Molecular evolution of the nuclear von Willebrand Factor gene in mammals and the phylogeny of rodents. Molecular Biology and Evolution 16: 577–589.

———. 2000. Variance of molecular datings, evolution of rodents and the phylogenetic affinities between Ctenodactylidae and Hystricognathi. Proceedings of the Royal Society of London B 267: 393–402.

Huchon, D., O. Madsen, M.J.J.B. Sibbald, K. Ament, M. J. Stanhope, F. M. Catzeflis, W. W. de Jong, and E.J.P. Douzery. 2002. Rodent phylogeny and a timescale for the evolution of Glires: Evidence from an extensive taxon sampling using three nuclear genes. Molecular Biology and Evolution 19: 1053–1065.

Hürzeler, J., 1936. Osteologie und Odontologie der Caenotheriden. Abhandlungen der Schweizerischen Palaeontologischen Gesellschaft 58: 1–89.

Jacobs, L. L., and W. R. Downs. 1994. The evolution of murine rodents in Asia. National Science Museum Monographs, Tokyo 8: 149–156.

Jacobs, L. L., and D. Pilbeam. 1980. Of mice and men: Fossil-based divergence dates and molecular "clock." Journal of Human Genetics 52: 152–166.

Jaeger, J.-J. 1988. Rodent phylogeny: New data and old problems; pp. 177–199 *in* M. J. Benton (ed.), The Phylogeny and Classification of the Tetrapods. Volume 2. Mammals. Clarendon Press, Oxford.

Jaeger, J.-J., H.-Y. Tong, and C. Denys 1986. Age de la divergence *Mus-Rattus* comparaison des données paléontologiques et moléculaires. Comptes Rendus de l'Académie des Sciences, Paris 302: 917–922.

Janke, A., G. Feldmaier-Fuchs, W. K. Thomas, A. von Haeseler, and S. Pääbo. 1994. The marsupial mitochondrial genome and the evolution of placental mammals. Genetics 137: 243–256.

Janke, A., O. Magnell, G. Wieczorek, M. Westerman, and Ú. Árnason. 2002. Phylogenetic analysis of 18S rRNA and the mitochondrial genomes of the wombat, *Vombatus ursinus,* and the spiny anteater, *Tachyglossus aculeatus.* Increased support for the Marsupionta hypothesis. Journal of Molecular Evolution 54: 71–80.

Kielan-Jaworowska, Z., T. M. Bown, and J. A. Lillegraven. 1979. Eutheria; pp. 221–258 *in* J. A. Lillegraven, Z. Kielan-

Jaworowska, and W. A. Clemens (eds.), Mesozoic Mammals. University of California Press, Berkeley.

Korth, W. W. 1984. Earliest Tertiary evolution and radiation of rodents in North America. Bulletin of the Carnegie Museum of Natural History 24: 1–71.

Kumar, S., and S. B. Hedges. 1998. A molecular timescale for vertebrate evolution. Nature 392: 917–920.

Landry, S. O., Jr.1999. A proposal for a new classification and nomenclature for the Glires (Lagomorpha and Rodentia). Mitteilungen aus dem Museum für Naturkunde in Berlin, Zoologische Reihe 75: 283–319.

Li, C.-K. 1977. Paleocene eurymyloids (Anagalida, Mammalia) of Quianshan, Anhui. Vertebrata PalAsiatica 15: 103–118.

Li, W.-H., W. A. Hide, and D. Graur. 1992. Origin of rodents and guinea-pigs. Nature 359: 277–278.

Li, C.-K., and S.-Y. Ting. 1985. Possible phylogenetic relationships of eurymylids and rodents, with comments on mimotonids; pp. 35–58 in W. P. Luckett and J.-L. Hartenberger (eds.), Evolutionary Relationships among Rodents: A Multidisciplinary Analysis. Plenum, New York.

———. 1993. New cranial and postcranial evidence for the affinities of the eurymylids (Rodentia) and mimotonids (Lagomorpha); pp. 151–158 in F. S. Szalay, M. J. Novacek, and M. C. McKenna (eds.), Mammal Phylogeny: Placentals. Springer-Verlag, New York.

Li, C.-K., R. W. Wilson, M. R. Dawson, and L. Krishtalka. 1987. The origin of rodents and lagomorphs; pp. 97–108 in H. H. Genoways (ed.), Current Mammalogy. Volume 1. Plenum, New York.

Li, C.-K., and D.-F. Yan. 1979. The systematic position of eurymylids (Mammalia) and the origin of Rodentia. Twelfth Annual Conference and 3rd National Congress of the Paleontological Society of China, Suzhou. Abstracts of Papers: 155–156.

Lillegraven, J. A. 1969. Latest Cretaceous mammals of upper part of Edmonton Formation of Alberta, Canada, and review of marsupial-placental dichotomy in mammalian evolution. The University of Kansas Paleontological Contributions, Article 50 (Vertebrata 12): 1–122.

Luckett, W. P. 1985. Superordinal and intraordinal affinities of rodents: Developmental evidence from dentition and placentation; pp. 227–276 in W. P. Luckett and J.-L. Hartenberger (eds.), Evolutionary Relationships among Rodents: A Multidisciplinary Analysis. Plenum, New York.

Luckett, W. P., and J.-L. Hartenberger. 1985. Evolutionary relationships among rodents: Comments and conclusions; pp. 685–712 in W. P. Luckett and J.-L. Hartenberger (eds.), Evolutionary Relationships among Rodents: A Multidisciplinary Analysis. Plenum, New York.

———. 1993. Monophyly or polyphyly of the order Rodentia: Possible conflict between morphological and molecular interpretations. Journal of Mammalian Evolution 1: 127–147.

Ma, D.-P., A. Zharkikh, D. Graur, J. L. VandeBerg, and W.-H. Li. 1993. Structure and evolution of opossum, guinea pig, and porcupine cytochrome b genes. Journal of Molecular Evolution 36: 327–334.

Madsen, O., M. Scally, C. J. Douady, D. J. Kao, R. W. DeBry, R. Adkins, H. M. Amrine, M. J. Stanhope, W. W. de Jong, and M. S. Springer. 2001. Parallel adaptive radiations in two major clades of placental mammals. Nature 409: 610–614.

Major, C.J.F. 1893. On some Miocene squirrels. Proceedings of the Zoological Society of London 1893: 179–215.

Martignetti, J. A., and J. Brosius. 1993. Neural BC1 RNA as an evolutionary marker: Guinea pig remains a rodent. Proceedings of the National Academy of Sciences USA 90: 9698–9702.

McKenna, M. C. 1961. A note on the origin of rodents. American Museum Novitates 2037: 1–5.

———. 1975. Toward a phylogenetic classification of the Mammalia; pp. 21–46 in W. P. Luckett and F. S. Szalay (eds.), Phylogeny of the Primates. Plenum, New York.

———. 1982. Lagomorpha interrelationships. Géobios Mémoire Spécial 6: 213–224.

———. 1994. Early relatives of flopsy, mopsy, and cottontail. Natural History 103: 56–58.

McKenna, M. C., and S. K. Bell. 1997. Classification of Mammals above the Species Level. Columbia University Press, New York.

McKenna, M. C., and J. Meng. 2001. A primitive relative of rodents from the Chinese Paleocene. Journal of Vertebrate Paleontology 21: 565–572

Meng, J. 2004. Phylogeny and divergence of basal Glires. Bulletin of the American Museum of Natural History 285: 97–109.

Meng, J., G. J. Bowen, J. Ye, P. L. Koch, S.-Y. Ting, Q. Li, and X. Jin. 2004. Gomphos elkema (Glires, Mammalia) from the Erlian Basin: Evidence for the Early Tertiary Bumbanian Land Mammal Age in Nei-Mongol, China. American Museum Novitates 3425: 1–24.

Meng, J., Y.-M. Hu., and C.-K. Li. 2003. The osteology of Rhombomylus (Mammalia, Glires): Implications for phylogeny and evolution of Glires. Bulletin of the American Museum of Natural History 275: 1–247.

Meng, J., and A. R. Wyss. 2001. The morphology of Tribosphenomys (Rodentiaformes, Mammalia): Phylogenetic implications for basal Glires. Journal of Mammalian Evolution 8: 1–71.

Meng, J., A. R. Wyss, M. R. Dawson, and R.-J. Zhai. 1994. Primitive fossil rodent from Inner Mongolia and its implications for mammalian phylogeny. Nature 370: 134–136.

Misawa, K., and A. Janke. 2003. Revisiting the Glires concept—phylogenetic analysis of nuclear sequences. Molecular Phylogenetics and Evolution 28: 320–327.

Miyamoto, M. M., and M. Goodman. 1986. Biomolecular systematics of eutherian mammals: Phylogenetic patterns and classification. Systematic Zoology 35: 230–240.

Moody, P. A., V. A. Cochran, and H. Drugg. 1949. Serological evidence on lagomorph relationships. Evolution 3: 25–33.

Mouchaty, S. K., F. Catzeflis, A. Janke, and Ú. Árnason. 2001. Molecular evidence of an African Phiomorpha–South American Caviomorpha clade and support for Hystricognathi based on the complete mitochondrial genome of the cane rat. Molecular Phylogenetics and Evolution 18: 127–135.

Murphy, W. J., E. Eizirik, W. E. Johnson, Y.-P. Zhang, O. A. Ryder, and S. J. O'Brien. 2001a. Molecular phylogenetics and the origins of placental mammals. Nature 409: 614–618.

Murphy, W. J., E. Eizirik, S. J. O'Brien, O. Madsen, M. Scally, C. J. Douady, E. Teeling, O. A. Ryder, M. J. Stanhope, W. W. de Jong, and M. S. Springer. 2001b. Resolution of the early placental mammal radiation using Bayesian phylogenetics. Science 294: 2348–2351.

Norell, M. A., 1992. Taxic origin and temporal diversity: The effect of phylogeny; pp. 89–118 in M. J. Novacek and Q. D. Wheeler (eds.), Extinction and Evolution. Columbia University Press, New York.

Novacek, M. J. 1990. Morphology, paleontology, and the higher clades of mammals; pp. 59–81 *in* H. H. Genoways (ed.), Current Mammalogy. Volume 2. Plenum, New York.

———. 1992. Mammalian phylogeny: Shaking the tree. Nature 356: 121–125.

———. 1993. Mammalian phylogeny: Morphology and molecules. Trends in Ecology and Evolution 8: 339–340.

Nowak, R. M. 1999. Walker's Mammals of the World. Sixth Edition. The Johns Hopkins University Press, Baltimore.

Osborn, H. F. 1902. American Eocene primates, and the supposed rodent family Mixodectidae. Bulletin of the American Museum of Natural History 17: 169–214.

Patterson, B., and A. E. Wood. 1982. Rodents from the Deseadan Oligocene of Bolivia and the relationships of the Caviomorpha. Bulletin of the Museum of Comparative Zoology, Harvard University 149: 371–543.

Penny, D., M. D. Hendy, and M. A. Steel. 1991. Testing the theory of descent; pp. 155–183 *in* M. M. Miyamoto and J. Cracraft (eds.), Phylogenetic Analysis of DNA Sequences. Oxford University Press, London.

Porter, C. A., M. Goodman, and M. J. Stanhope. 1996. Evidence on Mammalian phylogeny from sequences of Exon 28 of the von Willebrand Factor gene. Molecular Phylogenetics and Evolution 5: 89–101.

Rodriguez-Trelles, F., R. Tarrío, and F. J. Ayala. 2001. Erratic overdispersion of three molecular clocks: GPDH, SOD, and XDH. Proceedings of the National Academy of Sciences USA 98: 11405–11410.

Russell, L. S. 1959. The dentition of rabbits and the origin of lagomorphs. National Museum of Canada Bulletin 166: 41–45.

Scally, M., O. Madsen, C. J. Douady, W. W. de Jong, M. J. Stanhope, and M. S. Springer. 2001. Molecular evidence for the major clades of placental mammals. Journal of Mammalian Evolution 8: 239–277.

Shoshani, J., and M. C. McKenna. 1998. Higher taxonomic relationships among extant mammals based on morphology, with selected comparisons of results from molecular data. Molecular Phylogenetics and Evolution 9: 572–584.

Simpson, G. G. 1945. The principles of classification and a classification of mammals. Bulletin of the American Museum of Natural History 85: 1–350.

Smith, A. B., and K. J. Peterson. 2002. Dating the time of origin of major clades: Molecular clocks and the fossil record. Annual Review of Earth and Planetary Sciences 30: 65–88.

Springer, M. S., W. J. Murphy, E. Eizirik, and S. J. O'Brien. 2003. Placental mammal diversification and the Cretaceous-Tertiary boundary. Proceedings of the National Academy of Sciences USA 100: 1056–1061.

Stanhope, M. J., W. J. Bailey, J. Czelusniak, M. Goodman, J.-S. Si, J. Nickerson, J. G. Sgouros, G.A.M. Singer, and T. K. Kleinschmidt. 1993. A molecular view of primate supraordinal relationships from the analysis of both nucleotide and amino acid sequences; pp. 251–292 *in* R.D.E. MacPhee (ed.), Primates and Their Relatives in Phylogenetic Perspective. Plenum, New York.

Stanhope, M. J., M. A. Smith, V. G. Waddell, C. A. Porter, M. S. Shivji, and M. Goodman. 1996. Mammalian evolution and the interphotoreceptor retinoid binding protein (IRPB) gene: Convincing evidence for several superordinal clades. Journal of Molecular Evolution 43: 83–92.

Sullivan, J., and D. L. Swofford. 1997. Are guinea pigs rodents? The importance of adequate models in molecular phylogenetics. Journal of Mammalian Evolution 4: 77–86.

Szalay, F. S. 1977. Phylogenetic relationships and a classification of the eutherian Mammalia; pp. 317–374 *in* M. K. Hecht, P. C. Goody, and B. M. Hecht (eds.), Major Patterns in Vertebrate Evolution. Plenum, New York.

———. 1985. Rodent and Lagomorph morphotype adaptations, origins, and relationships: Some postcranial attributes analyzed; pp. 83–132 *in* W. P. Luckett and J.-L. Hartenberger (eds.), Evolutionary Relationships among Rodents: A Multidisciplinary Analysis. Plenum, New York.

Szalay, F. S., and M. C. McKenna. 1971. Beginning of the age of mammals in Asia: The late Paleocene Gashato fauna, Mongolia. Bulletin of the American Museum of Natural History 144: 269–318.

Van Valen, L. 1964. A possible origin for rabbits. Evolution 18: 484–491.

———. 1966. Deltatheridia, a new order of mammals. Bulletin of the American Museum of Natural History 132: 1–126.

———. 1971. Adaptive zones and the orders of mammals. Evolution 25: 420–428.

Wible, J. R., M. J. Novacek, and G. W. Rougier. 2004. New data on skull and dentition in the Mongolian Late Cretaceous eutherian mammal *Zalambdalestes*. Bulletin of the American Museum of Natural History 281: 1–144.

Wilson, R. W. 1949. Early Tertiary rodents of North America. Contribution to Palaeontology, Carnegie Institution of Washington 584: 67–164.

———. 1989. Rodent origin; pp. 3–6 *in* C. C. Black and M. Dawson (eds.), Papers on Fossil Rodents in Honor of Albert Elmer Wood. Natural History Museum of Los Angeles County Science Series 33. Natural History Museum of Los Angeles County, Los Angeles.

Wood, A. E. 1942. Notes on the Paleocene lagomorph, *Eurymylus*. American Museum Novitates 1162: 1–7.

———. 1957. What, if anything, is a rabbit? Evolution 11: 417–427.

———. 1962. The early Tertiary rodents of the family Paramyidae. Transactions of the American Philosophical Society of Philadelphia 52: 1–261.

———. 1968. Early Cenozoic mammalian faunas, Fayum Province, Egypt. Part II. The African Oligocene Rodentia. Peabody Museum of Natural History, Yale University Bulletin 28: 23–105.

———. 1977. The Rodentia as clues to the Cenozoic migrations between the Americas and Europe and Africa; pp. 95–109 *in* R. M. West (ed.), Paleontology and Plate Tectonics with Special Reference to the History of the Atlantic Ocean. Special Publication 2. Milwaukee Public Museum of Geology, Milwaukee.

Woodburne, M. O., and C. C. Swisher III. 1995. Land mammal high-resolution geochronology, intercontinental overland dispersals, sea level, climate, and vicariance; pp. 335–363 *in* W. A. Berggren, D. V. Kent, M.-P. Aubry, and J. Hardenbol (eds.), Geochronology, Time Scales and Global Stratigraphic Correlation. SEPM Special Publication 54. Society for Sedimentary Geology, Tulsa.

Wyss, A. R., J. J. Flynn, M. A. Norell, C. C. Swisher III, R. Charrier, M. J. Novacek, and M. C. McKenna. 1993. South America's earliest rodent and recognition of a new interval of mammalian evolution. Nature 365: 434–437.

Wyss, A. R., J. J. Flynn, M. A. Norell, C. C. Swisher III, M. J. Novacek, M. C. McKenna, and R. Charrier. 1994. Paleogene mammals from the Andes of central Chile: A preliminary taxonomic, biostratigraphic and geochronologic assessment. American Museum Novitates 3098: 1–31.

Wyss, A. R., and J. Meng. 1996. Application of phylogenetic taxonomy to poorly resolved crown clades: A stem-modified node-based definition of Rodentia. Systematic Biology 45: 559–568.

NANCY B. SIMMONS

11

CHIROPTERA

T HE ORDER CHIROPTERA (BATS) IS one of the oldest and most distinct groups of placental mammals. As currently recognized, Chiroptera includes 18 extant families and 6 extinct families (Table 11.1), all of which are characterized by specializations for powered flight, most notably well-developed wings. The oldest known bats come from early Eocene deposits in North America, and forelimb structure indicates that they had wings very similar to modern forms; the same is true of Eocene fossil bats from other continents. No fossils of "pre-bats" or morphological intermediates between bats and their non-volant ancestors have ever been found. The earliest known members of the bat lineage were already clearly bats, and the morphological gap between these and other mammals is quite large. Because phylogenetic analyses of different data sets have placed bats in several different positions in the mammalian family tree, the origins of the order Chiroptera remain unclear. Morphological data have suggested that the closest living relatives of bats lie among the archontans (gliding lemurs, primates, and tree shrews), but multiple molecular data sets have strongly supported placement of bats with either ferungulates (ungulates, whales, and carnivores) or eulipotyphlans (shrews, moles, and hedgehogs).

Bats are typically divided into two large groups, Microchiroptera (echolocating bats) and Megachiroptera (Old World fruit bats, which do not echolocate). These groups have been considered to be reciprocally monophyletic and are formally recognized as suborders in most classifications. However, recent analyses of molecular

Table 11.1   Extant and extinct families of bats

†Icaronycteridae	Nycteridae (slit-faced bats)
†Archaeonycteridae	Emballonuridae (sheath-tailed bats)
†Palaeochiropterygidae	Myzopodidae (sucker-footed bats)
†Hassianycteridae	Mystacinidae (New Zealand short-tailed bats)
†Tanzanycteridae[a]	Phyllostomidae (New World leaf-nosed bats)
†Philisidae	Mormoopidae (leaf-chinned bats)
Pteropodidae (Old World fruit bats, flying foxes)	Noctilionidae (bulldog bats; fishing bats)
Rhinolophidae (horseshoe bats)	Thyropteridae (disk-winged bats)
Hipposideridae (Old World leaf-nosed bats)	Furipteridae (smoky bats; thumbless bats)
Megadermatidae (false vampire bats)	Natalidae (funnel-eared bats)
Rhinopomatidae (mouse-tailed bats)	Molossidae (free-tailed bats)
Craseonycteridae (hog-nosed bat; bumblebee bat)	Vespertilionidae[b] (evening bats; vesper bats)

Notes: Classification above the family level is in a state of flux, due to major differences between phylogenies based on morphology and those based on gene sequence data. The taxa given here are presently recognized by most or all workers as distinct families (see Simmons, in press, and references cited therein). Classifications that recognize Megachiroptera and Microchiroptera as distinct, reciprocally monophyletic taxa would place Pteropodidae in Megachiroptera and all of the remaining extant families in Microchiroptera (e.g., Simmons and Geisler, 1998). Classifications that recognize Microchiroptera as a paraphyletic group would place Pteropodidae, Rhinolophidae, Hipposideridae, Megadermatidae, Rhinopomatidae, and Craseonycteridae in Yinpterochiroptera; the remaining extant families would be placed in Yangochiroptera (e.g., Hulva and Horacek, 2002; Teeling et al., 2002; Van Den Bussche et al., 2002a, 2003; and Hoofer et al., 2003).

[a] This family is a new taxon named by Gunnell et al. (2003) for *Tanzanycteris.* They spelled the family name Tanzanycteridiidae, but we follow Simmons and Geisler (1998: 133, footnote 13), who argued that all bat family group names based on generic epithets ending with the greek root *-nycteris* should be spelled the same way—that is, -nycteridae rather than -nycterididae.

[b] Includes Antrozoidae, following Hoofer and Van Den Bussche (2001) and Simmons (in press).

sequence data from several mitochondrial and nuclear genes have indicated that Microchiroptera is not monophyletic; instead, rhinolophids and their relatives appear to be more closely related to Megachiroptera than to the remaining microchiropteran families (see below). These new phylogenies suggest that echolocation may have evolved very early in the bat lineage, before diversification of the extant families (Springer et al., 2001). This hypothesis is congruent with the fossil record, as all Eocene bats known from well-preserved specimens are thought to have been echolocating (Simmons and Geisler, 1998).

In addition to amazing specializations for flight and echolocation, one of the most remarkable features of the order Chiroptera is its diversity and broad geographic distribution. In terms of species diversity, bats comprise more than 20% of extant mammals. The most recent survey of extant bat diversity recognizes 1,116 species in 202 genera (Simmons, in press), and more than 40 extinct genera are known from the fossil record (McKenna and Bell, 1997). These figures place bats as second only to rodents in taxonomic diversity. The geographic range of the order is enormous, covering all continents except Antarctica (Fig. 11.1). Species diversity of bats is greatest in the Neotropics, where more than 100 species may coexist in sympatry (Voss and Emmons, 1996). Diversity tends to progressively reduce as one moves from low to high latitudes on all continents. However, many bat species thrive at high latitudes, although they must hibernate or migrate to survive winter conditions of low temperatures and low food availability. Fossil bats are known from all continents except Antarctica (McKenna and Bell, 1997) and are most frequently found in deposits believed to have been formed in tropical or subtropical climatic conditions.

MORPHOLOGY OF BATS

Bats are placental mammals of small body size (1.5 g–1.5 kg; Rayner, 1981). All are characterized by the presence of wings, which consist of skin membranes (patagia) supported by a modified forelimb and hand with greatly elongated digits. A tail membrane (uropatagium) is present between the hind legs in many taxa, often supported by a bony or cartilaginous calcar (a neomorphic element that articulates with the calcaneum). Simmons and Geisler (1998) identified 33 morphological synapomorphies of Chiroptera; these are listed in Table 11.2. Unfortunately, many of these characters are soft-tissue features that are not preserved in fossils. However, the majority of the osteological traits in Table 11.2 have been observed in some or all of the better known Eocene fossil bats (*Icaronycteris, Archaeonycteris, Palaeochiropteryx,* and *Hassianycteris;* Fig. 11.2), confirming their presence in the earliest known chiropterans (Simmons and Geisler, 1998). For a review of bat anatomy and functional morphology of the wings, see Hill and Smith (1984).

Because many fossils of small mammals consist of little more than fragments of the cranium and dentition, there is also a practical need for dental characters that can be used to determine whether a fossil represents a bat or some other taxon. Most bats have a typical insectivorous dentition with tribosphenic molars, which makes characterization of unique dental features difficult, given the great diversity of early Tertiary small mammals. Indeed, several fossils have been the source of controversy over whether they are actually bats (e.g., *Wyonycteris* was originally described as a bat by Gingerich, 1987, but later removed from Chiroptera by Hand et al., 1994). Hand et al. (1994: 379) identified two dental characters as "probable synapomorphies of early bats": a

Fig. 11.1. Distribution of the order Chiroptera worldwide. The range of bats is indicated in black; dotted lines surround island groups inhabited by bats. Range data from Hill and Smith (1984) and Simmons (in press).

well-developed buccal cingulum on the lower molars and a marked reduction of the para- and metaconules on the upper molars. Other characteristics of Eocene bat teeth include a W-shaped ectoloph, poorly molariform p4, relatively high lingual cusps, and a small hypoconulid, but none of these features are unique to bats (Hand et al., 1994). Dental morphology of modern bats is more diverse (especially among fruit, flower, and blood-feeding forms), but the majority of extant species have teeth that are not much different from those of many Eocene bats (Slaughter, 1970).

An anatomical feature of great significance in bats is the relative size of the cochlea, which is correlated with the presence of sophisticated echolocation (Novacek, 1985, 1987; Habersetzer and Storch, 1992; Simmons and Geisler, 1998). Although there is some overlap in size between echolocating and non-echolocating bats, nearly all echolocating bats have a cochlea whose basal turn is proportionally larger than in non-echolocating bats and other mammals (Novacek, 1985, 1987; Habersetzer and Storch, 1992; Simmons and Geisler, 1998). Three Eocene fossil genera, *Palaeochiropteryx, Hassianycteris,* and *Tanzanycteris,* have cochleae that fall well within the range known for extant echolocating bats (outside the zone of overlap between microchiropterans and non-echolocating megachiropterans; Habersetzer and Storch, 1992; Simmons and Geisler, 1998; Gunnell et al.,

2003). There is little doubt that these bats were echolocators (Novacek, 1985, 1987; Habersetzer and Storch, 1992; Simmons and Geisler, 1998; Gunnell et al., 2003). Two other taxa, *Icaronycteris* and *Archaeonycteris,* have cochleae that fall within the narrow zone of overlap between echolocating and non-echolocating bats (Habersetzer and Storch, 1992; Simmons and Geisler, 1998). Because *Icaronycteris* and *Archaeonycteris* share with echolocating bats several unique basicranial features (e.g., an enlarged orbicular apophysis on the malleus, an expanded cranial tip on the stylohyal, loose attachment of the periotic to the basicranium) that are not seen in non-echolocating forms, these two taxa are also thought to have used echolocation (Simmons and Geisler, 1998).

Morphological specializations of extant bats are numerous, with different clades and species exhibiting a wide variety of wing forms, echolocation strategies, and dietary habits (for a review, see Norberg and Rayner, 1987). Bats with low-aspect-ratio wings tend to be specialized for slow, maneuverable flight near or within vegetation, whereas bats with high-aspect-ratio wings are capable of much faster flight but are less maneuverable, preferring open areas. Some bats hunt from perches (where they sit and wait for prey); others hawk insects on the wing. Dietary habits are diverse, with different groups specialized for feeding on different types of

Table 11.2 Morphological synapomorphies of Chiroptera

Number	Character	Number	Character
1	Deciduous dentition does not resemble adult dentition; deciduous teeth with long, sharp, recurved cusps	18	Modification of hip joint: 90° rotation of hindlimbs effected by reorientation of acetabulum and shaft of femur; neck of femur reduced; ischium tilted dorsolaterally; anterior pubes widely flared and pubic spine present; absence of m. obturator internus
2	Palatal process of premaxilla reduced; left and right incisive foramina fused in midsagital plane		
3	Postpalatine torus absent		
4	Jugal reduced and jugolacrimal contact lost	19	Sacrum terminates posterior to midpoint of acetabulum
5	Two entotympanic elements in the floor of the middle-ear cavity: a large caudal element and a small rostral element associated with the internal carotid artery	20	Absence of m. gluteus minimus
		21	Absence of m. sartorius
		22	Vastus muscle complex not differentiated
6	Tegmen tympani tapers to an elongate process that projects into the middle-ear cavity medial to the epitympanic recess	23	Modification of ankle joint: reorientation of upper ankle-joint facets on calcaneum and astragalus; trochlea of astragalus convex, lacks medial and lateral guiding ridges; tuber of calcaneum projects in plantolateral direction away from ankle and foot; peroneal process absent; sustentacular process of calcaneum reduced, calcaneo-astragalar and sustentacular facets on calcaneum and astragalus coalesced; absence of groove on astragalus for tendon of m. flexor digitorum fibularis
7	Proximal stapedial artery enters cranial cavity medial to the tegmen tympani; ramus inferior passes anteriorly dorsal to the tegmen tympani		
8	Enlarged fenestra rotundum		
9	Vomeronasal epithelial tube absent		
10	Accessory olfactory bulb absent		
11	Posterior laminae present on ribs		
12	Modification of scapula: reorientation of scapular spine and modification of shape of scapular fossae; reduction in of height of spine; presence of a well-developed transverse scapular ligament; presence of at least two facets in infraspinous fossa	24	Entocuneiform proximodistally shortened, with flat, triangular distal facet
		25	Elongation of proximal phalanx of digit I of foot
		26	Embryonic disc oriented toward tubo-uterine junction at time of implantation
13	Modification of elbow: reduction of olecranon process and humeral articular surface on ulna; presence of ulnar patella; absence of olecranon fossa on humerus	27	Differentiation of a free, glandlike yolk sac
		28	Preplacenta and early chorioallantoic placenta diffuse or horseshoe-shaped, with definitive placenta reduced to a more localized discoidal structure
14	Absence of supinator ridge on humerus	29	Definitive chorioallantoic placenta endotheliochorial
15	Absence of entepicondylar foramen in humerus	30	Baculum present
16	Occipitopollicalis muscle and cephalic vein present in leading edge of propatagium	31	Left central lobe of liver separate from other lobes or partially fused with right central lobe
17	Digits II–V of forelimb elongated with complex carpo metacarpal and intermetacarpal joints, support enlarged interdigital flight membranes (patagia); digits III–V lack claws	32	Caecum absent
		33	Cortical somatosensory representation of forelimb reverse of that in other mammals

Source: Simmons and Geisler (1998: table 7).

insects, fruit, flowers, small vertebrates, and even blood (for a summary, see Hill and Smith, 1984; also see Ferrarezzi and Gimenez, 1996). One family, Phyllostomidae, includes members that span the whole range of dietary habits, but most families are specialized for only one diet or a few diets (e.g., pteropodids eat primarily fruit and flower products, vespertilionids eat insects, megadermatids eat insects and small vertebrates). Most bats known from the fossil record are thought to have been insectivorous.

EOCENE BAT DIVERSITY AND BIOGEOGRAPHY

The oldest known bat is *Icaronycteris index* from North American early Eocene Green River Formation deposits, now thought to be approximately 53 million years old (Jepsen, 1966; Simmons and Geisler, 1998). Gingerich (1987) referred three isolated fragmentary molar teeth from Clarkforkian deposits in North America to cf. *Icaronycteris* sp., but it is not

clear whether these actually represent a bat. No undisputed Paleocene fossils of bats are presently known.

One of the most notable features of the bat fossil record is the nearly simultaneous appearance of multiple bat taxa on several continents (Table 11.3). Nine genera of bats first occur in early Eocene deposits, including taxa from North America (*Icaronycteris, Honrovits*), Europe (*Ageina, Archaeonycteris, Eppsinycteris, Hassianycteris, Palaeochiropteryx*), Africa (*Dizzya*), and Australia (*Australonycteris*). By the end of the middle Eocene, bats are known from every continent except South America, where the oldest chiropteran fossil is from Oligocene deposits (Legendre, 1984) and Antarctica, which presently has no fossil record for bats. Given that the early Tertiary record of South America is quite poor with respect to microfossil sites, it would not be surprising if future finds on that continent include Eocene bats. In this context, it seems clear that the geographic spread of bats across the globe was a very early feature of the chiropteran radiation.

The diversity of bats in the Eocene is also remarkable from a taxonomic standpoint. Twenty-seven genera of bats

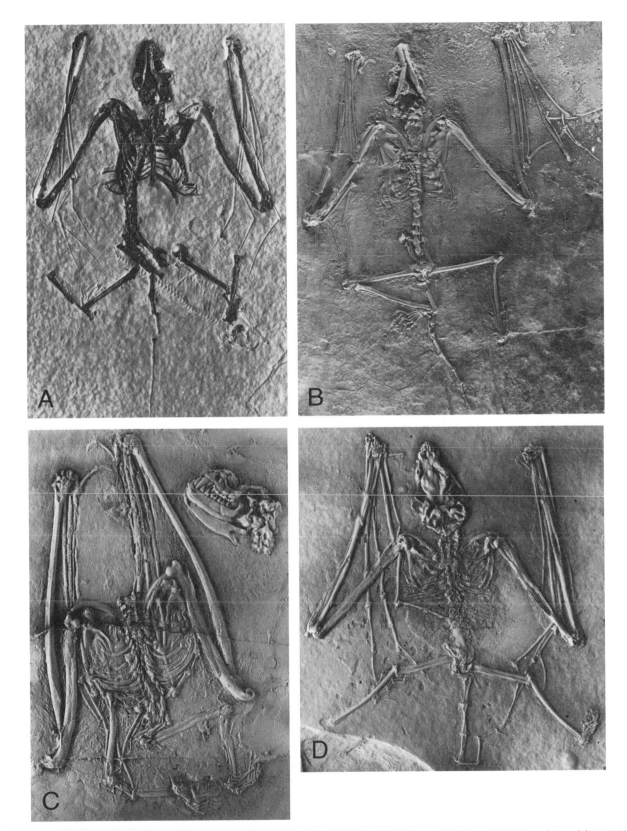

Fig. 11.2. Eocene fossil bats. Note the forelimb modifications associated with powered flight, including an elongation of the radius, ulna, and digits II–V. (A) *Icaronycteris index* (UW 21481a–b) from the Green River Formation, Wyoming; photograph by C. Tarka (American Museum of Natural History). (B) *Archaeonycteris trigonodon* (SMF 80/1379) from Messel, Germany; photograph by E. Pantak (Senckenbergmuseum). (C) *Hassianycteris messelensis* (SMF ME 1414a) from Messel, Germany; photograph by E. Pantak (Senckenbergmuseum). (D) *Palaeochiropteryx tupaiodon* (SMF ME 10) from Messel, Germany; photograph by E. Pantak (Senckenbergmuseum). Abbreviations: SMF ME, Senckenbergmuseum (Messel Collection), Frankfurt am Main, Germany; UW, University of Wyoming Museum of Geology, Laramie, Wyoming.

Table 11.3 Eocene bat diversity and biogeography

Taxon	Epoch	Location
First occurrence in early Eocene[a]		
Ageina (Family indet.)	Early Eocene	Europe and North America
Archaeonycteris (Archaeonycteridae)	Early Eocene–middle Eocene	Europe
Australonycteris (Family indet.)	Early Eocene	Australia
Dizzya (Philisidae)	Early Eocene	Africa
Eppsinycteris (Emballonuridae)	Early Eocene	Europe
Hassianycteris (Hassianycteridae)	?Early Eocene, middle Eocene	Europe
Honrovits (Vespertilionoidea, Family indet.)	Early Eocene	North America
Icaronycteris (Icaronycteridae)	Early Eocene, ?middle Eocene	North America
	?Early Eocene	Europe
Palaeochiropteryx (Palaeochiropterygidae)	Early Eocene–middle Eocene	Europe
First occurrence in Middle Eocene		
Cecilionycteris (Palaeochiropterygidae)	Middle Eocene	Europe
Hipposideros[b] (Hipposideridae)	Middle Eocene–early Pliocene	Europe
	Early Miocene–Recent	Africa
	?Pleistocene–Recent	Madagascar
	Middle Pleistocene–Recent	Asia
	Late Pleistocene–Recent	East Indies
	Miocene–Recent	Australia and New Guinea
Lapichiropteryx[c] (Palaeochiropterygidae)	Middle Eocene–late Eocene	China
Matthesia (Palaeochiropterygidae)	Middle Eocene	Europe
Palaeophyllophora (Hipposideridae)	Middle Eocene–early Miocene	Europe
Rhinolophus (Rhinolophidae)	Middle Eocene–Recent	Europe
	Middle Miocene–Recent	Africa
	Early Pleistocene–Recent	Asia
	Late Pleistocene–Recent	East Indies
	Late Pleistocene–Recent	Australia and New Guinea
Stehlinia[d] (Vespertilionidae)	Middle Eocene–late Oligocene	Europe
Tanzanycteris[e] (Tanzanycteridae)	Middle Eocene	Africa
Tachypteron[f] (Emballonuridae)	Middle Eocene	Europe
Vespertiliavus (Emballonuridae)	Middle Eocene–middle Oligocene	Europe
Wallia[g] (Molossidae)	Middle Eocene	North America
Genus and species indet.[h] (Nycteridae)	Middle Eocene	Africa
First occurrence in late Eocene		
Chadronycteris (Family indet.)	Late Eocene	North America
Cuvierimops (Molossidae)	Late Eocene–early Oligocene	Europe
Necromantis[i] (Megadermatidae)	Late Eocene	Europe
Paraphyllophora (Hipposideridae)	Late Eocene and/or early Oligocene, middle Miocene	Europe
Philisis (Philisidae)	Late Eocene	Africa
Vampyravus[j] (Philisidae)	Late Eocene	Africa
Vaylatsia[k] (Rhinolophidae)	Late Eocene–?early Miocene	Europe
Genus and species indet.[l] (Pteropodidae)	Late Eocene	Thailand

Sources: Modified and updated from Simmons and Geisler (1998: table 1). Unless otherwise noted, age ranges are from McKenna and Bell (1997).

Note: Isolated teeth of uncertain affinities have been referred to Chiroptera by a number of authors; see review in Simmons and Geisler (1998: footnotes to table 1).

[a]*Wyonycteris*, which was described as an early Eocene palaeochiropterygid bat by Gingerich (1987), is omitted from this list, based on arguments developed by Hand et al. (1994). Similarly, *Paradoxonycteris* from the late Eocene of Europe is omitted, based on arguments presented by Russell and Sigé (1970) and Sigé (1976). As with *Wyonycteris*, in the absence of more complete material, there seems little reason to refer *Paradoxonycteris* to Chiroptera. *Zanycteris* from the Wasatchian of Colorado is omitted because it is now thought to be a picrodontid primate (Simpson, 1945; Savage and Russell, 1983).

[b]Includes *Pseudorhinolophus* and *Alastor*. The range for *Hipposideros* (subgenus *Pseudorhinolophus*) is Middle Eocene–Middle Miocene, Europe (Hand and Kirsch, 1988).

[c]Age after Tong (1997), who described the genus.

[d]Includes *Nycterobius*, *Paleunycteris*, and *Revilliodia*; see Handley (1955) and Sigé (1974).

[e]Described by Gunnell et al. (2003); see that reference for information on age.

[f]Described by Storch et al. (2002) from the fossil beds of Messel, Germany.

[g]Originally described as a proscalopid insectivoran, but transferred to Molossidae by Legendre (1985).

[h]This record, which is based on a single lower molar tooth from Chambi, Tunisia, is significant because it apparently represents the earliest record of the family Nycteridae. It was originally described by Sigé (1991) as Rhinolophoidea genus and species indet., but subsequently reinterpreted as a nycterid after comparisons with *Chilbanycteris* (see Sigé et al., 1994: 42).

[i]Includes *Necromanter* and *Necronycteris* (McKenna and Bell, 1997).

[j]Includes *Provampyrus*, which is apparently an objective junior synonym of *Vampyravus* (Sigé, 1985). *Vampyravus* may be a senior synonym of *Philisis*, but this cannot be determined based on available material (see discussion in Sigé, 1985).

[k]Age range following Hand and Kirsch (1998).

[l]This record, which is based on a single premolar tooth, is significant because it represents the first record of the family Pteropodidae (Megachiroptera); see Ducrocq et al. (1993).

representing at least six extinct families (Icaronycteridae, Archaeonycteridae, Palaeochiropterygidae, Hassianycteridae, Tanzanycteridae, Philisidae) and eight extant families (Pteropodidae, Rhinolophidae, Hipposideridae, Megadermatidae, Nycteridae, Emballonuridae, Molossidae, Vespertilionidae) are known from Eocene deposits (Table 11.3). This list includes families with very different—and sometimes very specialized—echolocation and foraging strategies. Among extant family groups known from Eocene deposits, sophisticated echolocation is used by members of seven of eight families (the only exception being the megachiropteran Pteropodidae). Of the six extinct groups known from the Eocene, members of at least five families (Icaronycteridae, Archaeonycteridae, Palaeochiropterygidae, Hassianycteridae, and Tanzanycteridae) are thought to have been echolocating bats (Novacek, 1985, 1987; Habersetzer and Storch, 1992; Simmons and Geisler, 1998; Gunnell et al., 2003). Fossils representing the sixth family, Philisidae, are too incomplete to evaluate.

Analyses of body size, wing morphology, cochlear structure, and stomach contents of the best-preserved Eocene bats indicate that a considerable amount of ecological diversity was present among taxa (e.g., Habersetzer and Storch, 1987; Norberg, 1990; Habersetzer et al., 1994; Simmons and Geisler, 1998; Gunnell et al., 2003). Based on wing morphology, presence of a moderately enlarged cochlea, and absence of calcar, Simmons and Geisler (1998) suggested that *Icaronycteris* and *Archaeonycteris* were most likely perch-hunting gleaners rather than predators on aerial insects. These bats may have occasionally captured prey on the wing (aerial hawking), but their cochlear size falls somewhat below that characteristic of extant full-time aerial hawkers (Habersetzer and Storch, 1992; Simmons and Geisler, 1998). Instead, these bats may have used passive clues (e.g., the sounds of insects crashing into foliage) to locate their prey, reserving echolocation principally for orientation during flight (Simmons and Geisler, 1998). In contrast, *Hassianycteris* and *Palaeochiropteryx* have an enlarged cochlea proportionally similar to those found among extant bats that prey almost exclusively on flying insects captured on the wing (Habersetzer and Storch, 1992; Simmons and Geisler, 1998). Stomach contents, wing morphology, presence of a calcar, and dental morphology suggest that these animals had very different ecologies. *Palaeochiropteryx* species are thought to have foraged near vegetation or water surfaces, hawking small, scale-bearing insect prey (e.g., moths, flies, caddis flies) during slow flight, whereas *Hassianycteris* foraged by hawking larger, more heavily armored prey (beetles and cockroaches) during fast flight in forest gaps or well above the canopy (Habersetzer and Storch, 1987; Norberg, 1990; Habersetzer et al., 1994; Simmons and Geisler, 1998).

Another Eocene bat, *Tanzanycteris*, apparently used an even more specialized echolocation strategy. *Tanzanycteris* has an extremely large cochlea, similar to that seen today only among high-duty-cycle echolocators from the families Rhinolophidae, Hipposideridae, and Mormoopidae (Gunnell et al., 2003). High-duty-cycle echolocation involves rapid emission of long constant-frequency calls, and takes advantage of the Doppler shift to avoid self-deafening, because call pulses and Doppler-shifted echoes are separated in frequency. This system differs from that used by other echolocating bats, which emit shorter calls and listen for returning echoes in the spaces between calls (low-duty-cycle echolocation). High-duty-cycle echolocation appears to be particularly well suited for detecting fluttering insect prey in dense vegetation or close to the ground, situations in which low-duty-cycle echolocation is much less effective (Fenton et al., 1995). Although it seems very likely that Eocene rhinolophids and hipposiderids also used high-duty-cycle echolocation (because optimizations suggest that this feature was present in the most recent common ancestor of those two families; Simmons and Geisler, 1998), no fossil evidence for this has yet been found.

In summary, observations of wing morphology, cranial morphology, stomach contents, taxonomic diversity, and geographic distribution indicate that Eocene bats (1) were capable of powered flight, (2) used sophisticated echolocation like that of extant microchiropteran bats, (3) were taxonomically and ecologically diverse, and (4) were globally widespread, occurring on virtually all continents. Clearly, the radiation of bats was a major event in early Tertiary mammalian evolution. Given that *Archaeonycteris, Palaeochiropteryx, Hassianycteris,* and *Tachypteron* species all coexisted in the early middle Eocene at Messel, Germany, it also seems clear that complex chiropteran communities—with different species occupying distinct ecological niches—were already established at this time.

BAT MONOPHYLY

Until the late 1980s, most biologists assumed that Chiroptera was monophyletic and that all bats shared a common flying ancestor, implying that powered flight evolved only once in mammals. However, evidence from the morphology of the penis and nervous system led some authors to propose that bats are actually diphyletic (e.g., Smith and Madkour, 1980; Hill and Smith, 1984; Pettigrew, 1986, 1995; Pettigrew et al., 1989). Although several versions of bat diphyly have been suggested, the hypothesis most commonly cited suggests that megachiropterans (members of the family Pteropodidae) are more closely related to dermopterans (gliding lemurs) and primates than to echolocating microchiropteran bats (Pettigrew, 1986, 1995; Pettigrew et al., 1989). If this is true, Megachiroptera and Microchiroptera must have evolved from different non-flying ancestors, and therefore represent a remarkable case of convergence.

Although the bat diphyly hypothesis has been discussed extensively in the literature and still remains in some textbooks, an ever-growing body of evidence has provided enormous support for bat monophyly in recent years. Data supporting chiropteran monophyly include morphological data for many organ systems (reviewed in Simmons, 1994), DNA hybridization data (e.g., Kirsch, 1996), and nucleotide sequence data from numerous mitochondrial and nuclear genes (e.g., Miyamoto, 1996; Murphy et al., 2001; Árnason et al., 2002; Teeling et al., 2002 and references cited therein). That these diverse data sets all unambiguously support bat monophyly is remarkable, and most systematists agree that bat monophyly now represents one of the most strongly supported hypotheses in mammalian systematics (Simmons, 1994; Miyamoto, 1996).

PHYLOGENETIC RELATIONSHIPS OF BAT FAMILIES

A central challenge involved in understanding bat origins and diversification is resolving the phylogenetic relationships among families. Eighteen extant families and six extinct families must be considered—not an insignificant problem. The most comprehensive study of bat relationships published to date is that of Simmons and Geisler (1998), who included all but two of these taxa (the poorly known family Philisidae, and Tanzanycteridae, which was described in 2003) in a simultaneous analysis of 195 morphological characters, 12 rDNA restriction site characters, and one character based on the number of R-1 tandem repeats in the mtDNA d-loop region. This study, which used family and subfamily units as terminals, resulted in a single most-parsimonious tree. More recently, Simmons and Geisler (2002) reanalyzed the same data, using a variety of different methods for treating taxonomic polymorphisms (the original study used ambiguity coding, which may lead to biased results under some circumstances; Simmons, 2001). Although some relationships varied depending on the treatment of polymorphic characters, the big picture derived from these primarily morphological data was much the same. Microchiropteran monophyly was strongly supported, with Megachiroptera (Pteropodidae) as the sister group to this clade. On the microchiropteran side of the tree, *Icaronycteris, Archaeonycteris, Hassianycteris,* and *Palaeochiropteryx* occupied a "picket fence" series of consecutive branches that were sister taxa to the crown clade of extant echolocating bats. Within the microchiropteran crown clade, three clades emerged at or near the base: (1) Emballonuridae, (2) Yinochiroptera (= Rhinolophidae + Hipposideridae + Megadermatidae + Nycteridae + Rhinopomatidae + Craseonycteridae), and Yangochiroptera (Mystacinidae + Noctilionidae + Mormoopidae + Phyllostomidae + Myzopodidae + Thyropteridae + Furipteridae + Natalidae + Molossidae + Antrozoidae + Vespertilionidae; Simmons and Geisler, 1998, 2002).

Relationships within Yinochiroptera were not affected by different treatments of taxonomic polymorphisms in Simmons and Geisler's (2002) study; in every case, two clades were supported: a group traditionally called Rhinolophoidea (Rhinolophidae + Hipposideridae + Megadermatidae + Nycteridae) and another named Rhinopomatoidea by Simmons (1998; Rhinopomatidae + Craseonycteridae).

In contrast, relationships did vary within Yangochiroptera, depending on the methods used to deal with taxonomic polymorphisms. In Simmons and Geisler's (2002) most reductive analyses that included all characters, Yangochiroptera comprised three clades: (1) Noctilionoidea (including Mystacinidae + Noctilionidae + Mormoopidae + Phyllostomidae), (2) Nataloidea (Myzopodidae + Thyropteridae + Furipteridae + Natalidae), and (3) Vespertilionoidea (Molossidae + Antrozoidae + Vespertilionidae).

Few of the relationships found in Simmons and Geisler's (1998, 2002) studies were well supported by decay or bootstrap values. Those clades supported strongly (e.g., by bootstrap values greater than 90%) included only the following: (1) Microchiropteraformes (*Palaeochiropteryx* + *Hassianycteris* + all families of extant echolocating bats); (2) Rhinolophoidea (Rhinolophidae + Hipposideridae + Megadermatidae + Nycteridae); (3) a clade comprising Rhinolophidae + Hipposideridae + Megadermatidae; (4) a clade including Rhinolophidae + Hipposideridae; (5) Noctilionoidea excluding Mystacinidae (Noctilionidae + Mormoopidae + Phyllostomidae); (6) Nataloidea; and (7) Molossidae (including Tomopeatinae + Molossinae). Remaining groups were supported either moderately or weakly, suggesting that data were not highly informative for these groups.

As discussed by Simmons and Geisler (1998, 2002), one of the weaknesses of their studies was the use of higher-level taxa as terminals. Although some of these groups are very small or monotypic (e.g., Craseonycteridae, Mystacinidae, Noctilionidae, Myzopodidae, Furipteridae), others are very large (e.g., Vespertilionidae includes more than 300 species). Efforts are presently underway to develop a new morphological data set sampling at the species level across all bat families, but this study has not yet been completed. In the meantime, better-supported phylogenetic hypotheses based on molecular data have been published by several research groups (Hutcheon et al., 1998; Kirsch et al., 1998; Van Den Bussche and Hoofer, 2000, 2001; Teeling et al., 2000, 2002; Murphy et al., 2001; Springer et al., 2001; Hulva and Horacek, 2002; Van Den Bussche et al., 2002a,b, 2003; Hoofer et al., 2003).

Several molecular studies have seriously challenged the monophyly of Microchiroptera, suggesting instead that Yinochiroptera is the sister-group of Megachiroptera (Hutcheon et al., 1998; Teeling et al., 2000, 2002). Springer et al. (2001) created a new suborder Yinpterochiroptera for this clade, which now appears to include Pteropodidae, Rhinolophidae, Hipposideridae, Megadermatidae, Craseonycteridae, and Rhinopomatidae (Hulva and Horacek, 2002; Teeling et al., 2002). Nycteridae, usually included in Yinochiroptera within the superfamily Rhinolophoidea, now appears more closely related to yangochiropteran bats, as does Emballonuridae (Teeling et al., 2002). Monophyly of other superfamilies recognized by Simmons (1998) and Simmons and Geisler (1998) has also been seriously questioned, based on molecular data (e.g., Nataloidea and Molossoidea; Van Den Bussche and Hoofer, 2000, 2001; Hoofer and Van Den

Bussche, 2001; Van Den Bussche et al., 2002b, 2003; Hoofer et al., 2003). Most data sets now agree that Mystacinidae belongs in Noctilionoidea (Van Den Bussche and Hoofer, 2000, 2001; Simmons and Conway, 2001; Van Den Bussche et al., 2002a). However, relationships of Mystacinidae to the families traditionally placed in this superfamily (Noctilionidae, Mormoopidae, Phyllostomidae) and to others recently allied with Noctilionoidea based on molecular data (Thyropteridae and Furipteridae; Van Den Bussche and Hoofer, 2000, 2001; Van Den Bussche et al., 2002b, 2003) remain somewhat uncertain. Similarly, relationships among Myzopodidae, Emballonuridae, Nycteridae, and the two large yangochiropteran superfamilies (Noctilionoidea and Vespertilionoidea) recognized in the molecular studies remain unclear. The strength of the molecular sequence data supporting many of the novel clades noted above (e.g., Yinpterochiroptera) is increasingly compelling, and it seems likely that a new consensus view of higher-level bat phylogeny will soon emerge that contradicts many traditional arrangements.

Only one study (Springer et al., 2001) has attempted to integrate fossil and molecular data to investigate the position of the better-known fossil taxa with respect to Yinpterochiroptera and Yangochiroptera. These authors used the morphological data from Simmons and Geisler (1998) and ran parsimony analyses of the data using a molecular tree as a backbone scaffold (i.e., constraining the topology of extant branches to match the molecular tree). The results of this analysis are intriguing. *Icaronycteris* and *Archaeonycteris* occupied the two most basal branches of the resulting tree, appearing as successive sister taxa to the crown clade Chiroptera (the smallest clade including all extant families); *Hassianycteris* and *Palaeochiropteryx* nested up within the crown clade as successive sister taxa to Yangochiroptera (Springer et al., 2001). The structure of this tree implies that echolocation evolved right at the base of the bat family tree, and was subsequently lost in megachiropteran bats. This novel hypothesis contradicts Simmons and Geisler's (1998) contention that lack of echolocation was the primitive condition in bats (retained in megachiropterans), and that echolocation first evolved early in the microchiropteran part of the tree.

The need for updated phylogenies including morphological data has been growing as new fossil bats have been described. Using Simmons and Geisler's (1998) character set, Gunnell et al. (2003) attempted to place the new taxon *Tanzanycteris* with respect to fossil and extant lineages, and found that the result depended on whether they used a backbone constraint based on morphological data (i.e., with Microchiroptera monophyletic) or molecular trees (i.e., with Yinpterochiroptera monophyletic). *Tanzanycteris* appeared as either the sister group of *Hassianycteris* or Microchiroptera under the first scenario, but might be placed within the crown group as the sister taxon of rhinolophids if the molecular backbone is correct.

Unfortunately, there has not yet been a definitive analysis of family relationships combining both molecular and morphological data in a simultaneous analysis. Although Springer et al. (2001) used both types of data, they assumed that the molecular data were superior to the morphological data, and used topology of the shortest molecular tree to constrain the results (morphology was used only to place taxa for which there were no molecular data). Comparisons of the results of numerous molecular studies with Simmons and Geisler's (2002) trees reveal that, although many of the clades strongly supported by molecular data are not strongly contradicted by morphology, there are some important points in which these data conflict with one another (e.g., Microchiroptera and Nataloidea are strongly supported by morphology but strongly refuted by molecular studies). In the absence of simultaneous analyses including both molecular and morphological data, these points of incongruence cannot be resolved. However, it must be noted that combined analyses of five nuclear genes have produced very strong support for Yinpterochiroptera (Teeling et al., 2002), and combined analyses of two nuclear genes and one mitochondrial gene have produced very strong support for nataloid paraphyly (Hoofer et al., 2003). Given the large number of informative characters in these molecular data sets, it seems likely that future combined analyses—including morphological as well as molecular data—will produce trees that resemble the molecular trees more closely than the morphology trees in these areas.

DIVERSIFICATION OF BATS AS REVEALED BY GHOST LINEAGES

As noted above, approximately half of the 24 family groups of bats are known from Eocene fossils. This raises the question of when the remaining families originated. Was there a single great burst of bat diversification in the Eocene, or did families continue to arise throughout the Tertiary? In the absence of a complete fossil record, information from phylogenies can be used to reconstruct "ghost lineages," those parts of evolutionary lineages that must have existed, but for which we currently have no fossils (Norell, 1992; Padian et al., 1994; Novacek, 1996). Ghost lineage reconstruction provides a means of investigating both the absolute and relative timing of various divergence events.

The absence of a single definitive family tree for bats raises the problem of which phylogeny or phylogenies to use. For the purposes of investigating the pattern and timing of diversification of the major bat lineages, it seems reasonable at this time to consider both morphological and molecular trees. For a morphology tree, I have chosen to use that of Simmons and Geisler (2002: fig. 3A), which included all bat families except the extinct families Philisidae (which is known only from a few incomplete fossils) and Tanzanycteridae (which was described after their study was completed). The tree was derived from a modified version of the Simmons and Geisler (1998) data matrix, in which taxonomic polymorphisms were reduced to single states (as far as possible) using a combination of methods. The

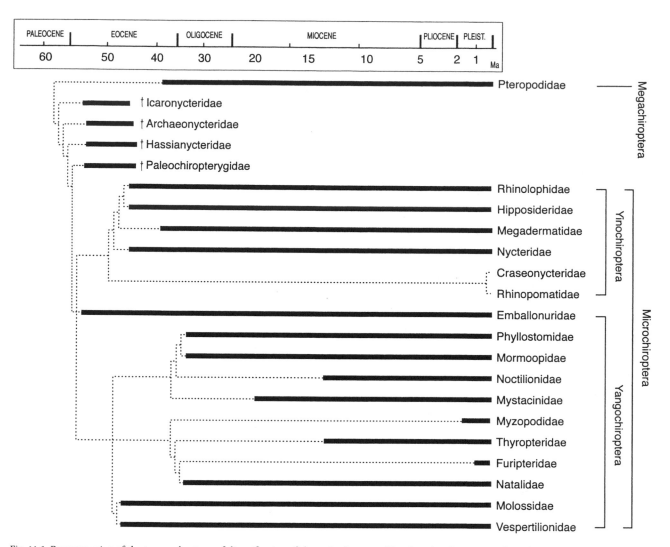

Fig. 11.3. Reconstruction of the temporal pattern of diversification of the major lineages of bats based on Simmons and Geisler's (1998) phylogenetic tree, which was derived principally from morphological data. Thick lines indicate time ranges documented by fossils; dotted lines represent ghost lineages that must have existed, given the presumed phylogenetic relationships. The time scale follows that of McKenna and Bell (1997: fig. 1). Data on fossil ranges are based on Ducrocq et al. (1993), Sigé et al. (1994), Czaplewski (1997), McKenna and Bell (1997), Simmons and Geisler (1998), Hand et al. (2001), Czaplewski and Morgan (2002), Morgan (2002), and references cited therein.

resulting tree is similar but not identical to that presented by Simmons and Geisler (1998), differing in the placement of Mystacinidae and the degree of resolution of a few nodes within Yangochiroptera. A ghost lineage reconstruction based on this tree topology is shown in Fig. 11.3. Sources for the dates of the oldest known fossils in each extant group were obtained from McKenna and Bell (1997), except as noted.

Producing a single molecular tree for bat families is complicated by the absence of any molecular study that includes representatives of all of the extant families of bats; in addition, analyses of different genes often result in trees with somewhat different topologies. For the purposes of this chapter, I built a composite molecular tree based on trees presented by Hulva and Horacek (2002), Teeling et al. (2002), Van Den Bussche et al. (2002a, 2003), and Hoofer et al. (2003). This composite tree is not a strict consensus,

but rather a conditional consensus containing only clades that were reasonably well supported in one or more studies (and not strongly refuted in any other). Eocene families were placed in the tree following Springer et al. (2001). A ghost lineage reconstruction based on the resulting tree topology is shown in Fig. 11.4.

Despite significant differences in topology of the phylogenetic trees on which they were based, the two ghost lineage reconstructions produced remarkably similar results concerning the timing of radiation of the major bat clades. As noted earlier, at least six extinct families (Icaronycteridae, Archaeonycteridae, Palaeochiropterygidae, Hassianycteridae, Tanzanycteridae, Philisidae) and eight extant families (Pteropodidae, Rhinolophidae, Hipposideridae, Megadermatidae, Nycteridae, Emballonuridae, Molossidae, Vespertilionidae) are known from Eocene deposits (see Table 11.3). In addition, the ghost lineage reconstructions indicate that

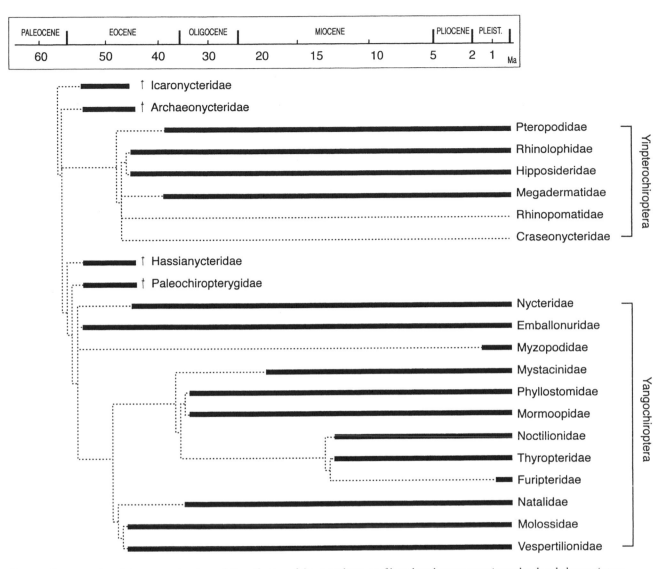

Fig. 11.4. Reconstruction of the temporal pattern of diversification of the major lineages of bats, based on a composite molecular phylogenetic tree derived from results published by Teeling et al. (2002), Hulva and Horacek (2002), Van Den Bussche et al. (2002a, 2003), and Hoofer et al. (2003). Thick lines indicate time ranges documented by fossils; dotted lines represent ghost lineages that must have existed, given the presumed phylogenetic relationships. The time scale follows that of McKenna and Bell (1997: fig. 1). Data on fossil ranges are based on Ducrocq et al. (1993), Sigé et al. (1994), Czaplewski (1997), McKenna and Bell (1997), Simmons and Geisler (1998), Hand et al. (2001), Czaplewski and Morgan (2002), Morgan (2002), and references cited therein.

three to five additional family-level lineages must have also been distinct at this time (Figs. 11.3, 11.4). The reconstruction based on the morphology tree (Fig. 11.3) indicates that three additional groups were distinct by the end of the Eocene: (1) a lineage leading to Rhinopomatoidea (Rhinopomatidae + Craseonycteridae), (2) a lineage leading to Noctilionoidea (Phyllostomidae + Mormoopidae + Noctilionidae + Mystacinidae), and (3) a lineage leading to Nataloidea (Myzopodidae + Thyropteridae + Furipteridae + Natalidae). The reconstruction based on the molecular tree (Fig. 11.4) produced quite similar results, indicating that five additional groups (beyond those known from Eocene fossils) were distinct by the end of the Eocene: (1) Rhinopomatidae, (2) Craseonycteridae, (3) Myzopodidae, (4) a lineage leading to an expanded Noctilionoidea clade (Phyllostomidae + Mormoopidae + Noctilionidae + Mystacinidae + Thy-

ropteridae + Furipteridae), and (5) Natalidae. In total, this means that 17–19 family lineages of bats (70–80% of the known family-level diversity of the clade) arose before the end of the Eocene.

Both reconstructions also agree that a second important radiation—diversification of the noctilionoid clade—took place in the Neotropics by the end of the early Oligocene. It is possible that noctilionoids diversified sooner (e.g., in the Eocene), but no bat fossils have yet been found in pre-Oligocene Neotropical deposits, so this remains conjecture. In the molecular-based reconstruction, the initial splits in Noctilionoidea (e.g, between Mystacinidae, Mormoopidae, Phyllostomidae, and the noctilionid branch) were subsequently followed by a third radiation, in which Noctilionidae, Thyropteridae, and Furipteridae diverged from one another. This third diversification event, which resulted in the establishment of the last of the chiropteran families, took place no later than the middle Miocene. If the morphological tree is correct, the diversification of Thyropteridae and Furipteridae must have taken place much earlier—in or before the early Oligocene.

Under either phylogenetic hypothesis, it seems clear that the family-level diversification of bats took place primarily in the Eocene, although some groups may have diversified later. Only one splitting event—the divergence of Rhinopomatidae and Craseonycteridae in the morphological tree—cannot be date-limited by fossils. But these two families (five species in total) represent a very small proportion of extant bat species (0.04%; Simmons, in press). The largest bat family, Vespertilionidae (>400 species), was distinct by the middle Eocene, as were many other diverse families (e.g., Pteropodidae, Hipposideridae, Rhinolophidae, Molossidae, Emballonuridae). Remarkably, some extant genera were even distinct in the middle Eocene (*Rhinolophus* and *Hipposideros;* see Table 11.3), and Eocene fossil species nest well up within family phylogenies in some groups (e.g., Hipposideridae; Hand and Kirsch, 1998). Clearly, the Eocene was a critical time for bats, and was the period in which they became established as one of the most diverse and successful mammalian orders.

ORIGINS OF FLIGHT AND ECHOLOCATION

The relative timing of the origins of powered flight and echolocation in bats has been the source of considerable controversy. Three theories emerged in the late twentieth century: the "echolocation first" theory (Hill and Smith, 1984; Fenton et al., 1995; Arita and Fenton, 1997), the "flight first" theory (Norberg, 1990, 1994; Simmons and Geisler, 1998), and the "tandem evolution" theory (Speakman, 1993). The echolocation-first theory suggests that primitive gliding representatives of the bat lineage evolved echolocation as a way to detect, track, and assess airborne prey that they captured during glides from perches. Specializations for powered flight evolved later to increase maneuverability and simplify the process of returning to the hunting perch after each foray (Fenton et al., 1995; Arita and Fenton, 1997). In contrast, the flight-first theory proposes that powered flight arose before sophisticated echolocation. This theory proposes that early bats were gliding nocturnal insectivores, and that powered flight evolved to improve mobility in the arboreal milieu and reduce the amount of time and energy required for foraging (Norberg, 1994). Under this scenario, after these animals achieved flight, they evolved sophisticated echolocation as a means of detecting, tracking, and evaluating airborne prey. Early gliding and flying taxa might have had a primitive echolocation system useful for orientation, but sophisticated echolocation evolved only after powered flight was achieved (Norberg, 1990, 1994). Finally, the tandem-evolution theory suggests that echolocation and powered flight evolved simultaneously (Speakman, 1993).

Simmons and Geisler (1998) argued that only one of the above theories—the flight-first theory—was compatible with bat monophly and microchiropteran monophyly, both of which appeared (at that time) to be enormously well-supported hypotheses. Given their phylogeny, the flight-first theory provided the simplest explanation of the observed distribution of these features, implying a single origin of powered flight (prior to the diversification of the bats) and a single origin of echolocation (after the split between megachiropterans and microchiropterans, but before diversification of the microchiropteran lineage). The other two theories would require loss of echolocation in megachiropterans, a scenario that Simmons and Geisler (1998) considered unnecessary, given the structure of the phylogenetic tree and unlikely, given the sensory system reorganization required.

However, new molecular phylogenies now being produced differ significantly from that of Simmons and Geisler (see discussion above). If Microchiroptera is indeed paraphyletic, as these studies strongly suggest, and the Eocene fossil families are related to extant forms as indicated by the analysis of Springer et al. (2001), then it seems very likely that both flight and echolocation evolved prior to the diversification of the known bat families. In this context, the echolocation-first, flight- first, and tandem-evolution theories are all equally viable. In the absence of fossils with wings but no evidence of echolocation abilities, or fossils showing evidence of echolocation but not flight, we cannot refute any of these scenarios. However, Simmons and Geisler's (1998) arguments that *Icaronycteris* and *Archaeonycteris*—two taxa clearly capable of powered flight—were perch-hunting gleaners rather than aerial hawkers suggests that the sophisticated echolocation abilities necessary to detect, track, and capture aerial prey may not have evolved until after powered flight was well established. The use of echolocation for orientation and obstacle avoidance could have evolved earlier, later, or simultaneously with the evolution of flight.

WHAT IS THE SISTER GROUP OF BATS?

Given that we do not have fossils of any evolutionary intermediates between bats and their non-flying ancestors, we must rely entirely on phylogenetic studies of interordinal relationships to indicate the place of bats in the family tree of mammals. Simmons (1994) reviewed evidence for various sister-group relationships between bats and other placental clades, and concluded that there was no consensus on this matter; the same is still true now, a decade later.

Morphological data have almost universally placed bats somewhere in the group Archonta, together with dermopterans, primates, and tree shrews (e.g., Wible and Novacek, 1988; Beard, 1993; Simmons, 1993, 1995; Szalay and Lucas, 1993; Miyamoto, 1996). Simmons (1995) provided a list of 17 morphological synapomorphies that apparently group bats with dermopterans in a monophyletic Volitantia within

Archonta. This arrangement is intuitively appealing, because dermopterans are gliding mammals and most researchers now believe that bats evolved from gliding ancestors (Simmons, 1995). However, this relationship has been strongly questioned in recent molecular studies.

Despite seemingly strong morphological evidence, bats have not appeared as a member of either Volitantia or Archonta in *any* of the more than two dozen molecular studies completed since the early 1990s. Regardless of the genes sampled or the phylogenetic methods used, bats never group with primates or dermopterans; multiple analyses of nuclear and mitochondrial gene sequences have resoundingly refuted the hypothesis that bats are archontan mammals. Instead, molecular studies uniformly place bats in a Laurasiatheria clade (e.g., Miyamoto et al., 2000; Murphy et al., 2001; Árnason et al., 2002; Douady et al., 2002; Van Den Bussche et al., 2002b). Within this group, bats most commonly appear as either the sister group or basal member of a ferungulate clade (which includes artiodactyls, cetaceans, perissodactyls, carnivores, and pholidotans) or as the sister group of a eulipotyphlan clade (including shrews, moles, and possibly hedgehogs). The evidence is fairly evenly split between these two arrangements, with different subsets of taxa or analysis methods sometimes producing alternate results for interrelationships even in the same study (e.g., Árnason et al., 2002). When bats are grouped with ferungulates, the sister group of that clade is Eulipotyphla; when bats are grouped with eulipotphylans, the sister group of that clade is ferungulates. Regardless, there is no molecular support for placing bats within Archonta.

WHERE DID BATS ORIGINATE?

Because the sister group of bats is uncertain—and most likely consists of a large clade distributed on multiple continents—it is not possible to isolate the geographic center of origin of bats based on relationships with other mammal groups. Similarly, the fossil record does not offer compelling evidence, as bat fossils seem to have appeared on several continents almost simultaneously in the early Eocene (North America, Europe, Africa, and Australia). Based on phylogenetic analyses, North America and Europe may appear to be the most likely areas for the geographic center of origin of bats, but the evidence is far from overwhelming. North America is indicated because it was the home of what may be the most primitive known bat, *Icaronycteris index,* which is also the oldest known bat species (Simmons and Geisler, 1998). Europe is implicated as a possible center of origin because it was home to a diverse group of early Eocene bats, including *Archaeonycteris* species, which may occupy the next branch up the tree from *Icaronycteris* (see Figs. 11.3, 11.4). However, African philisids and the early Eocene fossil *Australonycteris* have not been included in phylogenetic analyses, so their phylogenetic relationships remain unclear. In this context, it is not yet possible to determine with certainty even which hemisphere was the center of origin for the first chiropter-

ans. The probable laurasiatherian affinities of the order suggest a northern hemisphere origin for the group, but additional fossils and/or additional resolution of mammalian interordinal relationships will be required before any hypotheses of bat origins can be seriously evaluated.

SUMMARY

As currently recognized, Chiroptera includes 18 extant families and 6 extinct families, all of which are characterized by specializations for powered flight. The fossil record of bats begins in the early Eocene with *Icaronycteris index,* now thought to be approximately 53 million years old. Twenty-seven genera of bats, representing at least six extinct families (Icaronycteridae, Archaeonycteridae, Palaeochiropterygidae, Hassianycteridae, Tanzanycteridae, Philisidae) and eight extant families (Pteropodidae, Rhinolophidae, Hipposideridae, Megadermatidae, Nycteridae, Emballonuridae, Molossidae, Vespertilionidae) are known from Eocene deposits. Ghost lineage reconstructions indicate that three to five additional family-level lineages must have also been distinct at this time. This means that 17–19 family lineages of bats (70–80% of the known family-level diversity of the clade) arose before the end of the Eocene. Analyses of wing morphology, cranial morphology, and stomach contents indicate that Eocene bats used sophisticated echolocation like that of extant microchiropteran bats, and that they were taxonomically and ecologically diverse. Complex chiropteran communities—with different species occupying distinct ecological niches—were already established by the end of the early Eocene.

Monophyly of bats is supported by numerous morphological and molecular data sets, so it seems clear that bats evolved from a single volant ancestral lineage. Unfortunately, no fossils of morphological intermediates between bats and their non-volant ancestors have been found. Because analyses of different data sets have placed bats in different parts of the mammalian family tree, the origins of the order remain unclear. Morphological data have suggested that the closest living relatives of bats lie among the archontans (gliding lemurs, primates, and tree shrews), but multiple molecular data sets have strongly supported placement of bats in Laurasiatheria with either ferungulates (ungulates, whales, and carnivores) or eulipotyphlans (shrews, moles, and hedgehogs) as their closest relatives. The probable laurasiatherian origin of bats suggests that they may have originated in the northern hemisphere, but additional fossils and better resolution of mammalian interordinal relationships are needed before any hypotheses of bat origins can be evaluated further.

Bats are typically divided into two large groups, Microchiroptera (echolocating bats) and Megachiroptera (Old World fruit bats, which do not echolocate). Until recently, these groups were considered to be reciprocally monophyletic, and were formally recognized as suborders in most classifications. However, recent analyses of molecular sequence data from several mitochondrial and nuclear genes have indicated that Microchiroptera is not monophyletic; instead, rhinolophids and their relatives appear to be more closely related to Megachiroptera than to the remaining microchiropteran families. This new phylogenetic hypothesis suggests that echolocation may have evolved very early in the bat lineage, before diversification of the extant families. It is not yet clear if powered flight evolved first, echolocation evolved first, or these two key features evolved simultaneously. Regardless, bats have become one of the most successful of all mammalian lineages, presently accounting for approximately one-fifth of all extant mammal species.

ACKNOWLEDGMENTS

Sue Hand and Nick Czaplewski reviewed an earlier version of this chapter, and I thank them for their many helpful comments. Special thanks to Mark Mandica for illustrations, and to Jörg Habersetzer for making available the photographs of the Messel bats reproduced in Fig. 11.2. This study was supported in part by National Science Foundation Research Grant DEB-9873663.

REFERENCES

Arita, H. T., and M. B. Fenton. 1997. Flight and echolocation in the ecology and evolution of bats. Trends in Ecology and Evolution 12: 53–58.

Árnason, Ú., J. A. Adegoke, K. Bodin, E. W. Born, Y. B. Esa, A. Gullberg, M. Nilsson, R. V. Short, X. Xu, and A. Janke. 2002. Mammalian mitogenetic relationships and the root of the eutherian tree. Proceedings of the National Academy of Sciences USA 99: 8151–8156.

Beard, K. C. 1993. Phylogenetic systematics of Primatomorpha, with special reference to Dermoptera; pp. 129–150 in F. S. Szalay, M. J. Novacek, and M. C. McKenna (eds.), Mammal Phylogeny: Placentals. Springer-Verlag, New York.

Czaplewski, N. J. 1997. Chiroptera; pp. 410–431 in R. F. Kay, R. H. Madden, R. L. Cifelli, and J. J. Flynn (eds.), Vertebrate Paleontology in the Neotropics: The Miocene Fauna of La Venta, Colombia. Smithsonian Institution Press, Washington, D.C.

Czaplewski, N. J., and G. S. Morgan. 2002. Phyllostomid bats from the Oligocene and Early Miocene of Florida. Journal of Vertebrate Paleontology 22(3): 48A.

Douady, C. J., P. I. Chatelier, O. Madsen, W. W. de Jong, F. Catzeflis, M. S. Springer, and M. J. Stanhope. 2002. Molecular phylogenetic evidence confirming the Eulipotyphla concept and in support of hedgehogs as the sister group to shrews. Molecular Phylogenetics and Evolution 25: 200–209.

Ducrocq, S., J.-J. Jaeger, and B. Sigé. 1993. Un mégachiroptère dans l'Éocène supérieur de Thaïlande: incidence dans la discussion phylogénique du groupe. Neues Jahrbuch für Geologie und Paläontologie Mh 9: 561–575.

Fenton, M. B., D. Audet, M. K. Obrist, and J. Rydell. 1995. Signal strength, timing, and self-deafening: The evolution of echolocation in bats. Paleobiology 21: 229–242.

Ferrarezzi, H., and E. A. Gimenez. 1996. Systematic patterns and the evolution of feeding habits in Chiroptera (Archonta: Mammalia). Journal of Comparative Biology 1: 75–94.

Gingerich, P. 1987. Early Eocene bats (Mammalia, Chiroptera) and other vertebrates in freshwater limestones of the Willwood Formation, Clark's Fork Basin, Wyoming. Contributions of the Museum of Paleontology, University of Michigan 27: 275–320.

Gunnell, G. F., B. F. Jacobs, P. S. Herendeen, J. J. Head, E. Kowlaski, C. P. Msuya, F. A. Mizambwa, T. Harrison, J. Habersetzer, and G. Storch. 2003. Oldest placental mammal from sub-Saharan Africa: Eocene microbat from Tanzania—evidence for early evolution of sophisticated echolocation. Palaeontologia Electronica 5(3): 1–10.

Habersetzer, J., G. Richter, and G. Storch. 1994. Paleoecology of Early Middle Eocene bats from Messel, FRG. Aspects of flight, feeding and echolocation. Historical Biology 8: 235–260.

Habersetzer, J., and G. Storch. 1987. Klassifikation und funktionelle Flügelmorphologie paläogener Fledermäuse (Mammalia, Chiroptera). Courier Forschungsinstitut Senckenberg 1: 11–150.

———. 1992. Cochlea size in extant Chiroptera and Middle Eocene Microchiroptera from Messel. Naturwissenchaften 79: 462–466.

Hand, S. J., M. Archer, and H. Godthelp. 2001. New Miocene Icarops material (Microchiroptera: Mystacinidae) from Australia, with a revised diagnosis of the genus. Memoirs of the Association of Australasian Palaeontologists 25: 139–146.

Hand, S. J., and J.A.W. Kirsch. 1998. A southern origin for the Hipposideridae (Microchiroptera)? Evidence from the Australian fossil record; pp. 72–90 in T. H. Kunz and P. A. Racey (eds.), Bat Biology and Conservation. Smithsonian Institution Press, Washington, D.C.

Hand, S. J., M. Novacek, H. Godthelp, and M. Archer. 1994. First Eocene bat from Australia. Journal of Vertebrate Paleontology 14: 375–381.

Handley, C. O. 1955. Nomenclature of some Tertiary Chiroptera. Journal of Mammalogy 36: 128–130.

Hill, J. E., and J. D. Smith. 1984. Bats: A Natural History. University of Texas Press, Austin.

Hoofer, S. R., S. A. Reeder, E. W. Hansen, and R. A. Van Den Bussche. 2003. Molecular phylogenetics and taxonomic review of noctilionoid and vespertilionoid bats (Chiroptera, Yangochiroptera). Journal of Mammalogy 84: 809–821.

Hoofer, S. R., and R. A. Van Den Bussche. 2001. Phylogenetic relationships of plecotine bats and allies based on mitochondrial ribosomal sequences. Journal of Mammalogy 82: 131–137.

Hulva, P., and I. Horacek. 2002. Craseonycteris thonglongyai (Chiroptera: Craseonycteridae) is a rhinolophoid: Molecular evidence from cytochrome b. Acta Chiropterologica 4: 107–120.

Hutcheon, J. M., J.A.W. Kirsch, and J. D. Pettigrew. 1998. Base-compositional biases and the bat problem. III. The question of microchiropteran monophyly. Philosophical Transactions of the Royal Society of London B, 353: 607–617.

Jepsen, G. L. 1966. Early Eocene bat from Wyoming. Science 154: 1333–1339.

Kirsch, J.A.W. 1996. Bats are monophyletic; megabats are monophyletic; but are microbats also? Bat Research News 36: 78.

Kirsch, J.A.W., J. M. Hutcheon, D.G.P. Byrnes, and B. D. Lloyd. 1998. Affinities and historical zoogeography of the New Zealand short-tailed bat, *Mystacina tuberculata* Gray 1843, inferred from DNA-hybridization comparison. Journal of Mammalian Evolution 5: 33–64.

Legendre, S. 1984. Identification de deux sous-genres fossiles et compréhension phylogénique du genre *Mormopterus* (Molossidae, Chiroptera). Comptes Rendus de l'Académie des Sciences, Série II, Paris 298: 715–720.

———. 1985. Molossidés (Mammalia, Chiroptera) cénozoiques de l'Ancien et du Nouveau Monde; statut systématique; intégration phylogénique des données. Neues Jahrbuch für Geologie und Paläontologie Abhandlungen 170: 205–227.

McKenna, M. C., and S. K. Bell. 1997. Classification of Mammals above the Species Level. Columbia University Press, New York.

Miyamoto, M. M. 1996. A congruence study of molecular and morphological data for eutherian mammals. Molecular Phylogeny and Evolution 6: 373–90.

Miyamoto, M. M., C. Porter, and M. Goodman. 2000. *c-Myc* gene sequences and the phylogeny of bats and other eutherian mammals. Systematic Biology 49: 501–514.

Morgan, G. S. 2002. New bats in the Neotropical families Emballonuridae and Mormoopidae from the Oligocene and Miocene of Florida, and the biochronology of Florida Whitneyan, Arikareean, and Hemingfordian Faunas. Journal of Vertebrate Paleontology 22 (supplement to no. 3): 90A.

Murphy, W. J., E. Eizirik, W. E. Johnson, Y. P. Zhang, O. A. Ryder, and S. J. O'Brien. 2001. Molecular phylogenetics and the origins of placental mammals. Nature 409: 614–618.

Norberg, U. M. 1990. Ecological determinants of bat wing shape and echolocation call structure with implications for some fossil bats; pp. 197–211 *in* V. Hanák, I. Horácek, and J. Gaisler (eds.), European Bat Research 1987. Charles University Press, Prague.

———. 1994. Wing design, flight performance, and habitat use in bats; pp. 205–239 *in* P. C. Wainwright and S. M. Reilly (eds.), Ecological Morphology: Integrative Organismal Biology. University of Chicago Press, Chicago.

Norberg, U. M., and J.M.V. Rayner. 1987. Ecological morphology and flight in bats (Mammalia: Chiroptera): wing adaptations, flight performance, foraging strategy and echolocation. Philosophical Transactions of the Royal Society of London, Series B 316: 335–427.

Norell, M. A. 1992. Taxic origin and temporal diversity: The effect of phylogeny: pp. 89–118 *in* M. J. Novacek and Q. D. Wheeler (eds.), Extinction and Phylogeny. Columbia University Press, New York.

Novacek, M. J. 1985. Evidence for echolocation in the oldest known bats. Nature 315: 140–141.

———. 1987. Auditory features and affinities of the Eocene bats *Icaronycteris* and *Palaeochiropteryx* (Microchiroptera, *incertae sedis*). American Museum Novitates 2877: 1–18.

———. 1996. Paleontological data and the study of adaptation; pp. 311–359 *in* M. D. Rose and G. V. Lauder (eds), Adaptation. Academic Press, New York.

Padian, K., D. R. Lindberg, and P. D. Polly. 1994. Cladistics and the fossil record: The uses of history. Annual Review of Earth and Planetary Sciences 22: 63–91.

Pettigrew, J. D. 1986. Flying primates? Megabats have the advanced pathway from eye to midbrain. Science 231: 1304–1306.

———. 1995. Flying primates: Crashed, or crashed through? *in* P.A. Racey and S. M. Swift (eds.), Ecology, Evolution and Behavior of Bats. Symposium of the Zoological Society of London 67: 3–26.

Pettigrew, J. D., B.G.M. Jamieson, S. K. Robson, L. S. Hall, K. I. McAnally, and H. M. Cooper. 1989. Phylogenetic relations between microbats, megabats and primates (Mammalia: Chiroptera and Primates). Philosophical Transactions of the Royal Society of London, Series B 325: 489–559.

Rayner, J.M.V. 1981. Flight adaptations in vertebrates. Symposium of the Zoological Society of London 48: 137–172.

Russell, D. E., and B. Sigé. 1970. Révision des chiroptères lutétiens de Messel (Hesse, Allemagne). Palaeovertebrata 3: 83–182.

Savage, D. E., and D. E. Russell. 1983. Mammalian Paleofaunas of the World. Addison-Wesley, Reading, Pennsylvania.

Sigé, B. 1974. Données nouvelles sur le genre *Stehlinia* (Vespertilionoidea, Chiroptera) du Paléogène d'Europe. Palaeovertebrata 6: 253–272.

———. 1976. Insectivores primitifs de l'Eocène supérieur et Oligocène inférieur d'Europe occidentale. Nyctithériidés. Mémoires du Muséum National d'Histoire Naturelle, Série C, Sciences de la Terre 34: 1–140.

———. 1985. Les chiroptères oligocènes du Fayum, Egypte. Geologica et Paleontologica 19: 161–189.

———. 1991. Rhinolophoidea et Vespertilionoidea du Chambi (Eocène inférieur de Tunisie) et l'origine des chiroptères modernes. Neues Jahrbuch für Geologie und Paläontologie Abhandlungen 182: 355–376.

Sigé, B., H. Thomas, S. Sen, E. Gheerbrant, J. Roger, and Z. Al-Sulaimani. 1994. Les chiroptères de Taqah (Oligocène inférieur, Sultanat d'Oman). Premier inventaire systématique. Münchner Geowissenschaftliche Abhandlungen Reihe A Geologie und Paläontologie 26: 35–48.

Simmons, N. B. 1993. The importance of methods: Archontan phylogeny and cladistic analysis of morphological data; pp. 1–61 *in* R.D.E. MacPhee (ed.), Primates and Their Relatives in Phylogenetic Perspective. Plenum, New York.

———.1994. The case for chiropteran monophyly. American Museum Novitates 3103: 1–54.

———. 1995. Bat relationships and the origin of flight; *in* P. A. Racey and S. M. Swift (eds.), Ecology, Evolution and Behavior of Bats. Symposium of the Zoological Society of London 67: 27–43.

———. 1998. A reappraisal of interfamilial relationships of bats; pp. 1–26 *in* T. H. Kunz and P. A. Racey (eds.), Bat Biology and Conservation. Smithsonian Institution Press, Washington, D.C.

———. 2001. Misleading results from the use of ambiguity coding to score polymorphisms in higher-level taxa. Systematic Biology 50: 613–620.

———. In press. Chiroptera; *in* D. E. Wilson and D. M. Reeder (eds.), Mammal Species of the World: A Taxonomic and Geographic Reference. Third Edition. Smithsonian Institution Press, Washington, D.C.

Simmons, N. B., and T. Conway. 2001. Phylogeny of mormoopid bats (Chiroptera: Mormoopidae) based on morphological data. Bulletin of the American Museum of Natural History 258: 1–97.

Simmons, N. B., and J. H. Geisler. 1998. Phylogenetic relationships of *Icaronycteris, Archaeonycteris, Hassianycteris,* and *Palaeochiropteryx* to extant bat lineages, with comments on the evolution of echolocation and foraging strategies in

Microchiroptera. Bulletin of the American Museum of Natural History 235: 1–182.

———. 2002. Sensitivity analysis of different methods of coding taxonomic polymorphism: An example from higher-level bat phylogeny. Cladistics 18: 571–584.

Simpson, G. G. 1945. The principles of classification and a classification of mammals. Bulletin of the American Museum of Natural History 85: 1–350.

Slaughter, B. H. 1970. Evolutionary trends of chiropteran dentition; pp. 51–83 in B. H. Slaughter and W. D. Walton (eds.), About Bats. Fondren Science Series. Volume 11. Southern Methodist University Press, Dallas.

Smith, J. D., and G. Madkour. 1980. Penial morphology and the question of chiropteran phylogeny; pp. 347–365 in D. E. Wilson and A. L. Gardner (eds.), Proceedings of the Fifth International Bat Research Conference. Texas Tech Press, Lubbock.

Speakman, J. R. 1993. The evolution of echolocation for predation. Symposium of the Zoological Society of London 65: 39–63.

Springer, M. S., E. C. Teeling, O. Madsen, M. J. Stanhope, and W. W. de Jong. 2001. Integrated fossil and molecular data reconstruct bat echolocation. Proceedings of the National Academy of Sciences 98: 6241–6246.

Storch, G., B. Sigé, and J. Habersetzer. 2002. Tachypteron franzeni n. gen., n. sp., earliest emballonurid bat from the Middle Eocene of Messel (Mammalia, Chiroptera). Paläontologische Zeitschrift 76: 189–199.

Szalay, F. S., and S. G. Lucas. 1993. Cranioskeletal morphology of archontans, and diagnoses of Chiroptera, Volitantia, and Archonta; pp. 187–226 in R.D.E. MacPhee (ed.), Primates and Their Relatives in Phylogenetic Perspective. Plenum, New York.

Teeling, E. C., O. Madsen, R. A. Van Den Bussche, W. W. de Jong, M. J. Stanhope, and M. S. Springer. 2002. Microbat paraphyly and the convergent evolution of a key innovation in Old World rhinolophid microbats. Proceedings of the National Academy of Sciences USA 99: 1431–1436.

Teeling, E. C., M. Scully, D. J. Kao, M. L. Romagnoli, M. S. Springer, and M. J. Stanhope. 2000. Molecular evidence regarding the origin of echolocation and flight in bats. Nature 403: 188–192.

Tong, Y. 1997. Middle Eocene small mammals from Liguanqiao Basin of Henan Province and Yuanqu Basin of Shanxi Province, Central China. Palaeontologica Sinica, Series C 18: 189–256.

Van Den Bussche, R. A., and S. R. Hoofer. 2000. Further evidence for inclusion of the New Zealand short-tailed bat (Mystacina tuberculata) within Noctilionoidea. Journal of Mammalogy 81: 865–874.

———. 2001. Evaluating monophly of Nataloidea (Chiroptera) with mitochondrial DNA sequences. Journal of Mammalogy 83: 320–327.

Van Den Bussche, R. A., S. R. Hoofer, and E. W. Hansen. 2002b. Characterization and phylogenetic utility of the mammalian protamine P1 gene. Molecular Phylogenetics and Evolution 22: 333–341.

Van Den Bussche, R. A., S. R. Hoofer, and N. B. Simmons. 2002a. Phylogenetic relationships of mormoopid bats using mitochondrial gene sequences and morphology. Journal of Mammalogy 83: 40–48.

Van Den Bussche, R. A., S. A. Reeder, E. W. Hansen, and S. R. Hoofer. 2003. Utility of the dentin matrix protein 1 (DMP1) gene for resolving mammalian intraordinal relationships. Molecular Phylogenetics and Evolution 26: 89–101.

Voss, R. S., and L. H. Emmons. 1996. Mammalian diversity in Neotropical lowland rainforests: A preliminary assessment. Bulletin of the American Museum of Natural History 230: 1–115.

Wible, J. R., and M. J. Novacek. 1988. Cranial evidence for the monophyletic origin of bats. American Museum Novitates 2911: 1–19.

JOHN J. FLYNN AND
GINA D. WESLEY-HUNT

12

CARNIVORA

SINCE THE ADVENT OF CLADISTIC methodologies, several approaches have been applied to generate higher-level phylogenies of the Carnivoramorpha (including the crown-clade, Carnivora) and determine the relationships of this clade to other mammals. Initial studies were morphologically based, including both living and fossil taxa to better constrain ancestral conditions for modern clades and help overcome homoplasies introduced when analyzing only living forms (e.g., Flynn and Galiano, 1982; Flynn et al., 1988). The parsimony-based morphological phylogeny of Wyss and Flynn (1993; Flynn, 1996) represents the most recent such synthetic effort, including hypotheses of interrelationships among major crown clades. Wyss and Flynn concluded that the early Cenozic "miacoids" lay outside crown-clade Carnivora and thus, represented more basal Carnivoramorpha. Alternatively, Wang and Tedford's (1994) analysis of canids concluded that at least some "miacids" were caniform members of crown-clade Carnivora (see also Flynn and Galiano, 1982; Flynn et al., 1988). Hunt and Tedford (1993) provided a general discussion of carnivoran evolution, suggesting that the two major branches of Carnivora (the Feliformia and Caniformia) had independent roots in Late Cretaceous *Cimolestes*-like ancestors. Flynn (1998) characterized the morphological attributes and temporal-geographic distributions of North American "miacoids," encompassing most of the diversity and many of the specimens of early carnivoramorphans—but these "miacoids" were assumed to be members of the two major carnivoran clades, Feliformia and Caniformia. Wesley-Hunt and Flynn (in press) provide extensive character analyses and a phylogeny of basal carnivoramorphans that does not presuppose their inclusion within crown-clade Carnivora. These morphologically focused studies have been complemented by the molecular, congruence, and total evidence analyses of living Carnivora by Flynn and Nedbal (1998), as well as numerous similar analyses of carnivoran subgroups or higher-level relationships of Mammalia mentioned later in the chapter. Only a few of the many important recent studies of carnivoran phylogeny and evolution can be cited in a short review, so interested readers are encouraged to consult the more extensive references cited in such papers as Wyss and Flynn (1993), Flynn and Nedbal (1998), and Bininda-Emonds et al. (1999).

In contrast to character-based morphological and/or molecular phylogenetic analyses, some studies of living taxa have advocated a "supertree" approach (e.g., matrix representation using parsimony analysis for Carnivora, Bininda-Emonds et al., 1999; for Mammalia, Liu et al., 2001), wherein the topologies of individual phylogenetic analyses are used as the "characters" and the resultant parsimony analyses generate a phylogeny reflecting the number of studies in which particular clades or phylogenetic relationships were indicated. The supertree approach is a kind of meta-analysis, which has the advantage of being able to incorporate into the phylogeny virtually all species within a clade (at least those species that have been included in one or more prior phylogenetic analyses). But it also has many disadvantages (see also Flynn et al., 2000), not least of which is that the immediately underlying data are not character attributes of the taxa, but rather, derivative information obtained from previously suggested hypotheses of relationships. Each of these hypotheses is of unknown quality, is based on unspecified character information and uncertain taxon sampling, and has been analyzed with unnamed methods. In addition, the data may be interdependent, because the character data or analyses may overlap or be duplicated in multiple studies.

As with many other clades, the relationship of Carnivora to other major lineages of eutherian mammals has remained enigmatic, presumably due to a rapid diversification of Eutheria. Such a radiation, with brief internodes, would generate concordant problems with small numbers of diagnostic characters accumulated along these short branches, masking the effects of subsequent modification of the few higher-level clade synapomorphies (molecular or morphological) and creating false linkages supported by homoplasious similarities. The extinct "Creodonta" are generally considered to be the closest relatives of Carnivora (together considered Ferae; Simpson, 1945), yet even this commonly accepted conclusion must be viewed with caution, as this group has long been a "wastebasket" taxon, with many previously included forms having no special relationship to either Carnivora or to other "creodonts"; modern conceptions of the group have been limited to only Hyaenodontidae and Oxyaenidae.

In this chapter, we summarize the most robustly supported phylogenetic relationships of major clades within Carnivora, remaining areas of controversy or taxa of ambiguous relationships, congruence or discordance between morphological and molecular phylogenetic analyses, and available evidence for the closest relatives of Carnivora among living and fossil Eutheria (Figs. 12.1, 12.2; Table 12.1). Table 12.1 presents a tentative, unranked classification, consistent with the hierarchical structure of the summary phylogeny of Fig. 12.2. Recent phylogenetic analyses have provided the framework for the examination of evolutionary rates (morphological, molecular, and taxic), anatomical modification, morphological diversification, and biogeographic and temporal distribution. We also discuss representative studies.

MORPHOLOGICAL CHARACTERIZATION OF CREODONTS AND CARNIVORAMORPHAN CLADES

Intuitively, it might seem simple to characterize a carnivoran—a meat-eater, with correlated dental, cranial, and postcranial/locomotor features. But carnivoran is a genealogical term, including a broad range of living forms (and their extinct carnivoran and carnivoramorphan relatives), some of which eat little or no meat. In fact, the range of diets, body masses, locomotor specializations, and distributions (geographic, habitat, and ecologic) in the more than 260 living species of carnivorans (the fourth most speciose mammalian "order" today) is truly remarkable, and is even greater when one assesses fossil forms as well. Matthew (1909), Savage (1977), and Flynn (1998) provide representative references and discussions of the anatomy and adaptations of early carnivoramorphans and Carnivora more generally.

Accurately inferring the morphological attributes of particular clades (both the ancestral condition for the clade and the range in known living and fossil forms) will depend on a number of factors, including discovery of new fossils, reinterpretation of homologies (e.g., via developmental studies), phylogenetic relationships, character polarity assessment, and optimization criteria for inferring distributions of synapomorphic characters. In some cases, the hierarchical level to which a particular feature pertains as a synapomorphy, or the pathways of character transformation in cladogenesis, remain ambiguous. Nevertheless, the distributions of a number of features and stability of recognition of various clades permit some generalizations about the morphology of members of those clades, or of their common ancestor, as well as about patterns of morphological transformation within lineages.

There have only been a small number of recent higher-level phylogenetic studies of Carnivoramorpha; few of these have included explicit hypotheses of potential morphological synapomorphies diagnosing major clades. Many features cited in earlier studies have new homology assessments and

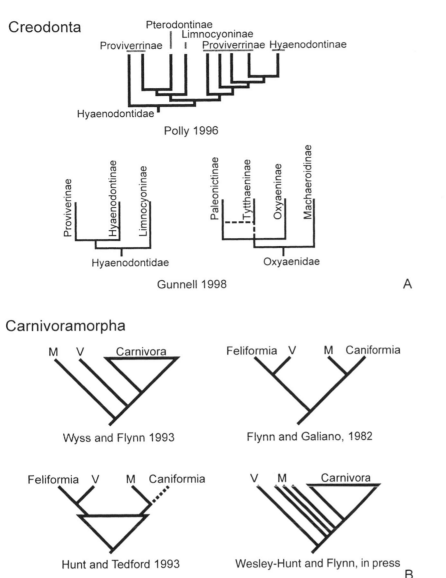

Fig. 12.1. Alternative hypotheses of phylogenetic relationships for Creodonta and early Cenozoic Carnivoramorpha. (A) Previously proposed relationships within Hyaenodontidae and Oxyaenidae (Creodonta). (B) Alternative relationships of early Cenozoic "Miacoidea" (Viverravidae and "Miacidae") relative to crown-group Carnivora.

pertain to different hierarchical levels than previously believed, due to reassessment of the morphology, or through inference of homoplasious occurrences in the parsimony analyses (e.g., Matthew, 1909; Flynn and Galiano, 1982; Flynn et al., 1988; Hunt and Tedford, 1993; Wyss and Flynn, 1993; Flynn, 1998). Table 12.2 and Fig. 12.3 include details of synapomorphies for the higher-level clades of Carnivoramorpha, Viverravidae, "miacids" plus Carnivora, Carnivora, Feliformia, and Caniformia. Creodonts have long been closely allied to carnivorans, and Wyss and Flynn (1993) provided potential synapomorphies for this grouping of Ferae. Gunnell (1998) cited some potential features shared by hyaenodontids and oxyaenids, the two major clades still recognized within Creodonta, but their monophyly remains contested by Polly (1996: 312), who noted:

If carnassialization is homologous in the two groups, then they should be characterized as primitively sharing carnassialization of the entire molar series (M1/m2 and M2/m3). Oxyaenids would be

united as a subgroup of Creodonta by the loss of the M2/m3 carnassial; hyaenodontids would be united by the primitive creodont characteristic of carnassialization of the molar series.

In addition to the carnivoramorphan synapomorphies listed in Table 12.2, we summarize here some anatomical features that historically have been considered synapomorphic or distinctive for these and other key clades within Ferae. We refer the reader to the references cited above for diagnostic features of other clades.

HYAENODONTIDAE. This clade is traditionally grouped with Oxyaenidae in Creodonta, although monophyly of Creodonta has been increasingly challenged. Some defining features of hyaenodontids are carnassialization of the molars (locus migrates during ontogeny, concentrated on M1/m2 and M2/m3 in adults), M3 generally present, m3 always present, skull narrow and long, basicranium narrow, tympanic bulla may be ossified or unossified, and frontals

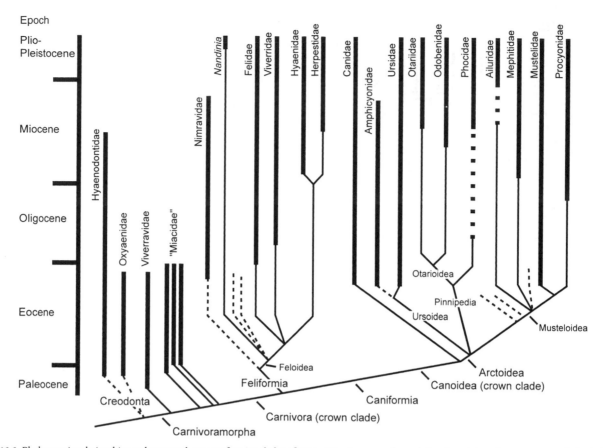

Fig. 12.2. Phylogenetic relationships and temporal ranges of major clades of Ferae (Carnivoramorpha and Creodonta), based on recent morphological and molecular studies. Dashed lines associated with named terminal taxa represent ambiguity in phylogenetic position for those lineages; for example, Amphicyonidae may be a caniform outgroup to Canoidea (Wesley-Hunt and Flynn, in press). Dashed lines in the Paleogene that are not associated with named taxa schematically represent basal fossil taxa with uncertain higher-level affinities. Broad dashed lines (Phocidae, Ailuridae) represent questionable range extensions into older temporal intervals. Creodonta (Hyaenodontidae, Oxyaenidae) may not be monophyletic, reflected in the tritomy of these taxa and Carnivoramorpha. "Miacidae" is shown as several distinct lineages, to reflect the sequentially nearer outgroup position to Carnivora of this paraphyletic array of species. Herpestidae represents both Herpestidae *sensu stricto* and their sister-taxon, a monophyletic clade of all native feloids in Madagascar, known only from the Recent. Mephitidae represents taxa traditionally placed within "Mephitinae" plus stink badgers.

concave between orbitals (see Gunnell, 1998, for further discussion).

OXYAENIDAE.   Some defining features of oxyaenids are carnassialization of the molars (locus migrates during ontogeny, concentrated on M1/m2 in adults), absence of M3 and m3, connate paracone and metacone on M1, M2 transverse and reduced or absent, braincase wide and low in skull, tympanic bulla unossified, nasals elongate (see Gunnell, 1998, for further discussion).

CARNIVORAMORPHA.   The classic diagnostic feature for this group is the P4/m1 carnassial pair in adults. Carnassial shear restricted to this tooth locus only, in adults, distinguishes this clade from all other Mammalia. Additional synapomorphies include the P4 protocone projecting far anteriorly (in front of the paracone), and a broad and far lingually projecting M1 parastyle (forming the posterior part of the embrasure for the shearing m1 trigonid).

VIVERRAVIDAE.   This monophyletic clade appears to be an outgroup to all other carnivoramorphans ("Miacidae" plus Carnivora). Both dental and cranial synapomorphies are present, including the lateral flange of the basioccipital nascent or small, subequal size of the M1 paracone and metacone, loss of m3, and posterior expansion of the m2 talonid and a very large and high m2 hypoconulid. Matthew's (1909; see also Flynn and Galiano, 1982; Flynn, 1998) original characterization of the group also noted their elongate skulls; digitigrady; humerus with flattened deltoid crest and high greater tuberosity; femur with small prominent third trochanter and long, narrow patellar trochlea; presence of a fibular facet on the calcaneum; and astragalus with narrow oblique head, moderately excavated trochlea, and well-developed keels.

"MIACIDAE".   As this is clearly is a paraphyletic assemblage of basal carnivoramorphan taxa, it cannot be diagnosed by synapomorphies. Species previously considered to

Table 12.1 Summary classification of Ferae (Carnivoramorpha and "Creodonta")

Ferae
  Creodonta (?)
    Oxyaenidae
    Hyaenodontidae
  Carnivoramorpha
    Viverravidae
    Unnamed clade ("Miacidae" and Carnivora)
      Various "Miacidae"
      Carnivora
        Feliformia
          Nimravidae
          Feloidea
            Nandinia
            Unnamed clade
              Felidae
              Unnamed clade
                Viverridae
                Unnamed clade
                  Hyaenidae
                  Unnamed clade
                    Herpestidae
                    Unnamed Malagasy feloid clade
        Caniformia
          Incertae sedis: Amphicyonidae[a]
          Canoidea
            Canidae
            Arctoidea
              Ursidae
              Pinnipedimorpha
                Enaliarctos
                Pinnipedia
                  Otarioidea
                    Otariidae
                    Odobenidae
                  Phocidae
              Musteloidea
              Ailuridae
              Mephitidae
              Unnamed clade
                Mustelidae[b]
                Procyonidae

Notes: The classification emphasizes the clades of living taxa and is arrayed dichotomously as an indented hierarchy to reflect the relationships shown in Fig. 12.2 (three taxa of equal identation indicate an unresolved polytomy). It does not detail various fossil stem taxa or all taxa of uncertain relationships, nor does it include all possible subclades within each grouping. The potential non-monophyly of Creodonta is noted by a (?); the "Miacidae" (various "Miacidae") are a paraphyletic array of taxa that are sequential outgroups between Viverravidae and crown-clade Carnivora. Clade names follow the phylogenetic definitions of Wyss and Flynn (1993), as emended by Bryant (1996a). Additional definitions provided in the text.

[a] Sister group to ursids within Canoidea, or possibly more basal carnivoramorphans; including *"Miacis" cognitus* and *Daphoenus*, of Wesley-Hunt and Flynn (in press).

[b] *Sensu stricto*, excluding mephitids and stink badgers.

be "miacids" appear to be arrayed between viverravids and Carnivora. Thus, the series of nodes on the phylogeny document sequential acquisition of diagnostic synapomorphies for nested clades and of features characteristic of later crown-clade Carnivora (Wesley-Hunt and Flynn, in press). Matthew (1909; see also Flynn and Galiano, 1982; Flynn, 1998) noted their relatively short skulls; plantigrady; humerus with high

ridged deltoid crest and small greater tuberosity; femur with low, broad third trochanter and wide, short patellar trochlea; and astragalus with wide, flat head and flat trochlea lacking an inner keel. As might be expected in this paraphyletic series of outgroups, many of these features are primitive with respect to crown-clade carnivorans. The "miacids" plus Carnivora clade shares a round infraorbital foramen, a rugose surface on the anteromedial promontorium (and rostral entotympanic likely present), and a deep fossa for the tensor tympani muscle in the middle-ear cavity.

CARNIVORA. Although restricted P4/m1 shear is usually considered distinctive of carnivorans, it is clear that this is a much more "ancient" feature present in all carnivoramorphans. Crown-clade carnivorans all possess a fused scaphoid and lunar (scapholunar) bone in the wrist, an expanded braincase, loss of M3, and a well-developed flange on the lateral edge of the basioccipital (for attachment of a caudal or combined rostral-caudal entotympanic).

FELIFORMIA. Included within this stem-based clade are nimravids, feloids, and various other fossil taxa, although viverravids appear to be excluded. All feliforms lack m3 (convergent with the earlier diverging viverravids and some later caniforms), lack or have reduced postglenoid foramen in the basicranium, and have the anterior palatine canal opening through the maxilla.

CANIFORMIA. Members of the stem-based clade of caniforms are quite distinctive in their possession of large and branching maxilloturbinals in the nasal passage, and of an internal carotid artery that is extrabullar, medially-situated, and runs inside a bony canal in the entotympanic. Nimravids are now known to lack complex maxilloturbinals (Joeckel et al., 2002), consistent with their recent exclusion from Caniformia and placement as early-diverging feliforms. If amphicyonids (represented in the analysis of Wesley-Hunt and Flynn, in press, by *Daphoenus* and *"Miacis" cognitus;* Wang and Tedford, 1994) are an outgroup to the remaining caniforms or crown-clade Canoidea, then the canoids share an elongate and stylized baculum and several distinctive basicranial features. The latter characters include an ossified entotympanic (also in the clade of all feloids, excluding their nearest outgroup *Nandinia*); an elongate anterior promontorium with a broad, flat apron extension; the epitympanic wing of the petrosal forming a bony ventral roof to the fossa for the tensor tympani muscle; the margin of the middle (medial) lacerate foramen defined anteriorly and positioned at or posterior to the basisphenoid/basioccipital suture.

PHYLOGENETIC TAXONOMY

Wyss and Flynn (1993) provided the first phylogenetic taxonomic definitions for the Carnivoramorpha (a name originally coined by Gureev, 1979, for a broad and peculiar mixture of eutherian taxa) and major subclades, including

Table 12.2   Diagnoses of major clades of Carnivoramorpha

Carnivoramorpha	M1 metastyle projection ≤ parastyle* [43(1)] M1 parastyle broad* [51(1)] **P4/m1 carnassial shear present*** [54(1)] **P4 protocone anterior to paracone** [82(1)]
Viverravidae	Flange on basioccipital later edge small or nascent [34(1)] M1 protocone ≅ paracone [42(1)] P4 parastyle cusp well-developed, defined [55(1)] Loss of m3 [88(1)]
Carnivoramorpha, excluding Viverravidae	**Infraorbital foramen round*** [3(1)] Mastoid process: blunt, rounded, does not protrude significantly [13(1)] **Surface of anteromedial promontorium rugose or rostral entotympanic present*** [30(1)] **Fossa for tensor tympani muscle defined and deep*** [39(1)] m$_2$ talonid not elongate, no enlarged hypoconulid [59(1)]
"Miacid" series	Numerous transformations, leading to various synapomorphies retained by crown-clade Carnivora
Carnivora	**Flange on basioccipital lateral edge well-developed (entotympanic attachment)*** [34(2)] **Loss of M3** [53(1)] Expanded braincase, fronto-parietal suture anteriorly located* [66(1)] **Scaphoid and lunate fused** [92(1)]
Caniformia	**Internal carotid artery medial, extrabullar, inside bony canal formed by entotympanic** [25(2)] Shelf between mastoid and paroccipital process laterally wide, surface rugose or bulbous [33(1)] **Maxilloturbinals large and branching*** [62(1)]
Canoidea	Anterior promontorium elongate, apron is broad flat extension* [28(3)] **Epitympanic wing of the petrosal forms ventral floor to fossa for the tensor tympani muscle*** [38(1)] Middle lacerate foramen margin defined anteriorly; positioned at or posterior to basisphenoid/basioccipital suture [40(1)] **Entotympanic fully ossified** (also in all post-*Nandinia* feloids) [68(1)] **Baculum long, stylized*** [89(1)]
Feliformia	**Anterior palatine canal opening through maxilla** [6(2)] **Postglenoid foramen greatly reduced or absent** [12(1)] Anterior promontorium blunt* [28(2)] Loss of M1 protocone lingual cingulum [−41(0)] Loss of m3 [88(1)]

Notes: Selected distinctive features are in bold; bracketed numbers are diagnostic synapomorphies from Wesley-Hunt and Flynn (in press), in both ACCTRAN and DELTRAN optimizations; asterisk indicates a character unique to the clade in the present analysis; minus sign indicates character reversal.

Carnivora, Feliformia, Caniformia, Feloidea, Arctoidea, Ursoidea, Pinnipedia, and Phocoidea. Several less inclusive, or less robustly supported, clades were not named at the time. Wolsan (1993) subsequently provided phylogenetic definitions for many of the same clades, although most were not based on explicitly defined clades (Bryant, 1996a). Bryant (1996a) noted important philosophical and methodological problems in the Wyss and Flynn (1993) and Wolsan (1996) studies, and provided amended node- or stem-based phylogenetic definitions for the names of various carnivoramorphan clades. Except where noted, we follow Bryant's (1996a: table 2; in quotes below) emendations of Wyss and Flynn's (1993) phylogenetic definitions for the names of key clades of Carnivora (see Table 12.1). We also indicate whether these are stem-, node-, or apomorphy-based (in parentheses after each name). Note that Carnivoramorpha, Feliformia, and Caniformia represent stem-based counterparts to the node-based crown clades of Carnivora, Feloidea, and Canoidea, respectively. In addition, we provide brief morphological characterizations of several of these carnivoran clades, as well as the most common creodont and "miacoid" groups.

Some of the clades recognized by Wyss and Flynn (1993), and other novel associations recognized more recently, remain unnamed here, as they will be formally named in manuscripts in preparation or when better support for their relationships is supplied (Table 12.1; e.g., various feliform clades recognized by Flynn and Nedbal, 1998; Yoder et al., 2003; Yoder and Flynn, 2003; Flynn et al., in prep.). In contrast, some clades receiving additional evidence of monophyly, or others that have been newly recognized (and strongly supported) in later phylogenetic analyses, herein receive formal phylogenetic definitions for the clade names, noted below as "new." As unranked taxonomies are becoming more widely accepted, clade name endings do not necessarily reflect rank or equivalent hierarchical categories, as specified in the International Code of Zoological Nomenclature (1999), but have been selected to most closely reflect existing usage or inclusiveness relative to previously named taxa (e.g., Pinnipedimorpha, applied as a stem-based name for the more inclusive clade including the previously named Pinnipedia clade).

The name Ursida (Wyss and Flynn, 1993; Bryant, 1996a) was applied to the clade including the last common ancestor of ursoids (or species of Ursidae) and pinnipeds, and all of its descendants. That name, however, may be redundant with Arctoidea (which would have priority), depending on how the currently ambiguous relationships among various arctoid subgroups are resolved in the future. The name Musteloidea has been widely, yet variably, applied in the past, but has not yet been defined phylogenetically. As prior usages have included some or all taxa traditionally considered to be Mustelidae (Wilson and Reeder, 1993), we chose to apply this name broadly, to a clade encompassing all those taxa. The more restrictive clade of Mustelidae (*sensu stricto*, excluding skunks and stink badgers) and Procyonidae, or the "Musteloidea *sensu stricto*" of Flynn and Nedbal (1998) and Flynn et al. (2000) currently remains without a name applied with a phylogenetic definition. Mustelidae, as defined by Bryant (1996a), is problematic, because most recent studies clearly document polyphyly of traditional "Mustelidae" and only a distant relationship between mephitids (skunks) and other Mustelidae *sensu stricto* (see below).

In this chapter, we use the following clade definitions:

Carnivoramorpha (stem-based): "Carnivora and all members of Mammalia (Rowe, 1988) that are more closely related to Carnivora than to taxa referred to Creodonta by Carroll (1988)" (Bryant, 1996a: table 2).

Carnivora (node-based): "most recent common ancestor of Feloidea, all species referred to Canidae by Wilson and Reeder (1993), and Arctoidea and all of its descendants" (Bryant, 1996a: table 2).

Feliformia (stem-based): "Feloidea and all members of Carnivora that are more closely related to Feloidea than to Canoidea" (Bryant, 1996a: table 2).

Feloidea (node-based): "most recent common ancestor of the species referred to Felidae, Viverridae, Herpestidae, and Hyaenidae by Wilson and Reeder (1993) and all of its descendants" (Bryant, 1996a: table 2).

Caniformia (stem-based): "Canoidea and all members of Carnivora that are more closely related to Canoidea than to Feloidea" (Bryant, 1996a: table 2).

Canoidea (node-based): "most recent common ancestor of Arctoidea and the species referred to Canidae by Wilson and Reeder (1993) and all of its descendants" (Bryant, 1996a: table 2).

Arctoidea (node-based): "most recent common ancestor of the species referred to Procyonidae, Mustelidae, and Ursidae by Wilson and Reeder (1993), of *Ailurus*, and of Pinnipedia and all of its descendants" (Bryant, 1996a: table 2).

Pinnipedimorpha (new; stem-based): Pinnipedia and all members of Arctoidea that are more closely related to Pinnipedia than to any species of Ursidae (of Wilson and Reeder, 1993) or Musteloidea.

Pinnipedia (node-based): "most recent common ancestor of all species referred to Otariidae, Phocidae, and Odobenidae by Wilson and Reeder (1993) and all of its descendants" (Bryant, 1996a: table 2).

Otarioidea (new, node-based): Most recent common ancestor of *Odobenus* and *Otaria* and all other species referred to Otariidae by Wilson and Reeder (1993) and all of its descendants. This new taxon also represents the traditional conception of a closer relationship of walruses to otariids than to phocids. If the alternative of closer phylogenetic relationship of *Odobenus* to Phocidae, initially suggested by Wyss (1987), is correct, then this name would be equivalent to the Pinnipedia, which would have priority.

Mustelida (node-based): "most recent common ancestor of all species referred to Procyonidae and Mustelidae by Wilson and Reeder (1993) and of Pinnipedia and all of its descendants" (Bryant, 1996a: table 2). Most recent phylogenies have been unable to definitively resolve the tritomy among Pinnipedia, Ursidae, and Musteloidea; thus, we do not use this name in the figures or the tentative classification of Table 12.1. Bryant's (1996a) definition reflects a phylogeny in which pinnipeds are a nearer outgroup to Musteloidea than are ursids among living taxa. If ursids are more closely related to Musteloidea, then this node-based name would be redundant with Arctoidea.

Musteloidea (new, node-based): Most recent common ancestor of *Ailurus, Mustela, Mephitis, Procyon,* and all other species referred to Procyonidae by Wilson and Reeder (1993) and all of its descendants. A slightly expanded usage of the name often was applied to a grouping of procyonids, mustelids (*sensu stricto*), and mephitids, but previously excluding *Ailurus*. This definition is equivalent to "Musteloidea *sensu lato*" of Flynn and Nedbal (1998) and Flynn et al. (2000). There is strong support for monophyly of this clade in recent phylogenies, although ambiguity remains as to resolution of the basal tritomy of mephitids-*Ailurus*-procyonids/mustelids. Musteloidea thus includes all the taxa previously considered to be mustelids (mephitines, mustelines, lutrines, melines, mellivorines, and gulonines; Bryant et al., 1993), although mephitids no longer appear to be the closest relatives of other Mustelidae (*sensu stricto;* Wayne et al., 1989; Dragoo and Honeycutt, 1997; Flynn et al., 2000). Thus, Mustelidae as defined by Bryant (1996a) certainly would be more inclusive than originally intended (to apply to a clade including only those species traditionally considered mustelids), as it likely also includes procyonids, and possibly would even apply to this entire broader musteloid clade, if the tritomy is ultimately resolved as (mephitids (*Ailurus* (procyonids, mustelids *sensu stricto*))).

PHYLOGENY

Comparing multiple lines of phylogenetic evidence (e.g., nucleotide sequences, morphology, fossil record) generally yields congruent phylogenetic results, and combining them can greatly increase robustness and resolution. Recent

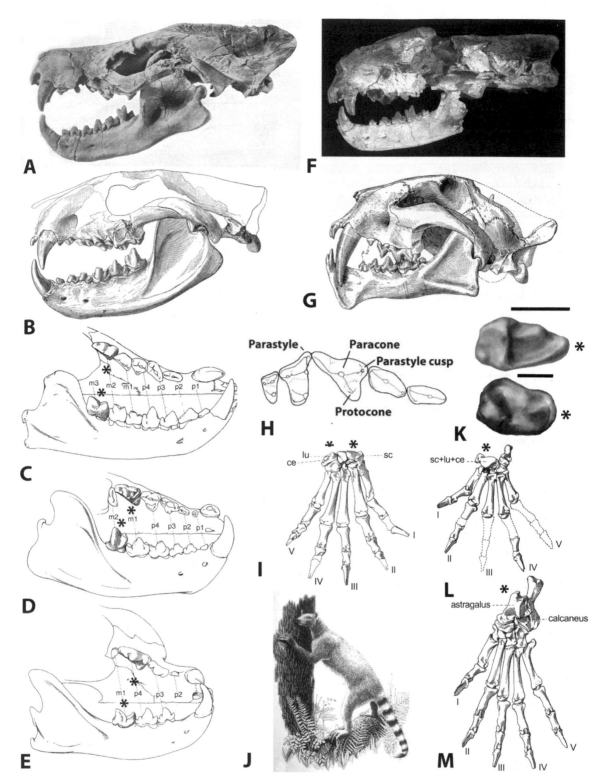

Fig. 12.3. Representative specimens and important morphological characteristics of Ferae (Carnivoramorpha and Creodonta). (A) Skull and jaw of Hyaenodontidae (*Limnocyon*). (B) Skull and jaw of Oxyaenidae (*Oxyaena*). (C–E) Upper and lower dentitions of Ferae, with asterisks marking the primary carnassial pair in adults. (C) Hyaenodontidae, M2/m3. (D) Oxyaenidae, M1/m2. (E) Carnivoramorpha, P4/m1. (F) Skull and jaw of early Carnivoramorpha ("miacid," *Tapocyon*). (G) skull and jaw of early Feliformia (Nimravidae, *Dinictis*). (H) *Viverravus* upper tooth row (anterior to right), illustrating derived characters of Carnivoramorpha (M1 metastyle projection less than or equal to the parastyle, P4 protocone anterior to the paracone [relative to long axis of tooth], P4 carnassial with extensive metastyle shearing blade) and Viverravidae (P4 parastyle cusp well developed, M1 protocone approximately equal to the paracone). (I) Manus of *Tritemnodon* (Hyaenodontidae), showing fissured claws, and the primitive condition of separate scaphoid and lunate bones (two asterisks) in the creodont. (J) Life reconstruction of middle Eocene *Vulpavus ovatus* standing on and grasping sycamore trunks, illustrating scansorial-arboreal features evident in the skeleton of this early carnivoramorphan. (K) The m2 of *Viverravus* (above, Viverravidae) and *Tapocyon* (below, "miacid," early Carnivoramorpha), with talonids drawn to similar scale, illustrating the elongate talonid (bar) and large, high hypoconulid (asterisk) of Viverravidae (m2 is the last molar in this clade, with loss of m3 a synapomorphy), and short talonid and weak or absent hypoconulid of "miacids" and Carnivora (which retain m3, although this and other molars are lost in various later clades). (L) Manus of *Vulpavus* ("miacid," early Carnivoramorpha), showing a single fused scapholunar bone (asterisk) characteristic of crown-clade Carnivora and possibly some earlier carnivoramorphans. (M) Pes of *Vulpavus*. "Miacids" retain the primitive condition of plantigrady and the astragalus (asterisk) has a wide, flat head and a flat trochlea lacking an inner keel (Matthew, 1909); Ferae (Carnivoramorpha and creodonts) share a uniquely derived condition of enlarged medial border of the astragalar tibial trochlea (Flynn et al., 1988; Wyss and Flynn, 1993). (N) Basicranium of *Protictis schaffi* (Viverravidae). There is a distinct Glasserian fissure (GF) forming a deep cleft for passage of the chorda tympani, and a strong sigmoid curvature to the ventral occipital condyle (possible synapomorphy of Carnivoramorpha, creodonts,

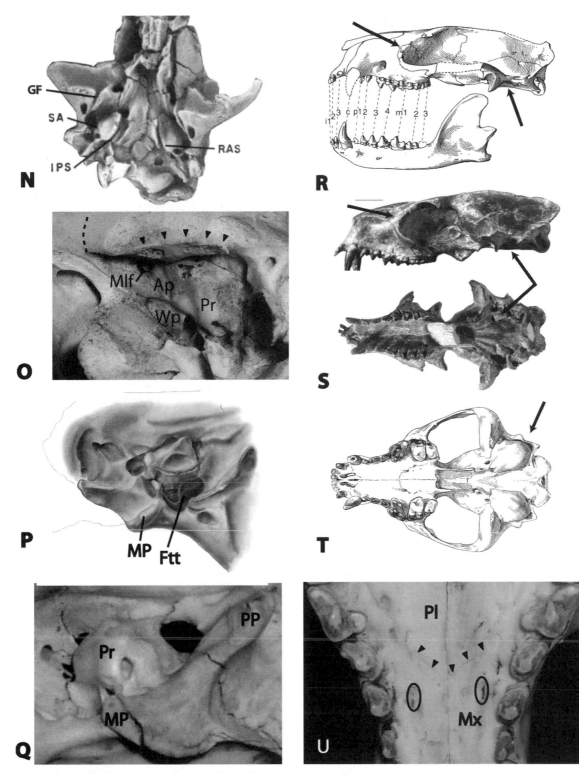

and lipotyphlans; Wyss and Flynn, 1993); the inferior petrosal sinus (IPS), carrying jugular venous drainage (rather than a third branch of the internal carotid artery, as assumed by Matthew, 1909), is indistinct and does not substantially excavate the surrounding petrosal and basioccipital (primitive for Carnivoramorph); the presence of a groove for the stapedial artery (SA) is also primitive for Carnivoramorpha (this branch of the internal carotid artery is absent in Carnivora and likely some more basal Carnivoramorphans). (O) *Canis* basicranium (canid, Caniformia; anterior to left), illustrating derived features of Canoidea: epitympanic wing of the petrosal forming a floor to the fossa for the tensor tympani (Wp), the anterior promontorium apron (Ap) is a broad flat extension (Pr, promontorium), flange on basioccipital lateral edge is well developed (arrows), the middle lacerate foramen (Mlf) has a defined anterior margin and lies at or posterior to the basioccipital/basisphenoid suture (indicated by a dashed line). (P) *Oödectes* basicranium (anterior to right), showing derived features of the clade of all Carnivoramorpha, excluding Viverravidae: fossa for tensor tympani (Ftt) muscle defined and deep, and mastoid process (MP) blunt, rounded, and does not protrude significantly. (Q) *Nandinia* basicranium (anterior to left), illustrating the blunt, truncated anterior edge of the promontorium, a synapomorphy of Feliformia (Pr; promontorium, PP; paroccipital process, MP; mastoid process). (R, S) Skull and jaws of *Vulpavus* (R) and *Oödectes* (S). Arrows indicate exposure of lacrimal bone on facial region (left arrow) and bony bulla not fused to skull (right arrow), primitive features retained in early carnivoramorphans (although the presence of an entotympanic likely was synapomorphic for most or all Carnivoramorpha). (T) Palate/basicranium of *Phlaocyon* (borophagine canid, Caniformia), showing a fully ossified composite ctotympanic/entotympanic bulla fused to the basicranium and covering the entire middle ear cavity (independently derived in Canoidea and post-*Nandinia* Feloidea). (U) *Nandinia* (Feloidea) palate, illustrating Feliformia synapomorphy of anterior palatine canal opening (circles) through the maxilla; arrows indicate the suture between maxilla (Mx) and palatine (Pl). Illustrations are modified from Matthew (1909: A, I, L, M, R), Wortman (1899: B), Osborn (1907: C–E, G, T), and Gingerich and Winkler (1985: N). The remaining panels were prepared by the authors or by Marlene Donnelly (Field Museum, Chicago).

molecular or combined analyses of living Carnivora and outgroups have revealed exceptionally strong support for many carnivoran crown clades identified in previous morphological phylogenies. But the interrelationships of most modern eutherian "orders" (including which are the nearest outgroups to Carnivora) remain more enigmatic, as do those of some fossil taxa, and the identity of the closest relatives of Carnivora among both fossil and living clades remains controversial. In contrast to the general congruence (or at least the lack of significant discordance) within integrative phylogenetic analyses, the inverse problem of using phylogenetic, fossil occurrence, and character (especially nucleotide divergences) data to calibrate divergence ages often yields contradictory results between fossil-based and molecular clock methods. In addition, there is variation between "strict" (e.g., single calibration, linearized tree, or quartet methods; Kimura, 1980; Takezaki et al., 1995; Rambaut and Bromham, 1998) and "relaxed" clock methods and divergence-age-estimating techniques (e.g., Bayesian methods; Thorne et al., 1998). We note here both robustly supported phylogenetic hypotheses and remaining areas of ambiguity. Identification of these areas should prove useful in providing constraints on the "reliability" of derivative evolutionary inferences (e.g., identifying groups for which phylogeny-dependent assessments of adaptation or transformation, such as independent contrasts, should be more or less robust) and targets for more intensive future investigations. Within this review of higher-level phylogeny of the Carnivora, we are not able to discuss the many recent advances in understanding lower-level interrelationships (e.g., within traditional families or subfamilies).

Carnivora within Eutheria

Among higher-level clades (orders) with living representatives, lipotyphlan "insectivores" most frequently have been cited as the closest relatives of Carnivora and creodonts (morphological: Wyss and Flynn, 1993; implicit in MacPhee and Novacek, 1993; molecular: Faith and Cranston, 1991, protein amino-acid sequences, with Permutation Tail Probability (PTP) test). Some or all "ungulates," especially various artiodactyls, commonly appear as close relatives of Carnivora. Together, these linkages are reminiscent of Simpson's (1945) clustering of "Ferungulata." This association may be fortuitous, however, as the taxa included differ and the justification for their relationships are actually based on widely different lines of evidence. Simpson's (1945) initial clustering assumed that various "creodonts" possessed primitive features that were intermediate between those in carnivorans and ungulates, and thus likely represented a common ancestral stock. This assumption is now known to be poorly substantiated or incorrect, and although there is some molecular support for an ungulate-Ferae clade, few, if any, morphological synapomorphies have been proposed (see also "Creodonta" below).

There have been widely discordant results over the past two decades, however, and many recent molecular phylogenies arrive at very different conclusions. Recent alternative phylogenies range from poorly resolved polytomies (e.g., morphology: Simpson, 1945; Novacek, 1992a; strict consensus, amino-acid sequences for 10 polypeptides: Honeycutt and Adkins, 1993; amino-acid sequences, with vespertilionid bats and pholidotans violating carnivoran monophyly by linking to particular carnivoran species: Czelusniak et al., 1990) to strongly conflicting topologies linking Carnivora to a wide array of other eutherians. Among the various eutherian clades proposed as the nearest relatives of Carnivora are:

1. Pholidota (e.g., Miyamoto and Goodman, 1986; Shoshani, 1986; Delsuc et al., 2001, 2002; Liu et al., 2001; Murphy et al., 2001). "Ungulates" are the nearest outgroup to this clade in both Liu et al. (2001) and Murphy et al. (2001);
2. Various "ungulates" (Árnason and Johnsson, 1992; Honeycutt and Adkins, 1993; Krettek et al., 1995; non-monophyly of Cetartiodactyla, Eulipotyphla, and Chiroptera: Jow et al., 2002);
3. Primates (Shoshani, 1986);
4. Chiroptera (Stanhope et al., 1992);
5. Tubulidentata/Insectivora (Novacek, 1992b); and
6. Perissodactyla (having a polytomous relationship to four other major clades: Stanhope et al., 1998; Van Dijk et al., 1998).

The most comprehensive recent molecular datasets most consistently appear to support a Pholidota-Carnivora clade, which do share (with creodonts) at least one potential morphological synapomorphy (osseous tentorium, Wyss and Flynn, 1993). It is often difficult to rigorously compare or assess congruence among various studies, however. Problems contributing to this include incomplete or variable taxon sampling of the major eutherian clades, different underlying data, and different analytical methods. Although some phylogenetic analyses are synthetic, integrating and simultaneously analyzing qualitatively different primary data, most focus on a single class of data. Some end members of the variability include molecular versus morphological, phenetic (immunodiffusion or DNA hybridization) versus character-based (nucleotide or protein sequences), nuclear versus mitochondrial or coding versus non-coding genes, or partially or entirely different suites of genes sequenced. Just as importantly, phylogenetic algorithms with radically different assumptions and methods are often applied. Although parsimony is most widespread, especially in analyses of morphological data, commonly applied alternatives range across phenetic clustering (e.g., neighbor joining), varying likelihood models, Bayesian posterior-probability methods, and supertrees. Resolution of the precise interrelationships between Carnivora and other eutherians is thus likely to remain in flux for at least some time.

Creodonta, and Relationships to Carnivora

Creodonta was first proposed by Cope in 1875. Since its inception, the composition of Creodonta has undergone numerous revisions. Within five years, Cope (1880) had already significantly expanded it to contain the Miacidae, Arctocyonidae, Leptictidae, Oxyaenidae, Ambloctonidae, and Mesonychidae. Hyaenodontidae, one of only two families now remaining in Creodonta, was not added until Matthew's (1909) landmark monograph. Over the subsequent decades, various groups were removed from the Creodonta, but it was not until 1969, when Van Valen removed the mesonychids and arctocyonids from the creodonts, that Creodonta appeared as it does today, approximating Matthew's (1909) Pseudocreodi, consisting only of the Hyaenodontidae and the Oxyaenidae.

Although there is a long history of studies (e.g., Matthew, 1909; Simpson, 1945) citing the extinct creodonts as the closest relatives of Carnivora, detailed studies of the group suggest that Creodonta may not be monophyletic, even in its more modern narrowly conceived configurations (regardless of whether one or both major clades of creodonts are closely related to Carnivora). Even the monophyly of each of the two most widely accepted creodont subgroups, Hyaenodontidae and Oxyaenidae, has been seriously questioned (see Fig. 12.1; Polly, 1996; Gunnell, 1998). Still, morphological analyses generally indicate similarities—and possible synapomorphies—shared between various creodont clades and carnivorans, suggesting that some or all of them may be closest relatives among living and fossil Eutheria (e.g., Wyss and Flynn, 1993).

The most recent phylogenetic studies of creodonts are those by Polly (1996) and Gunnell (1998; see Fig. 12.1). In Polly's (1996) revision of the Hyaenodontidae, the Proviverrinae were still considered a paraphyletic stem group within Hyaenodontidae, with various taxa previously included within the Proviverrinae giving rise to (or being closest relatives of) the monophyletic clades Hyaenodontinae, Limnocyoninae, Machaeroidinae, and the newly proposed Pterodontinae. Gunnell (1998) assessed the relationships of taxa within both Hyaenodontidae and Oxyaenidae, although he reviewed only North American forms. In that study, Gunnell placed the enigmatic Machaeroidinae tentatively within Oxyaenidae. Although Gunnell noted some "defining" features of creodonts, neither Polly (1996) nor Gunnell (1998) advocated or provided compelling evidence supporting monophyly of the Creodonta. Thus, we conservatively treat the two major clades (Hyaenodontidae, Oxyaenidae) as separate basal taxa in an unresolved polytomy with respect to Carnivoramorpha.

Carnivoramorpha Versus Crown-Clade Carnivora, and the "Miacoid" Problem

Matthew (1909) provided the first explicit discussion of the interrelationships of early Cenozoic "miacoids" and living groups of Carnivora (termed "Fissipeda," as his concept included only the terrestrial Carnivora, to the exclusion of the marine pinniped carnivorans; see also Simpson, 1945). Since then, debate has centered on a series of questions, including whether (1) "miacoids" are monophyletic or paraphyletic, (2) the two traditional subgroups of "miacoids"

(Viverravidae and Miacidae) are each monophyletic, and (3) the various "miacoids" are members of the crown-clade Carnivora or stem-group carnivoramorphans (see Fig. 12.1). In early cladistic studies, the "miacid" and viverravid subgroups (or at least some individual taxa within them, if they are paraphyletic) were considered basal members of the Caniformia and Feliformia, and thus were defined as crown-clade carnivorans (e.g., Flynn and Galiano, 1982; Flynn et al., 1988; Hunt and Tedford, 1993; Wang and Tedford, 1994). More recent parsimony analyses alternatively place them as carnivoramorphan stem-groups to the Carnivora, with either viverravids (Wyss and Flynn, 1993) or "miacids" (Wesley-Hunt and Flynn, in press; see also discussion in Bryant, 1991) as the nearest outgroups to crown-clade Carnivora.

"Miacoids" first appear in the Puercan of North America, at least 63–64.5 million years ago (Flynn, 1998). These early (Paleocene) forms all are either viverravids (e.g., *Protictis, Didymictis, Bryanictis, Intyrictis, Raphictis, Simpsonictis*) or incertae sedis (e.g., *Ictidopappus, Pristinictis, Ravenictis*). Fox and Youzwyshyn (1994) alternatively considered *Pristinictis* a viverravid, and *Ravenictis* to be of uncertain affinities but most closely resembling such viverravids as *Pristinictis*. "Miacids" do not appear until much later (Clarkforkian, latest Paleocene), whereas species that definitively are members of crown-clade Carnivora first appear in the middle Eocene. Thus, although the Carnivoramorpha (and possibly the Carnivora, depending on the phylogenetic position of these early "miacoids") have one of the earliest fossil records for a modern placental order or lineage, the time at which crown-clade Carnivora originated has remained uncertain. This origination time, the minimum divergence age of Caniformia/Feliformia, has been considered to lie somewhere before either about 65 or about 45–50 million years ago (permissive and conservative estimates of Flynn, 1996, respectively).

Hunt and Tedford (1993) concluded that Carnivora might be diphyletic, with the Miacidae (and caniform Carnivora) and Viverravidae (and feliform Carnivora) evolved from separate lineages within Cretaceous and possibly Paleocene species of *Cimolestes*, although no specific "ancestors" or closest relatives were definitively proposed. Fox and Youzwyshyn (1994), in contrast, assigned all the early "miacoids" to Carnivora, alternatively invoking origins of this monophyletic group from an unspecified ancestral stock rather than *Cimolestes*. In particular, Hunt and Tedford (1993) argued that the viverravids are aeluroid (feloid) ancestors, based on dental evidence, extending the Feloidea (Aeluroidea in that paper) lineage into the Paleocene. Although Hunt and Tedford never explicitly stated a hypothesis of the relationships of miacids, it is implicit that they are ancestors of the Caniformia, stemming from a different *Cimolestes* ancestor. Hunt and Tedford's (1993) diphyly concept actually differs less significantly from earlier monophyly hypotheses (e.g., Flynn and Galiano, 1982) than it might at first appear, as both retain monophyly of living crown-clade Carnivora with respect to any other living Mammalia, and both placed "miacoids" with the Carnivora as stem taxa either to living clades of Caniformia or Feliformia

(Aeluroidea of Hunt and Tedford, 1993). It differs substantively only in drawing the root of the Carnivora back in time, by implicitly including various older *Cimolestes* taxa within the base of the crown clade. This topology, however, also would significantly modify reconstructions of ancestral morphologies for the Carnivora (or even the Caniformia and Feliformia clades), including requiring that long-accepted synapomorphies of Carnivora or Carnivoramorpha (e.g., see the older non-cladistic hypotheses of diagnostic characters, as in Matthew, 1909; and the cladistic phylogenies of Flynn and Galiano, 1982, Flynn et al., 1988; Wyss and Flynn, 1993; Table 12.2) be considered convergences instead. For example, the restriction of adult carnassials to P4/m1 or the reduction of posterior molars could not be synapomorphies of Carnivora, but rather would have been independently derived somewhere within the base of the Feliformia and the Caniformia, as the hypothesized *Cimolestes* "ancestors" of each of these carnivoran subclades lack these features.

The most recent fossil-focused studies, including descriptions of many new and well-preserved specimens and more comprehensive anatomical (especially basicranial) and phylogenetic analyses, indicate that there is an even more complex pattern of relationships of early Cenozoic carnivoramorphans ("miacoids") to the living clades than had been proposed in any previous phylogenies. Most prominently, viverravids (Didymictida plus Viverravidae of Flynn and Galiano, 1982), appear to be monophyletic, but "miacoids" are not (Wesley-Hunt and Flynn, in press). Furthermore, all "miacoid" taxa appear to be basal members of the Carnivoramorpha ("conservative," Miacoid 1 pattern of Flynn, 1996), and thus are outgroups to the crown-clade Carnivora rather than early representatives of the Feliformia or Caniformia ("permissive," Miacoid 2 pattern of Flynn, 1996), although "missing data" leave open the possibility that some may later be determined to belong within the crown clade. The current interrelationships of Carnivoramorpha suggest that a monophyletic Viverravidae is nearest outgroup to all remaining taxa, with various taxa of the paraphyletic "Miacidae" arrayed as sequential outgroups to crown-clade Carnivora (see Fig. 12.2). This result yields a middle Eocene (about 45–50 Ma) minimum estimate for the origin of the Carnivora and divergence of its two major subgroups (Feliformia/Caniformia), based on the earliest occurrence of a fossil that clearly is a member of the crown clade, thus favoring the "conservative" rather than "permissive" divergence age estimates of Flynn (1996).

Higher-Level Phylogeny of the Carnivora

Carnivoran crown clades identified in early morphological phylogenies and strongly supported in all or most subsequent studies include Carnivora, Feliformia (Feloidea), Caniformia, Arctoidea, Musteloidea, and all traditional families (except Viverridae and Mustelidae; see Fig. 12.2). Robust support for these carnivoran subclades emerges in analyses as varied as molecular analyses and combined character analyses (details and references in Flynn and Nedbal, 1998;

Flynn et al., 2000; see also Vrana et al., 1994), phenetic similarity linkages (DNA hybridization, Wayne et al., 1989), and phylogeny-based supertree approaches (Bininda-Emonds et al., 1999). Several longstanding controversies or areas of ambiguity appear to be better resolved and some unexpected results have been produced in recent morphological, molecular, and/or integrative analyses. Examples include:

1. Documenting the monophyly of Pinnipedia, although the relationship of this clade to other arctoids remains uncertain;
2. Supporting an Otariidae-Odobenidae pairing (Otarioidea) within Pinnipedia;
3. Determining that skunks (mephitines, traditionally) are not closely related to other mustelids;
4. Placing the red panda (*Ailurus*) as a basal musteloid (in a currently unresolved tritomy with mephitines and a procyonid-mustelid clade);
5. Documenting monophyly of the Madagascar carnivorans; and
6. Ascertaining that *Nandinia* is not a viverrid, but rather is an outgroup to all other living Feloidea.

Monophyly of the living catlike carnivorans has long been robustly supported by morphological evidence (especially basicranial, with some dental and soft-anatomy synapomorphies) and more recently by molecular data. Various fossil taxa have been assigned as stem groups to the living clade, and others as early members of the crown clade or familial lineages within it. There is some debate over terminology for the crown clade, with Feliformia, Aeluroidea, and Feloidea representing the most frequently cited naming options. As fossils (many of them likely to be stem taxa) played key roles in Kretzoi's (1943) original conceptualization of Feliformia and Caniformia, and the phylogenetic position of many early fossil taxa remains uncertain, application of a stem-based definition of Feliformia seems most appropriate (Wyss and Flynn, 1993; Bryant, 1996a). The more commonly recognized name "Feloidea" (not "Aeluroidea," following Simpson, 1945) has been applied to the crown clade (Wyss and Flynn, 1993; Bryant, 1996a). The placement of various fossils within or basal to this clade thus has no effect on its conceptualization, although they will be crucial for calibrating or constraining divergence-age estimates for both the caniform-feliform split and for major feloid subclades.

Although monophyly of Feliformia and Feloidea is well supported by all available data, the biogeographic and temporal distributions are more enigmatic (Fig. 12.4). Hunt and Tedford (1993) and Hunt (1996b, 1998, 2001) analyzed these distributions, suggesting an Old World focus for diversification of the crown-clade Feloidea (Aeluroidea in those papers). But they also emphasized the ambiguity in the biogeographic origin for the clade and a temporal gap between occurrences of putative ancestors and the first definitive crown-clade feloids. Well-preserved cranial specimens of viverravid "miacoids," considered members of the feliform

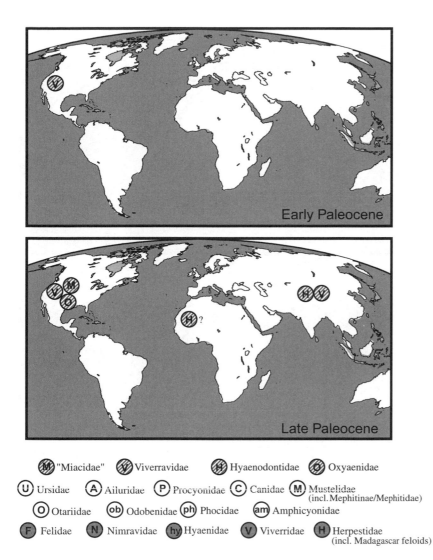

Fig. 12.4. Geographic and temporal distribution of traditional carnivoramorphan and creodont families through the Cenozoic. Question mark (?) indicates a questionable temporal-geographic occurrence of the taxon. The time slices are early Paleocene, late Paleocene, early Eocene, mid-late Eocene, Oligocene, Miocene, and Pliocene-Pleistocene. (*Continued on following pages.*)

lineage or closer relatives of feloids than are any other carnivoramorphans, last occur in the middle Eocene of North America. In contrast, forms that certainly represent crown-clade Feloidea, or are definitively assignable to modern feloid families, are early Oligocene Old World taxa, such as *Anictis, Paleoprionodon,* and other forms from Quercy in Europe and similarly aged taxa from Hsanda Gol in Asia. If one accepts viverravids as basal feliforms, within crown-clade Carnivora, then the fossil record would clearly suggest a North American diversification for the group. This early diversification would have been followed by Holarctic distributions of viverravid feliforms, then a rapid late Eocene or early Oligocene radiation of crown-clade Feloidea, potentially centered in the Old World. Only felids (excepting a single Plio-Pleistocene hyaenid lineage) successfully back-dispersed to the New World (first appearances in the middle Miocene of North America, and about 2 Ma—latest Pliocene—in South America), with a diverse in-situ late Cenozoic radiation of felids in both North and South America. Nimravids were not considered in the analyses of feliforms by Hunt and Tedford (1993) and Hunt (1996b, 1998), as they were of uncertain affinities. Recent phylogenetic

analyses clearly support nimravids as basal feliforms, as an outgroup to crown-clade Feloidea. The earliest nimravids occur in the late Eocene of North America and the earliest Oligocene of Europe (Hunt, 1998; Janis et al., 1998; Joeckel et al., 2002), indicating a caniform-feliform split by that time and requiring only slight modifications to the previous temporal-biogeographic scenario.

If the relatively generalized viverravid "miacoids" are basal carnivoramorphans rather than members of the crown-clade Carnivora, then no obvious potential early Cenozoic (pre-late Eocene to early Oligocene) ancestors for the crown-clade Feloidea have yet been recovered anywhere in the world. Eocene nimravids are the only pre-Oligocene feliforms, and they already are quite specialized. They do represent the earliest definitive feliforms known, however, helping to constrain a much younger minimum age for the caniform-feliform split than the early Paleocene age inferred previously, when "miacoids" were considered members of the crown clade. But the presence of early nimravids in both the New and Old World does not help resolve whether one or the other region served as a locus of early diversification for feliforms. Basal feloids are absent from the well-sampled

Fig. 12.4. (*Continued*)

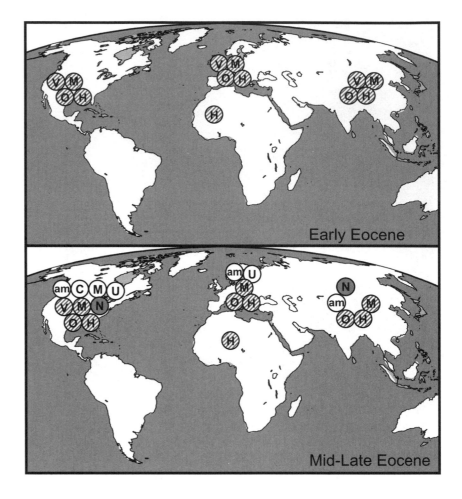

North American early Cenozoic record, and the only early Cenozoic feliforms (nimravids) known are much too autapomorphic to represent early feloids. The earliest definitive feloids thus occur in the early Oligocene of the Old World, suggesting an Old World origin for the Feloidea, and possibly the Feliformia. The more robust phylogenies, combined with observed temporal distributions, are yielding better constraints on divergence ages and biogeographic histories of major carnivoramorphan clades. In general, there is increasing concordance between the relative order of cladogenetic events and observed temporal ranges. Much older origin remains possible for either the Feloidea or the Carnivora, but it would require ad hoc invocation of major gaps in the fossil record.

Within Feloidea, Hunt and Tedford (1993) indicated first occurrences for both viverrids and felids in the later Oligocene to early Miocene (Europe and Asia; late early Miocene appearance in Africa), herpestids in the late early Miocene (Europe and Asia; late Miocene appearance in Africa), and hyaenids in the late early Miocene (Europe and Asia; middle Miocene appearance in Africa). Such distributions, refined by analyses within particular groups, can be used to infer minimum divergence ages for feloid lineages (see Fig. 12.2; Flynn, 1996), reconstruct diversity histories within such clades as Hyaenidae (Werdelin and Solounias, 1991; Flynn, 1996; but see Wagner, 2000), and alter reconstructions

of ancestral character states for modern clades from those based only on distributions in living taxa (Werdelin, 1996; Flynn and Nedbal, 1998). These inferred temporal and geographic distributions may be contradicted, or at least made more complex, by recent phylogenetic hypotheses for "viverrids" and herpestids. These hypotheses also would invalidate some putative synapomorphies, and thus call into question assignment of various early fossil taxa (e.g., the rich diversity of fossil feliforms from the European Quercy deposits) to specific families rather than more conservatively to Feloidea incertae sedis (see also Wesley-Hunt and Flynn, in press).

The most striking changes from traditional taxonomic and phylogenetic arrangements within the feliforms are (1) consistent placement of the "viverrid" *Nandinia* (African palm civet) as an outgroup to all other Feloidea, based on morphological data (Hunt, 1987; Flynn et al., 1988; Véron, 1995) and molecular or combined analyses (Flynn, 1996; Flynn and Nedbal, 1998; Yoder et al., 2003; Yoder and Flynn, 2003; Flynn et al., in press); and (2) robust identification of several subclades within Feloidea by recent molecular analyses (Yoder et al., 2003; Yoder and Flynn, 2003; Flynn et al., in prep.), in contrast to the longstanding difficulties in resolving higher-level feloid interrelationships because of conflicting morphological data (Flynn et al., 1988). Key among these changes are monophyly of all of Madagascar's carnivorans,

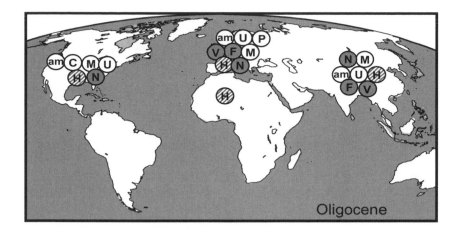

Fig. 12.4. (*Continued*)

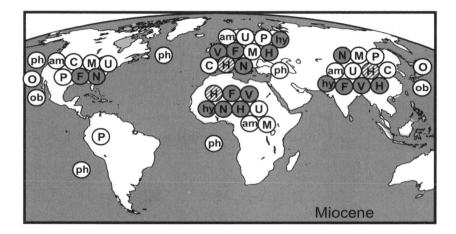

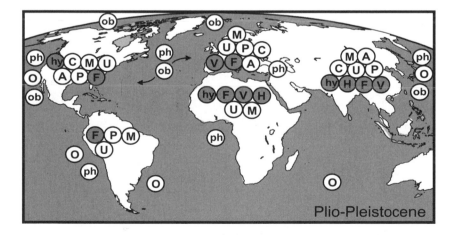

which previously had been placed in two or three different families ("Viverridae," Herpestidae, and sometimes, Felidae); a monophyletic clade of Herpestidae *sensu stricto* as the nearest relatives of the Malagasy clade; and a mid-late Cenozoic African origin for the group. These results clearly indicate paraphyly for "Viverridae" and polyphyly for "Herpestidae" as traditionally configured (Flynn and Nedbal, 1998; Flynn et al., in prep.). Based on those studies, the best available assessment of interrelationships among modern Feloidea (see Fig. 12.2) is a topology of (*Nandinia* (Felidae (Viverridae (Hyaenidae (Herpestidae [*sensu stricto*], Malagasy

feloids))))), although the relative positions of Felidae and Viverridae remain problematic, at least in parsimony analyses. The extinct nimravids, often controversial in their phylogenetic placement (Flynn and Galiano, 1982; Neff, 1983; Bryant, 1988; Flynn et al., 1988), in our recent study are a near outgroup to the Feloidea within the Feliformia, and thus are members of crown-clade Carnivora (Wesley-Hunt and Flynn, in press).

Among arctoids, the most significant areas of controversy remain the relationships of Pinnipedia to other arctoids, and the interrelationships of living and fossil musteloids.

Proposals of pinniped diphyly achieved some notoriety several decades ago, with large taxa like walruses (Odobenidae) and sea lions and fur seals (Otariidae) allied to bears (Ursidae), and true seals (Phocidae) related to otters (lutrine Mustelidae). Virtually all the more recent phylogenetic analyses (beginning with Wyss, 1987), however, support pinniped monophyly. Within Pinnipedia, Wyss (1987, 1988a, 1989) proposed novel associations of odobenids with phocids, rather than with otariids, and of various taxa within Phocidae, challenging the commonly accepted notions of interrelationships among the monophyletic pinniped taxa. This also led to markedly different interpretations of evolutionary transformations in body size and the skeleton, such as limb "retrogression" within Phocinae (Wyss, 1988b). Analysis of the important basal fossil outgroup to crown-clade pinnipeds, the pinnipedimorph *Enaliarctos* (Berta et al., 1989; Berta, 1991), and more recent molecular and combined molecular/morphological analyses robustly support pinniped monophyly (e.g., Árnason et al. 1995; Lento et al., 1995; Flynn and Nedbal, 1998). Together, this series of important analyses led to a refined understanding of the biogeographic history and morphological transitions in the evolution of this marine clade. A number of recent studies, both molecular and integrative, appears to have overturned Wyss's proposed interrelationships within the group (supporting the traditional sister-group relationship of odobenids and otariids; Lento et al., 1995; Flynn and Nedbal, 1998; Bininda-Emonds et al., 1999; Flynn et al., 2000). Other studies, however, still yield the non-traditional odobenid-phocid link (e.g., Berta and Wyss, 1994; Berta and Sumich, 1999; Berta, 2001). These discordances, marked differences in the geometry of relationships of taxa within both Otariidae and Phocidae, and the striking inability of all recent analyses to generate a robust hypothesis for the closest relatives of Pinnipedia within Arctoidea suggest that much fruitful work remains to be done on these groups.

Monophyly of Musteloidea was initially based on, but only weakly supported by, morphological data (Flynn et al., 1988). That clade included living and fossil mustelids and procyonids (Musteloidea *sensu stricto*), and possibly, the *Ailurus* lineage (Musteloidea *sensu lato*), but excluded phocids, which had been considered to be linked to lutrine mustelids in earlier diphyly propositions. Although numerous subsequent morphological and molecular studies of fossils and living taxa suggested drastic rearrangements of traditional concepts of various musteloid subclades, support for the monophyly of Musteloidea has continued to grow. For example, in character-based analyses (Flynn and Nedbal, 1998; Flynn et al., 2000), Musteloidea (*sensu lato*) was strongly supported by individual and combined data sets. In Flynn and Nedbal (1998), however, Musteloidea (*sensu stricto*) was weakly supported by transthyretin data alone, but received robust support in parsimony analyses of combined molecular (three genes) and morphological data, as well as in likelihood analyses of all the molecular data. Supertree analyses (Bininda-Emonds et al., 1999) yielded similar, but less robustly supported, results for both clades. The less robust results presumably reflect the strong discordance among the various input trees (Flynn et al., 2000).

Relationships within the Musteloidea are another story, however, with a great deal of flux and new controversy, due to both morphological and molecular analyses. The phylogenetic relationships of the red panda (*Ailurus*), although clearly an arctoid, have long been contested, as have the precise relationships of procyonids (e.g., Flynn et al., 1988, 2000; Wyss and Flynn, 1993; Wolsan and Lange-Badré, 1996). As noted above, there is also recent, growing evidence that skunks (mephitines or mephitids) are only distantly related to other mustelids, making traditional Mustelidae paraphyletic (DNA hybridization: Árnason and Widegren, 1986; Wayne et al., 1989; nucleotide sequences: Dragoo and Honeycutt, 1997; Flynn et al., 2000; but see Wolsan, 1993, 1999). To date, there is no compelling reason to reject monophyly of the remaining Mustelidae (*sensu stricto,* including mustelines, lutrines, and various badgers—mellivorines, melines, gulonines; see Bryant et al., 1993) or of the Procyonidae. The morphological phylogeny of mustelids by Bryant et al. (1993) concluded that both mephitines and lutrines are monophyletic (with the stink badger, *Mydaus,* closely related to mephitines), but that meline badgers are polyphyletic and mustelines are probably paraphyletic. As with phylogenetic hypotheses for other parts of the carnivoran evolutionary tree, some of the conflict and controversy for interrelationships among musteloids has arisen because of taxonomic sampling problems. Compelling molecular evidence now exists supporting the musteloid affinities of *Ailurus* (Flynn and Nedbal, 1998) and an early fossil that possibly represents the first occurrence of the lineage (*Simocyon;* Wang, 1997). The most recent analysis of living taxa indicated a tritomy of *Ailurus,* mephitines, and a mustelid (*sensu stricto*)/procyonid clade (Flynn et al., 2000; see also Slattery and O'Brien, 1995, for an *Ailurus*/procyonid link, but with sparse arctoid taxonomic sampling). Together, these results suggest that it may be difficult to determine the precise phylogenetic affinities of various early fossil musteloids. This is probably because there are few morphological synapomorphies characterizing the bases of the major sublineages, and some of these are homoplasious relative to other caniforms, a situation analogous to that discussed earlier for basal feloids.

One major group of extinct caniforms, the amphicyonids, has been variously allied with canids or ursids (as reflected in their common name, "bear-dogs"). Wyss and Flynn (1993) indicated a sister-group relationship to ursids, as did Hunt (1996a), who most recently reviewed the entire group. Wolsan (1993) alternatively suggested that amphicyonids were basal arctoids. In the parsimony analyses of Wesley-Hunt and Flynn (in press) the amphicyonids (represented by "*Miacis*" *cognitus* and *Daphoenus*) are sister taxa within Caniformia, as the nearest outgroup to the crown-clade Canoidea, although this topology is not supported in the bootstrap analysis.

Congruence and Discordance of Molecular and Morphological Phylogenies

Perceptions commonly center on dramatic conflict between phylogenetic hypotheses based on molecular data and those based on morphological data. Undoubtedly, some such disagreements do exist. Nevertheless, the bulk of these disagreements pertains to situations in which there is weak support for the competing hypotheses from one or the other (or both) of the data sets, and only rarely are the conflicting nodes robustly supported in the alternative analyses. Disagreement may even arise within partitions of the main types of morphological or molecular data, such as different phylogenetic signals in different genes, yielding conflict among topologies based on those individual genes. Even in cases in which some conflict is evident (determined using any of several tests for heterogeneity among data partitions), combining conflicting data can yield exactly the same topology derived from the other partition or from the combination of several homogeneous data partitions. In fact, combining these data often yields an overall increase in phylogenetic signal, such as increased support for various nodes, providing additional support for including all data in total evidence analyses (e.g., Flynn and Nedbal, 1998).

For Carnivora, many of the clades long recognized as monophyletic based on morphological features receive similarly robust support in molecular or combined analyses (e.g., Carnivora, Feliformia, Caniformia, Feloidea, Arctoidea, Musteloidea, Pinnipedia). Similarly, many clades of ambiguous or controversial relationships in morphological studies (e.g., those with few synapomorphies shared with other clades, or with many homoplasies between different clades) remain of uncertain relationship, even with the addition of extensive molecular data (e.g., Pinnipedia within Arctoidea, *Ailurus* relative to other musteloids). Presumably this arises because of long-branch effects, or rapid radiations, which similarly limit the resolving power of both morphological and molecular data. Ultimately, increasing taxonomic and character data (living or fossil, morphological or molecular) —and assessing the phylogenetic signals from partitions separately and in combined total evidence analyses—can identify and help resolve problems with prior character analyses or homology assumptions. Combining molecular and morphological data has already yielded more stable, better-resolved, and more robustly supported hypotheses of higher-level relationships of the Carnivora; permitted resolution of conflict among the signal in partitions of earlier data sets; and suggested some new and perhaps unexpected hypotheses of interrelationships (see Fig. 12.2). Subsets of morphological synapomorphies supporting very different hypotheses had made it notoriously difficult to discern the interrelationships among feloid families (Flynn et al., 1988). This problem has been at least partially overcome by additional data and character analyses (e.g., Wyss and Flynn, 1993; Flynn and Nedbal, 1998; Yoder et al., 2003; Yoder and Flynn, 2003). The previous discussion of the relationships of *Ailurus, Nan-*

dinia, and taxa within Pinnipedia provide additional examples. New fossil discoveries, anatomical studies, and integrative phylogenetic analyses promise to further enhance our understanding of carnivoran diversification and evolution.

PATTERNS OF EVOLUTION WITHIN CARNIVORA

Wide interest in Carnivora, coupled with better understanding of their phylogeny in recent years, has resulted in a wealth of new insights into the evolution of the group (see Bininda-Emonds et al., 1999). Recent accomplishments include:

1. The study of fossils from exceptional preservation sites (e.g., early feloid diversity in the Phosphorites du Quercy; de Bonis and Cirot, 1995; Peigné and de Bonis, 1998, 1999);
2. The analyses of rates of evolution (taxic, anatomical, molecular; e.g., Flynn, 1996; Werdelin, 1996); and
3. The inference of transformations and adaptations in life history strategies (Gittleman, 1994a); brain size (e.g., Gittleman, 1991, 1994b, Gittleman and Van Valkenburgh, 1997; see also pioneering studies by Radinsky, 1971, 1975, 1982; Jerison, 1973); social structure (Creel and Macdonald, 1995; Geffen et al., 1996); and physiology (Lee et al., 1991, McNab, 1995).

Space constraints enable us to touch briefly on only three aspects of the pattern of carnivoramorphan evolution: rates of molecular evolution, anatomical evolution, and biogeography and temporal patterns.

Rates of Molecular Evolution

Of particular relevance to integrating molecular and morphological data in phylogenetic analyses, marked rate heterogeneity—differences of up to 1.5 orders of magnitude in nucleotide changes per unit time across the carnivoran radiation—has been documented in several recent studies. Heterogenous rates, violating molecular clock models or assumptions, have been observed in many genes, both locally (among lineages within Carnivora) and globally (between carnivorans and other taxa). Both statistical (likelihood ratio and relative rate) tests and fossil/phylogeny-calibrated absolute-rate assessments have been used to determine the degree of rate variability in Carnivora (e.g., Flynn, 1996; Jenks and Werdelin, 1998; Flynn and Zehr, 2001; Yoder et al., 2003). This variability may be a general pattern in many groups, but it has been documented for only a few. Although there clearly is a general correlation between time and sequence divergence (for most genes and groups), the rejection of clock models in various tests and the observed high rates of variability both suggest that clocks must be applied with caution in phylogenetic analyses or molecular-clock inverse calculations of divergence ages from nucleotide differences among taxa.

Anatomical Evolution

EAR REGION.   The ear region has been an important character complex in understanding carnivoran phylogeny since the pioneering work of Turner (1848) and Flower (1869). In the past, researchers lumped together the early Cenozoic "miacoids," largely based on the shared (symplesiomorphic) lack of a preserved auditory bulla, whereas all living Carnivora (except *Nandinia*) have a fully ossified bulla, consisting of ectotympanic and rostral and caudal entotympanic elements, with bullar details distinguishing the feloid, canid, and arctoid subgroups. Various workers have since looked more closely at the basicrania of "miacoids" and generally agree that these taxa must have had something resembling a bulla, and more recently, researchers have documented morphological evidence that a cartilaginous or poorly attached ossified entotympanic and ectotympanic bulla did, in fact, exist on these early carnivoramorphans (e.g., Wesley and Flynn, 2003; Wesley-Hunt and Flynn, in press; Polly, pers. comm.). Significant contributions have been made to understanding the evolutionary transformations and phylogenetic implications of the ear region in recent years by many workers, including Hunt (e.g., 1974, 1989, 1991, 1998, 2001), Schmidt-Kittler (1981), Hunt and Tedford (1993), Wang and Tedford (1994), Wolsan (1999), Ivanoff (2000, 2001), Wesley and Flynn (2003), and Wesley-Hunt and Flynn (in press).

SKELETON.   Ken Rose and colleagues have undertaken the most extensive recent efforts to reconstruct skeletal evolution and infer locomotor mode across a wide variety of Paleogene mammals (including Carnivoramorpha and creodonts). These studies are complemented by a variety of analyses by these and other workers, including studies of specific subgroups (e.g., hyaenodontids:, Gebo and Rose, 1993; "miacoids," Heinrich and Rose, 1995, 1997; canids, Wang, 1993; pinnipeds, Wyss, 1988a,b; Berta and Adam, 2001) or anatomical features (e.g., retractile claws). Retractile claws are now known to be widely distributed across carnivoramorphans, among both living and fossil taxa, suggesting that this feature may be a synapomorphy for the entire clade (Flynn et al., 1988; Flynn, 1998; Wesley and Flynn, 2003). Wang's (1993) analysis of canid locomotor adaptations indicates loss of retractile claws within Canidae during transition from arboreal plantigrade "miacoids" to later terrestrial digitigrade canids (with *Hesperocyon* intermediate), implying that retractile claws were primitive for caniforms. Hunt and Tedford (1993), however, proposed that retractile claws were not present primitively in feloids (or in the viverravid *Didymictis*, considered a close relative of feloids, following Flynn and Galiano, 1982) and that living felids and viverrids probably evolved retraction independently. MacLeod and Rose (1993) built on earlier functional morpho-

logic studies by Van Valkenburgh (1987) and Rose (1990), by using eigenshape analysis of ungual phalanges and proximal radial morphology to infer locomotor mode in various Paleogene fossils, including carnivoramorphans. Although strong functional signal is present in skeletal morphology of the fossils, distinctions that are strong between modern locomotor guilds are not as distinct or obvious in fossils. MacLeod (2001) applied phylogenetic autocorrelation methods to determine that, although these skeletal elements clearly evolved in response to functional constraints in carnivoramorphans, phylogeny can mask or override functional signal. Integrating both sources of information may permit more accurate assessment/reconstruction of locomotor mode in fossil taxa, particularly those belonging to clades with close living relatives.

Biogeography and Temporal Patterns

Hunt (1996a) provided the most comprehensive recent overview of carnivoran biogeography. We summarize his analyses—supplemented by pinniped data from Deméré et al. (2003) and our own literature review—as a series of temporal slices with biogeographic distributions (by occurrence on a continent) for the major carnivoramorphan and creodont clades (Fig. 12.4). Briefly, Paleocene occurrences of Ferae (see Table 12.1) are limited to viverravids in Asia and North America (with an earliest Paleocene first appearance in North America), the first appearance of oxyaenodontids in the latest Paleocene of Asia (*Prolimnocyon;* Meng et al., 1998) and possibly the late Paleocene of Africa. The only Paleogene Ferae recorded from the southern hemisphere are African hyaenodontids.

The most extensive fossil record of early Ferae (Carnivoramorpha, Hyaenodontidae, and Oxyaenidae) is centered in North America. During the Eocene, both creodont groups, viverravids, "miacids," and caniform carnivorans are known from all three Laurasian continents, and nimravid feliforms were present in North America and Europe. Crown-clade Carnivora did not appear until the mid-late Eocene, and all basal carnivoramorphans were extinct by the late Eocene. Caniforms always were far more diverse in North America than on any other continent. As noted above, Eocene nimravids are known from Europe and North America, but the earliest crown-clade feloids are known only from the early Oligocene of Europe. Noteworthy among these are taxa from the early Oligocene portions of the Quercy deposits, such as *Palaeoprionodon, Haplogale, Stenoplesictis, Stenogale,* and *Proailurus* (Hunt, 1996b, 1998, 2001). Hunt considered some of these taxa to be early viverrids and felids, but perhaps they should all be placed more conservatively as incertae sedis within Feloidea, pending a comprehensive phylogenetic analysis of Feloidea. Oxyaenids were extinct by the Oligocene, although hyaenodontids persisted across Laurasia and in Africa through the Miocene. Later Cenozoic feliforms (basal feloids, felids, and putative "viverrids") diversified in Eurasia, beginning in the Oligocene, but only nimravids were present in North America during the Oligocene.

Caniforms also diversified in Eurasia and remained common in North America throughout the later Cenozoic. Carnivorans flourished during the Miocene, which marked the first appearances of all the remaining major clades, including hyaenids, herpestids, all three pinniped clades, and possibly ailurids. In contrast to the early Cenozoic, the later diversity of Ferae centered on the Old World, with extremely diverse faunas in Europe, Asia, and most notably, Africa. North American faunas remained dominated by caniforms, although by the Miocene, felid feloids also joined the more basal feliform nimravids. Late Miocene procyonids represent the first incursion of carnivorans into South America. These basic distribution patterns were maintained throughout the Plio-Pleistocene, with some modifications (see Fig. 12.4), although the late Pliocene "Great American Biotic Interchange" resulted in profound modification to the South American carnivore fauna, with the extinction of many native marsupial carnivores and the arrival of mustelids, ursids, felids, and canids to complement the earlier immigrant procyonids.

As the biogeography of basal Ferae and the Nimravidae have not been as thoroughly reviewed as those of the Carnivora, we here provide a somewhat more detailed overview of key points for creodonts, basal carnivoramorphans, and nimravids. Creodonts are unknown prior to the late Paleocene, and biogeographic distribution patterns have not yet helped elucidate their phylogenetic origins. Oxyaenids first appeared in the late Paleocene of North America, and later diversified there extensively, whereas the group did not appear in Europe and Asia until the Eocene. Oxyaenids became extinct after the mid-Eocene in North America, and by the late Eocene in Europe and Asia (Gunnell, 1998; Janis et al., 1998). Hyaenodontidae appeared first in the late Paleocene in Asia, and in Africa possibly as early as the late Paleocene, but certainly by the Eocene (Meng et al., 1998). They were extinct before the end of the Miocene. Hyaenodontids underwent extensive diversification in Africa and were the top mammalian predators there until the Miocene, which brought members of the Carnivora and the extinction of hyaenodontids, once the Tethys Seaway no longer isolated Africa (Ginsburg, 1999). Hyaenodontidae arrived in North America and Europe in the early Eocene, where the group also experienced successful radiations, but no taxa survived past the Oligocene. Due to the close, if not coincident, first appearance of the Oxyaenidae and Hyaenodontidae on separate continents, neither temporal nor biogeographic evidence provides support for their monophyly, which is otherwise based only on relatively weak morphological evidence (Gingerich, 1980; Fox and Youzwyshyn, 1994; Polly, 1996). In addition, when hyaenodontids do appear in North America, they are already taxonomically diverse, and their arrival is coincident with the arrival of numerous other taxa, such as perissodactyls, artiodactyls, and primates, suggesting a major immigration event rather than in situ cladogenesis (Gunnell, 1998).

The fossil record of early carnivoramorphans is richest in North America, in both the number of specimens and taxonomic diversity; it is poor in Asia and Europe (barring a few exceptionally well-preserved specimens from the middle Eocene Lagerstätten of Messel, Germany), and there is no record of early carnivoramorphans on any southern continent (e.g., Matthew, 1909; Flynn and Galiano, 1982; Gingerich and Winkler, 1985; Flynn, 1986, 1998; Janis et al., 1998). The earliest members of Carnivoramorpha to appear in the fossil record are the Viverravidae in the early Paleocene of North America. In Asia (China), viverravids occur in the late Paleocene or early Eocene, but the record is poor and limited to the enigmatic taxon *Pappictidops* (Qiu and Li, 1977; Flynn and Galiano, 1982). In Europe, cf. *Viverravus* from the early Eocene is the only viverravid known (Rich, 1971). The taxonomic diversity of viverravids is greatest in the late Paleocene of North America. In the early Eocene, the diversity drops, and viverravids do not survive through the late Eocene (Flynn, 1998).

Taxa traditionally assigned to the paraphyletic "Miacidae" first occur in the late Paleocene of North America, but do not diversify until the Eocene (Flynn, 1998). In Asia and Europe, the earliest occurrence of "miacids" is in the early Eocene, and they were extinct worldwide in the late Eocene. The European record of "miacids" is not as diverse as the North American, with only five genera (*Miacis, Paramiacis, Messelogale, Paroödectes,* and *Quercygale*) recorded; in Asia, there is only limited material of *Miacis* reported (Gingerich, 1983).

The Nimravidae are an extinct group of catlike carnivorans, which appear to be early feliforms. They are Holarctic in distribution, with the exception of a limited later invasion of Northern Africa (Bryant, 1996b; Hunt, 1996b; Ginsburg, 1999; Morales et al., 2001; Joeckel et al., 2002). Nimravidae is typically divided into the Nimravinae and Barbourofelinae clades (following Bryant, 1991). In North America, nimravines appear in the late Eocene. With six reported genera, they are relatively diverse and extend through the late Oligocene. Only one barbourofeline (*Barbourofelis*) is present in the fossil record of North America, and its range was limited to the late Miocene. In Europe, the Nimravinae do not appear until the early Oligocene and are extinct by the Miocene. The barbourofelines first occur in the early Miocene in Europe, but do not extend into the Pliocene. Nimravids are known from Asia in the late Eocene through the Miocene. The Asian Oligocene fossils referred to Nimravidae are primarily *Nimravus*. Barbourofelinae (*Sansanosmilus*) are not known to occur in Asia until the middle Miocene. Barbourofelines reached North Africa during the early Miocene. The record of Nimravinae in Africa is limited, and they are absent from the record before the end of the Miocene.

SUMMARY

Comparing multiple lines of phylogenetic evidence (e.g., nucleotide sequences, morphology, fossil record) generally yields congruent phylogenetic results, and combining them can greatly increase robustness and resolution. Recent molecular or combined analyses of living Carnivora and outgroups have revealed exceptionally strong support for many clades identified in previous morphological phylogenies, including Carnivora, Feliformia (includes crown-clade Feloidea and fossil stem outgroups), Caniformia (includes crown-clade Canoidea and fossil stem outgroups), Arctoidea, Musteloidea, a clade of Procyonidae and Mustelidae (*sensu stricto*), and all families (except Viverridae and Mustelidae). Integrative analyses have better resolved several longstanding controversies or areas of ambiguity, but also have yielded some unexpected results. Examples include documenting the monophyly of Pinnipedia; pairing of Otariidae and Odobenidae (Otarioidea); determining that skunks (mephitines or mephitids) are not closely related to other mustelids; placing the red panda (*Ailurus*) as a basal musteloid; documenting the monophyly of the Madagascar carnivorans; and ascertaining that *Nandinia* is not a "viverrid," but rather a sister-group to all other living Feloidea.

Fossil-focused studies, which include descriptions of many new and well-preserved specimens and more comprehensive phylogenetic analyses, suggest that there is a complex pattern of relationships of early Cenozoic carnivoramorphans ("miacoids") to the living clades. Viverravidae is likely to be both monophyletic and basal to all other Carnivoramorpha. In contrast, the "Miacidae" is paraphyletic, but taxa traditionally assigned to this group are carnivoramorphans more closely related to Carnivora than are viverravids. Both Viverravidae and "Miacidae" appear to be outside of the crown-clade Carnivora; if these phylogenetic relationships are correct, then the oldest known member of the Carnivora is Middle Eocene in age. Nimravidae likely are early feliforms, rather than caniforms or more basal carnivoramorphans. The relationship of amphicyonids remains uncertain.

A reasonable amount of morphologic evidence supports a relationship between all or some Creodonta (Hyaenodontidae and Oxyaenidae) and Carnivoramorpha, together forming the clade Ferae, although there is little evidence suggesting that the creodonts themselves are monophyletic. The identity of the closest relatives of Carnivora among both fossil and living clades remains controversial. Available morphological evidence supports Lipotyphla (the more restrictive Eulipotyphla of some authors) as the closest relatives of the Carnivora among living placental clades, although some of the most recent molecular analyses indicate that Pholidota are the closest relatives of Carnivora, and lipotyphlans are more distant to very distantly related. There is growing evidence of a link between Carnivora and ungulates, especially from molecular data, providing somewhat surprising validation of Simpson's (1945) Ferungulata.

These better-resolved and more robust phylogenies are yielding significant enhancements of our understanding of clade divergence ages, rates of taxic and molecular evolution, ancestral morphotypes, character distributions and transformations, biogeography, and ecomorphological and diversity changes through time. We presented summaries of the biogeography, temporal distributions, and phylogenetic relationships of the major groups of Carnivora and their Ferae relatives, and highlighted a few recent studies that used these groups to address important evolutionary questions.

ACKNOWLEDGMENTS

We thank Marlene Donnelly (Field Museum, Chicago) for her exceptional drawings, photographic illustrations, and assistance in preparation of Fig. 12.3. Dick Tedford, Lars Werdelin, and Ken Rose provided reviews, greatly clarifying prose and strengthening content. Annalisa Berta and David Polly gave us access to their manuscripts and unpublished works in progress, for which we are very grateful. JJF was originally inspired to work on this group by William Diller Matthew's pioneering monograph on Bridger Basin mammals—Matthew generated the first "phylogenetic" (in a modern conception) insights into carnivoramorphan relationships, building on the excellent foundation of prior anatomists and paleontologists like Turner, Flower, Cope, and Wortman. GDW thanks Ann Elder and Scott Madsen of Dinosaur National Monument, Utah, for taking the time to introduce her to paleontology. Although we may have inadvertently overlooked many of the colleagues that have contributed to our interest in and knowledge of the Ferae, we acknowledge the support of Pam Austin, Annalisa Berta, Harold Bryant, Mary Dawson, Jerry Dragoo, Henry Galiano, Steve Goodman, Greg Gunnell, Ron Heinrich, Rodney Honeycutt, Bob Hunt, Julian Kerbis, Michael Morlo, Mike Nedbal, David Polly, Len Radinsky, Bill Stanley, Dick Tedford, Xiaoming Wang, Lars Werdelin, Andy Wyss, Anne Yoder, and Sarah Zehr.

REFERENCES

Árnason, Ú., K. Bodin, A. Gullberg, C. Ledje, and S. Mouchaty. 1995. A molecular view of pinniped relationships with particular emphasis on the true seals. Journal of Molecular Evolution 40: 78–85.

Árnason, Ú., and E. Johnsson. 1992. The complete mitochondrial DNA sequence of the harbor seal, *Phoca vitulina*. Journal of Molecular Evolution 34: 493–505.

Árnason, Ú., and B. Widegren. 1986. Pinniped phylogeny enlightened by molecular hybridizations using highly repetitive DNA. Molecular Biology and Evolution 3: 356–365.

Berta, A. 1991. New *Enaliarctos* (Pinnipedimorpha) from the Oligocene and Miocene of Oregon and the role of "enaliarctids" in pinniped phylogeny. Smithsonian Contributions to Paleobiology 69: 1–33.

———. 2001. Pinnipedia, overview; Pinniped evolution; Systematics, overview; pp. 903–911, 921–929; 1222–1226 *in* W. Perrin, J.G.M. Thewissen, and B. Wursig (eds.), Encyclopedia of Marine Mammals 2002. Academic Press, San Diego.

Berta, A., and P. J. Adam. 2001. Evolutionary biology of pinnipeds; pp. 235–260 *in* J.-M. Mazin and V. de Buffrénil (eds.), Secondary Adaptations of Tetrapods to Life in Water. Verlag Dr. Friedrich Pfeil, Munich.

Berta, A., C. E. Ray, and A. R. Wyss. 1989. Skeleton of the oldest known pinniped, *Enaliarctos mealsi.* Science 244: 60–62.

Berta, A., and J. Sumich. 1999. Marine Mammals: Evolutionary Biology. San Diego: Academic Press.

Berta, A., and A. R. Wyss. 1994. Pinniped phylogeny; pp. 33–56 *in* A. Berta and T. A. Demeré (eds.), Contributions in Marine Mammal Paleontology Honoring Frank C. Whitmore, Jr. Proceedings of the San Diego Society of Natural History 29, San Diego Society of Natural History, San Diego.

Bininda-Emonds, O.R.P., J. L. Gittleman, and A. Purvis. 1999. Building large trees by combining phylogenetic information: A complete phylogeny of the extant Carnivora (Mammalia). Biological Reviews 74: 143–175.

Bonis, L. de, and E. Cirot. 1995. Le Garouillas et les sites contemporains (Oligocène, MP 25) des phosphorites du Quercy (Lot, Tarn-et-Garonne, France) et leurs faunes de Vertébrés. 7. Carnivores. Palaeontographica Abteilung A (Stuttgart) 236: 135–149.

Bryant, H. N. 1988. The anatomy, phylogenetic relationships and systematics of the Nimravidae (Mammalia: Carnivora). Ph.D. Dissertation, University of Toronto.

———. 1991. Phylogenetic relationships and systematics of the Nimravidae (Carnivora). Journal of Mammalogy 72: 56–78.

———. 1996a. Explicitness, stability, and universality in the phylogenetic definition and usage of taxon names: A case study of the phylogenetic taxonomy of the Carnivora (Mammalia). Systematic Biology 45: 174–189.

———. 1996b. Nimravidae; pp. 453–475 *in* D. R. Prothero and R. J. Emry, (eds.), The Terrestrial Eocene-Oligocene Transition in North America. Cambridge University Press, New York.

Bryant, H. N., A. P. Russell, and W. D. Fitch. 1993. Phylogenetic relationships within the extant Mustelidae (Carnivora): Appraisal of the cladistic status of the Simpsonian subfamilies. Zoological Journal of the Linnean Society 108: 301–334.

Carroll, R. L. 1988. Vertebrate Paleontology and Evolution. Freeman, New York.

Cope, E. D. 1875. On the supposed Carnivora of the Eocene of the Rocky Mountains. Proceedings of the Academy of Natural Sciences, Philadelphia 27: 444–449.

———. 1880. On the genera of the Creodonta. Proceedings of the American Philosophical Society 1882: 76–82.

Creel, S., and D. Macdonald. 1995. Sociality, group size, and reproductive suppression in carnivores. Advances in the Study of Behavior 24: 203–257.

Czelusniak, J., M. Goodman, B. Koop, D. Tagle, J. Shoshani, G. Braunitzer, T. Kleinschmidt, W. W. de Jong, and G. Matsuda. 1990. Perspectives from amino acid and nucleotide sequences on cladistic relationships among higher taxa of Eutheria; pp. 545–572 *in* H. Genoways (ed.), Current Mammalogy. Volume 2. Plenum, New York.

Delsuc, F., F. M. Catzeflis, M. J. Stanhope, and E.J.P. Douzery, 2001. The evolution of armadillos, anteaters and sloths depicted by nuclear and mitochondrial phylogenies: Implications for the status of the enigmatic fossil *Eurotamandua.* Proceedings of the Royal Society Biological Sciences, Series B 268: 1605–1615.

Delsuc, F., M. Scally, O. Madsen, M. J. Stanhope, W. W. de Jong, F. M. Catzeflis, M. S. Springer, and E.J.P. Douzery. 2002. Molecular phylogeny of living xenarthrans and the impact of character and taxon sampling on the placental tree rooting. Molecular Biology and Evolution 19: 1656–1671.

Deméré, T. A., A. Berta, and P. J. Adam. 2003. Pinnipedimorph evolutionary biogeography. Bulletin of the American Museum of Natural History 279: 32–76.

Dragoo, J. W., and R. L. Honeycutt. 1997. Systematics of mustelid-like carnivores. Journal of Mammalogy 78: 426–443.

Faith, D. P., and P. S. Cranston. 1991. Could a cladogram this short have arisen by chance alone? On permutation tests for cladistic structure. Cladistics 7: 1–28.

Flower, W. H. 1869. On the value of the characters of the base of the cranium in the classification of the Order Carnivora. Proceedings of the Zoological Society of London: 4–37.

Flynn, J. J. 1986. Faunal provinces and the Simpson Coefficient; pp. 317–338 *in* K. Flanagan and J. A. Lillegraven (eds.), Vertebrates, Phylogeny, and Philosophy. George Gaylord Simpson Memorial Volume. University of Wyoming Contributions to Geology Special Paper 3. University of Wyoming, Laramie, Wyoming.

———. 1996. Phylogeny and rates of evolution: Morphological, taxic and molecular; pp. 542–581 *in* J. Gittleman (ed.), Carnivore Behavior, Ecology, and Evolution. Volume 2. Cornell University Press, Ithaca, New York.

———. 1998. Early Cenozoic Carnivora ("Miacoidea"); pp. 110–123 *in* C. M. Janis, K. M. Scott, and L. L. Jacobs (eds.), Evolution of Tertiary Mammals of North America. Volume 1. Terrestrial Carnivores, Ungulates, and Ungulatelike Mammals. Cambridge University Press, Cambridge.

Flynn, J. J., J. A. Finarelli, S. M. Zehr, J. Hsu, and M. A. Nedbal. In press. Molecular phylogeny of the Carnivora (Mammalia): Assessing the impact of increased sampling on resolving enigmatic relationships. Systematic Biology.

Flynn, J. J., and H. Galiano. 1982. Phylogeny of Early Tertiary Carnivora, with a description of a new species of *Protictis* from the Middle Eocene of Northwestern Wyoming. American Museum Novitates 2725: 1–64.

Flynn, J. J., S. M. Goodman, and A. D. Yoder. In prep. Systematics and evolution of Malagasy Carnivora.

Flynn, J. J., and M. A. Nedbal. 1998. Phylogeny of the Carnivora (Mammalia): Congruence vs incompatibility among multiple data sets. Molecular Phylogenetics and Evolution 9: 414–426.

Flynn, J. J., M. A. Nedbal, J. W. Dragoo, and R. L. Honeycutt. 2000. Whence the red panda? Molecular Phylogenetics and Evolution 17: 190–199.

Flynn, J. J., N. A. Neff, and R. H. Tedford. 1988. Phylogeny of the Carnivora; pp. 73–116 *in* M. Benton (ed.), The Phylogeny and Classification of the Tetrapods. Volume 2. Mammals. Clarendon Press, Oxford.

Flynn, J. J., and S. M. Zehr. 2001. Testing clocks: Assessing morphological, molecular, and paleontological data on the timing and rate of evolution in mammals. Paleobios 21(supplement to no. 2): 52.

Fox, R. C., and G. P. Youzwyshyn. 1994. New primitive carnivorans (Mammalia) from the Paleocene of western Canada, and their bearing on relationships of the order. Journal of Vertebrate Paleontology 14: 382–404.

Gebo, D. L., and K. D. Rose. 1993. Skeletal morphology and locomotor adaptation in *Prolimnocyon atavus,* an early Eocene hyaenodontid creodont. Journal of Vertebrate Paleontology 13: 125–144.

Geffen, E., M. E. Gompper, J. L. Gittleman, H.-W. Luh, D. W. Macdonald, and R. K. Wayne. 1996. Size, life-history traits, and social organization in the Canidae: A reevaluation. American Naturalist 147: 140–160.

Gingerich, P. D. 1980. *Tytthaena parrisi,* oldest known oxyaenid (Mammalia, Creodonta) from the late Paleocene of Western North America. Journal of Paleontology 54: 570–576.

———. 1983. Systematics of early Eocene Miacidae (Mammalia, Carnivora) in the Clark's Fork Basin, Wyoming. Contributions from the Museum of Paleontology 26: 197–225.

Gingerich, P. D., and D. A. Winkler. 1985. Systematics of Paleocene Viverravidae (Mammalia, Carnivora) in the Bighorn Basin and Clark's Fork Basin, Wyoming. Contributions Museum of Paleontology University of Michigan 27: 87–128.

Ginsburg, L. 1999. Order Creodonta; pp. 105–108 *in* G. E. Roessner, K. Heissig, and V. Fahlbusch (eds.), The Miocene Land Mammals of Europe. Verlag Dr. Friedrich Pfeil, Munich.

Gittleman, J. L. 1991. Carnivore olfactory bulb size: Allometry, phylogeny, and ecology. Journal Zoology, London 225: 253–272.

———. 1994a. Are the pandas successful specialists or evolutionary failures? BioScience 44: 456–464.

———. 1994b. Female brain size and parental care in carnivores. Proceedings of the National Academy of Sciences USA 91: 5495–5497.

Gittleman, J. L., and B. Van Valkenburgh. 1997. Sexual dimorphism in the canines and skulls of carnivores: The effects of size, phylogeny, and behavioral ecology. Journal of Zoology 242: 97–117.

Gunnell, G. F. 1998. Creodonta; pp. 91–109 *in* C. M. Janis, K. M. Scott, and L. L. Jacobs (eds.), Evolution of Tertiary Mammals of North America. Volume 1. Terrestrial Carnivores, Ungulates, and Ungulatelike Mammals. Cambridge University Press, Cambridge.

Gureev, A. A. 1979. [Fauna of the USSR, Mammals]. Volume 4, part 2. [Insectivores (Mammalia, Insectivora, Erinaceidae, Talpidae, Soricidae)]. Moskva Nauka, Leningrad. (In Russian.)

Heinrich, R. E., and K. D. Rose. 1995. Partial skeleton of the primitive carnivoran *Miacis petilus* from the Early Eocene of Wyoming. Journal of Mammalogy 76: 148–162.

———. 1997. Postcranial morphology and locomotor behavior of two early Eocene miacoid carnivorans, *Vulpavus* and *Didymictis.* Palaeontology 40: 279–305.

Honeycutt, R. L., and R. M. Adkins. 1993. Higher level systematics of eutherian mammals: An assessment of molecular characters and phylogenetic hypotheses. Annual Review of Ecology and Systematics 24: 279–305.

Hunt, R. M., Jr. 1974. The auditory bulla in Carnivora: An anatomical basis for reappraisal of carnivore evolution. Journal of Morphology 143: 21–76.

———. 1987. Evolution of the aeluroid Carnivora: Significance of auditory structure in the nimravid cat *Dinictis.* American Museum Novitates 2886: 1–74.

———. 1989. Evolution of the aeluroid Carnivora: Significance of the ventral promontorial process of the petrosal, and the origin of basicranial patterns in the living families. American Museum Novitates 2930: 1–32.

———. 1991. Evolution of the aeluroid Carnivora: Viverrid affinities of the Miocene carnivoran *Herpestides.* American Museum Novitates 3023: 1–34.

———. 1996a. Amphicyonidae; pp. 476–485 *in* D. Prothero and R. J. Emry (eds.), The Terrestrial Eocene-Oligocene Transition in North America. Cambridge University Press, Cambridge.

———. 1996b. Biogeography of the order Carnivora; pp. 485–541 *in* J. Gittleman (ed.), Carnivore Behavior, Ecology, and Evolution. Volume 2. Cornell University Press, Ithaca, New York.

———. 1998. Evolution of the aeluroid Carnivora: Diversity of the earliest aeluroids from Eurasia (Quercy, Hsanda-Gol) and the origin of felids. American Museum Novitates 3252: 1–65.

———. 2001. Basicranial anatomy of the living linsangs *Prionodon* and *Poiana* (Mammalia, Carnivora, Viverridae), with comments on the early evolution of aeluroid carnivorans. American Museum Novitates 3330: 1–24.

Hunt, R. M., Jr., and R. H. Tedford. 1993. Phylogenetic relationships within the aeluroid Carnivora and implications of their temporal and geographic distribution; pp. 53–73 *in* F. S. Szalay, M. J. Novacek, and M. C. McKenna (eds.), Mammal Phylogeny: Placentals. Springer-Verlag, New York.

International Commission on Zoological Nomenclature. 1999. International Code of Zoological Nomenclature. Fourth Edition, adopted by the International Union of Biological Sciences. International Trust for Zoological Nomenclature, London.

Ivanoff, D. V. 2000. Origin of the septum in the canid auditory bulla: Evidence from morphogenesis. Acta Theriologica 45: 253–270.

———. 2001. Partitions in the carnivoran auditory bulla: Their formation and significance for systematics. Mammal Review 31: 1–16.

Janis, C. M., J. A. Baskin, A. Berta, J. J. Flynn, G. F. Gunnell, R. M. Hunt, Jr., L. D. Martin, and K. Munthe. 1998. Carnivorous mammals; pp. 73–90 *in* C. M. Janis, K. M. Scott, and L. L. Jacobs (eds.), Evolution of Tertiary Mammals of North America. Volume 1. Terrestrial Carnivores, Ungulates, and Ungulatelike Mammals. Cambridge University Press, Cambridge.

Jenks, S., and L. Werdelin. 1998. Taxonomy and systematics of living hyaenas (family Hyaenidae); pp. 8–17 *in* M.G.L. Mills and H. Hofer (eds.), Hyaenas: Status Survey and Conservation Action Plan. The World Conservation Union (IUCN), Gland, Switzerland.

Jerison, H. J. 1973. Evolution of the Brain and Intelligence. Academic Press, New York.

Joeckel, R. M., S. Peigné, R. M. Hunt, Jr., and R. I. Skolnick. 2002. The auditory region and nasal cavity of Oligocene Nimravidae (Mammalia: Carnivora). Journal of Vertebrate Paleontology 22: 830–847.

Jow, H., C. Hudelot, M. Rattray, and P. G. Higgs. 2002. Bayesian phylogenetics using an RNA substitution model applied to early mammalian evolution. Molecular Biology and Evolution 19: 1591–1601.

Kimura, M. 1980. A simple method for estimating evolutionary rates of base substitutions through comparative studies of nucleotide sequences. Journal of Molecular Evolution 16: 111–120.

Krettek, A., A. Gullberg, and Ú. Árnason. 1995. Sequence analysis of the complete mitochondrial DNA molecule of the hedgehog, *Erinaceus europaeus,* and the phylogenetic position of the Lipotyphla. Journal of Molecular Evolution 41: 952–957.

Kretzoi, M. 1943. *Kochitiis centenii* n.g. n.sp., ein altertümlicher Creodonte aus dem Oberoligozän Siebenbürgens. Földtany Közlöny 73: 10–17, 190–195.

Lee, P. C., P. Majluf, and I. J. Gordon. 1991. Growth, weaning and maternal investment from a comparative perspective. Journal of Zoology 225: 99–114.

Lento, G. M., R. E. Hickson, G. K. Chambers, and D. Penny. 1995. Use of spectral analysis to test hypotheses on the origin of pinnipeds. Molecular Biology and Evolution 12: 28–52.

Liu, F.-G.R., M. M. Miyamoto, N. P. Freire, P. Q. Ong, M. R. Tennant, T. S. Young, and K. F. Gugel. 2001. Molecular and morphological supertrees for eutherian (placental) mammals. Science 291: 1786–1789.

MacLeod, N. 2001. The role of phylogeny in quantitative paleobiological data analysis. Paleobiology 27: 226–240.

MacLeod, N., and K. D. Rose. 1993. Inferring locomotor behavior in Paleogene mammals via eigenshape analysis. American Journal of Science 293-A: 300–355.

MacPhee, R.D.E., and M. J. Novacek. 1993. Definition and relationships of Lipotyphla; pp. 13–31 in F. S. Szalay, M. J. Novacek, and M. C. McKenna (eds.), Mammal Phylogeny: Placentals. Springer-Verlag, New York.

Matthew, W. D. 1909. The Carnivora and Insectivora of the Bridger basin, middle Eocene. Memoirs of the American Museum of Natural History 9: 289–567.

McNab, B. K. 1995. Energy expenditure and conservation in frugivorous and mixed-diet carnivorans. Journal of Mammalogy 76: 206–222.

Meng, J., R. Zhai, and A. R. Wyss. 1998. The late Paleocene Bayan Ulan fauna of Inner Mongolia, China; pp. 145–185 in C. K. Beard and M. R. Dawson (eds.), Dawn of the Age of Mammals in Asia. Bulletin of the Carnegie Museum of Natural History 43.

Miyamoto, M. M., and M. Goodman. 1986. Biomolecular systematics of eutherian mammals: Phylogenetic patterns and classification. Systematic Zoology 35: 230–240.

Morales, J., M. J. Salesa, M. Pickford, and D. Soria. 2001. A new tribe, new genus and two new species of Barbourofelinae (Felidae, Carnivora, Mammalia) from the early Miocene of East Africa and Spain. Transactions of the Royal Society of Edinburgh, Earth Sciences 92: 97–102.

Murphy, W. J., E. Eizirik, S. J. O'Brien, O. Madsen, M. Scally, C. J. Douady, E. Teeling, O. A. Ryder, M. J. Stanhope, W. W. de Jong, and M. S. Springer. 2001. Resolution of the early placental mammal radiation using Bayesian phylogenetics. Science 294: 2348–2351.

Neff, N. A. 1983. The basicranial anatomy of the Nimravidae (Mammalia: Carnivora): Character analyses and phylogenetic inferences. Ph.D. Thesis, City University of New York, New York.

Novacek, M. J. 1992a. Mammalian phylogeny: Shaking the tree. Nature 356: 121–125.

———. 1992b. Fossils as critical data for phylogeny; pp. 46–88 in M. J. Novacek and Q. D. Wheeler (eds.), Extinction and Phylogeny. Columbia University Press, New York.

Osborn, H. F. 1907. Evolution of mammalian molar teeth. MacMillan, New York.

Peigné, S., and L. de Bonis. 1998. The genus Stenoplesictis Filhol (Mammalia, Carnivora) from the Oligocene deposits of the Phosphorites of Quercy, France. Journal of Vertebrate Paleontology 19: 566–575.

———. 1999. Le premier crâne de Nimravus (Mammalia, Carnivora) d'Eurasie et ses relations avec N. brachyops d'Amérique du Nord. Revue de Paléobiologie, Genève 18: 57–67.

Polly, P. D. 1996. The skeleton of Gazinocyon vulpeculus gen. et comb. nov. and the cladistic relationships of Hyaenodontidae (Eutheria, Mammalia). Journal of Vertebrate Paleontology 16: 303–319.

Qiu, Z.-X., and C.-K. Li. 1977. Miscellaneous mammalian fossils from the Paleocene of Qianshan Basin, Anhui. Vertebrata PalAsiatica 15: 94–102. (In Chinese.)

Radinsky, L. B. 1971. An example of parallelism in carnivore brain evolution. Evolution 25: 518–522.

———. 1975. Viverrid neuroanatomy: Phylogenetic and behavioral implications. Journal of Mammalogy 56: 130–150.

———. 1982. Evolution of skull shape in carnivores. 3. The origin and early radiation of the modern carnivore families. Paleobiology 8: 177–195.

Rambaut, A., and L. Bromham. 1998. Estimating divergence dates from molecular sequences. Molecular Biology and Evolution 15: 442–448.

Rich, T.H.V. 1971. Deltatheridia, Carnivora, and Condylarthra (Mammalia) of the early Eocene, Paris Basin, France. University of California Publication in Geological Sciences 88: 1–72.

Rose, K. D. 1990. Postcranial skeletal remains and adaptations in early Eocene mammals from the Willwood Formation, Bighorn Basin, Wyoming; pp. 107–133 in T. M. Bown and K. D. Rose (eds.), Dawn of the Age of Mammals in the Northern Part of the Rocky Mountain Interior, North America. Special Paper 243. Geological Society of America, Boulder, Colorado.

Rowe, T. 1988. Definition, diagnosis, and origin of Mammalia. Journal of Vertebrate Paleontology. 8: 241–264.

Savage, R.J.G. 1977. Evolution in carnivorous mammals. Palaeontology 20: 237–271.

Schmidt-Kittler, N. 1981. Zur Stammesgeschichte der marderverwandten Raubtiergruppen (Musteloidea, Carnivora). Eclogae Geologicae Helvetiae 74: 753–801.

Shoshani, J. 1986. Mammalian phylogeny: Comparison of morphological and molecular results. Molecular Biology and Evolution 3: 222–242.

Simpson, G. G. 1945. The principles of classification and a classification of mammals. Bulletin of the American Museum of Natural History 85: 1–350.

Slattery, J. P., and S. J. O'Brien. 1995. Molecular phylogeny of the red panda (Ailurus fulgens). Journal of Heredity 86: 413–422.

Stanhope, M. J., J. Czelusniak, J. S. Si, J. Nickerson, and M. Goodman. 1992. A molecular perspective on mammalian evolution from the gene encoding interphotoreceptor retinoid binding protein, with convincing evidence for bat monophyly. Molecular Phylogenetics and Evolution 1: 148–160.

Stanhope, M. J., V. C. Waddell, O. Madsen, W. W. de Jong, S. B. Hedges, G. C. Cleven, D. Kao, and M. S. Springer. 1998. Molecular evidence for multiple origins of Insectivora and for a new order of endemic African insectivore mammals. Proceedings of the National Academy of Sciences USA 95: 9967–9972.

Takezaki, N., A. Rzhetsky, and M. Nei. 1995. Phylogenetic test of the molecular clock and linearized trees. Molecular Biology and Evolution 12: 823–833.

Thorne, J. L., H. Kishino, and I. S. Painter. 1998. Estimating the rate of evolution of the rate of evolution. Molecular Biology and Evolution 15: 1647–1657.

Turner, H. N. 1848. Observations relating to some of the foramina at the base of the skull in Mammalia, and on the classification of the Order Carnivora. Proceedings of the Zoological Society of London: 63–88.

Van Dijk, M.A.M., O. Madsen, F. M. Catzeflis, M. J. Stanhope, W. W. de Jong, and M. Pagel. 1998. Protein sequence signatures support the African clade of mammals. Proceedings of the National Academy of Sciences USA 98: 188–193.

Van Valen, L. 1969. The multiple origins of the placental carnivores. Evolution 23: 118–130.

Van Valkenburgh, B. 1987. Skeletal indicators of locomotor behavior in living and extinct carnivores. Journal of Vertebrate Paleontology 7: 162–182.

Véron, G. 1995. La position systématique de *Cryptoprocta ferox* (Carnivora). Analyse cladistique des caractères morphologiques de carnivores Aeluroidea actuels et fossiles. Mammalia 59: 551–582.

Vrana, P. B., M. C. Milinkovitch, J. R. Powell, and W. C. Wheeler. 1994. Higher level relationships of the arctoid Carnivora based on sequence data and "total evidence." Molecular Phylogenetics and Evolution 3: 47–58.

Wagner, P. J. 2000. Exhaustion of morphologic character states among fossil taxa. Evolution 54: 365–386.

Wang, X. 1993. Transformation from plantigrady to digitigrady: Functional morphology of locomotion in *Hesperocyon* (Canidae: Carnivora). American Museum Novitates 3069: 1–23.

———. 1997. New cranial material of *Simocyon* from China, and its implications for phylogenetic relationship to the red panda (*Ailurus*). Journal of Vertebrate Paleontology 17: 184–198.

Wang, X., and R. H. Tedford. 1994. Basicranial anatomy and phylogeny of primitive canids and closely related miacids (Carnivora: Mammalia). American Museum Novitates 3092: 1–34.

Wayne, R. K., R. E. Benveniste, D. N. Janczewski, and S. J. O'Brien. 1989. Molecular and biochemical evolution of the Carnivora; pp. 465–494 in J. Gittleman (ed.), Carnivore Behavior, Ecology, and Evolution. Volume 1. Cornell University Press, Ithaca, New York.

Werdelin, L. 1996. Carnivoran ecomorphology: A phylogenetic perspective; pp. 582–624 in J. Gittleman (ed.), Carnivore Behavior, Ecology, and Evolution. Volume 2. Cornell University Press, Ithaca, New York.

Werdelin, L., and N. Solounias. 1991. The Hyaenidae: Taxonomy, systematics and evolution. Fossils and Strata 30: 1–104.

Wesley, G. D., and J. J. Flynn. 2003. A revision of *Tapocyon*, including analysis of the first cranial specimens, identification of a new species, and implications for the phylogeny of basal Carnivora. Journal of Paleontology 77: 769–783.

Wesley-Hunt, G. D., and J. J. Flynn. In press. Phylogeny of the Carnivora: Basal relationships among the carnivoramorphans, and assessment of the position of "Miacoidea" relative to Carnivora. Journal of Systematic Palaeontology.

Wilson, D. E., and D. M. Reeder (eds.). 1993. Mammal Species of the World. Second Edition. Smithsonian Institution Press, Washington, D.C.

Wolsan, M. 1993. Phylogeny and classification of early European Mustelida (Mammalia: Carnivora). Acta Theriologica 38: 345–384.

———. 1999. Oldest mephitine cranium and its implications for the origin of skunks. Acta Palaeontologica Polonica 44: 223–230.

Wolsan, M., and B. Lange-Badré. 1996. An arctomorph carnivoran skull from the Phosphorites du Quercy and the origin of procyonids. Acta Palaeontologica Polonica 41: 277–298.

Wortman, J. L. 1899. Restoration of *Oxyaena lupina* Cope, with descriptions of certain new species of Eocene creodonts. Bulletin of the American Museum of Natural History 12(7): 139–148.

Wyss, A. R. 1987. The walrus auditory region and the monophyly of pinnipeds. American Museum Novitates 2871: 1–31.

———. 1988a. Evidence from flipper structure for a single origin of pinnipeds. Nature 334: 427–428.

———. 1988b. On "retrogression" in the evolution of the Phocinae and phylogenetic affinities of the monk seals. American Museum Novitates 2924: 1–38.

———. 1989. Flippers and pinniped phylogeny: Has the problem of convergence been overrated? Marine Mammal Science 5: 343–360.

Wyss, A. R., and Flynn, J. J. 1993. A phylogenetic analysis and definition of the Carnivora; pp. 32–52 in F. S. Szalay, M. J. Novacek, and M. C. McKenna (eds.), Mammal Phylogeny: Placentals. Springer-Verlag, New York.

Yoder, A. D., M. M. Burns, S. Zehr, T. Delefosse, G. Véron, S. M. Goodman, and J. J. Flynn. 2003. Single origin of Malagasy Carnivora from an African ancestor. Nature 421: 734–737.

Yoder, A. D., and J. J. Flynn. 2003. Origin of Malagasy Carnivora; pp. 1253–1256 in S. M. Goodman and J. Benstead (eds), The Natural History of Madagascar. University of Chicago Press, Chicago. [See also references in composite Mammals reference list on pp. 1390–1422.]

JEREMY J. HOOKER

13

PERISSODACTYLA

THE ORDER PERISSODACTYLA IS REPRESENTED today by a relict assemblage of 17 species: eight horses (family Equidae), four tapirs (family Tapiridae) and five rhinoceroses (family Rhinocerotidae). In contrast, the fossil record demonstrates a highly diverse group extending back unequivocally to the beginning of the Eocene Epoch 55.5 million years ago.

The order is universally regarded as being monophyletic. The form of the astragalus, with its saddle-shaped navicular facet and small cuboid facet (Fig. 13.1A) is unique to the clade and found in every taxon in which this bone is known (e.g., Radinsky, 1966). Other synapomorphies include an upper molar protoloph, which is primitively stepped or kinked just lingual of the paraconule and occludes with a twinned lower molar metaconid (Fig. 13.2A–B, E–F; Hooker, 1984, 1989, 1994); nasals that are laterally flared toward the posterior end and that form an essentially transverse suture with the frontals (Fig. 13.1D; Holbrook 2001); and an entocuneiform that is rotated to a nearly horizontal orientation (Fig. 13.1E; Radinsky, 1963; Holbrook, 2001).

INFRAORDINAL CLASSIFICATION

When only the three modern perissodactyl families are considered, there is unanimous agreement on their interrelationships. Rhinocerotidae and Tapiridae are more closely related to one another than either is to Equidae, and they are combined to form the clade Ceratomorpha, which is usually treated as a suborder or infraorder.

Fig. 13.1. (A) Right astragalus in dorsal view of *"Hyracotherium."* Not to scale. (B) Right astragalus in dorsal view of *Phenacodus.* Not to scale. (A) and (B) are redrawn from Radinsky (1966: fig. 5b,c). (C) Right astragalus in dorsal view of a hyracoid (*Thyrohyrax* or *Saghatherium*) from the Oligocene of the Fayum, Egypt. Not to scale. Redrawn reversed from Rasmussen et al. (1990: fig. 1d). (D) Dorsal view of anterior part of cranium of holotype of the primitive equid *Pliolophus vulpiceps* Owen (Natural History Museum, London, BMNH.44115), from the London Clay, Harwich, showing posteriorly flared nasals. Scale bar = 10 mm. (E) Posterior view of part of right pes of *Tapirus indicus* Desmarest, Recent, showing orientation of entocuneiform and metatarsal I. modified from Radinsky (1963). Scale bar = 20 mm.

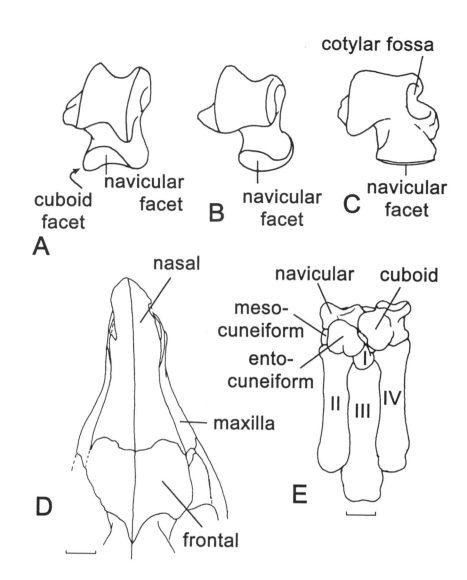

However, there are two extinct groups, the superfamilies Brontotherioidea and Chalicotherioidea, whose relationships to Equidae and Ceratomorpha have varied greatly in recent decades. The Brontotherioidea has usually been treated as monofamilial, although one of the subfamilies, Lambdotheriinae, was raised to family rank by Prothero and Schoch (1989b) to reflect the opinion that it was remote from typical brontotheres (Mader, 1989). The Chalicotherioidea has usually been divided into two families, although the Eomoropidae is accepted as being paraphyletic (Coombs, 1989; Hooker and Dashzeveg, 2004) and was subsumed into the Chalicotheriidae by Prothero and Schoch (1989b). Similar controversies surround the precise relationships of various extinct families related to either Equidae (e.g., Palaeotheriidae) or Ceratomorpha (e.g., Hyrachyidae, Isectolophidae, Lophialetidae, Rhodopagidae, Deperetellidae). For details and discussion of these, see especially Hooker (1984, 1989, 1994), Franzen (1989, 1995), Dashzeveg and Hooker (1997), Froehlich (1999, 2002), Remy (2001), Holbrook (2001), Hooker and Dashzeveg (2003) and references therein. Schoch (1989) has summarized the earlier history of perissodactyl

classification and I shall only be concerned with relatively recent developments, mainly at a level higher than family.

In common with most earlier workers, Radinsky (1964) regarded brontotheres as sister group to equoids, but he separated chalicotheres as a distinct third group (suborder), for which he resurrected a modified version of Cope's (1889) Ancylopoda (Fig. 13.3). The 1960s saw essentially the last pre-cladistic treatments of perissodactyl phylogeny (e.g., Savage et al., 1965, 1966; Radinsky, 1969). Hooker (1984, 1989), on the basis of derived character states, linked Ancylopoda (to which he added the previously ceratomorph, extinct family Lophiodontidae) with Ceratomorpha as sister group and resurrected a modified version of Haeckel's Tapiromorpha for this clade (Fig. 13.4). He also separated brontotheres either as the sister group to all other perissodactyls or as the third member of an unresolved trichotomy with Equoidea and Tapiromorpha. These interrelationships were followed by Janis et al. (1998b), but substituting Moropomorpha Schoch, 1984, for Tapiromorpha. In their classification, McKenna and Bell (1997) resurrected Borissiak's (1945, 1946) concept of a sister-group relationship

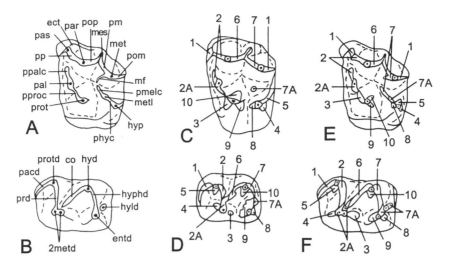

Fig. 13.2. Crown views of left upper and right lower first or second molars of a primitive perissodactyl and a phenacodontid, showing dental terminology or wear facets numbered according to Butler (1952) and Hooker (1984). Mesial is to the left, distal to the right. Not to scale. (A) The perissodactyl *Lambdotherium*, upper molar. (B) *Lambdotherium*, lower molar. (C) The phenacodontid *Ectocion*, upper molar. (D) *Ectocion*, lower molar. (E) *Lambdotherium*, upper molar. (F) *Lambdotherium*, lower molar. The postparacrista plus the premetacrista are known as the centrocrista. The protocone, preprotocrista, paraconule, preparaconule crista ± the preparacrista make up the protoloph. The hypocone, prehypocrista, metaconule, premetaconule crista ± the mesiolingual metacone fold compose the metaloph. Abbreviations: co, cristid obliqua (metalophid); ect, ectocingulum; entd, entoconid; hyd, hypoconid; hyld, hypoconulid; hyp, hypocone; hyphd, hypolophid (may be divided into posthypocristid and postentocristid); mes, mesostyle; met, metacone; 2 metd, twinned metaconid; metl, metaconule; mf, mesiolingual metacone fold; pacd, paracristid (paralophid); par, paracone; pal, paraconule; pas, parastyle; phyc, prehypocrista; pm, premetacrista; pmelc, premetaconule crista; pom, postmetacrista; pop, postparacrista; pp, preparacrista; ppalc, preparaconule crista; pproc, preprotocrista; prd, protocristid (protolophid); prot, protocone; protd, protoconid.

between brontotheres and chalicotheres, to which clade they gave the name "Selenida." They placed Selenida as sister group to Tapiroidea plus Rhinocerotoidea, and switched the earlier usages of Ceratomorpha and Tapiromorpha. They also returned the Lophiodontidae to the Tapiroidea. In these ways, they did not follow any earlier cladistic studies or the classification of Prothero and Schoch (1989b) and at the same time, provided no evidence in support of their new arrangement (Fig. 13.5). Froehlich's (1999) cladistic study also found a sister-group relationship between Ancylopoda (including Lophiodontidae) and Ceratomorpha, but linked Brontotheriidae once again with Equoidea, either as sister groups (using all characters unordered) or with brontotheres nested within a paraphyletic Equoidea (using relevant multistate characters ordered; Fig. 13.6).

The interrelationships of these major infraordinal groups, whose origins probably extend to the beginning of the perissodactyl record as currently known, are clearly not yet fully resolved. For instance, evidence against the integrity of the Tapiromorpha is beginning to emerge (e.g., Holbrook, 2001; Hooker and Dashzeveg, 2004). The latter reference provides evidence that isectolophids are stem ancylopods, which, in turn, have a sister relationship to a ceratomorph-equoid clade. This also conforms with the recent finding that the postcranials of the isectolophid *Homogalax* are more primitive than those of the most primitive equoids and approach the phenacodont condition (Rose, 1996). It is therefore not yet possible to provide even a higher-rank classification that represents a consensus view of recent cladistic phylogenies, if extinct groups are to be included. Table 13.1 represents a recommended, approximately majority view of recently published work.

PHYLOGENETIC ORIGINS

The answer to the question of what is the nearest sister group to perissodactyls remains elusive, although on morphological grounds, it is either the order Hyracoidea or the superorder Paenungulata, which today comprises the Hyracoidea plus the Tethytheria, the latter comprising the living orders Proboscidea and Sirenia, with or without the extinct orders Embrithopoda and Desmostylia (Prothero et al., 1988; Court, 1990, 1992; Thewissen and Domning, 1992; Fischer and Tassy, 1993; Janis et al., 1998a). When originally named (Simpson, 1945), Paenungulata also included Pantodonta, Dinocerata, and Pyrotheria, extinct orders now universally excluded from the superorder. An alternative name for Paenungulata that avoids the exclusion problem is Uranotheria McKenna and Bell, 1997.

Despite these uncertainties, most authors accept a more inclusive grouping for all four of these modern orders, plus

Fig. 13.3. Cladogram representing the
relationships of the major groups of
perissodactyls and phenacodontid
"condylarths," according to Radinsky (1964).

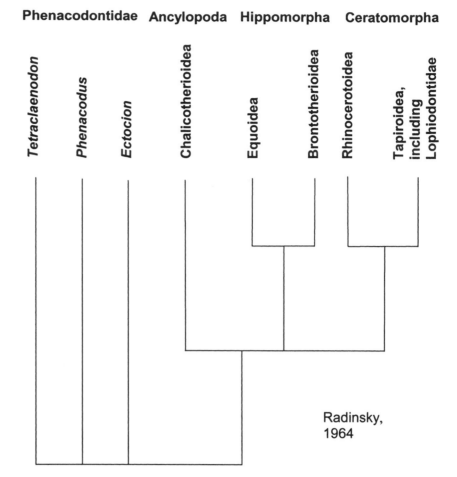

or minus the two extinct ones. The name for this clade is either "Pantomesaxonia Franz, 1924," or "Altungulata Prothero and Schoch, 1989a." The Pantomesaxonia was resurrected by Fischer (1986), who modified it by excluding from the original content the South American ungulates, phenacodontids, and meniscotheriids, and by including the sirenians. The Altungulata was erected because the extent of modification from the original concept of the emended Pantomesaxonia was regarded as extreme. Subsequently, Pantomesaxonia has been used by Thewissen and Domning (1992), Hooker (1994), Froehlich (1999), Gheerbrant et al. (2001), Thewissen and Simons (2001), and Hooker and Dashzeveg (2003). Altungulata, however, has been used by McKenna and Bell (1997). Fischer and Tassy (1993) simply referred to the group informally as "PSHM," after the initials of the included groups. On usage grounds, Pantomesaxonia seems to win.

Synapomorphies of the Pantomesaxonia are the presence of an upper molar metaloph and lower molar hypolophid; upper molar hypocone and metaloph extending onto M3, and m3 elongated with separate hypoconulid lobe; palatine fissures fused (Thewissen and Domning, 1992); and foramen stylomastoideum primitivum close to the vestibular window (Thewissen and Domning, 1992). A few additional characters used by Fischer and Tassy (1993) to support their PHSM clade are either not exclusive or do not characterize all members.

The nearest sister group to Pantomesaxonia is usually considered to be the extinct stem ungulate ("condylarth") family Phenacodontidae, with or without the Meniscotheriidae, sometimes incorporated as Phenacodonta (e.g., Thewissen and Domning, 1992; Fischer and Tassy, 1993). Phenacodonta, whether monophyletic or not, together with Pantomesaxonia, has been incorporated into a more inclusive clade Taxeopoda Cope, 1882 (e.g., Gheerbrant et al., 2001). This taxon was resurrected by Archibald (1998) and, like Pantomesaxonia, modified from its original meaning. Cope's definition included "Condylarthra," Proboscidea, and possibly some notoungulates and hyracoids, although he subsequently added and subtracted various other orders now considered remote from ungulates.

Hyracoid Relationships

When the order Perissodactyla was first named, Owen (1848) included taxa that now constitute the order Hyracoidea. For

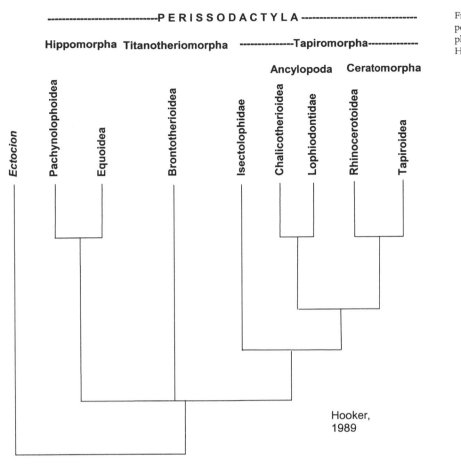

Fig. 13.4. Cladogram of the major groups of perissodactyls and the most closely related phenacodontid "condylarth," following Hooker (1989).

much of the twentieth century, hyracoids were almost universally separated from the Perissodactyla and allied with the Proboscidea (for a review, see Fischer, 1989). Most morphologic characters that favored that link have been reinterpreted as either primitive or convergent, but Fischer (1986, 1989) proposed a new set of synapomorphies for a hyrax-perissodactyl sister relationship and returned to Owen's concept of the Perissodactyla (i.e., including hyracoids), referring the Perissodactyla exclusive of Hyracoidea to Marsh's (1884) Mesaxonia. Because of the long usage of the ordinal name "Perissodactyla" excluding the Hyracoidea and because of the major doubts surrounding their interrelationships, Perissodactyla rather than Mesaxonia is used here for the group in question.

The extinct order Embrithopoda is also relevant here, because it was originally thought (Andrews, 1906) to be most closely related to Hyracoidea, these two groups together therefore potentially forming the sister group to perissodactyls. Court (1990, 1992), in a cladistic analysis using cranial (including petrosal) and postcranial characters, found instead a sister-group relationship between Embrithopoda and Proboscidea within the Tethytheria. His phylogenies show hyracoids as sister taxon to Perissodactyla plus Tethytheria, although he recognized that a sister-group relationship between Hyracoidea and Perissodactyla was only

slightly less parsimonious. He represented embrithopods only with *Arsinoitherium,* the best known—albeit the most derived—taxon. The extinct family Phenacolophidae has also been included in the Embrithopoda by McKenna and Manning (1977), and the more primitive nature of the phenacolophid dentition helps us to compare embrithopods with primitive perissodactyls that are known mainly from teeth. On this basis, the phenacolophid *Radinskya* has been thought by some to be most closely related to the Perissodactyla (see discussion below on outgroups).

Novacek and Wyss (1986) and Novacek (1992a,b) discounted some of Fischer's hyracoid-perissodactyl synapomorphies on the grounds of doubtful homology and patchy distribution in the groups concerned, some features being known or recognizable only in modern animals. They also listed alternative synapomorphies that, instead, favored a hyracoid-tethythere sister relationship. Fischer and Tassy (1993: 220) relinquished some of the characters that were earlier considered to unite hyracoids and perissodactyls and argued for others that outnumbered those supporting the Paenungulata. Their numbered characters are:

22. Upper molar parastyle enlarged;
27. Astragalar foramen absent;
31(1). Astragalar navicular facet flattened (coded

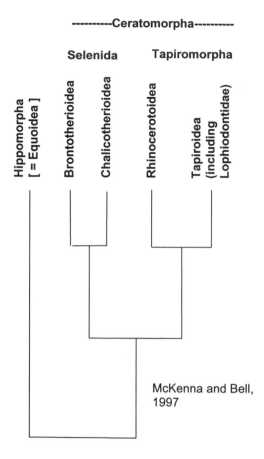

Fig. 13.5. Cladogram representing the relationships of the major groups of perissodactyls, according to the classification of McKenna and Bell (1997).

as an intermediate stage between the convex condylarth state and the saddle-shaped perissodactyl state);

64. Eustachian sac present;
65. Internal carotid extrabullar;
66. Stapedial system absent;
67. Tuber maxillae present, forming floor of orbit;
68. Metatarsals I and V reduced;
69. M. sternoscapularis inserts at the superior cranial edge of the scapula;
71. Hoof, mesial side with horned lamellae, even fused.

Subsequent work has shown that some of these characters also occur in different phenacodontids, whereas others either do not occur in primitive members of the relevant groups or are difficult to recognize in fossils. Thus, characters 22 and 27 also occur in the phenacodontid *Ectocion*, and 66 in *Phenacodus* (Thewissen, 1990). Moreover, character 27 also occurs in proboscideans and embrithopods.

The presence of an eustachian sac or diverticulum (character 64) is an important soft-part character unknown outside Perissodactyla and Hyracoidea. It has been argued that its absence in rhinoceroses is secondary, because, as in the taxa that have the sac, the rhino internal carotid artery is displaced, so that its course is extrabullar (character 65). This extrabullar course means that there is no trace of the internal carotid on the surface of the promontorium, unlike those taxa where it is perbullar. Absence of such a promontorial groove appears to be the only way in fossils of recognizing an extrabullar course for the internal carotid and, by extrapolation, presence (or at least former presence) of an eustachian sac. Thewissen and Simons (2001) noted the absence of promontorial grooves in *Phenacodus*, implying that this taxon also had an eustachian sac.

The presence of horned lamellae on the hooves of modern hyraxes and perissodactyls is thought to be recognizable in fossils by the fissuring of the ungual phalange (Fischer and Tassy, 1993). If so, the slight fissuring in *Phenacodus* too (Radinsky, 1966: fig.4) is suggestive of character 71 being present in this taxon.

The sharing of characters 22, 65, 66, and 71 with phenacodontids does not, in itself, invalidate the hyracoid-perissodactyl relationship, however, as it simply suggests that some phenacodontids are more closely related to these orders than to other pantomesaxonians. Nevertheless, other characters of phenacodontids would not be congruent with them.

The reduction of metatarsals I and V (character 68) is shared by all hyracoids and perissodactyls in which these bones are known. However, the importance of this synapomorphy is weakened by the poor knowledge of hyracoid postcranials prior to the Oligocene (Thomas et al., 2004).

Character 69, insertion of the sternoscapularis muscle at the superior cranial edge of the scapula, is recognized in fossils by the loss of the acromion process. As noted by Novacek and Wyss (1986), the presence of an acromion in a number of primitive perissodactyls—in particular, primitive equoids (Kitts, 1956) and the tapiroid *Heptodon* (Radinsky, 1965)—means that this character is not a synapomorphy of hyracoids and perissodactyls.

The presence of a tuber maxillae, forming the floor of the orbit (character 67) has been observed by Thewissen and Simons (2001) to occur also in *Phenacodus, Meniscotherium,* and derived Proboscidea. However, these authors considered that a modification of this character, namely, "proximity of maxilla and pterygoid," could be maintained as a hyracoid-perissodactyl synapomorphy.

Character 31, state 1 (astragalar navicular facet flattened), was coded as an intermediate stage between the convex condylarth state (31, state 0) and the saddle-shaped perissodactyl state (31, state 2). Perissodactyls share with phenacodontids a proximal extension, or wrap-around, of the navicular facet (Fig. 13.1A,B), so that it is partly visible in anterior view (Radinsky, 1966). No such extension is present in hyracoids, the flat or slightly mediolaterally convex navicular facet being entirely restricted to the distal face (Fig.

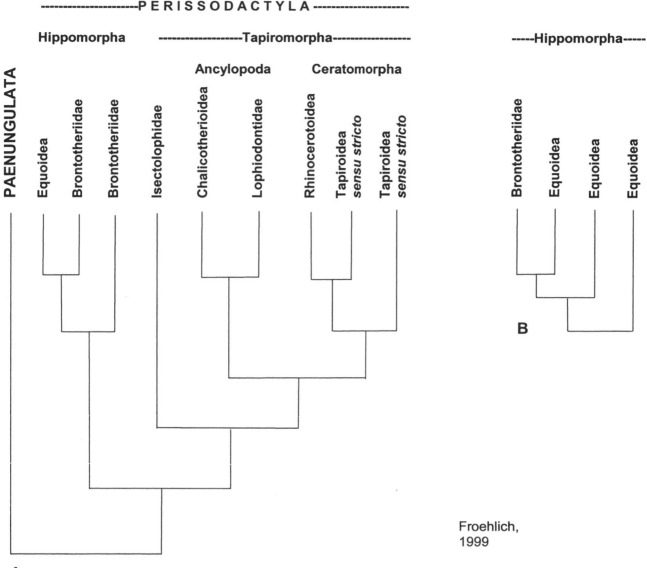

A

B

Froehlich, 1999

Fig. 13.6. Cladogram of the major groups of perissodactyls and their nearest sister group, Paenungulata, following Froehlich (1999). (A) The entire cladogram with multistate characters entered as unordered. (B) The hippomorph clade analyzed from the same matrix, but with relevant multistate characters entered as ordered. Taxa shown multiple times in the cladograms are interpreted to be paraphyletic.

13.1C; Rasmussen et al., 1990). This state in hyracoids is thus clearly derived and unconnected with the saddle shape of the perissodactyl navicular facet, which is thus independently derived. Interestingly, the hyracoid state is also shared with primitive proboscideans, as is a distinct cotylar fossa (Fig. 13.1C) for articulation with the medial malleolus of the tibia (Andrews, 1906; Rasmussen et al., 1990). This cotylar fossa is not present in perissodactyls.

This survey of the most recently published synapomorphies of hyracoids and perissodactyls suggests that only Fischer and Tassy's characters 67 (modified) and, tentatively, 68 can be strictly used to favor a hyrax-perissodactyl sister relationship, thus weakening this hypothesis. Discussion of alternative characters favoring the Paenungulata grouping is outside the scope of this chapter and is dealt with elsewhere (Gheerbrant et al., chapter 7, this volume).

Potential Solutions: Establishing Basal Characters

The biggest problem with the conflicting characters involved in pantomesaxonian ingroup relationships is that they occur or at least potentially occur only in derived members. As has often been pointed out (e.g., Hooker, 1989; Holbrook, 2001), it is important to recognize primitive members of clades to establish their basal characters.

Perissodactyls have a good record back to their earliest occurrence, although cranial and postcranial remains are still patchy for the earliest records. Hyracoids, however, have a very poor record prior to the latest Eocene, and their postcranials are only just beginning to be recovered from rocks of this age (Tabuce et al., 2001). The diversity of latest Eocene and early Oligocene hyracoids, as well as the known derived nature of the contemporaneous perissodactyls,

Table 13.1　Higher-rank classification and distribution of perissodactyls

Order PERISSODACTYLA Owen, 1848
　Suborder †TITANOTHERIOMORPHA Hooker, 1989
　　　Superfamily Brontotherioidea Marsh, 1873
　　　　Family †Lambdotheriidae Cope, 1889 (e–mEoc, NA, Asia)
　　　　Family †Brontotheriidae Marsh, 1873 (e–lEoc, NA; m–lEoc, Asia; ?lEoc, Eu (SE only))
　Suborder HIPPOMORPHA Wood, 1937
　　　Superfamily Equoidea Gray, 1821
　　　　Family Equidae Gray, 1821 (eEoc–Pleist, NA; eEoc, eMioc–R, Eu; eMioc–R, Asia; lMioc–R, Af; lPlioc—Pleist, SA)
　　　　Family †Palaeotheriidae Bonaparte, 1850 (eEoc–lOlig, Eu)
　Suborder TAPIROMORPHA Haeckel, 1866
　　　　Family †"Isectolophidae" Peterson, 1919 (e–mEoc, NA, Asia)
　　　　Family †Pachynolophidae Pavlow, 1888 (e–lEoc, Eu, ?eOlig, Asia)
　　Infraorder †ANCYLOPODA Cope, 1889
　　　　Family †Chalicotheriidae Gill, 1872 (mEoc–Pleist, Asia; mEoc, e–mMioc, NA; eOlig–ePlioc, Eu; mMioc–Pleist, Af)
　　　　Family †Lophiodontidae Gill, 1872 (e–mEoc, Eu; e—?m, ?lEoc, NA)
　　Infraorder CERATOMORPHA Wood, 1937
　　　　Family †Lophialetidae Matthew and Granger, 1925 (mEoc—?lEoc, Asia)
　　　Superfamily Tapiroidea Gray, 1821
　　　　Family † "Helaletidae" Osborn, 1892 (eEoc–eOlig, NA; mEoc–eOlig,?lOlig, Asia; eOlig, Eu)
　　　　Family †Deperetellidae Radinsky, 1965 (m–lEoc, ?eOlig, Asia)
　　　　Family Tapiridae Gray, 1821 (mEoc–Pleist, NA; Olig-Pleist, Eu; mMioc–R, Asia; lPlioc–R, SA)
　　　Superfamily Rhinocerotoidea Gray, 1821
　　　　Family †"Hyrachyidae" Osborn, 1892 (e–mEoc, NA, Asia, Eu)
　　　　Family †Hyracodontidae Cope, 1879 (mEoc–eMioc, Asia; mEoc–lOlig, NA; e–lOlig, Eu, ?eMioc, SE Eu)
　　　　Family †Rhodopagidae Reshetov, 1975 (mEoc, Asia)
　　　　Family †Amynodontidae Scott and Osborn, 1883 (mEoc–eOlig, ?lOlig, Asia; mEoc–eOlig, NA; Olig–?eMioc, Eu)
　　　　Family Rhinocerotidae Gray, 1821 (mEoc–R, Asia; mEoc–ePlioc, NA; mEoc (SE), eOlig–Pleist, Eu; eMioc–R, Af)

Notes: Quote marks denote an unresolved, but commonly used, paraphyletic group. Epoch abbreviations: Pal, Paleocene; Eoc, Eocene; Olig, Oligocene; Mio, Miocene; Plio, Pliocene; Pleist, Pleistocene; R, Recent; e, m, or l preceding epochs designate early, middle or late.

suggests that our current knowledge of hyracoid character states may well not represent basal ones for the order. Early Eocene *Seggeurius* is usually considered the most primitive hyracoid (Court and Mahboubi, 1993), but the meager sample shows a few features that appear more derived than some Oligocene genera, and the premolars are very poorly known.

Early equoids that have classically been referred to the genus *Hyracotherium* have long been regarded as the most primitive perissodactyls and applied to investigations into the origins of the order (e.g., Novacek, 1992b,c; Thewissen and Domning, 1992). An otherwise paraphyletic genus, *Hyracotherium* has recently been restricted to the European type species *H. leporinum* Owen, 1841, which is now interpreted as a stem palaeotheriid (Hooker, 1989, 1994; Froehlich 1999, 2002), and other species from Europe and North America are referred to other genera (Hooker, 1984, 1994; Froehlich, 2002). It is the members of these other genera that are the source of the original concept of the primordial perissodactyl. Thus, Radinsky (1966) argued that perissodactyls must be derived from the most primitive phenacodontid *Tetraclaenodon*, as only this genus lacked an upper molar mesostyle like "*Hyracotherium*." The other phenacodontid genera that bear mesostyles—*Phenacodus* and *Ectocion*—were regarded as offshoots from the perissodactyl stem and not ancestral to the order Perissodactyla. Two assumptions

were being made: first, that the earliest known perissodactyl was automatically the most primitive and second, that the mesostyle character was irreversible. Neither is an established fact, but can be tested for parsimony by cladistic analysis. Results of analyses can vary, depending on the choice of outgroup, whether multistate characters are ordered or not, and exactly how characters are defined and coded.

Hooker (1989) hypothesized that perissodactyls like "*Homogalax* sp." (a species now placed in the related genus *Cardiolophus* Gingerich, 1991), although functionally bilophodont, bore a small, apparently functionless mesostyle, and that this mesostyle must be vestigial and therefore indicative of an origin in a mesostyle-bearing group of mammals. He envisaged the primitive perissodactyl dental state as essentially resembling the primitive equoid *Cymbalophus* Hooker, 1984, which, unlike "*Homogalax*," was less derived for bilophodonty in retaining intermediate conules on the upper molars and a small hypoconulid lobe on m3. In his analysis, Hooker used a hypothetical outgroup; his analysis did not support *Cymbalophus* as the stem perissodactyl.

CHOICE OF OUTGROUP.　Subsequent studies of perissodactyl phylogeny have used specific outgroups. All have used at least *Phenacodus*, and most have employed multiple outgroups, as implicitly advocated by Maddison et al. (1984). Thus, Fischer and Tassy (1993) used the hyopsodontid *Hyopsodus,* the phenacodontans *Phenacodus* and *Meniscotherium,* and the phenacolophid *Radinskya,* but included only one perissodactyl (*"Hyracotherium"*), as their study was dealing broadly with pantomesaxonian relationships. In his analysis of basal perissodactyls, Froehlich (1999) also used multiple outgroups; namely, the Phenacodontidae (based on *Phenacodus* and *Ectocion*), the phenacolophids *Radinskya* and *Phenacolophus,* the Tethytheria (based on *Numidotherium*), and the Hyracoidea (based on *Seggeurius,* plus various Fayum taxa for the postcranials). He identified no basal taxon. The problem with these outgroups is the diversity of form, especially dental, including varying degrees of bilophodonty, selenolophodonty, and bunodonty, which provides conflicting character polarity. However, multiple outgroups are by no means a necessity, and it is recommended that the outgroup should possess "more inclusive synapomorphies shared with the ingroup" (Nixon and Carpenter, 1993: 421).

Hooker (1994), in an analysis of primitive equoids, found that opposite polarities were obtained by using non-equoid perissodactyls vs. phenacodontids as the outgroup. As non-equoid perissodactyls were apparently as derived as the group he was analyzing, Hooker opted to use the more primitive phenacodontid *Phenacodus* as outgroup and include the more derived phenacodontid *Ectocion* in the ingroup. *Phenacodus* was perceived to be sufficiently generalized among placental mammals to provide primitive character states, yet shared enough derived similarities with perissodactyls to facilitate the homologizing of character states across the matrix. The result was that Brontotheriidae (*sensu lato*) split off first and represented the most primitive perissodactyl structural type. A broader study of primitive perissodactyls also used *Phenacodus* as outgroup and two other phenacodontids, together with perissodactyls, in the ingroup (Hooker and Dashzeveg, 2003).

The use of *Radinskya* as an outgroup taxon often goes beyond its concept as a primitive embrithopod. Instead, it is often regarded as the non-perissodactyl most closely related to perissodactyls following the original assessment of McKenna et al. (1989). However, it shares with phenacolophids (themselves usually accepted as primitive embrithopods) the following derived character states on its upper molars: a crestiform parastyle and a tall metaconule close to the hypocone, lying exactly on a line drawn between the metacone and hypocone (Hooker and Dashzeveg, 2003). The perissodactyl-like bilophodonty is unlike that in primitive perissodactyls, in which the metaloph is formed around a small, mesially-shifted metaconule. The π-shaped loph structure more closely resembles that found in more derived ceratomorphs and must therefore be convergently evolved. This distances *Radinskya* from the Perissodactyla—although, according to other characters (McKenna et al., 1989), it is likely to share with this order a derivation from within the phenacodontid "condylarths."

CHARACTER TYPES. Froehlich's (1999) study provided results using multistate characters as both ordered and unordered. He preferred the ordered version, because it accorded better with stratigraphic order. The results from each choice were rather different (see Fig. 13.6). Unordered, the brontotheres are stem members of an equoid clade, and some palaeotheres are stem equids. Ordered, the palaeotheres all plot together but have the brontotheres nested within them. Using states unordered avoids assuming prior information about a given transformation sequence. However, if the various states show morphological continuity, not ordering the states could weight transformations that are not the most parsimonious. The second of these problems seems to be the more serious.

Holbrook (2001) treated most of his multistate characters as unordered. He did this both when analyzing his own matrix, which included only one non-tapiromorph perissodactyl, and when reanalyzing the emended matrices of Thewissen and Domning (1992) and Court (1992). In the case of the reanalysis of Thewissen and Domning (1992), it is telling that the choice between a perissodactyl-hyrax vs. a perissodactyl-paenungulate sister relationship depended on whether Holbrook followed Thewissen and Domning in scoring basal perissodactyl character states based on *"Hyracotherium,"* or whether he interpreted these states from the observed variation in a number of taxa.

Hooker (1994) used multistate characters ordered in nearly every case. Hooker and Dashzeveg (2003) also added two stepmatrices to reflect certain dichotomizing characters (see Lipscomb, 1992). The stepmatrices prevent the duplication of primitive states, which, if reversed, would create overweighting.

CHARACTER DEFINITION AND CODING. When defining characters, it is important to avoid inadvertently weighting some characters over others. Tooth characters linked by occlusal function is an example of such erroneous preferential weighting. Combining such tooth characters will reduce the inappropriate weighting (e.g., Hooker, 1989). Examining wear facets can also help to understand cusp homology.

The formation of a step in the upper molar protoloph of perissodactyls divides the lower molar metaconid into two parts (Fig. 13.2A,B,E,F). The distal metaconid is therefore distinct from the metastylid, which occludes with the protocone at facet 3 (Fig. 13.2C,D), and with which it has been long confused. If the step in the protoloph is treated as a character distinct from the twinned metaconid, it will double the weight of the character. Examination of the wear facets in this case also shows that the more distal of the twinned metaconid cusps is not homologous with the metastylid, which occurs in many groups, including phenacodontids (Hooker, 1994).

New Possibilities

The previous sections outline some of the problems encountered in various attempts to achieve a consensus on

Fig. 13.7. Cladogram of primitive perisso-dactyls and phenacodontid "condylarths," following Hooker and Dashzeveg (2003). Letters appended to taxa indicate their occurrence in the continents of Asia (A), Europe (E), and/or North America (NA).

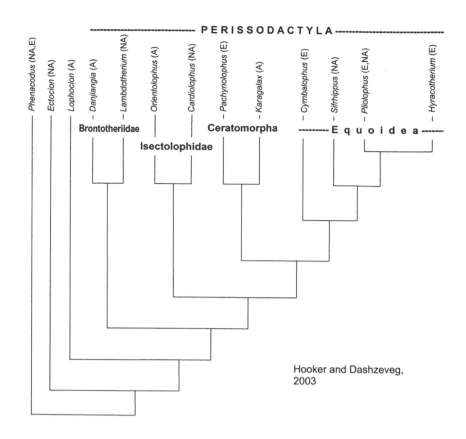

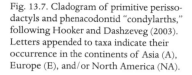

perissodactyl phylogeny. In this section, I consider how some of the conflicts might be addressed.

By adding newly discovered Asian taxa (the perissodactyl *Orientolophus* Ting, 1993, and the phenacodontid *Lophocion* Wang and Tong, 1997), by using matrices for which the multistate characters are ordered according to morphological continuity and for which functionally linked characters are combined, and by employing *Phenacodus* as the outgroup (with the phenacodont *Ectocion* also included in the analysis), Hooker and Dashzeveg (2003) obtained the result shown in Fig. 13.7.

Their analysis results in Brontotheriidae *sensu lato* (represented here by *Lambdotherium* and *Danjiangia*), not equoids, emerging as the most primitive perissodactyl group. As a result, the presence of an upper molar mesostyle is primitive for the order and bilophodonty does not evolve until the next node up, which accounts for some level of bilophodonty in primitive members of all three remaining groups: Isectolophidae, Ceratomorpha, and Equoidea. This bilophodonty in tapiromorphs and equoids was subsequently overprinted by a secondary selenodonty, resulting in the selenolophodont dentitions of derived equoids and chalicotheres.

A new analysis that also includes primitive chalicotheres (Hooker and Dashzeveg, 2004) has similar topology. However, Radinsky's hypothesis that *Tetraclaenodon,* rather than any other phenacodont, is more closely related to perissodactyls has not been tested. To this end, therefore, *Tetraclaenodon* was added to the Hooker and Dashzeveg matrix, replacing *Phenacodus* as the outgroup, and characters were

recoded to attribute primitive states to *Tetraclaenodon*. The result was that, as before, *Phenacodus, Ectocion,* and *Lophocion* branched from successive nodes below Perissodactyla, falsifying Radinsky's hypothesis and further supporting the idea that the presence of the upper molar mesostyle is primitive for perissodactyls.

Dental similarities between perissodactyls and hyraxes have often been remarked upon, and these are greater if brontotheres, rather than equoids, are stem perissodactyls. However, very few dental characters have been employed in most analyses supporting a hyracoid-perissodactyl relationship (Fischer and Tassy, 1993, is an exception), even though the majority of characters in the perissodactyl cladograms are of necessity dental. To see where hyraxes would fit, a primitive hyracoid, based mainly on *Seggeurius,* was added to the matrix without adding more characters. The result was that hyracoids were placed as the sister group to perissodactyls. However, this result does not mean that hyracoids are necessarily the nearest sister group to perissodactyls. Tethytheres, if included, might well plot out at around the same node, tentatively suggesting an origin for the Pantomesaxonia from a derived phenacodontid and making this family paraphyletic.

A few characters of hyraxes are apparently more primitive than those of *Tetraclaenodon* and may be biasing the result. These characters are:

1. The absence in most hyraxes of an upper premolar postprotocrista;

2. The P4 paracone and metacone close together but subequal in size (this might indicate a separate origin from a condylarth more primitive than *Tetraclaenodon* by splitting the paracone into two, instead of the metacone appearing as a new cusp on the postparacrista, as apparently happened in *Tetraclaenodon*); and

3. The absence of an entoconid on all lower premolars (the entoconid is often reduced to a crest in perissodactyls with nonmolariform premolars, but is constant in all phenacodontids; in most early hyraxes, it is missing along with a basined talonid).

Adding to the matrix an arctocyonid "condylarth" (*Protogonodon*) with similarities to phenacodontids (and whose states for the above characters are at least as primitive as those of hyraxes) and replacing *Tetraclaenodon* as outgroup still resulted in hyraxes and perissodactyls being sister taxa. This result thus indicates a reversal for the above listed characters in hyraxes. The absence of a p4 entoconid in hyraxes was regarded as primitive by Tabuce et al. (2001), but polarity was not explicitly argued, and no outgroup was specified by these authors.

The biggest problem, however, for assessing perissodactyl-hyracoid relationships is that many potential characters are missing from the fragmentary earliest members of both groups. This will clearly only be resolved by finding more complete material of early fossil forms. The description of more pre-Miocene hyrax postcranials (recently begun for articulated skeletons from the Oligocene; Thomas et al., 2004), as well as a rigorous cladistic analysis of the group using both dental and osteological characters, should also produce more evidence to support or refute a morphological relationship with Perissodactyla.

Whether the grouping Pantomesaxonia (i.e., Perissodactyla, Hyracoidea, Proboscidea, Sirenia, plus or minus Embrithopoda, with an origin distinct from a monophyletic Phenacodontidae) or the more inclusive Taxeopoda (i.e., these groups with origins within a paraphyletic Phenacodontidae) more accurately reflects reality depends on whether *Ectocion* and *Lophocion* are closer to Perissodactyla (with or without Hyracoidea) than to the other groups. The analyses of Fischer and Tassy (1993) and Thewissen and Domning (1992) found a monophyletic Pantomesaxonia. However, the former used only one phenacodontid, *Phenacodus,* and combined characters of ancient and modern perissodactyls, whereas the latter used "*Hyracotherium*" as the primitive perissodactyl, and evidence for a monophyletic Phenacodontidae (Phenacodonta) was weak.

Current molecular phylogenies could hardly be more different from the morphology-based ones. The former place hyracoids with the Proboscidea and Sirenia (Tethytheria) as Paenungulata, not with the Perissodactyla (e.g., Liu and Miyamoto, 1999; Springer et al., 1999). Scally et al. (2001) placed Perissodactyla closest to Carnivora, with or without Pholidota, or to Cetartiodactyla, together combined with Chiroptera and some Lipotyphla (Eulipotyphla) in a major northern hemisphere clade, Laurasiatheria. Thus, the Perissodactyla were considered remote from the Tethytheria, which Scally et al. included in a major southern hemisphere clade, the Afrotheria. The analysis of Jow et al. (2002) showed Perissodactyla as either nested within Cetartiodactyla or as polyphyletic, split by Carnivora and with Equidae as sister taxon to Cetartiodactyla. There is much inconsistency in the detail of these results, although the proximity of Perissodactyla to Cetartiodactyla and Carnivora and its remoteness from Paenungulata is a recurring pattern. If these results are broadly correct, a large number of morphological characters hitherto recognized as congruent would have to be homoplastic. However, problems in the mtDNA sequences have raised some doubts over the integrity of the clade Afrotheria (Corneli, 2003).

GEOGRAPHIC ORIGIN

The place of origin of perissodactyls has long been disputed. Members of the order appear suddenly by dispersal at the beginning of the Eocene in North America, Asia, and Europe. Theories of geographic origin have centered on four main areas: Central America, Africa, India, and non-Indian Asia.

Sloan (1969) and Gingerich (1976) suggested an origin in Central America. This was based on the theory that Paleocene cooling pushed various Torrejonian taxa south, removing them from the North American Tiffanian record (their descendants reappearing in North America in the Eocene, when temperatures recovered). Support for this conjecture is provided by the North American continent being the main area of diversity for phenacodontid "condylarths." Arguing against this, however, is that the earliest perissodactyls are of low diversity in North America.

Gingerich (1986, 1989) and Franzen (1989) suggested an African origin, supported by the existence of the possible sister group Hyracoidea in that continent. However, in this scenario, perissodactyls would have to have become extinct in Africa immediately following their dispersal to Northern Hemisphere continents, as no pre-Miocene perissodactyls have ever been recorded in the African continent. Although not impossible, this is an unlikely scenario that is not backed up by fossil evidence.

Krause and Maas (1990) suggested an Indian "Noah's Ark" origin, the docking of India with Asia providing for the simultaneous appearance of this and several other major groups of mammals. In view of the long history (extending into the Mesozoic) of India as an island and the close relationship to perissodactyls of exclusively Paleocene phenacodontids, this center of origin has little support.

Beard (1998) suggested Asia (excluding India) as the origin, based on his view that *Radinskya* represents a stem perissodactyl and on fragmentary teeth of a *Lambdotherium*-like perissodactyl in the Chinese late Paleocene site of Bayan Ulan. McKenna et al. (1989), however, regarded *Radinskya*

Fig. 13.8. Cladogram from Figure 13.7 plus *Tetraclaenodon,* with nodes constrained by the geomagnetic polarity timescale (Berggren et al., 1995). Taxonomic ranges are indicated by broad bars, accurately (solid) or tentatively (hatched) calibrated. *Karagalax* and *Danjiangia* are too imprecisely dated for their range to be indicated, although they are probably both late early Eocene (Wang, 1995; Maas et al., 2001). These ranges are from Hooker (1994), Janis et al. (1998a,b), Ting (1998), Bowen et al. (2002), and Froehlich (2002). North American Land Mammal Ages (NALMA) divided into numbered biozones are calibrated to the timescale according to Butler et al. (1987), Clyde et al. (1994, 2001), and Clemens (2002). Abbreviations: BR, Bridgerian; CF, Clarkforkian; PU, Puercan; Torrejon, Torrejonian; Wasatch, Wasatchian.

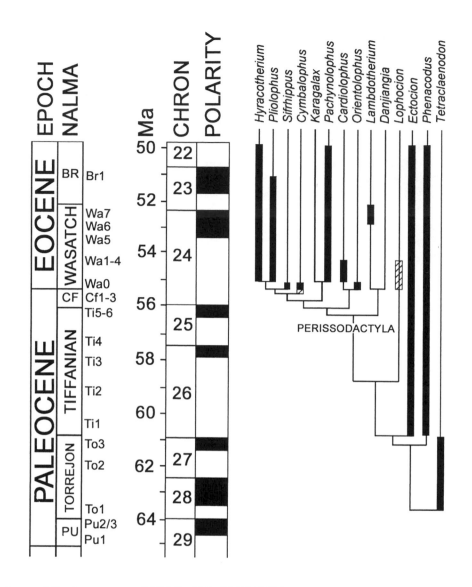

as a primitive embrithopod, and Hooker and Dashzeveg (2003) have listed characters establishing the genus as a primitive phenacolophid embrithopod with apomorphies distinct from any perissodactyl (see above). Moreover, the Bayan Ulan perissodactyl is from early collections. The taxon has not been recollected at Bayan Ulan and might represent faunal mixing from another locality and horizon (Meng et al., 1998).

The finding in China of a phenacodontid "condylarth" (*Lophocion;* Wang and Tong, 1997) that shares more characters with perissodactyls than does *Ectocion* provides the strongest support so far for the origin of the order. Using Hooker and Dashzeveg's (2003) cladogram, a combination of estimated node dates (Fig. 13.8) and of continental locations of terminal taxa (see Fig. 13.7) is strongly suggestive of an Asian origin for perissodactyls. This origin was followed by a colonization of Europe by horses and other perissodactyls, via the Turgai Straits, then of North America via the Greenland land bridge. North America was colonized via the Bering land bridge by brontotheres, ancylopods and ceratomorphs, which then dispersed to Europe (minus the brontotheres) via the Greenland land bridge.

TIME OF ORIGIN

The time ranges added to Hooker and Dashzeveg's (2003) cladogram constrain the possibilities for time of origin. There is essentially a good correspondence between order of nodes and stratigraphic sequence (Fig. 13.8). The early Eocene age of *Lophocion,* however, is an exception and suggests an earlier ghost range. Such a range extension would not be unexpected as the closely related North American phenacodontid *Ectocion* ranged from the beginning of the late Paleocene (Tiffanian) at 60.9 million years ago until the latest early Eocene (Bridgerian) at 50 million years ago. The earliest *Lophocion* is unlikely to be earlier than the earliest record of *Ectocion,* so perissodactyl origination should fall within the six-million-year span of the Tiffanian perhaps plus early Clarkforkian (56–60.9 Ma).

Whether this represents the divergence time from the nearest modern sister group depends on whether the relationship to other pantomesaxonian orders is closer than to members of the Phenacodontidae. If it does represent the divergence time, the implied rapid diversification might explain the degree of homoplasy and thus, the difficulties encountered in trying to resolve the interordinal relationships of the Pantomesaxonia.

The estimate of 87.6 million years ago (Coniacian, Late Cretaceous; Waddell et al., 1999) as the divergence time of the Perissodactyla, based on mitochondrial DNA, is much older. Nevertheless, Waddell and coworkers' estimate of 55 million years ago for the horse-rhino split is close to that provided by the fossil record. These results accord with the findings of Corneli (2003) that mtDNA genes, although producing highly resolved intraordinal divergence times, are inadequate for defining interordinal ones.

SUMMARY

The monophyly of the order Perissodactyla is well established. However, based on morphology, it remains unresolved as to whether perissodactyls are more closely related to hyracoids or to paenungulates. Molecular studies support the grouping of hyracoids in Paenungulata within the major clade Afrotheria and place perissodactyls remote from this group, suggesting instead a sister-group relationship with either Carnivora or Cetartiodactyla in the major clade Laurasiatheria.

A monophyletic Pantomesaxonia is tentatively supported by morphology, as is a paraphyletic Phenacodontidae. Molecular studies, however, split Pantomesaxonia between Afrotheria and Laurasiatheria.

Which one of the morphological cladograms the reader chooses to follow depends on his or her preference for the interpretation and coding of characters. An interpretation of characters that reflects morphological continua and minimizes inadvertent, unequal weighting gives the Brontotheriidae *sensu lato* as stem perissodactyls and provides basal character states for the order, on which broader analyses can be based. Cladistic analysis of primitive perissodactyls plus phenacodontids, combined with palaeogeography, points to Asia as the center of origin for the order Perissodactyla.

The fossil record, where stratigraphy is concordant with order of nodes on the morphology-based cladogram, is indicative of an origin in the late Paleocene—56. to 60.9 million years ago. This may also represent the divergence time from either Hyracoidea or Paenungulata, but molecular evidence gives a much older divergence time of 87.6 million years ago (Late Cretaceous).

To resolve the question of phylogenetic origin, we should concentrate on finding out much more about the teeth and skeleton of the most primitive hyracoids to evaluate their relationships to perissodactyls and tethytheres; undertake more detailed cladistic analysis of all primitive pantomesaxonians, together with relevant "condylarths;" learn more about the relationships of the various "condylarth" groups; and attempt to find unquestionable Paleocene perissodactyls—probably in Asia. These efforts will put the morphological arguments on a more secure footing and help to test more rigorously the phylogenies provided by the molecular data.

ACKNOWLEDGMENTS

I thank many colleagues for illuminating discussions over the years on the subject of perissodactyl relationships and origins; especially Percy Butler, Bill Clemens, Matt Colbert, Margery Coombs, Nick Court, Demberel Dashzeveg, Cyrille Delmer, Daryl Domning, Martin Fischer, Jens Franzen, David Froehlich, Emmanuel Gheerbrant, Phil Gingerich, Marc Godinot, Luke Holbrook, Spencer Lucas, Malcolm McKenna, David Polly, Don Prothero, the late Len Radinsky, Jean Remy, Ken Rose, Don Russell, Robert Schoch, Pascal Tassy, Hans Thewissen and Suyin Ting. Critical comments by Luke Holbrook, Don Prothero, and Ken Rose improved the chapter. I am also grateful to David Archibald and Ken Rose for inviting me to contribute to the symposium commemorating Simpson's birth, and to this volume.

REFERENCES

Andrews, C. W. 1906. A Descriptive Catalogue of the Tertiary Vertebrata of the Fayum, Egypt. British Museum (Natural History), London.

Archibald, J. D. 1998. Archaic ungulates ("Condylarthra"); pp. 292–331 in C. M. Janis, K. M. Scott, and L. L. Jacobs (eds.), Evolution of Tertiary Mammals of North America. Volume 1. Terrestrial Carnivores, Ungulates, and Ungulate-like Mammals. Cambridge University Press, Cambridge.

Beard, K. C. 1998. East of Eden: Asia as an important center of taxonomic origination in mammalian evolution. Bulletin of the Carnegie Museum of Natural History 34: 5–39.

Berggren, W. A., D. V. Kent, C. C. Swisher III, and M.-P. Aubry. 1995. A revised Cenozoic geochronology and chronostratigraphy; pp. 129–212 in W. A. Berggren, D. V. Kent, M.-P. Aubry, and J. Hardenbol (eds.), Geochronology, Time Scales, and Global Stratigraphic Correlation. SEPM Special Publication 54. Society for Sedimentary Geology, Tulsa.

Bonaparte, C.-L. 1850. Conspectus Systematis Mastozoologiae. Editio altera reformata. In: Conspectus Systematum. Lugduni Batavorum with E. J. Brill, Leiden.

Borissiak, A. 1945. Chalicotheres as a biological type. American Journal of Science 243: 667–679.

———. 1946. Novyi predstavitel' khalikoteriev iz tretichnykh otlozhenii Kazakhstana. [A new chalicothere from the Tertiary of Kazakhstan.] Trudy Paleontologicheskogo Instituta 13(3): 1–135, 14 plates. (In Russian.)

Bowen, G. B., W. C. Clyde, P. L. Koch, S.-Y. Ting, J. Alroy, T. Tsubamoto, Y.-Q. Wang, and Y. Wang. 2002. Mammalian dispersal at the Paleocene-Eocene boundary. Science 295: 2062–2065.

Butler, P. M. 1952. The milk-molars of Perissodactyla, with remarks on molar occlusion. Proceedings of the Zoological Society of London 121: 777–817.

Butler, R. F., D. W. Krause, and P. D. Gingerich. 1987. Magnetic polarity stratigraphy and biostratigraphy of middle-late Paleocene continental deposits of south-central Montana. Journal of Geology 95: 647–657.

Clemens, W. A. 2002. Evolution of the mammalian fauna across the Cretaceous-Tertiary boundary in northeastern Montana and other areas of the Western Interior; pp. 217–245 *in* J. H. Hartman, K. R. Johnson, and D. J. Nichols (eds.), The Hell Creek Formation and the Cretaceous-Tertiary Boundary in the Northern Great Plains: An Integrated Continental Record of the End of the Cretaceous. Special Paper 361. Geological Society of America, Boulder, Colorado.

Clyde, W. C., N. D. Sheldon, P. L. Koch, G. F. Gunnell, and W. S. Bartels. 2001. Linking the Wasatchian/Bridgerian boundary to the Cenozoic global climate optimum: New magneto-stratigraphic and isotopic results from South Pass, Wyoming. Palaeogeography, Palaeoclimatology, Palaeoecology 167: 175–199.

Clyde, W. C., J. Stamatakos, and P. D. Gingerich. 1994. Chronology of the Wasatchian Land-Mammal Age (early Eocene): Magnetostratigraphic results from the McCullough Peaks section, northern Bighorn Basin, Wyoming. Journal of Geology 102: 367–377.

Coombs, M. C. 1989. Interrelationships and diversity in the Chalicotheriidae; pp. 438–457 *in* D. R. Prothero and R. M. Schoch (eds.), The Evolution of Perissodactyls. Oxford University Press, New York.

Cope, E. D. 1879. On the extinct species of the Rhinoceridae of North America and their allies. Bulletin of the United States Geological and Geographical Survey of the Territories 5: 227–237.

———. 1882. On the Taxeopoda, a new order of Mammalia. American Naturalist 16: 522–523.

———. 1889. The Vertebrata of the Swift Current River, II. American Naturalist 23: 151–155.

Corneli, P. S. 2003. Complete mitochondrial genomes and eutherian evolution. Journal of Mammalian Evolution 9: 281–305.

Court, N. 1990. Periotic anatomy of *Arsinoitherium* (Mammalia, Embrithopoda) and its phylogenetic implications. Journal of Vertebrate Paleontology 10: 170–182.

———. 1992. The skull of *Arsinoitherium* (Mammalia, Embrithopoda) and the higher order interrelationships of ungulates. Palaeovertebrata 22: 1–43.

Court, N., and M. Mahboubi. 1993. Reassessment of lower Eocene *Seggeurius amourensis:* Aspects of primitive dental morphology in the mammalian order Hyracoidea. Journal of Paleontology 67: 889–893.

Dashzeveg, D., and J. J. Hooker. 1997. New ceratomorph perissodactyls (Mammalia) from the Middle and Late Eocene of Mongolia: Their implications for phylogeny and dating. Zoological Journal of the Linnean Society 120: 105–138.

Fischer, M. S. 1986. Die Stellung der Schliefer (Hyracoidea) im phylogenetischen System der Eutheria. Courier Forschungsinstitut Senckenberg 84: 1–132.

———. 1989. Hyracoids, the sister-group of perissodactyls; pp. 37–56 *in* D. R. Prothero and R. M. Schoch (eds.), The Evolution of Perissodactyls. Oxford University Press, New York.

Fischer, M. S., and P. Tassy. 1993. The interrelation between Proboscidea, Sirenia, Hyracoidea, and Mesaxonia: The morphological evidence; pp. 217–234 *in* F. S. Szalay, M. J. Novacek, and M. C. McKenna (eds.), Mammal Phylogeny: Placentals. Springer-Verlag, New York.

Franz, V. 1924. Die Geschichte der Organismen. G. Fischer, Jena.

Franzen, J. L. 1989. Origin and systematic position of the Palaeotheriidae; pp. 102–108 *in* D. R. Prothero and R. M. Schoch (eds.), The Evolution of Perissodactyls. Oxford University Press, New York.

———. 1995. Die Equoidea des europäischen Mitteleozäns (Geiseltalium). Hallesches Jahrbuch für Geowissenschaften B17: 31–45.

Froehlich, D. J. 1999. Phylogenetic systematics of basal perissodactyls. Journal of Vertebrate Paleontology 19: 140–159.

———. 2002. Quo vadis eohippus? The systematics and taxonomy of the early Eocene equids (Perissodactyla). Zoological Journal of the Linnean Society 134: 141–256.

Gheerbrant, E., J. Sudre, M. Iarochene, and A. Moumni. 2001. First ascertained African "condylarth" mammals (primitive ungulates: cf. Bulbulodentata and cf. Phenacodonta) from the earliest Ypresian of the Ouled Abdoun Basin, Morocco. Journal of Vertebrate Paleontology 21: 107–118.

Gill, T. 1872. Arrangement of the families of mammals with analytical tables. Smithsonian Miscellaneous Collections 11 (230, Article 1): 1–98.

Gingerich, P. D. 1976, Cranial anatomy and evolution of early Tertiary Plesiadapidae (Mammalia, Primates). University of Michigan Papers on Paleontology 15: 1–141.

———. 1986. Early Eocene *Cantius torresi*—oldest primate of modern aspect from North America. Nature 320: 319–321.

———. 1989. New earliest Wasatchian mammalian fauna from the Eocene of northwestern Wyoming: Composition and diversity in a rarely sampled high-floodplain assemblage. University of Michigan Papers on Paleontology 28: 1–97.

———. 1991. Systematics and evolution of early Eocene Perissodactyla (Mammalia) in the Clarks Fork Basin, Wyoming. Contributions from the Museum of Paleontology, the University of Michigan 28: 181–213.

Gray, J. E. 1821. On the natural arrangement of vertebrose animals. London Medical Repository 15: 296–310.

Haeckel, E. 1866. Generelle Morphologie der Organismen. Allgemeine Entwicklungsgeschichte der Organismen. Kritische Grundzüge der mechanischen Wissenschaft von der enstehenden formen der Organismen, begründet durch die Descendenz-Theorie. Georg Reimer, Berlin.

Holbrook, L. T. 2001. Comparative osteology of early Tertiary tapiromorphs (Mammalia, Perissodactyla). Zoological Journal of the Linnean Society 132: 1–54.

Hooker, J. J. 1984. A primitive ceratomorph (Perissodactyla, Mammalia) from the early Tertiary of Europe. Zoological Journal of the Linnean Society 82: 229–244.

———. 1989. Character polarities in early Eocene perissodactyls and their significance for *Hyracotherium* and infraordinal relationships; pp. 79–101 *in* D. R. Prothero and R. M. Schoch (eds.), The Evolution of Perissodactyls. Oxford University Press, New York.

———. 1994. The beginning of the equoid radiation. Zoological Journal of the Linnean Society 112: 29–63.

Hooker, J. J., and D. Dashzeveg. 2003. Evidence for direct mammalian faunal interchange between Europe and Asia near the Paleocene-Eocene boundary; pp. 479–500 *in* S. L. Wing, P. D. Gingerich, B. Schmitz, and E. Thomas (eds), Causes and Consequences of Globally Warm Climates in the Early Paleogene. Special Paper 369. Geological Society of America, Boulder, Colorado.

———. 2004. The origin of chalicotheres (Perissodactyla, Mammalia). Palaeontology 47(6): 1363–1386.

Janis, C. M., J. D. Archibald, R. L. Cifelli, S. G. Lucas, C. R. Schaff, R. M. Schoch, and T. E. Williamson. 1998a. Archaic ungulates and ungulatelike mammals; pp. 247–259 *in* C. M. Janis, K. M. Scott, and L. L. Jacobs (eds.), Evolution of Tertiary

Mammals of North America. Volume 1. Terrestrial Carnivores, Ungulates, and Ungulatelike Mammals. Cambridge University Press, Cambridge.

Janis, C. M., M. W. Colbert, M. C. Coombs, W. D. Lambert, B. J. MacFadden, B. J. Mader, D. R. Prothero, R. M. Schoch, J. Shoshani, and W. P. Wall. 1998b. Perissodactyla and Proboscidea; pp. 511–524 in C. M. Janis, K. M. Scott, and L. L. Jacobs (eds.), Evolution of Tertiary Mammals of North America. Volume 1. Terrestrial Carnivores, Ungulates, and Ungulatelike Mammals. Cambridge University Press, Cambridge.

Jow, H., C. Hudelot, M. Ratray, and P. G. Higgs. 2002. Bayesian phylogenetics using an RNA substitution model applied to early mammalian evolution. Molecular Biology and Evolution 19: 1591–1601.

Kitts, D. B. 1956. American *Hyracotherium* (Perissodactyla, Equidae). Bulletin of the American Museum of Natural History 110: 1–60.

Krause, D. W., and M. C. Maas. 1990. The biogeographic origins of late Paleocene–early Eocene mammalian immigrants to the Western Interior of North America; pp. 71–105 in T. M. Bown and K. D. Rose (eds.), Dawn of the Age of Mammals in the Northern Part of the Rocky Mountain Interior, North America. Special Paper 243. Geological Society of America, Boulder, Colorado.

Lipscomb, D. L. 1992. Parsimony, homology and the analysis of multistate characters. Cladistics 8: 45–65.

Liu, F.-G.R., and M. M. Miyamoto. 1999. Phylogenetic assessment of molecular and morphological data for eutherian mammals. Systematic Biology 48: 54–64.

Maas, M. C., S. T. Hussain, J.J.M. Leinders, and J.G.M. Thewissen. 2001. A new isectolophid tapiromorph (Perissodactyla, Mammalia) from the early Eocene of Pakistan. Journal of Paleontology 75: 407–417.

Maddison, W. P., M. J. Donoghue, and D. R. Maddison. 1984. Outgroup analysis and parsimony. Systematic Zoology 33: 83–103.

Mader, B. J. 1989. The Brontotheriidae: A systematic revision and preliminary phylogeny of North American genera; pp. 458–484 in D. R. Prothero and R. M. Schoch (eds.), The Evolution of Perissodactyls. Oxford University Press, New York.

Marsh, O. C. 1873. Notice of new Tertiary mammals. American Journal of Science, Series 3, 5: 407–410, 485–488.

———. 1884. The gigantic mammals of the order Dinocerata; pp. 245–302 in Fifth Annual Report of the United States Geological Survey. U.S. Government Printing Office, Washington, D.C.

Matthew, W. D., and W. Granger. 1925. The smaller perissodactyls of the Irdin Manha Formation, Eocene of Mongolia. American Museum Novitates 199: 1–9.

McKenna, M. C., and S. K. Bell. 1997. Classification of Mammals above the Species Level. Columbia University Press, New York.

McKenna, M. C., M. Chow, S.-Y. Ting, and Z. Luo. 1989. *Radinskya yupingae,* a perissodactyl-like mammal from the late Paleocene of China; pp. 24–36 in D. R. Prothero and R. M. Schoch (eds.), The Evolution of Perissodactyls. Oxford University Press, New York.

McKenna, M. C., and E. Manning. 1977. Affinities and paleobiogeographical significance of the Mongolian Paleocene genus *Phenacolophus.* Géobios Mémoire spécial 1: 61–85.

Meng, J., R.-J. Zhai, and A. R. Wyss. 1998. The late Paleocene Bayan Ulan fauna of Inner Mongolia, China. Bulletin of the Carnegie Museum of Natural History 34: 148–185.

Nixon, K. C., and J. M. Carpenter. 1993. On outgroups. Cladistics 9: 413–426.

Novacek, M. J. 1992a. Mammal phylogeny: Shaking the tree. Nature 356: 121–125.

———. 1992b. Fossils, topologies, missing data, and the higher level phylogeny of eutherian mammals. Systematic Biology 41: 58–73.

———. 1992c. Fossils as critical data for phylogeny; pp. 46–88 in M. J. Novacek and Q. D. Wheeler (eds.), Extinction and Phylogeny. Columbia University Press, New York.

Novacek, M. J., and A. R. Wyss. 1986. Higher-level relationships of the recent eutherian orders: Morphological evidence. Cladistics 2: 257–287.

Osborn, H. F. 1892. The classification of the Perissodactyla. In H. F. Osborn and J. L. Wortman, Fossil mammals of the Wahsatch and Wind River Beds. Collection of 1891. Bulletin of the American Museum of Natural History 4: 81–147.

Owen, R. 1841. Description of the fossil remains of a mammal (*Hyracotherium leporinum*) and of a bird (*Lithornis vulturinus*) from the London Clay. Transactions of the Geological Society of London 6: 203–208.

———. 1848. Description of teeth and portions of jaws of two extinct Anthracotherioid quadrupeds (*Hyopotamus vectianus* and *Hyop. bovinus*) discovered by the Marchioness of Hastings in the Eocene deposits on the N.W. coast of the Isle of Wight; with an attempt to develop Cuvier's idea of the classification of pachyderms by the number of their toes. Quarterly Journal of the Geological Society of London 4: 103–141.

Pavlow, M. 1888. Etudes sur l'histoire paléontologique des ongulés. II. Le développement des Equidae. Bulletin de la Société des Naturalistes de Moscou (NS) 2: 135–182.

Peterson, O. A. 1919. Report upon the material discovered in the upper Eocene of the Uinta Basin by Earl Douglass in the years 1908–1909, and by O. A. Peterson in 1912. Annals of the Carnegie Museum 12: 40–168.

Prothero, D. R., E. M. Manning, and M. Fischer. 1988. The phylogeny of the ungulates; pp. 201–234 in M. J. Benton (ed.), The Phylogeny and Classification of the Tetrapods. Volume 2: Mammals. Systematics Association Special Volume 35B. Clarendon Press, Oxford.

Prothero, D. R., and R. M. Schoch. 1989a. Origin and evolution of the Perissodactyla: Summary and synthesis; pp. 504–529 in D. R. Prothero and R. M. Schoch (eds.), The Evolution of Perissodactyls. Oxford University Press, New York.

———. 1989b. Classification of the Perissodactyla; pp. 530–537 in D. R. Prothero and R. M. Schoch (eds.), The Evolution of Perissodactyls. Oxford University Press, New York.

Radinsky, L. B. 1963. The perissodactyl hallux. American Museum Novitates 2145: 1–8.

———. 1964. *Paleomoropus,* a new early Eocene chalicothere (Mammalia, Perissodactyla), and a revision of Eocene chalicotheres. American Museum Novitates 2179: 1–28.

———. 1965. Evolution of the tapiroid skeleton from *Heptodon* to *Tapirus.* Bulletin of the Museum of Comparative Zoology 134: 69–106.

———. 1966. The adaptive radiation of the phenacodontid condylarths and the origin of the Perissodactyla. Evolution 20: 408–417.

———. 1969. The early evolution of the Perissodactyla. Evolution 23: 308–328.

Rasmussen, D. T., M. Gagnon, and E. L. Simons. 1990. Taxeopody in the carpus and tarsus of Oligocene Pliohyracidae (Mammalia: Hyracoidea) and the phyletic position of hyraxes. Proceedings of the National Academy of Sciences USA 87: 4688–4691.

Remy, J. A. 2001. Sur le crâne de *Propalaeotherium isselanum* (Mammalia, Perissodactyla, Palaeotheriidae) de Pépieux (Minervois, Sud de la France). Geodiversitas 23: 105–127.

Reshetov, V. Yu. 1975. Obzor rannetretichnykh tapiroobraznykh Mongolii i SSSR. In N. N. Kramarenko (ed.), Iskopaemaya fauna i flora Mongolii. Trudy sovmestnaya Sovetsko-Mongolskaya Paleontologicheskaya Ekspeditsya 2: 19–53. (In Russian.)

Rose, K. D. 1996. Skeleton of early Eocene *Homogalax* and the origin of Perissodactyla. Palaeovertebrata 25: 243–260, 1 plate.

Savage, D. E., D. E. Russell, and P. Louis. 1965. European Eocene Equidae (Perissodactyla). University of California Publications in Geological Science 56: 1–94.

———. 1966. Ceratomorpha and Ancylopoda (Perissodactyla) from the lower Eocene Paris Basin, France. University of California Publications in Geological Science 66: 1–38.

Scally, M., O. Madsen, C. J. Douady, W. W. de Jong, M. J. Stanhope, and M. S. Springer. 2001. Molecular evidence for the major clades of placental mammals. Journal of Mammalian Evolution 8: 239–277.

Schoch, R. M. 1984. Two unusual specimens of *Helaletes* in the Yale Peabody Museum collections, and some comments on the ancestry of the Tapiridae (Perissodactyla, Mammalia). Postilla 193: 1–20.

———. 1989. A brief historical review of perissodactyl classification; pp. 13–23 *in* D. R. Prothero and R. M. Schoch (eds.), The Evolution of Perissodactyls. Oxford University Press, New York.

Scott, W. B., and H. F. Osborn. 1883. On the skull of the Eocene rhinoceros *Orthocynodon,* and the relation of the genus to other members of the group. Contributions from the E. M. Museum of Geology and Archaeology, Princeton College Bulletin 3: 1–22.

Simpson, G. G. 1945. The principles of classification and a classification of mammals. Bulletin of the American Museum of Natural History 85: 1–350.

Sloan, R. E. 1969. Cretaceous and Paleocene terrestrial communities of western North America. Proceedings of the North American Paleontological Convention, Part E: 427–453.

Springer, M. S., H. M. Amrine, A. Burke, and M. J. Stanhope. 1999. Additional support for Afrotheria and Paenungulata, the performance of mitochondrial versus nuclear genes, and the impact of data partitions with heterogeneous base composition. Systematic Biology 48: 65–75.

Tabuce, R., M. Mahboubi, and J. Sudre. 2001. Reassessment of the Algerian Eocene hyracoid *Microhyrax*. Consequences on the early diversity and basal phylogeny of the order Hyracoidea (Mammalia). Eclogae Geologicae Helvetiae 94: 537–545.

Thewissen, J.G.M. 1990. Evolution of Paleocene and Eocene Phenacodontidae (Mammalia, Condylarthra). University of Michigan Papers on Paleontology 29: 1–107.

Thewissen, J.G.M., and D. P. Domning. 1992. The role of phenacodontids in the origin of the modern orders of ungulate mammals. Journal of Vertebrate Paleontology 12: 494–504.

Thewissen, J.G.M., and E. L. Simons. 2001. Skull of *Megalohyrax eocaenus* (Hyracoidea, Mammalia) from the Oligocene of Egypt. Journal of Vertebrate Paleontology 21: 98–106.

Thomas, H., E. Gheerbrant, and J.-M. Pacaud. 2004. Découverte de squelettes subcomplets de mammifères (Hyracoidea) dans le Paléogène d'Afrique (Libye). Comptes Rendus Palevol 3: 209–217.

Ting, S.-Y. 1993. A preliminary report on an early Eocene mammalian fauna from Hengdong, Hunan Province, China. Kaupia 3: 201–207.

———. 1998. Paleocene and early Eocene land mammal ages of Asia. Bulletin of the Carnegie Museum of Natural History 34: 124–147.

Waddell, P. J., N. Okada, and M. Hasegawa. 1999. Towards resolving the interordinal relationships of placental mammals. Systematic Biology 48: 1–5.

Wang, J.-W., and Y.-S. Tong. 1997. A new phenacodontid condylarth (Mammalia) from the early Eocene of the Wutu Basin, Shandong. Vertebrata PalAsiatica 35: 283–289.

Wang, Y. 1995. A new primitive chalicothere (Perissodactyla, Mammalia) from the early Eocene of Hubei, China. Vertebrata PalAsiatica 33: 138–159.

Wood, H. E. 1937. Perissodactyl suborders. Journal of Mammalogy 18: 106.

JESSICA M. THEODOR,
KENNETH D. ROSE,
AND JÖRG ERFURT

14

ARTIODACTYLA

ARTIODACTYLA ARE EVEN-TOED UNGULATES, character-
ized by cursorially adapted skeletons with paraxonic feet and a double-
trochleated astragalus. Recent discoveries of double-trochleated astragali in
fossil whales from Pakistan show that this diagnostic criterion of Artiodactyla is
shared by early cetaceans. Molecular data have called into question the monophyly
of Artiodactyla unless Cetacea is included; despite these new finds, most morphological studies still support the monophyly of the Artiodactyla as traditionally assumed (but see Geisler and Uhen, 2003).

However, the question of artiodactyl origins remains as complex as it was a decade
ago. The oldest known fossil artiodactyls, referred to the paraphyletic genus *Diacodexis,* appear suddenly in the earliest Eocene, roughly contemporaneously in North
America (early Wasatchian), Pakistan (Bumbanian), and Europe (Sparnacian). The
traditional view has been that Artiodactyla evolved from the "Condylarthra" during
the Paleocene, probably from an arctocyonid, such as *Chriacus* or a similar form, or
from Hyopsodontidae (*sensu lato*). But the most recent morphological cladistic
analyses place whales either as the sister taxon to Artiodactyla, with mesonychids as
a paraphyletic stem assemblage (Thewissen et al., 2001), or place whales + mesonychids as the sister taxon to Artiodactyla (Geisler, 2001).

The nearly simultaneous appearance of artiodactyls on three continents also com-
plicates our understanding of their biogeography. The sudden appearance of a
large number of artiodactyls in North America and Europe suggests that they are immigrants to these regions, and that their origins lie in Asia, particularly the Indian

subcontinent, or perhaps in Africa, but the evidence is very scant. Clearly, more work remains to be done in the earliest Eocene and late Paleocene to better understand the origin of the order. There is no evidence to suggest that Artiodactyla originated prior to the Paleocene.

WHAT IS AN ARTIODACTYL?

Traditional Definition

The order Artiodactyla was established for the even-toed ungulates, and includes ten extant families: Suidae (pigs), Tayassuidae (peccaries), Hippopotamidae (hippos), Camelidae (camels and llamas), Tragulidae (chevrotains), Moschidae (musk deer), Antilocapridae (pronghorn antelope), Giraffidae (giraffe and okapi), Cervidae (deer) and Bovidae (cows, goats, sheep, and antelope) (McKenna and Bell, 1997). Although Linnaeus (1758) recognized the ruminants and camels as a natural group on the basis of foregut rumination, it was not until Blainville (1816) that the pigs, peccaries, and hippopotamids were added to the group, based on the even number of digits on the foot ("au doigts pair"). Owen (1848) formally united the extant families in the Artiodactyla, citing paraxonic symmetry of the manus and pes, in which the axis of symmetry runs between the third and fourth digits, as a defining character. Early discoveries of fossil anoplotheres (Cuvier, 1822) and anthracotheres (Owen, 1848) were recognized as belonging to the order. Until very recently, there has not been much debate about which taxa belong in Artiodactyla, with the exception of

some primitive fossil taxa for which skeletons were not well known.

Current Phylogenetic Definition

Artiodactyla has always been defined using a character-based definition that has been based on the postcranial skeleton. Owen (1848) formally based Artiodactyla on paraxonic symmetry of the feet. Schaeffer (1947, 1948) placed more emphasis on the unique derivation of the double-trochleated astragalus, and this character has come to be regarded as the artiodactyl hallmark (Fig. 14.1). More recently, Luckett and Hong (1998) added the character of a six-cusped deciduous fourth premolar (dp4) as a potential synapomorphy of Artiodactyla, in response to molecular evidence challenging the monophyly of the order (Sarich, 1993; Gatesy et al., 1996; Gatesy 1997, 1998; Milinkovitch and Thewissen, 1997; Milinkovitch et al., 1998; Ursing and Árnason, 1998).

The use of the term Artiodactyla as traditionally conceived is rendered ambiguous by the discovery of double-trochleated astragali in early archaeocete whales (Gingerich et al., 2001; Thewissen et al., 2001). Based on the traditional definition of paraxonic feet and a double-trochleated astragalus, Cetacea would belong in Artiodactyla. The six-cusped dp4 cannot be used as a substitute character on which to base a definition for the traditional Artiodactyla until dp4 is known for key early cetacean taxa.

The new fossil whale evidence provides no new data supporting a specific relationship between hippos and whales (as suggested by some molecular data), and hence, does not contradict artiodactyl monophyly (Thewissen et al., 2001).

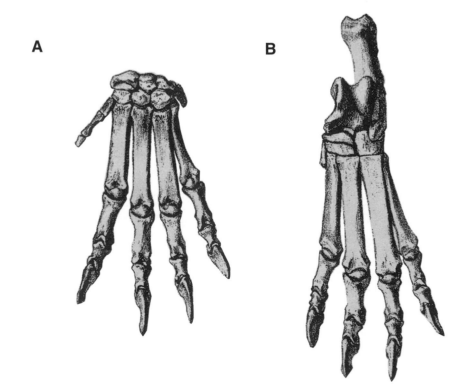

A   **B**

Fig. 14.1. Foot morphology of artiodactyls, as illustrated by Oligocene *Merycoidodon*. (A) Forefoot. (B) Hind foot. From Scott (1940).

If whales are the sister taxon to hippos, the traditional taxon definitions fail, and Cetacea are members of Artiodactyla for all potential definitions. However, if whales are in fact the sister taxon to all taxa in Artiodactyla as traditionally defined, there are two nomenclatural possibilities. Without an explicit node-based definition, under the current character-based definition, whales belong in Artiodactyla. Usage of the name Artiodactyla for the traditional grouping could be retained if it were redefined using a node-based definition (Rowe, 1988; de Queiroz and Gauthier, 1992). A node-based definition would have the advantage of retaining a familiar, stable grouping that is in wide use. The most straightforward node-based definition for Artiodactyla would include the living crown-group taxa Ruminantia, Tylopoda, Suiformes, the extinct stem taxon *Diacodexis,* and their most recent common ancestor. A strict crown-group concept is simpler, but would exclude stem taxa that have long been included in the traditional concept of the order, which would defeat the goal of stabilizing current usage. For this chapter, we retain the traditional membership of Artiodactyla for clarity, using the above node-based definition. This also potentially leaves the Cetacea as a traditional order—clearly, the sister taxon to the Artiodactyla.

Current Classification

The families included in Artiodactyla are listed in the classification given in Table 14.1. McKenna and Bell (1997) followed Simpson (1945) in placing the dichobunoids, entelodonts, anthracotheres, oreodonts, mixtotheres, cainotheres, and anoplotheres in the Suiformes. However, recent phylogenetic analyses (Fig. 14.2; Gentry and Hooker, 1988; Norris, 2000) cluster the oreodonts, mixtotheres, cainotheres, and anoplotheres more closely with the Tylopoda, and the classification adopted in this chapter reflects those analyses. We also follow recent studies that place the dichobunoids and entelodonts in basal positions among artiodactyls (Janis et al., 1998; Stucky, 1998), as there is little evidence to indicate that they are linked with suoids by anything more than primitive characters.

MORPHOLOGICAL CHARACTERIZATION

General Features

The diagnostic features of Artiodactyla are the paraxonic, hoofed fore- and hind feet, the double-trochleated astragalus with a modified calcaneus, the elongate, six-cusped deciduous p4, and the great reduction or loss of the clavicle and the third trochanter of the femur (Schaeffer, 1948; Rose, 1985; Thewissen and Hussain, 1990; Luckett and Hong, 1998).

However, there are a number of other morphological features typical of artiodactyls, several of which have been listed as putative synapomorphies for Artiodactyla. They include narrow molar trigonids with closely appressed paraconid and metaconid, an enlarged M3 metacone, a small P4 protocone crista and a long P3, a wide epitympanic recess, confluent foramen lacerum and foramen rotundum, an enlarged pars facialis of the lacrimal, and an enlarged orbitosphenoid separating the frontal and alisphenoid (Prothero et al., 1988). To this list, Thewissen et al. (2001) added upper molars with a small trigon basin, M2 paraconule and metaconule present and similar in size, m2 trigonid similar in height to the talonid, and m2 hypoconulid present. Additional features include the absence of an entepicondylar foramen on the distal humerus; a deeply grooved tibial surface of the astragalus; loss of the astragalar canal; and astragalar contact, with the cuboid large and oriented perpendicular to the sagittal plane (Geisler, 2001). However, note that the distribution of some of these characters is not known for many early artiodactyl taxa, and some of the cranial characters are also poorly known for many primitive ungulates. Lack of a hypocone has also been listed as a synapomorphy for Artiodactyla, but the polarity and distribution of this character (which is not limited to artiodactyls) is difficult to interpret. A hypocone is variably present in populations of some early artiodactyl taxa (e.g., *Diacodexis, Protodichobune, Eygalayodon*) and also among some archaic ungulate taxa (the arctocyonid *Chriacus*).

The earliest artiodactyls in the fossil record, immediately recognizable because of their fully developed diagnostic tarsal morphology, are generally assumed to display the primitive condition (Fig. 14.3). They were small, relatively long-legged animals, similar in overall morphology to the chevrotains or small antelope, such as dik-dik (see the reconstruction of *Anthracobunodon weigelti,* Fig. 14.3C,D), but lacking cranial ornamentation and certain postcranial specializations, and retaining a long tail. The oldest known artiodactyl, early Eocene *Diacodexis* (Rose, 1982, 1985) was smaller than *Anthracobunodon* or any extant species, comparable in size to the rabbit *Sylvilagus*. In these primitive artiodactyls, the molars are tritubercular, bunodont to bunoselenodont, and relatively unspecialized. In contrast, the postcranial morphology appears relatively "derived." Before good postcranial skeletons were found, paleontologists traditionally assumed that the long torso and shorter limbs found in a number of artiodactyls (including the modern suoids) was the ancestral artiodactyl condition (e.g., Romer, 1966). More recently, skeletons of several early and middle Eocene artiodactyl taxa have been discovered (Franzen, 1981, 1983, 1988; Rose, 1982, 1985, 1990; Thewissen et al., 1983; Tobien, 1985; Erfurt and Haubold, 1989; Thewissen and Hussain, 1990). These fossils indicate that the oldest and, likely, most primitive morphology among artiodactyls is highly cursorial compared with other Wasatchian mammals, much more closely resembling *Sylvilagus* than contemporary *Hyopsodus* in limb proportions. Early artiodactyls had long limbs, the hind limbs much longer than the forelimbs, and a relatively specialized tarsus, yet retained a small third trochanter on the femur and a small clavicle in the shoulder in the oldest taxa. However, the earliest artiodactyls retain

Table 14.1 Comparison of classifications of Artiodactyla

This Chapter	McKenna and Bell	Gentry and Hooker
Artiodactyla	Artiodactyla	Artiodactyla
Dichobunoidea	Suiformes	Bunodontia
†Diacodexeidae eE-mE; As. eE-mE; Eu. eE-mE, NA.	Raoellidae	Suiformes
†Dichobunidae eE-lO; Eu. mE; As.	Choeropotamidae	Suoidea
†Homacodontidae mE; NA. mE; As.	Suoidea	Suidae
†Leptochoeridae mE-lO; NA.	Suidae	Tayassuidae
†Cebochoeridae eE-eO; Eu.	Tayassuidae	Entelodontoidea
†Helohyidae mE; As. mE; NA.	Sanitheriidae	Entelodontidae
Entelodontoidea	Hippopotamidae	Choeropotamoidea
†Entelodontidae mE-eM; As. mE-eM; NA. eO-lO; Eu.	Dichobunoidea	Choeropotamidae
Anthracotherioidea	Dichobunidae (includes diacodexeids, homacodonts, leptochoerids and dichobunids *sensu stricto*)	Cebochoeridae
†Anthracotheriidae mE-?ePle; As. mE-lM; Eu. mE-eM; NA. lE-?ePli; Af.	Cebochoeridae	Hippopotamoidea
Suiformes	Mixtotheriidae	Anthracotheriidae
†Raoellidae eE-mE; As.	Helohyidae	Hippopotamidae
†Choeropotamidae eE-lE; Eu.	Anthracotherioidea	Raoellidae
Suoidea	Haplobunodontidae	Helohyidae
Suidae ?lE-R; Eu. lO-R; As. eM-R; Af.	Anthracotheriidae	Selenodontia
Tayassuidae lE; As. lE-R; NA. lPli-R; SA.	Anoplotherioidea	Tylopoda
†Sanitheriidae eM-mM; Af, As. mM; Eu.	Dacrytheriidae	Diacodexis pakistanensis[a]
Hippopotamidae lM-R; Af. lM-lPle; As.	Anoplotheriidae	Bunomerycidae
Tylopoda	Cainotheriidae	Anoplotherioidea
Cainotherioidea	Oreodontoidea	Anoplotheriidae
†Mixtotheriiidae e-lE; Eu. ?lE; Af.	Agriochoeridae	Dacrytheriidae
†Cainotheriidae lE-mM; Eu.	Oreodontidae = Merycoidodontidae	Xiphodontidae
Anoplotherioidea	Entelodontoidea	Cameloidea
†Anoplotheriidae eE-lO; Eu.	Entelodontidae	Camelidae
†Xiphodontidae m-lE, ?eO; Eu.	Tylopoda	Oromerycidae
Oreodontoidea	Xiphodontidae	Protoceratidae
†Agriochoeridae mE-eM; NA.	Cameloidea	Oreodonta
†Merycoidodontidae mE-lM; NA.	Camelidae	Mixtotheriidae
†Protoceratidae mE-ePli; NA.	Oromerycidae	Merycoidodontoidea
Cameloidea	Protoceratoidea	Agriochoeridae
Camelidae mE-lPle; NA. LM-R; Eu. ePli-R; As. lPli-R; Af. lPli-R; SA.	Protoceratidae	Merycoidodontidae
†Oromerycidae mE-lE; NA.	Ruminantia	Cainotheriidae
Ruminantia	Tragulina	Merycotheria
†Amphimerycidae eE-eO; Eu.	Traguloidea	Ruminantia
†Hypertragulidae mE-eM; NA. eM-mM; Eu.	Tragulidae	
Tragulidae eM-R; Af. eM-R; As. eM-lM; Eu.	Leptomerycidae	
†Leptomerycidae mE-eO; As. mE-mM; NA.	Amphimerycoidea	
†Bachitheriidae eO-lO; Eu.	Amphimerycidae	
†Lophiomerycidae lE-lO; As. eO-lO; Eu.	Hypertraguloidea	
†Gelocidae lE-lO; As. lE-lO; Eu. eM; Af. lM; NA.	Hypertragulidae	
Cervoidea	Pecora	
Moschidae eO-lM; Eu. eM-R; As. eM-lPli; NA.	Moschoidea	
Antilocapridae eM-R; NA.	Gelocidae	
†Palaeomerycidae eM-mM; Eu. mM-?lM; As.	Moschidae	
Cervidae eM-R; As. eM-R; Eu. ePli-R; NA. ePle-R; SA. lPle-R; Af.	Giraffoidea	
Giraffidae eM-R; Af. eM-Ple; As. mM-?lPle; Eu.	Palaeomerycidae	
Bovoidea	Giraffidae	
Bovidae O-R; As. eM-R; Af. eM-R; Eu. lM-R; NA.	Climacoceridae	
	Cervoidea	
	Dromomerycidae	
	Cervidae	
	Hoplitomerycidae	
	Bovoidea	
	Bovidae	
	Antilocapridae	

Source: The alternate classifications shown in columns 2 and 3 are from McKenna and Bell (1997) and Gentry and Hooker (1988).

Note: Abbreviations following current classification indicate the time ranges in which groups are found in continental regions: e, early; m, middle; l, late; E, Eocene; O, Oligocene; M, Miocene; Pli, Pliocene; Ple, Pleistocene; R, Recent; Af, Africa; As, Asia; Eu, Europe; NA, North America; SA, South America.

[a] Gentry and Hooker did not specify placement of other species of *Diacodexis*.

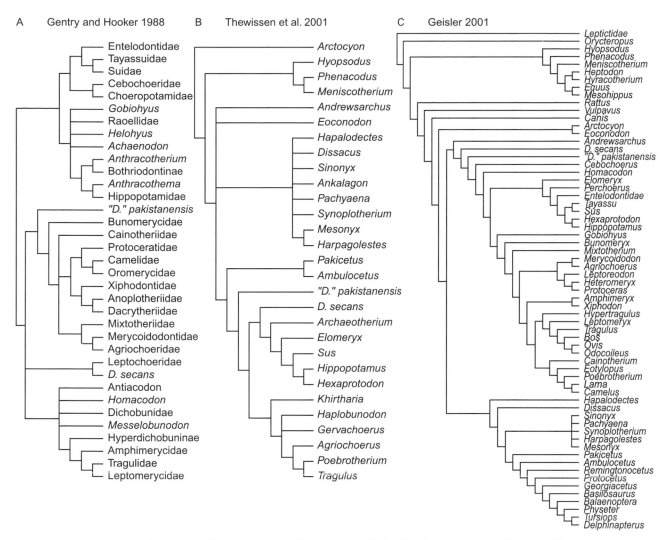

Fig. 14.2. Current phylogenies of Artiodactyla. (A) After Gentry and Hooker (1988). (B) After Thewissen et al. (2001). (C) After Geisler (2001).

remarkably primitive dentition, very similar to that of some arctocyonids and other condylarths.

Artiodactyla presents a diversity of body size and morphologies through the fossil record. There are two major body morphologies or "growth types" (*sensu* Erfurt, 2000) seen among living and fossil artiodactyls: a piglike morphology, with a large head, long heavy body, and relatively short legs, as seen in anthracotheres, pigs, peccaries, and hippos; and the more gracile, long-legged forms found among the small diacodexeids, early cebochoerids, cainotheriids, and the much larger ruminants and tylopods.

Artiodactyls are primitively bunodont (Fig. 14.4), but selenodonty, a condition in which the tooth cusps join into mesiodistal, crescentiform crests, has evolved—probably repeatedly—in numerous groups in the order.

Extant ruminants and tylopods have evolved foregut rumination, a system involving symbiotic gut bacteria in a series of chambers pouched off from the anterior stomach. All living members of these clades share foregut rumination, although the details of stomach morphology differ slightly among camelids, tragulids, and other extant mem-

bers of Ruminantia. The anterior chambers (rumen, reticulum, and omasum in ruminants; rumen and reticulum in camelids) are modified for bacterial fermentation and sieving of particulate material. The posterior chamber, the abomasum, is the functional equivalent of the stomach of other mammals, secreting acid and digestive enzymes. The stomach in the suiforms is somewhat less complex. Suid stomachs range from relatively simple to somewhat more elaborate subdivisions. The tayassuids show further subdivision of the stomach into three chambers. Hippos represent the most complex stomach among suiforms. Although hippos eat grasses, they do not chew the cud; neither do pigs and peccaries.

To explore the origin of the Artiodactyla, it is instructive to review the diversity of primitive early groups.

Basal Groups

DIACODEXEIDS. *Diacodexis,* the most primitive known artiodactyl genus, was first discovered in the early Eocene

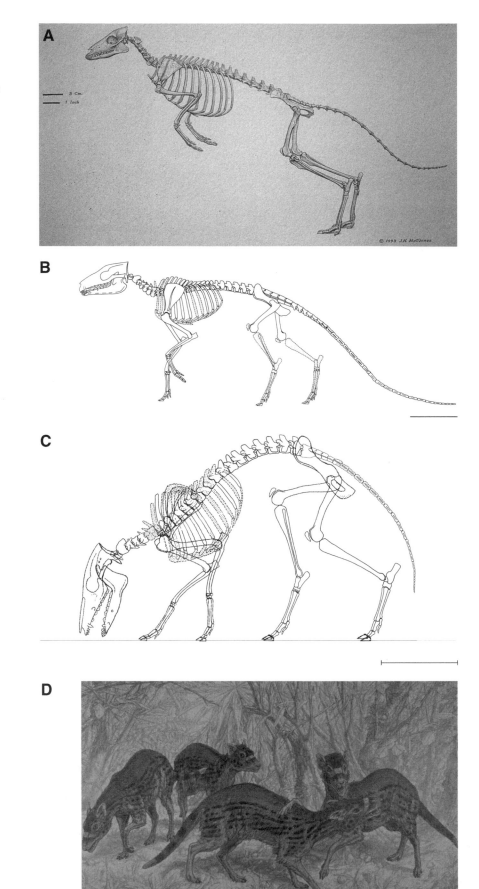

Fig. 14.3. Skeletal morphology and reconstruction of early artiodactyls. (A) Diacodexeid, *Diacodexis secans,* by Jay Matternes, © 1993, reprinted with permission. (B) Cebochoerid, *Gervachoerus jaegeri.* (C) Choeropotamid, *Anthracobunodon weigelti.* (D) Reconstruction of *Anthracobunodon weigelti.* Scale bar = 5 cm in (A); scale bars = 100 mm in (B), (C). Panel (B) from Erfurt and Haubold (1989); panel (C) modified from Erfurt (2000). Panels (B)–(D) courtesy of the Geiseltal Museum, Halle, Germany.

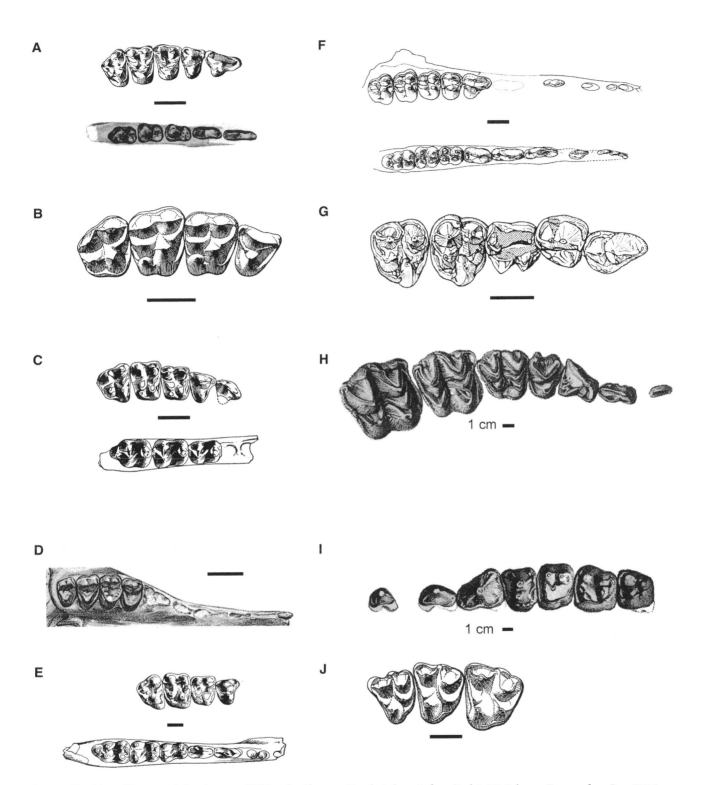

Fig. 14.4. Dentitions of Eocene artiodactyl groups. (A) Diacodexeid, uppers *Diacodexis chacensis,* from Sinclair (1914); lowers *D. secans,* from Rose (1996). (B) Dichobunid, *Dichobune robertiana,* from Sudre (1972). (C) Homacodontid, *Homacodon vagans,* from Sinclair (1914). (D) Leptochoerid upper dentition, *Stibarus* sp., from Scott (1940). (E) Helohyid, *Helohyus plicodon,* from Sinclair (1914). (F) Cebochoerid, *Gervachoerus jaegeri,* from Erfurt (1988). (G) Choeropotamid upper dentition, *Anthracobunodon weigelti,* from Erfurt and Haubold (1989). (H) Anthracothere upper dentition, *Bothriodon americanus,* from Scott (1940). (I) Entelodont upper dentition, *?Entelodon magnus,* from Scott (1940). (J) Mixtothere upper dentition, *Mixtotherium gresslyi,* from Sudre (1972). Scale bars = 0.5 cm for all panels except in (H) and (I); scale bars = 1.0 cm in (H), (I).

of North America (Cope, 1881; Gazin, 1952; Rose, 1982; Krishtalka and Stucky, 1985). In the past two decades, findings from contemporaneous deposits in Europe have enlarged our knowledge of this and other primitive artiodactyl genera (Godinot, 1981; Sudre et al., 1983a,b; Estravís and Russell, 1989; Sudre and Marandat, 1993; Smith et al., 1996; Sudre and Erfurt, 1996; Checa-Soler, 2004). Diacodexeids have also been discovered in Asia (Gabunia, 1973; Thewissen et al., 1983; Averianov, 1994; Métais et al., in press). *Diacodexis* itself is known from Europe and Asia, as well as North America, but there is little clue as to the ultimate phylogenetic or geographic origins of the family. *Diacodexis* is basal within Artiodactyla, and because one or more of the species seem to be more closely related to selenodont artiodactyls than to the others, the genus is considered to be paraphyletic (Gentry and Hooker, 1988; Stucky, 1998). Some authors have included diacodexeids in the Dichobunidae *sensu lato* (Simpson, 1945; Romer, 1966; Rose, 1982; McKenna and Bell, 1997), but there is general agreement that diacodexeids are the most primitive known artiodactyls.

Diacodexeids (*Diacodexis, Neodiacodexis, Bunophorus* [including *Wasatchia*], *?Tapochoerus, ?Lutzia,* and *?Eolantianius*) are small artiodactyls with primitive, tritubercular upper molars, which display three main cusps with small para- and metaconules (Fig. 14.4A). The main cusps may be blunt (bunodont) or crescentiform. The lower molars have moderately elevated trigonids that are anteroposteriorly compressed, narrower than the talonids, and have reduced paraconids. In some forms, the entocristid is shallow, leaving the talonid basin open lingually. The p2–p4 tend to become elongate, with one main cusp and a low talonid; they may also bear a parastylid or small paraconid.

Postcranial morphology is substantially established in *Diacodexis* and, to a lesser extent, in some other diacodexeids that appear to be very similar. *Diacodexis* is specialized for cursorial locomotion, yet it retains several primitive features, such as a vestigial clavicle and a small third trochanter of the femur. It has long limbs, the hind limbs longer than the forelimbs and the intermediate and distal segments relatively elongate; it has digitigrade feet with small hooves. The manus is pentadactyl and mesaxonic, whereas the pes is paraxonic, with large central digits (III and IV), somewhat reduced lateral digits, and a vestigial hallux (Rose, 1982, 1985; Thewissen et al., 1983; Thewissen and Hussain, 1990).

DICHOBUNIDS. Dichobunids (*Protodichobune, Buxobune, Messelobunodon, Eygalayodon, Aumelasia, Neufferia, Eurodexis, Parahexacodus, Dichobune, Meniscodon, Metriotherium, Synaphodus, Hyperdichobune, Mouillacitherium, Pakibune,* and *Chorlakkia*) are primarily European in distribution and early Eocene–early Oligocene in age. There are three named subfamilies in Dichobunidae: Gentry and Hooker (1988) established the Hyperdichobuninae (*Hyperdichobune* and *Mouillacitherium*); Erfurt and Sudre (1996) followed Krishtalka and Stucky (1985) in distinguishing the Eurodexinae (*Eurodexis, Parahexacodus, Eygalayodon*) from the Dichobuninae

(*Protodichobune, Buxobune, Messelobunodon, Aumelasia, Neufferia, Dichobune, Meniscodon, Metriotherium,* and *Synaphodus*) within the restricted family Dichobunidae. They regarded the Eurodexinae to have evolved in parallel to the North American homacodontids. *Chorlakkia* and *Pakibune* are currently known from poor material from Pakistan and are not placed in a subfamily. New lantianiine dichobunids from Asia (Métais et al., in press) indicate that there may yet be more morphological diversity warranting a fourth subfamily among the dichobunids.

As in diacodexeids, the dentition of dichobunids is generalized and primitive (Fig. 14.4B), but dichobunids are more bunodont than are diacodexeids, and their upper molars often bear a hypocone. The presence of a hypocone is variable such in primitive genera as *Protodichobune* (from Mas de Gimel, France; Sudre et al., 1983b) and in diacodexeids (*Diacodexis* from Silveirinha, Portugal; Estravís and Russell, 1989), ranging from complete absence, to a small cuspule, to a full hypocone. This character distribution may reflect variability in the early Eocene that became functionally important and taxonomically significant in the middle Eocene. The hypocone is absent in the very small eurodexine artiodactyls (*Eurodexis*) of the European middle Eocene, suggesting that they belong to an insectivorous/frugivorous lineage distinct from the bigger omnivorous and/or herbivorous dichobunines (e.g., *Meniscodon*), all of which have a hypocone. The origin of these eurodexines probably lies within a lineage of *Diacodexis* with more crescentiform teeth, older than European reference level MP (Mammal Paleogene)10. Other dichobunine genera are thought to be descended from bunodont lineages of *Diacodexis* through *Protodichobune* or a similar form. Cebochoerids (e.g., *Gervachoerus*) and choeropotamids (e.g., *Hallebune*) are also thought to be descended from *Protodichobune* around the end of the early Eocene (MP 10/11). Skeletons of *Messelobunodon* and *Aumelasia* from Messel, Germany (Franzen, 1981, 1983, 1988), are very similar to those of *Diacodexis*, differing only slightly in proportions.

HOMACODONTIDS. Homacodontidae includes Bunomerycinae (*Bunomeryx, Hylomeryx, Mesomeryx, Mytonomeryx,* and *Pentacemylus*) and Homacodontinae (*Hexacodus, Antiacodon, Homacodon, Auxontodon, Microsus,* and *Texodon*), and is known from the early-middle Eocene of North America. The family contains bunodont artiodactyls with a basic triangular arrangement of the main cusps on the upper molars, but they differ from diacodexeids in having more selenodont cusps (Fig. 14.4C). The protocone is conical, and the hypocone is usually well developed (at least on M1 and M2) but is sometimes replaced by a large, posterolingual metaconule (e.g., *Pentacemylus*). The lower molars display a hypoconulid on the postcingulid, and the entoconid is generally larger than in diacodexeids. Homacodontidae are generally similar to diacodexeids in their postcranial skeletons (West, 1984). Consequently, Homacodontidae are regarded as North American descendants of a diacodexeid stock. Craniodental features suggest that bunomerycines lie in or near the ancestry of tylopods (Gentry and Hooker, 1988; Stucky, 1998; Norris, 1999).

LEPTOCHOERIDS.   The Leptochoeridae (*Stibarus, Ibarus, Laredochoerus,* and *Leptochoerus*) are small to medium-sized bunodont artiodactyls with enlarged premolars: the p2–p4 are robust and longer than m1–m3, and P4 reaches the size of M1. The upper molars lack a hypocone and show a marked decrease in size from M1 to M3 (Fig. 14.4D). These genera, known only from the middle Eocene–Oligocene of North America, are not clearly referable to the diacodexeid or dichobunid clades mentioned above, so their classification as an independent family seems appropriate.

RAOELLIDS.   The Raoellidae (*Indohyus, Kirtharia, Kunmunella, Raoella, Metkatius,* and *Haqueina*) are a poorly known early and middle Eocene family, endemic to Indo-Pakistan. They are known mainly from dental remains. The lower molars are broad, low-crowned, and weakly bilophodont or bunolophodont, with four main cusps; they lack a paraconid (Sahni and Khare, 1971; Thewissen et al., 1987, 2001). Some raoellids have wrinkled enamel. The relationships and paleoecology of this group are poorly understood. Although they could be related to some later suiform artiodactyls, this has yet to be established. Their derived dental morphology, however, reduces their relevance to the origin of Artiodactyla.

HELOHYIDS.   The Helohyidae (*Helohyus, Achaenodon, Parahyus, Dyscritochoerus, Lophiohyus, Pakkokuhyus, Gobiohyus,* and *Apriculus*) are found in the middle Eocene of Asia and North America. They range in body size from dichobunid-sized to much larger, pig-sized animals. Cranially, they have a much larger sagittal crest than do dichobunids, and there is considerable variation in snout length. In some helohyids, there are diastemata separating the anterior premolars from the molars and from one another. The cheek teeth have inflated crowns and bunodont cusps, the canines are prominent, and some taxa have wrinkled enamel (Fig. 14.4E). The upper molars have an enlarged, posterolingually displaced metaconule, but a small hypocone is often present as well. The paraconule is reduced and there is no mesostyle. The lower molars increase in size toward the back of the jaw. The paraconid is small or absent.

Helohyids are craniodentally derived relative to the families described above, but they remain primitive in having four-toed fore- and hind feet, with unfused metapodials. *Achaenodon,* with a skull length of 35 cm, was the largest North American middle Eocene artiodactyl. Like other primitive artiodactyls, it had short forelimbs and longer hindlimbs.

CEBOCHOERIDS.   This family, consisting of the genera *Cebochoerus, Acotherulum, Moiachoerus,* and *Gervachoerus,* is known from the early Eocene to early Oligocene of Europe (Sudre, 1978; Erfurt and Haubold, 1989). In contrast to other Eocene artiodactyls of Europe (except choeropotamids), the upper molars of cebochoerids display four bulbous cusps, generally quadratically arranged (Fig. 14.4F). The lingual cusps are basally inflated and point buccally. The paraconule is small or absent, the mesostyle is absent, and the metaconule is posterolingually situated and as big as the protocone. The canines are incisiform, and the first upper and lower premolars are enlarged and often caniniform. The P4 is triangular, and P3 is elongate; p4 is shorter than p2–3, and lacks a paraconid. The lower molars are narrow, with slight crest development. The paraconid is separate from the metaconid, and the hypoconulid is posteriorly displaced. These dental characters indicate an advanced stage compared with diacodexeids or dichobunids. Cebochoerids have short snouts, and the lower jaw deepens posteriorly.

Among artiodactyls, cebochoerids display unspecialized postcranial morphology. In *Gervachoerus* cf. *jaegeri* from Geiseltal (Fig. 14.3B), the length of the proximal and intermediate segments in each limb are about the same length (in contrast to the relatively longer tibia in diacodexeids, dichobunids, and homacodontids), and metapodials II and V are only slightly shorter than metapodials III and IV. As in other middle Eocene taxa, the metapodials remain separate. The tail was unusually long, about 30% of the total body length.

The oldest cebochoerid, *Gervachoerus dawsoni,* from the middle Eocene (MP 13, Bouxwiller, France), is similar in the size and form of the molars to dichobunids like *Protodichobune* (from Mas de Gimel, France), as well as to the choeropotamid *Hallebune* (Geiseltal, Germany). Erfurt and Sudre (1996: fig. 3) derived cebochoerids from a cebochoerid/choeropotamid stock, which they, in turn, derived from undifferentiated early diacodexeids. The geographic origin of cebochoerids probably lies in Europe, but the age of their origin remains open because of the poor fossil record between MP 10 and 12.

CHOEROPOTAMIDS.   This formerly monotypic family has recently been redefined to include the haplobunodontids, and now includes *Thaumastognathus, Haplobunodon, Lophiobunodon, Tapirulus, Rhagatherium, Hallebune, Amphirhagatherium, Anthracobunodon, Masillabune, Cuisitherium, Parabunodon,* and probably *Diplopus,* in addition to *Choeropotamus* (Hooker and Weidmann, 2000; Hooker and Thomas, 2001). Choeropotamids are known from early to late Eocene deposits in Europe and Asia Minor. These small to medium-sized artiodactyls are characterized by bunodont or bunoselenodont quadricuspate molars, lacking a hypocone but possessing an enlarged lingual metaconule (Fig. 14.4G). There is a small paraconule and a well-developed mesostyle (Sudre, 1978). In the lower jaw, the first premolar is primitively incisiform, separated from the incisiform canine by a diastema. In more derived genera, the first lower premolar is caniniform, and another diastema separates it from the second premolar. Compared to cebochoerids, Choeropotamidae display a stronger differentiation of the buccal cusps, including the mesostyle and centrocrista. As noted above, choeropotamids probably evolved from diacodexeids.

The skull is slender and elongated, with an exposed mastoid, and there is no evidence of any cranial ornamentation. The mandible deepens posteriorly, with no significant enlargement of the angular process. The teeth are brachydont and sometimes deeply worn, indicating a more abrasive diet than in contemporary eurodexines. The bunoselenodont cusp pattern implies a mixed diet, including fruit and leaves, a prediction supported by preserved stomach contents in *Masillabune* (Tobien, 1980; Erfurt, 2000; Hooker and Thomas, 2001).

The postcranial anatomy of choeropotamids is best known from *Masillabune* from Messel (Tobien, 1985), and *Haplobunodon? mülleri* and *Anthracobunodon weigelti* from Geiseltal (Erfurt and Haubold, 1989; Erfurt, 2000). Choeropotamids were cursorially adapted, similar to dichobunids and diacodexeids. The limbs are long and slender, with the forelimbs shorter than the hind, producing a dorsal curvature in the lumbar spine. Muscle attachment crests and processes are reduced, and the femur lacks a third trochanter. The fibula is large and not fused with the tibia, and the cuboid and navicular are separate. The tail, at least in *Anthracobunodon*, was relatively long, about 25% of body length. They have strongly paraxonic feet, with a reduced pollex in the forefoot and possibly a vestigial first metatarsal in the hind foot, at least in *Anthracobunodon* (Erfurt, 2000). The unguals are narrow and hoofed, but as in *Diacodexis*, the articular surfaces face caudally, suggesting a digitigrade foot posture.

ANTHRACOTHERIIDS.

The anthracotheres (comprised of 37 genera, according to McKenna and Bell, 1997) first appeared in the middle Eocene of Laurasia and the late Eocene of Africa. They became extinct in North America in the early Miocene, but survived longer in Africa and Eurasia. The high diversity of anthracothere taxa in Asia in the middle–late Eocene suggests an Asian origin, perhaps from helohyids (Coombs and Coombs, 1977a,b), but the precise phylogeny is poorly understood.

Anthracotheres were large animals, with a piglike body morphology. They have amastoid skulls (one of the derived characters that has been cited to link them with suiforms), with an incomplete postorbital bar and a prominent sagittal crest. Most species have long, narrow snouts with diastemata, although this is not the case for one of the earliest anthracotheres, Duchesnean-Whitneyan *Heptacodon*. The mandible has an expanded angle and an elongated symphysis. The dental formula is complete (3.1.4.3 / 3.1.4.3), with large canines in many taxa.

The molar teeth of anthracotheres are brachydont and bunoselenodont, with bunodont lingual cusps and selenodont labial ones, and they increase in size posteriorly (Fig. 14.4H). The upper molars are square, with a large paraconule, V-shaped labial cusps and a large posterolingual metaconule, features common to many early selenodont artiodactyls. The postprotocrista is reduced or absent, such

that the molar is bisected by a transverse valley. The lower molars lack a paraconid. The premolars are simple, and p4 is strongly crested between the paraconid and metaconid (Kron and Manning, 1998). Tooth wear patterns indicate an abrasive diet, based on the pattern of heavy wear of the first molars by the time of eruption of the third molars.

Anthracotheres have relatively short, robust limbs comparable to those of suoids and oreodonts. The fore and hind limbs are similar in length. The radius and ulna are not fused, but are closely appressed, and the fibula is large and separate. The hind foot is four-toed, whereas the forefoot retains a pollex. Both fore- and hind feet have short, unfused metapodials. The feet are generally paraxonic, although in *Elomeryx* (early Oligocene), the third digit is elongated and the manus is mesaxonic.

Despite their relatively generalized limb structure, anthracotheres are craniodentally specialized, compared to early Eocene artiodactyls. Colbert (1935) suggested that anthracotheres were probably ancestral to the Hippopotamidae, a hypothesis that has been followed uncritically by most authors since then, although there is relatively little evidence to support it. Pickford (1983) advocated an alternative derivation of hippos from Old World peccaries, probably in the early Miocene, based on numerous craniodental resemblances.

ENTELODONTIDS.

Entelodonts were large, piglike animals with long bodies and relatively long skulls. The dentition and postcranial morphology is generally primitive, although the feet are didactyl. The diagnostic features of the group include a large head with an elongate snout, a large ventral flange on the jugal, a dorsal flange on the temporal, a postorbital bar, and one or two tubercles along the ventral edge of the mandible. They have a complete tooth formula (3.1.4.3 / 3.1.4.3), with variable diastemata and low-crowned, bunodont molars, with a lobed cristid obliqua on the lower molars and a cingulum on the anterior external margin of the M2 (Foss, 2001). The molars have six main cusps, except M3, which has five or occasionally four (Fig. 14.4I). A hypocone is present on the upper molars, and the paraconid is closely appressed to the metaconid on the lower molars, so that it appears absent on worn specimens (Foss, 2001). The hypoconulid is variable in size, and unlike that of some bunodont artiodactyls, is never found isolated on a posterior lobe, but is always part of the talonid (Foss, pers. comm.).

Of the 23 genera of entelodonts in McKenna and Bell (1997), Foss (2001) recognized only seven genera as valid: *Entelodon, Eoentelodon, Paraentelodon, Daeodon, Archaeotherium, Brachyhyops,* and *"Archaeotherium" coarctatum.* The age of the first appearance of entelodonts is middle Eocene, but whether the oldest record is Asian or North American is equivocal. *Eoentelodon* occurs in the lower Lunan Formation of China (Chow, 1958). Zhang et al. (1983) regarded the lower Lunan fauna to be comparable to the Irdin Manha fauna in age, which is middle Eocene, according to Meng and McKenna (1998). The oldest North American entelodont is *Brachyhyops,* for which the oldest known specimen is from the upper Uinta Formation (late middle Eocene), and the majority of specimens are Duchesnean-Chadronian. The family persisted in North America (Foss and Fremd, 1988; Hunt, 1990) and Asia until the early Miocene, and appeared in Europe only during the Oligocene.

A bunodont lower dentition from the early Eocene Wutu local fauna of China has been referred questionably to Entelodontidae (Tong and Wang, 1998). Whether it indicates an even earlier origin of Entelodontidae or pertains to some other early artiodactyl group must await further study.

MIXTOTHERES.   The Mixtotheriidae is a family of selenodont artiodactyls found in the early-late Eocene of Europe. The family includes only the genus, *Mixtotherium*. The trapezoidal upper molars are wide transversely, bear five V-shaped cusps with large styles, and increase in size posteriorly. The P4 is partially molarized; p4 is molariform and elongated, with large protoconid and metaconid forming a transverse lobe. Sudre and Ginsburg (1993) discuss some postcranial material from La Défense, tentatively referred to *Mixtotherium,* showing separate facets for the navicular and cuboid on the astragalus, with oblique proximal and distal trochleas of the astragalus, both primitive features among artiodactyls.

RUMINANTS.   The oldest fossils assigned to the Ruminantia are the exceptionally long-legged amphimerycids, although there is controversy about their inclusion in Ruminantia. They are included in Ruminantia because the navicular and cuboid are fused, although it has been argued that these elements are not well associated with the dental remains, and that the group is more closely related to the xiphodontids (Viret, 1961). There are two genera, *Amphimeryx* and *Pseudamphimeryx,* known from the early Eocene through early Oligocene of Europe. The upper molars are five cusped and selenodont. The lower dentition is complete, with a small canine, single-rooted elongate p2, elongate tricuspid p3, and a diastema separating p3 from p4. The p1 of *Amphimeryx* is elongated and bears three laterally compressed cusps arranged anteroposteriorly, but *Pseudamphimeryx* bears a simpler p1 with a single main cusp (Hooker and Weidmann, 2000).

ORIGIN, TIMING, AND RELATIONSHIPS

Phylogenetic Origin

The preceding review of early artiodactyl groups reaffirms the conclusion that the most primitive known artiodactyls are the diacodexeids, which are also the oldest known. Several of the most primitive artiodactyl families (e.g., dichobunids, homacodontids, cebochoerids, choeropotamids) seem clearly derived from diacodexeids. The others are plausibly derived from diacodexeids, but the evidence of their precise origins is less clear, and it remains possible that

one or more of them (particularly raoellids, anthracotheres, and enteledonts) originated from unknown, more primitive artiodactyls.

The question of artiodactyl ancestry has proved to be a vexing one; in part, because no transitional stages show how the artiodactyl tarsus evolved. Various authors have pointed to particular lineages from the paraphyletic assemblage "Condylarthra" (archaic ungulates) as ancestral to the artiodactyls, primarily based on dental evidence. Dentally, early artiodactyls show strong resemblances to certain hyopsodontids, mioclaenids, and arctocyonids, and several genera now placed in the Hyopsodontidae (*sensu lato*) were originally classified as artiodactyls (e.g., *Phenacodaptes;* Jepsen, 1930; Simpson, 1937).

Prothero et al. (1988) considered Artiodactyla to be the sister taxon of all other ungulates, including arctocyonids and other condylarths, which suggests a long ghost lineage (at least all of the Paleocene), and early divergence in the Cretaceous. Although this would accord with the timing suggested by some molecular phylogenies, no fossil evidence currently supports such an early origin of Artiodactyla.

Potential Ancestral Groups

MESONYCHIANS.   Matthew (1909) suggested that unknown creodonts, similar to the Mesonychidae (now excluded from creodonts and referred to suborder Acreodi, order Cete; McKenna and Bell, 1997) might have been ancestral to Artiodactyla. Mesonychians precede and overlap with early artiodactyls in the fossil record, being known from the early Paleocene and Eocene of North America and Asia, and the late Paleocene and Eocene of Europe. Dentally, there is little resemblance between mesonychids (which have very specialized cheek teeth) and artiodactyls, but mesonychids have paraxonic feet with hooflike unguals, and an astragalus with an oblique cuboid articulation and a shallowly grooved navicular articulation (e.g., Matthew, 1915; O'Leary and Rose, 1995). This astragalar structure has been interpreted as foreshadowing the double-trochleated astragalus of artiodactyls (Wortman, 1901; Gregory, 1910), but Schaeffer (1947) argued that this is unlikely, because of the relatively broad calcaneus and primitive orientation of the posterior astragalocalcaneal facet of mesonychids.

Even if mesonychians are excluded from artiodactyl ancestry, the early Paleocene triisodontids from North America, generally regarded as the sister taxon or ancestral stock of Mesonychia, could still lie near the origin of Artiodactyla. Triisodontids have less specialized cheek teeth than do mesonychians, more conducive to relationship with artiodactyls; but their postcranial skeleton is poorly known and provides little support for a special relationship to artiodactyls.

HYOPSODONTIDS.   Simpson (1937) postulated that artiodactyls could be derived from Paleocene members of the Hyopsodontidae *sensu lato* (including taxa now separated as the Mioclaenidae), based on general dental resemblance between Paleocene hyopsodontids and early Eocene artiodactyls. Presumably on Simpson's authority, Schaeffer (1947) used the astragalus of early Paleocene *Choeroclaenus* (then placed in Hyopsodontidae, now considered a mioclaenid) to show how the artiodactyl tarsus could have originated; but he acknowledged that the astragalus of *Choeroclaenus* was much closer to those of arctocyonids than to those of artiodactyls. Nevertheless, the view of a hyopsodontid ancestry for Artiodactyla predominated until the mid-1960s. However, the postcranial skeleton of middle Eocene *Hyopsodus* (Matthew, 1909, 1915; Gazin, 1968), unlike that of artiodactyls, is short-limbed, pentadactyl, mesaxonic, and clawed. In addition, the limb bones have more prominent crests and processes than do those of artiodactyls. There is a prominent low third trochanter, and the astragalus has an almost flat trochlea and a convex head, contrary to expectations for the structural ancestor of the long-limbed and cursorial artiodactyls. The postcranial skeleton is poorly known in other hyopsodontids, and what little is known does not support a special relationship to Artiodactyla.

MIOCLAENIDS.   Several of the taxa that have been suggested as potential ancestors of Artiodactyla, based on dental resemblance, are North American genera now referred to the Paleocene family Mioclaenidae. Postcranial comparisons, however, show little that would support mioclaenid ancestry of Artiodactyla, with the possible exception of early Paleocene *Protoselene*. Early Paleocene *Ellipsodon* is postcranially similar to *Hyopsodus* (Matthew, 1937). The astragalus of early Paleocene *Choeroclaenus* (which Matthew referred to *Mioclaenus*), although used by Schaeffer (1947) as a structural antecedent for the artiodactyl condition, is narrower than that of *Hyopsodus*, with a convex head and shallowly grooved trochlea, unlike artiodactyls. Late Paleocene *Pleuraspidotherium* from Europe has moderately robust postcrania, with shorter and broader astragalus and calcaneus than those of *Hyopsodus*, and its femur has a medially projecting lesser trochanter and broad, flat patellar trochlea, in contrast to artiodactyls (Thewissen, 1991).

Protoselene, however, shows postcranial features that more closely approach the artiodactyl condition than does *Hyopsodus* or any mioclaenid. These include a relatively long humerus and femur—the humerus distally narrow and with a low deltoid crest, and the femur with a narrow trochlea and a high third trochanter (Matthew, 1937).

ARCTOCYONIDS.   Early serological data suggested a relationship between artiodactyls and cetaceans (Boyden and Gemeroy, 1950). Van Valen (1966) postulated that this could be explained by common ancestry, with both Artiodactyla and Mesonychia—which he considered ancestral to whales—being derived from within the "condylarths." He proposed the small Paleocene arctocyonid *Tricentes* (=*Metachriacus*, now referred to *Chriacus*) as the ancestor of Artiodactyla, based on close dental similarity, as the postcranial skeletons of *Chriacus* and the earliest artiodactyls were then largely unknown (Van Valen, 1971, 1978).

Rose (1987) described the first relatively complete postcranial skeleton of *Chriacus*, based on early Eocene specimens that clearly lacked the features expected in a form ancestral to the cursorial Artiodactyla, instead displaying extensive specializations for climbing. These features include pentadactyl manus and pes with a slightly divergent hallux and clawed digits, large deltopectoral crest and entepicondyle on the humerus, an elbow joint allowing substantial forearm supination, an abducted femur with a medially directed lesser trochanter and a relatively shallow patellar groove, and very mobile ankle joints. Such features appeared to provide strong evidence against an arctocyonid ancestry for Artiodactyla.

An early Paleocene (Torrejonian) arctocyonid specimen from New Mexico, attributed to *"Chriacus truncatus,"* weakened that conclusion, however, and revived the possibility of an arctocyonid-artiodactyl relationship. The specimen, represented by associated dentaries and fragmentary hindlimb bones, closely approaches diacodexeids in dental and hindlimb anatomy (Rose, 1996). In particular, the limb bones lack the arboreal features of Wasatchian *Chriacus*, which had led to the conclusion that *Chriacus* was unlikely to be near the ancestry of Artiodactyla. Instead, the tibia of *"C. truncatus"* is gracile, with a distinct tibial crest (but longer than in *Diacodexis*), and the femur has a narrow distal end, with a narrow, elevated patellar groove and laterally compressed condyles, features approaching the cursorial locomotor adaptations of primitive artiodactyls. Thus, arctocyonids remain a possible source for Artiodactyla, although no transitional forms are known.

CETACEA.   Recently, our understanding of artiodactyl relationships has been altered by new molecular and morphological data that suggest a special relationship to whales. Studies from molecular biology (Graur, 1993; Sarich, 1993; Springer and Kirsch, 1993; Graur and Higgins, 1994; Gatesy, 1997, 1998; Shimamura et al., 1997; Ursing and Árnason, 1998; Gatesy et al., 1999; Nikaido et al., 1999; Madsen et al., 2001; Murphy et al., 2001) have not only supported a sister-group relationship between Cetacea and Artiodactyla, as suggested by Van Valen (1971), but have also been used to argue that the whales are, in fact, deeply nested within a paraphyletic Artiodactyla (using the traditional definition), and most closely related to the Hippopotamidae.

New data from the postcranial skeleton of the early–middle Eocene archaeocete whales *Artiocetus*, *Rodhocetus*, *Ichthyolestes*, and *Pakicetus* show that the double trochleated head of the astragalus and several details of the morphology of the calcaneus and cuboid—which have long been considered unique to Artiodactyla—are shared with archaeocetes (Gingerich et al., 2001; Thewissen et al., 2001). The tarsus of all these taxa confirms that the earliest cetaceans had an astragalus that was double-trochleated, although the morphology shares some primitive features with mesonychids (retention of a remnant of the astragalar foramen and a shallower tibial trochlea; Rose, 2001). Although these discoveries remove the double-trochleated astragalus as a synapomorphy of Artiodactyla *sensu stricto,* it does seem to be a synapomorphy of the Cetartiodactyla. Phylogenetic analyses of morphological data of extant mammals generally support a sister-group relationship between artiodactyls and whales, but they have thus far not supported inclusion of whales in Artiodactyla (Geisler and Luo, 1998; O'Leary and Geisler, 1999; Geisler, 2001). When extinct taxa are included, however, some analyses find Mesonychia to be the sister group of Cetacea (O'Leary and Geisler, 1999). Total evidence analysis (Geisler and Uhen, 2003, in press) and new dental data from fossil deciduous teeth (Theodor and Foss, in press) may provide some support for inclusion of Cetacea within Artiodactyla.

IMPLICATIONS OF COMPARISONS FOR ANCESTRY AND CHARACTER TRANSFORMATIONS

Postcranial remains of Eocene cetaceans show that several features of the tarsus are shared with artiodactyls. At the same time, early cetaceans are more primitive than artiodactyls and resemble mesonychids in retaining the astragalar canal and a shallower astragalar trochlea with more rounded edges than is found among artiodactyls. This suggests that Cetacea diverged from a common ancestor with Artiodactyla at a more primitive stage than is known in any artiodactyl.

There are also a number of derived morphological features linking mesonychians to cetaceans. They include details of dental morphology (e.g., mesiodistal alignment of lower molar cusps, loss of the m3 paraconid, reduction of the metacone and loss of the lingual cingulum of M2, presence of proximal reentrant grooves in the lower cheek teeth), as well as basicranial anatomy (a small postglenoid foramen, position of the foramen ovale, and enlargement of the tegmen tympani; Geisler and Luo, 1998; O'Leary and Geisler, 1999; Geisler, 2001).

There are two plausible character reconstructions that could account for the known fossil evidence. Either one would seem to require substantial convergence in one anatomical system. If the shared features of the ankle joint in artiodactyls and early whales are synapomorphies, then craniodental resemblances between whales and mesonychids must be convergent. This seems to be the current consensus. Alternatively, the cranial and dental resemblances between mesonychids and early cetaceans are synapomorphies (Geisler and Luo, 1998; O'Leary, 1998; 1999; O'Leary and Geisler, 1999; Geisler, 2001) and there has been extensive convergence in cetacean and artiodactyl ankle joints.

IMPLICATIONS FOR TIMING OF ORIGIN

The oldest artiodactyls known in the fossil record are all assigned to *Diacodexis,* which first appeared 55 million years ago, at the beginning of the Eocene (Fig. 14.5). *Diacodexis ilicis* is found in North America in the earliest Eocene, Wa-0 subage of the Wasatchian NALMA (North American Land Mammal Age), and this occurrence is associated with the best (Gingerich, 1989, 2001) biostratigraphically constrained age estimate. Several species of *Diacodexis* appear suddenly in the earliest Eocene (Sparnacian or Neustrian, reference level MP 7) at Rians, France; Silveirinha, Portugal; and Dormaal, Belgium. *"Diacodexis" pakistanensis* appears in the Bumbanian LMA (Land Mammal Age) in Pakistan, where it is thought to be slightly younger in age (Gingerich, 1989).

In North America and Europe, the oldest artiodactyls appear abruptly and contemporaneously, and without obvious direct ancestors—evidence that their first appearances are a result of immigration rather than in situ evolution. All taxa that have been considered plausibly ancestral to Artiodactyla are present in the early Paleocene and are more primitive dentally, skeletally, or both, than early artiodactyls. Clearly, a significant part of the record of the origin of Artiodactyla has not been found, but the existing evidence strongly suggests that the transition took place during the Paleocene, and not earlier.

The oldest cetaceans are slightly younger than the oldest artiodactyls, being known from the Ypresian and Lutetian stages, in shallow marine deposits of the eastern Tethys Sea. *Himalayacetus* dates from around 53.5 million years ago, *Pakicetus* at 48 million years ago, and *Artiocetus* around 47 million years ago (Gingerich et al., 2001). Middle Eocene protocetids were the first cetaceans to disperse widely outside the Indo-Pakistani region. O'Leary and Uhen (1999) postulated a cetacean ghost lineage back to the early Paleocene (Torrejonian), based on a common origin with Mesonychia. However, if cetaceans and artiodactyls are sister groups to the exclusion of Mesonychia, they might have shared a common ancestor later in the Paleocene. A later Paleocene origin of Cetacea is also possible if the group evolved from a mesonychian.

The molecular hypothesis that Cetacea are the sister group of Hippopotamidae is difficult to evaluate from fossil evidence, because the oldest known hippopotamid is *Kenyapotamus,* from Miocene deposits of Kenya, dated between 15.6 and 15.8 million years ago (Pickford, 1983; Behrensmeyer et al., 2002). Thus, if whales are the sister taxon to the Hippopotamidae, there is a gap of over 37 million years between their earliest known representatives. There are two schools of thought regarding the ancestry of hippopotamids within artiodactyls: one holds that they are descended from Old World tayassuids; the other, that hippos are descended from the anthracotheres. Kron and Manning (1998) argue that anthracotheres are outside the Suoidea, and hence, not directly ancestral to hippos, based on the anthracothere synapomorphy of a crest joining the p4 protoconid to the hypoconid. Several morphological phylogenetic analyses support a relationship between Suoidea and anthracotheres but do not indicate that they are sister groups (Gentry and Hooker, 1988; Thewissen et al., 2001).

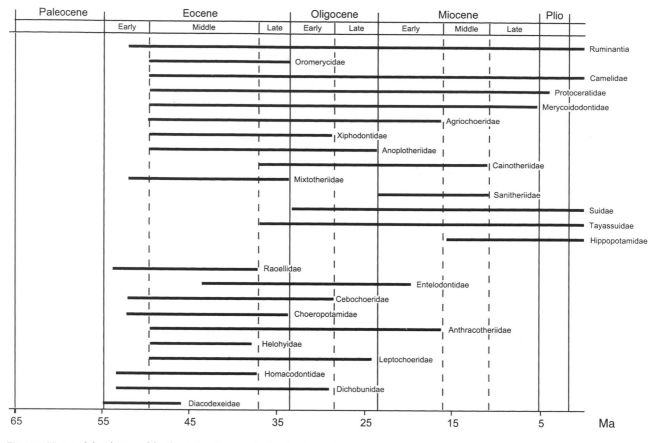

Fig. 14.5. Temporal distribution of fossil artiodactyl groups during the Cenozoic.

If hippos are a sister taxon to anthracotheres, or anthracotheres are directly ancestral to hippos, the common lineage would extend back to the origin of the Anthracotheriidae, about 40 million years ago. Even with this extension, which is by no means strongly supported, there is still a gap of 13.5 million years after the appearance of cetaceans and before the appearance of any fossil even potentially closely related to the hippopotamids. Note also that if hippos and anthracotheres share common ancestry with whales, the lineage as a whole would not share aquatic ancestry, as there is little evidence to indicate that anthracotheres were aquatic. In summary, although the precise relationships of hippopotamids within suiforms are controversial, there is little extant or fossil morphological evidence to support a close relationship between hippopotamids and cetaceans.

IMPLICATIONS FOR BIOGEOGRAPHY

Once they appear in North America and Europe, artiodactyls were present in some abundance, and we would expect them to have been preserved in the Paleocene record in at least low numbers if they had originated on one of those continents (Fig 14.6). Gingerich (1986, 1989) has suggested that artiodactyls entered North America and Europe via Asia, from a center of origin in Africa. There is, as yet, no evidence that there were artiodactyls in Africa before the

middle Eocene, although the record is poorly sampled (Gheerbrant, 1997). It seems to us that, although it cannot be ruled out, if Africa were the birthplace of Artiodactyla, we would expect to find some trace of them prior to the middle Eocene. However, in the absence of better sampling, Gingerich's hypothesis cannot be falsified.

Geographically, we note the coincidence in Pakistan between the earliest records of amphibious Cetacea and one of the earliest artiodactyls, *"Diacodexis" pakistanensis* (raoellids as well). This may indicate rather precisely the region in which the Cetartiodactyla originated.

The occurrence of *"Diacodexis" pakistanensis* in the Bumbanian LMA of Pakistan, and the diversity of artiodactyl taxa found in slightly younger deposits (Métais, 2002) indicate that it is probably to Asia—perhaps the Indian subcontinent—where we should look for the biogeographic origin of Artiodactyla. The Paleogene record of Asia is also relatively poorly sampled to date, but thus far, there are many more artiodactyl taxa reported from Asia than from Africa, and it is here that we might expect to find evidence leading to the ancestry of Artiodactyla and bearing on their relationship to the Cetacea.

SUMMARY

The preceding discussion might lead one to ask what we do know about artiodactyl origins. Early artiodactyl morphol-

■ Agriochoeridae	
▪ Anthracotheriidae	◉ Helohyidae
▫ Anoplotheriidae	○ Homacodontidae
● Cainotheriidae	✷ Mixtotheriidae
◉ Camelidae	★ Merycoidodontidae
○ Cebochoeridae	☆ Oromerycidae
▲ Choeropotamidae	✳ Protoceratidae
▲ Diacodexeidae	✚ Raoellidae
△ Dichobunidae	▼ Ruminantia
⬣ Entelodontidae	▽ Xiphodontidae

Fig. 14.6. Middle Eocene continental distribution of artiodactyl families. Each point indicates only the presence of that taxon on the continent, not its biogeographic distribution within the continent.

ogy is consistent: all of the earliest artiodactyls in the fossil record are small, with relatively long limbs (hind limbs longer than forelimbs) and paraxonic feet, double-trochleated astragalus, a vestigial (or absent) clavicle, greatly reduced or absent third trochanter on the femur, and a six-cusped dp4. The basal artiodactyl must have been similar and, like *Diacodexis,* would have had a mesaxonic manus and bunodont molars. Compared to *Diacodexis,* it probably had shorter limbs, a larger clavicle, a more prominent third trochanter on the femur, and possibly a more primitive astragalus— double-trochleated, but with a shallower tibial trochlea and an astragalar foramen, as in the early whales *Artiocetus* and *Rodhocetus.*

The Artiodactyla probably did not originate in North America or Europe; more likely, they arose in the late Paleocene of Asia, possibly the Indian subcontinent or Africa, well after the Cretaceous, probably in the mid-late Paleocene. Both morphological and molecular data indicate that Artiodactyla is the sister taxon to Cetacea. The most probable source group for both orders is the Arctocyonidae, but the evidence is still relatively weak, and new fossil discoveries could result in significant changes in this hypothesis. In spite of their excellent fossil record, we have much to learn about the origins of this group, particularly from new localities in Asia and Indo-Pakistan.

ACKNOWLEDGMENTS

We thank Scott Foss for informative discussion, and Christine Janis and Dave Webb for reading the chapter and offering constructive suggestions for its improvement.

REFERENCES

Averianov, A. 1994. Artiodactyla from the early Eocene of Kyrgyzstan. Palaeovertebrata 25(2–4): 359–369.

Behrensmeyer, A. K., A.L. Deino, A. Hill, J. D. Kingston, and J. J. Saunders. 2002. Geology and geochronology of the middle Miocene Kipsaramon site complex, Muruyur Beds, Tugen Hills, Kenya. Journal of Human Evolution 42: 11–38.

Blainville, H.M.D. de. 1816. Prodrome d'une nouvelle distribution systématique du règne animal. Bulletin des Sciences Philomatiques, Série 3, Paris, 3: 105–124.

Boyden, A., and D. Gemeroy. 1950. The relative position of Cetacea among the orders of Mammalia as indicated by precipitin tests. Zoologica 35: 145–151.

Checa-Soler, L. 2004. Revisión del género *Diacodexis* (Artiodactyla, Mammalia) en el Eoceno inferior del Noreste de España. Géobios 37(3): 325–335.

Chow, M. 1958. *Eoentelodon*—a new primitive entelodont from the Eocene of Lunan, Yunan. Vertebrata Palasiatica 2(1): 30–36.

Colbert, E. H. 1935. Distributional and phylogenetic studies on Indian fossil mammals. IV. The phylogeny of the Indian Suidae and the origin of the Hippopotamidae. American Museum Novitates 799: 1–24.

Coombs, M. C., and W. P. Coombs, Jr. 1977a. Dentition of *Gobiohyus* and a reevaluation of the Helohyidae (Artiodactyla). Journal of Mammalogy 58(3): 291–308.

Coombs, W. P., Jr., and M. C. Coombs. 1977b. The origin of anthracotheres. Neues Jahrbuch für Geologie und Paläontologie, München 10: 584–599.

Cope, E. D. 1881. The systematic arrangement of the order Perissodactyla. Proceedings of the American Philosophical Society 19: 377–401.

Cuvier, G. 1822. Recherches sur les ossemens fossiles, où l'on rétablit les caractères de plusieurs espèces d'animaux que les révolutions du globe ont détruit les espèces. G. Dufour and E. d'Ocagne, Paris.

de Queiroz, K., and J. Gauthier. 1992. Phylogenetic taxonomy. Annual Reviews in Ecology and Systematics 23: 449–480.

Erfurt, J. 1988. Systematik, Paläoökologie und Stratigraphische Bedeutung der Artiodactyla des Geiseltales. Ph.D. Dissertation, Martin-Luther-Universität, Halle-Wittenberg, Germany.

———. 2000. Rekonstruktion des Skelettes und der Biologie von *Anthracobunodon weigelti* (Artiodactyla, Mammalia) aus dem Eozän des Geiseltales. Hallesches Jahrbuch für Geowissenschaften B 12: 57–141.

Erfurt, J., and H. Haubold. 1989. Artiodactyla aus den Eozänen Braunkohlen des Geiseltales bei Halle (DDR). Palaeovertebrata 19(3): 131–160.

Erfurt, J., and J. Sudre. 1996. Eurodexeinae, eine neue Unterfamilie der Artiodactyla (Mammalia) aus dem Unter- und Mitteleozän Europas. Palaeovertebrata 25(2–4): 371–390.

Estravís, C., and D. E. Russell. 1989. Découverte d'un nouveau Diacodexis (Artiodactyla, Mammalia) dans l'Eocène inférieur de Silveirinha, Portugal. Palaeovertebrata 19: 29–44.

Foss, S. E. 2001. Systematics and Paleobiology of the Entelodontidae (Mammalia, Artiodactyla). Ph.D. Dissertation, Northern Illinois University, DeKalb.

Foss, S. E., and T. Fremd. 1988. A survey of the species of entelodonts (Mammalia, Artiodactyla) of the John Day Basin, Oregon. Dakoterra 5: 63–72.

Franzen, J. L. 1981. Das erste Skelett eines Dichobuniden (Mammalia, Artiodactyla), geborgen aus mitteleozänen Ölschiefern der "Grube Messel" bei Darmstadt (Deutschland, S-Hessen). Senckenbergiana Lethaea 61: 299–353.

———. 1983. Ein zweites Skelett von *Messelobunodon* (Mammalia, Artiodactyla, Dichobunidae), aus der "Grube Messel" bei Darmstadt (Deutschland, S-Hessen). Senckenbergiana Lethaea 64: 403–445.

———. 1988. Skeletons of *Aumelasia* (Mammalia, Artiodactyla, Dichobunidae) from Messel (M. Eocene, W. Germany). Courier Forschungsinstitut Senckenberg 107: 309–321.

Gabunia, L. K. 1973. On the presence of the Diacodexinae in the Eocene of Asia. Bulletin of the Academy of Sciences of the Georgian SSR 71: 741–744. (In Russian.)

Gatesy, J. 1997. More DNA support for a Cetacea/Hippopotamidae clade: The blood clotting protein gene γ-fibrinogen. Molecular Biology and Evolution 14(5): 537–543.

———. 1998. Molecular evidence for the phylogenetic affinities of Cetacea; pp. 63–111 in J.G.M. Thewissen (ed.), The Emergence of Whales: Evolutionary Patterns in the Origin of Cetacea. Plenum, New York and London.

Gatesy, J., C. Hayashi, M. A. Cronin, and P. Arctander. 1996. Evidence from milk casein genes that cetaceans are close relatives of hippopotamid artiodactyls. Molecular Biology and Evolution 13(7): 954–963.

Gatesy, J., M. Milinkovitch, V. Waddell, and M. Stanhope. 1999. Stability of cladistic relationships between Cetacea and higher-level artiodactyl taxa. Systematic Biology 48(1): 6–20.

Gazin, C. L. 1952. The Lower Eocene Knight Formation of western Wyoming and its mammalian faunas. Smithsonian Miscellaneous Collections 117(18): 1–82.

———. 1968. A study of the Eocene condylarthran mammal *Hyopsodus*. Smithsonian Miscellaneous Collections 153(4): 1–90.

Geisler, J. H. 2001. New morphological evidence for the phylogeny of Artiodactyla, Cetacea, and Mesonychidae. American Museum Novitates 3344: 1–53.

Geisler, J. H., and Z.-X. Luo. 1998. Relationships of Cetacea to terrestrial ungulates and the evolution of cranial vasculature in Cete; pp. 163–212 in J.G.M. Thewissen (ed.), The Emergence of Whales: Evolutionary Patterns in the Origin of Cetacea. Plenum, New York and London.

Geisler, J. H., and M. D. Uhen. 2003. Morphological support for a close relationship between hippos and whales. Journal of Vertebrate Paleontology 23(4): 991–996.

Geisler, J. H., and M. D. Uhen. In press. Phylogenetic relationships of extinct cetartiodactyls: results of simultaneous analyses of molecular, morphological and stratigrpahic data, Journal of Mammalian Evolution.

Gentry, A. W., and J. J. Hooker. 1988. The phylogeny of the Artiodactyla; pp. 235–72 in M. J. Benton (ed.), The Phylogeny and Classification of the Tetrapods. Volume 2. Mammals. Systematics Association Special Volume 35B. Clarendon Press, Oxford.

Gheerbrant, E. 1997. The oldest known proboscidean and the role of Africa in the radiation of modern orders of placentals. Bulletin of the Geological Society of Denmark 44: 181–185.

Gingerich, P. D. 1986. Early Eocene *Cantius torresi*—oldest primate of modern aspect from North America. Nature 320: 319–321.

———. 1989. New earliest Wasatchian mammalian fauna from the Eocene of Northwestern Wyoming: Composition and diversity in a rarely sampled high-floodplain assemblage. University of Michigan Papers on Paleontology 28: 1–97.

———. 2001. Biostratigraphy of the continental Paleocene-Eocene boundary interval on Polecat Bench in the northern Bighorn Basin; pp. 37–73 in P. D. Gingerich (ed.), Paleocene-Eocene Stratigraphy and Biotic Change in the Bighorn and Clarks Fork Basins, Wyoming. University of Michigan Papers on Paleontology 33.

Gingerich, P. D., M. ul Haq, I. S. Zalmout, I. H. Khan, and M. S. Malkani. 2001. Origin of whales from early artiodactyls: Hands and feet of Eocene Protocetidae from Pakistan. Science 293: 2239–2242.

Godinot, M. 1981. Les mammifères de Rians (Eocène inférieur, Provence). Palaeovertebrata 10: 43–126.

Graur, D. 1993. Molecular phylogeny and the higher classification of Eutherian mammals. Trends in Ecology and Evolution 8(4): 141–147.

Graur, D., and D. G. Higgins. 1994. Molecular evidence for the inclusion of cetaceans within the Order Artiodactyla. Molecular Biology and Evolution 11(3): 357–364.

Gregory, W. K. 1910. The orders of mammals. Bulletin of the American Museum of Natural History 27: 1–524.

Hooker, J. J., and K. M. Thomas. 2001. A new species of *Amphirhagatherium* (Choeropotamidae, Artiodactyla, Mammalia) from the Late Eocene Headon Hill Formation of Southern England and phylogeny of endemic European "anthracotherioids." Palaeontology 44(5): 827–853.

Hooker, J. J., and M. Weidmann. 2000. The Eocene mammal faunas of Mormont, Switzerland. Schweizerische Paläontologische Abhandlungen 120: 1–143.

Hunt, R. M., Jr. 1990. Taphonomy and sedimentology of Arikaree (lower Miocene) fluvial, eolian, and lacustrine paleoenvironments, Nebraska and Wyoming; A paleobiota entombed in fine-grained volcaniclastic rocks; pp. 69–111 in M. G. Lockley and A. Rice (eds.), Volcanism and Fossil Biotas. Geological Society of America Special Paper 244.

Janis, C. M., M. E. Ahearn, J. A. Effinger, J. A. Harrison, J. G. Honey, D. G. Kron, B. Lander, E. Manning, D. R. Prothero, M. S. Stevens, R. K. Stucky, S. D. Webb, and D. B. Wright. 1998. Artiodactyla; pp. 337–357 in C. M. Janis, K. M. Scott, and L. L. Jacobs (eds.), Evolution of Tertiary Mammals of North America. Volume 1. Terrestrial Carnivores, Ungulates, and Ungulatelike Mammals. Cambridge University Press, Cambridge.

Jepsen, G. L. 1930. New vertebrate fossils from the Lower Eocene of the Bighorn Basin, Wyoming. Proceedings of the American Philosophical Society 69: 117–131.

Krishtalka, L., and R. K. Stucky. 1985. Revision of the Wind River faunas, early Eocene of central Wyoming. Part 7. Revision of *Diacodexis* (Mammalia, Artiodactyla). Annals of the Carnegie Museum 54(14): 413–486.

Kron, D. G., and E. Manning. 1998. Anthracotheriidae; pp. 381–388 In C. M. Janis, K. M. Scott, and L. L. Jacobs (eds.), Evolution of Tertiary Mammals of North America. Volume 1. Terrestrial Carnivores, Ungulates, and Ungulatelike Mammals. Cambridge University Press, Cambridge.

Linnaeus, C. 1758. Systema Naturae per Regna tria Naturae, secundum Classes, Ordines, Genera, Species, cum Characteribus, Differentiis, Synonymis, Locis. Volume I. Tenth Edition. Impensis Direct, Laurentii Salvii, Stockholm.

Luckett, W. P., and N. Hong. 1998. Phylogenetic relationships between the orders Artiodactyla and Cetacea: A combined assessment of morphological and molecular evidence. Journal of Mammalian Evolution 5(2): 127–182.

Madsen, O., M. Scally, C. J. Douady, D. J. Kao, R. W. DeBry, R. Adkins, H. M. Armine, M. J. Stanhope, W. W. de Jong, and M. S. Springer. 2001. Parallel adaptations in two major clades of placental mammals. Nature 409: 610–614.

Matthew, W. D. 1909. The Carnivora and Insectivora of the Bridger Basin, middle Eocene. Memoirs of the American Museum of Natural History 9: 289–567.

———. 1915. A revision of the lower Eocene Wasatch and Wind River faunas, Part II: Order Condylarthra, family Hyopsodontidae. Bulletin of the American Museum of Natural History 34: 311–328.

———. 1937. Paleocene faunas of the San Juan Basin, New Mexico. Transactions of the American Philosophical Society, n.s. 30: 1–510.

McKenna, M. C., and S. K. Bell. 1997. Classification of Mammals Above the Species Level. Columbia University Press, New York.

Meng, J., and M. C. McKenna. 1998. Faunal turnovers of Palaeogene mammals from the Mongolian Plateau. Nature 394: 364–367.

Métais, G. 2002. Evolution des artiodactyls proto-selenodontes du Paleogène d'Asie: Nouvelles données sur l'origine du clade Ruminantia. Ph.D. Dissertation, Université Montpellier II Sciences et Techniques du Languedoc.

Métais, G., G. Jianwei, and K. C. Beard. In press. A new small dichobunid artiodactyl from Shanghuang (Middle Eocene, Eastern China). Bulletin of the Carnegie Museum of Natural History.

Milinkovitch, M. C., M. Bérubé, and P. J. Palsbøll. 1998. Cetaceans are highly derived artiodactyls; pp. 113–131 in J.G.M. Thewissen (ed.), The Emergence of Whales: Evolutionary Patterns in the Origin of Cetacea. Plenum, New York and London.

Milinkovitch, M. C., and Thewissen, J.G.M. 1997. Even-toed fingerprints on whale ancestry. Nature 388: 622–624.

Murphy, W. J., E. Eizirik, W. E. Johnson, Y. P. Zhang, O. A. Ryder, and S. J. O'Brien. 2001. Molecular phylogenetics and the origin of mammals. Nature 409: 614–618.

Nikaido, M. A., P. Rooney, and N. Okada. 1999. Phylogenetic relationships among cetartiodactyls based on insertions of short and long interspersed elements: Hippopotamuses are the closest extant relatives of whales. Proceedings of the National Academy of Sciences USA 96: 10261–10266.

Norris, C. A. 1999. The cranium of *Bunomeryx* (Artiodactyla: Homacodontidae) from the Upper Eocene Uinta deposits of Utah and its implications for tylopod systematics. Journal of Vertebrate Paleontology 19(4): 742–754.

———. 2000. The cranium of *Leptotragulus,* a hornless protoceratid (Artiodactyla: Protoceratidae) from the Middle Eocene of North America. Journal of Vertebrate Paleontology 20(2): 341–348.

O'Leary, M. A. 1998. Phylogenetic and morphometric reassessment of the dental evidence for a mesonychian and cetacean clade; pp. 133–161 in J.G.M. Thewissen (ed.), The Emergence of Whales: Evolutionary Patterns in the Origin of Cetacea. Plenum, New York and London.

———. 1999. Parsimony analysis of total evidence from extinct and extant taxa, and the cetacean-artiodactyl question. Cladistics 15: 315–330.

O'Leary, M. A., and J. H. Geisler. 1999. The position of Cetacea within Mammalia: Phylogenetic analysis of morphological data from extinct and extant taxa. Systematic Biology 48: 455–490.

O'Leary, M. A., and K. D. Rose. 1995. Postcranial skeleton of the Early Eocene mesonychid *Pachyaena* (Mammalia: Mesonychia). Journal of Vertebrate Paleontology 15(2): 401–430.

O'Leary, M. A., and M. D. Uhen. 1999. The time of origin of whales and the role of behavioral changes in the terrestrial-aquatic transition. Paleobiology 25(4): 534–556.

Owen, R. 1848. Description of teeth and proportion of jaws of two extinct anthracotherioid quadrupeds (*Hyopotamus vectianus* and *Hyop. bovinus*) discovered by the Marchioness of Hastings in the Eocene deposits on the N.W. coast of the Isle of Wight: With an attempt to develope Cuvier's idea of the classification of pachyderms by the number of their toes. Quarterly Journal of the Geological Society of London 4: 103–141, plates 7–8.

Pickford, M. 1983. On the origins of the Hippopotamidae together with descriptions of two new species, a new genus and a new subfamily from the Miocene of Kenya. Géobios 16: 193–217.

Prothero, D. R., E. M. Manning, and M. Fischer. 1988. The phylogeny of the Ungulates; pp. 201–234 in M. J. Benton (ed.), The Phylogeny and Classification of the Tetrapods. Volume 2. Mammals. Systematics Association Special Volume 35B. Clarendon Press, Oxford.

Romer, A. S. 1966. Vertebrate Paleontology. University of Chicago Press, Chicago.

Rose, K. D. 1982. Skeleton of *Diacodexis,* oldest known artiodactyl. Science 216: 621–623.

———. 1985. Comparative osteology of North American dichobunid artiodactyls. Journal of Paleontology 59(5): 1203–1226.

———. 1987. Climbing adaptations in the early Eocene mammal *Chriacus* and the origin of Artiodactyla. Science 236: 314–316.

———. 1990. Postcranial skeletal remains and adaptations in early Eocene mammals from the Willwood Formation, Bighorn Basin; pp. 107–133 in T. M. Bown and K. D. Rose (eds.), Dawn of the age of mammals in the northern part of the Rocky Mountain Interior, North America. Geological Society of America Special Paper 243.

———. 1996. On the origin of the Order Artiodactyla. Proceedings of the National Academy of Sciences USA 93: 1705–1709.

———. 2001. The ancestry of whales. Science 293: 2216–2217.

Rowe, T. 1988. Definition, diagnosis, and origin of Mammalia. Journal of Vertebrate Paleontology 8(3): 241–264.

Sahni, A., and S. K. Khare. 1971. Three new Eocene mammals from Rajauri District, Jammu and Kashmir. Journal of the Palaeontological Society of India 169: 41–53.

Sarich, V. M. 1993. Mammalian systematics: Twenty-five years among their albumins and transferrins; pp. 103–114 in F. S. Szalay, M. J. Novacek, and M. C. McKenna (eds.), Mammal Phylogeny: Placentals. Springer-Verlag, New York.

Schaeffer, B. 1947. Notes on the origin and function of the artiodactyl tarsus. American Museum Novitates 1356: 1–24.

———. 1948. The origin of a mammalian ordinal character. Evolution 2: 164–175.

Scott, W. B. 1940. The mammalian fauna of the White River Oligocene, Part IV. Artiodactyla. Transactions of the American Philosophical Society 28(4): 363–746.

Shimamura, M., H. Yasue, K. Ohshima, H. Abe, H. Kato, T. Kishiro, M. Goto, I. Munechika, and N.Okada. 1997. Molecular evidence from retroposons that whales form a clade within even-toed ungulates. Nature 388: 666–670.

Simpson, G. G. 1937. The Fort Union of the Crazy Mountain Field, Montana and its mammalian faunas. Bulletin of the U.S. National Museum 169: 1–287.

———. 1945. The principles of classification and a classification of mammals. Bulletin of the American Museum of Natural History 85: 1–350.

Sinclair, W. J. 1914. A revision of the bunodont Artiodactyla of the Middle and Lower Eocene of North America. Bulletin of the American Museum of Natural History 33: 267–295.

Smith, R., T. Smith, and J. Sudre. 1996. *Diacodexis gigasei* n. sp., le plus ancien Artiodactyle (Mammalia) belge, proche de la limite Paléocène-Eocène. Bulletin de l'Institut Royal des Sciences Naturelles de Belgique, Sciences de la Terre 66: 177–186.

Springer, M. S. and J.A.W. Kirsch. 1993. A molecular perspective on the phylogeny of placental mammals based on mitochondrial 12S rDNA sequences, with special reference to the problem of the Paenungulata. Journal of Mammalian Evolution 1: 149–166.

Stucky, R. K. 1998. Eocene bunodont and bunoselenodont Artiodactyla; pp. 358–374 in C. M. Janis, K. M. Scott, and L. L. Jacobs (eds.), Evolution of Tertiary Mammals of North America. Volume 1. Terrestrial Carnivores, Ungulates, and Ungulatelike Mammals. Cambridge University Press, Cambridge.

Sudre, J. 1972. Révision des Artiodactyles de l'Eocène moyen de Lissieu (Rhône). Palaeovertebrata 5(4): 111–156.

———. 1978. Les artiodactyles de l'Eocène moyen et supérieur d'Europe occidentale; systématique et évolution. Mémoires et Travaux de l'École Pratique des Hautes Études, Institut de Montpellier 7: 1–229.

Sudre, J., and J. Erfurt. 1996. Les artiodactyles du gisement Ypresien terminal de Premontre (Aisne, France). Palaeovertebrata 25(2–4): 391–414.

Sudre, J., and L. Ginsburg. 1993. La faune de mammifères de la Défense (Calcaire Grossier; Lutétien Supérieur) à Puteaux près Paris; artiodactyles et *Lophiodon parisiense* Gervais, 1848–1852. Bulletin du Muséum National d'Histoire Naturelle, Série 4, Paris, 15 section C (1–4): 155–181.

Sudre, J., and B. Marandat. 1993. First discovery of an Homacodontinae (Artiodactyla, Dichobunidae) in the Middle Eocene of Western Europe: *Eygalayodon montenati* new genus, new species. Considerations on the evolution of primitive artiodactyls. Kaupia: Darmstädter Beiträge zur Naturgeschichte 3: 157–164.

Sudre, J., D. E. Russell, P. Louis, and D. E. Savage. 1983a. Les artiodactyles de l'Eocène inférieur d'Europe (Première partie). Bulletin du Muséum National d'Histoire Naturelle, Paris, Série 4, Paris, 5 section C (3): 281–333.

———. 1983b. Les artiodactyles de l'Eocène inférieur (Deuxième partie). Bulletin du Muséum National d'Histoire Naturelle, Série 4, Paris, 5 section C (4): 339–365.

Theodor, J. and Foss, S. E. In press. Deciduous dentitions of Eocene cebochoerid artiodactyls and cetartiodactyl relationships. Journal of Mammalian Evolution.

Thewissen, J.G.M. 1991. Limb osteology and function of the primitive Paleocene ungulate *Pleuraspidotherium* with notes on *Tricuspidon* and *Dissacus* (Mammalia). Géobios 24(4): 483–495.

Thewissen, J.G.M., P. D. Gingerich, and D. E. Russell. 1987. Artiodactyla and Perissodactyla from the Early-Middle Eocene Kuldana Formation of Kohat (Pakistan). Contributions from the Museum of Paleontology, University of Michigan 27(10): 247–274.

Thewissen, J.G.M., and S. T. Hussain. 1990. Postcanial osteology of the most primitive artiodactyl *Diacodexis pakistanensis* (Dichobunidae). Anatomia, Histologia, Embryologia. Zentralblatt für Veterinarmedizin 19: 37–48.

Thewissen, J.G.M., D. E. Russell, P. D. Gingerich, and S. T. Hussain. 1983. A new dichobunid artiodactyl (Mammalia) from the Eocene of North-West Pakistan. Proceedings of the Koninklijke Nederlandse Akademie van Wetenschapen, Series B 86(2): 153–180.

Thewissen, J.G.M., E. M. Williams, L. J. Roe, and S. T. Hussain. 2001. Skeletons of terrestrial cetaceans and the relationship of whales to artiodactyls. Nature 413: 277–281.

Tobien, H. 1980. Ein anthracotherioider Paarhufer (Artiodactyla, Mammalia) aus dem Eozän von Messel be Darmstadt (Hessen). Geologisches Jahrbuch Hessen 108: 11–22.

———. 1985. Zur Osteologie von *Masillabune* (Mammalia, Artiodactyla, Haplobunodontidae) aus dem Mitteleozän der Fossilfundstätte Messel bei Darmstadt (S-Hessen, Bundesrepublik Deutschland). Geologisches Jahrbuch Hessen 113: 5–58.

Tong, Y., and J. Wang. 1998. A preliminary report on the early Eocene mammals of the Wutu fauna, Shandong Province, China; pp. 186–193 *in* K. C. Beard and M. R. Dawson (eds.), Dawn of the Age of Mammals in Asia. Bulletin of Carnegie Museum of Natural History 34.

Ursing, B. M., and Ú. Árnason. 1998. Analyses of mitochondrial genomes strongly support a hippopotamus-whale clade. Proceedings of the Royal Society of London, Series B, Biological Science 265: 2251–2255.

Van Valen, L. 1966. Deltatheridia, a new order of mammals. Bulletin of the American Museum of Natural History 132: 1–126.

———. 1971. Toward the origin of artiodactyls. Evolution 25(3): 523–529.

———. 1978. The beginning of the Age of Mammals. Evolutionary Theory 4: 45–80.

Viret, J. 1961. Artiodactyla; pp. 887–1021 *in* J. Piveteau (ed.), Traité de Paléontologie. Volume VI(1). Masson, Paris.

West, R. M. 1984. Paleontology and geology of the Bridger Formation, southern Green River Basin, southwestern Wyoming. Part 7. Survey of Bridgerian Artiodactyla, including description of a skull and partial skeleton of *Antiacodon pygmaeus*. Contributions in Biology and Geology, Milwaukee Public Museum, 56: 1–47.

Wortman, J. L. 1901. Studies of Eocene Mammalia in the Marsh Collection, Peabody Museum. Part 1. Carnivora. American Journal of Science 12: 421–432.

Zhang, Y., H. J. Long, and S. Ding. 1983. The Cenozoic Deposits of the Yunnan Region. Geological Publishing House, Peking.

PHILIP D. GINGERICH

CETACEA

GEORGE GAYLORD SIMPSON'S 1945 *Classification of Mammals* marks a turning point in the study of the origin and relationships of Cetacea. This is principally because Simpson so clearly characterized the state of knowledge at the time, but also because he unknowingly set the stage for a new approach to the study of whale relationships. First, Simpson (1945: 213) pointed out that the phylogenetic interpretation of the anatomy of cetaceans in comparison to that of other living mammals was ambiguous:

Because of their perfected adaptation to a completely aquatic life, with all its attendant conditions of respiration, circulation, dentition, locomotion, etc., the cetaceans are on the whole the most peculiar and aberrant of mammals. Their place in the sequence of cohorts and orders [of mammalian classification] is open to question and is indeed quite impossible to determine in any purely objective way.

And second, Simpson (1945: 214) noted that the fossil record known to him did little to constrain either the timing of whale origins or the relationships of suborders:

It is clear that the Cetacea are extremely ancient as such. . . . They probably arose very early and from a relatively undifferentiated eutherian ancestral stock. . . . Throughout the order Cetacea there is a noteworthy absence of annectent types, and nothing approaching a unified structural phylogeny can be suggested at present. Successive grades of structure appear in waves without any known origin for each. This is strikingly true in many orders, not only of mammals but of all animals, but within the Mammalia it is perhaps most striking among the Cetacea. Thus the Archaeoceti . . . are definitely the most primitive of cetaceans, but they can hardly have given rise to the other suborders.

Reaction was swift, and five years later, Boyden and Gemeroy (1950: 150–151) published the first biochemical study of cetacean relationships, indicating "a greater similarity in the serum proteins of representative Cetacea and Artiodactyla than between the Cetacea and any other orders tested, all existing orders but the Lagomorpha being included in the comparisons" (lagomorphs were excluded because the antigens were tested in rabbits). Boyden and Gemeroy thus corroborated the much earlier comparative anatomy by John Hunter (1787) and William H. Flower (1883), who emphasized resemblances of cetaceans to ungulates rather than carnivores. Flower's (1883: 376) scenario for the origin of Cetacea is particularly interesting:

We may conclude by picturing to ourselves some primitive generalised, marsh-haunting animals with scanty covering of hair like the modern hippopotamus, but with broad, swimming tails and short limbs, omnivorous in their mode of feeding, probably combining water-plants with mussels, worms, and freshwater crustaceans, gradually becoming more and more adapted to fill the void place ready for them on the aquatic side of the borderland on which they dwelt, and so by degrees being modified into dolphin-like creatures inhabiting lakes and rivers, and ultimately finding their way into the ocean.

Following Boyden and Gemeroy, Van Valen (1966: 90) reconciled both the newly hypothesized relationship of Cetacea to Artiodactyla and derivation of predaceous whales from herbivores by tracing the origin of whales through Paleocene members of the condylarth family Mesonychidae:

Only two known families need to be considered seriously as possibly ancestral to the archaeocetes and therefore to recent whales. These are the Mesonychidae and Hyaenodontidae (or just possibly some hyaenodontid-like palaeoryctid). No group that differentiated in the Eocene or later need be considered, since the earliest known archaeocete, *Protocetus atavus,* is from the early middle Eocene and is so specialized in the archaeocete direction that it is markedly dissimilar to any Eocene or earlier terrestrial mammal. It is also improbable that any strongly herbivorous taxon was ancestral to the highly predaceous archaeocetes.

In effect, Van Valen's synthesis ranked Mesonychidae as the closer extinct sister taxon of Cetacea, and Artiodactyla as the more distant extant sister taxon of Cetacea.

Van Valen's hypothesis deriving cetaceans from mesonychians initiated a slowly accelerating dance of disagreement between paleontologically trained morphologists committed to a carnivorous mesonychid origin on one hand, and biochemically trained systematists advocating direct derivation from or within herbivorous artiodactyls on the other. The principal cladistic studies of morphology favoring a relationship to mesonychids were by McKenna (1975), Pro-

thero et al. (1988), Thewissen (1994), Geisler and Luo (1998), O'Leary (1998), Luo and Gingerich (1999), O'Leary and Geisler (1999), and Geisler (2001). Some of the principal cladistic studies of molecular sequences favoring close a relationship to artiodactyls were by Irwin and Árnason (1994), Gatesy et al. (1996), Montgelard et al. (1997), Gatesy (1998), Milinkovitch et al. (1998), Liu and Miyamoto (1999), Nikaido et al. (1999), and Árnason et al. (2000). Thus, it was both a surprise and a relief when early archaeocete skeletons were discovered in 2000, enabling the issue to be resolved (Gingerich et al., 2001a; see below).

Here I review the general morphological characteristics of Cetacea, living and extinct. Tracing characteristics of living cetaceans back through geological time enables us to identify traits that are primitive in Cetacea and, hence, to recognize the origin of the order and its broader phylogenetic relationships. Relative likelihoods of various hypotheses of temporal distribution constrain a narrow range of credible times for the origin of whales. Finally, the stratigraphic record of early whale evolution enables discussion of the origin and diversification of Cetacea in a paleoenvironmental context.

MORPHOLOGICAL CHARACTERIZATION

The mammalian order Cetacea is interesting from an evolutionary point of view, because it represents entry into and eventual mastery of a new aquatic adaptive zone markedly different from that of its terrestrial ancestors. It is true, as Simpson wrote in 1945, that adaptation to a predaceous aquatic life, with all its attendant modifications of morphology and physiology, has made living cetaceans markedly different from other mammals. This obscures phylogenetic relationships. However, on the positive side, life in water gives cetaceans excellent potential for preservation as fossils. Whale bones are large and hard to overlook, and remarkable progress has been made in recent years tracing modern groups of whales back through time in the fossil record.

Living cetaceans fall naturally into two groups, distinguished by their feeding apparatus. These are the baleen whales or Mysticeti (Fig. 15.1B), and the toothed whales or Odontoceti (Fig. 15.1C). Living odontocetes and mysticetes are so different from one another that distinguished authors argued for years that each evolved independently from a different terrestrial ancestor (e.g., Kükenthal, 1893, 1922; Miller, 1923; Slijper, 1936; Yablokov, 1964). Others argued that odontocetes and mysticetes are too similar not to be related (e.g., Flower and Lydekker, 1891; Weber, 1904; Winge, 1921; Kellogg, 1936; Simpson, 1945; Van Valen, 1968). Kellogg (1936: 339) summarized the controversy:

Contradictions of statement in regard to identical anatomical structures subserving the same function show how difficult it is even for such eminent experts as Kükenthal and Weber to come to some sort of an agreement in regard to the relative importance

Fig. 15.1. Comparison of skeletons in three suborders of Cetacea. Note the very different skull sizes and forms related to different modes of feeding, and retention of hindlimbs with feet and toes in the oldest suborder, represented by *Dorudon*. All skeletons are shown in right lateral view and drawn to approximately the same length (not to scale). (A) The middle-late Eocene archaeocete *Dorudon atrox*. (~5 m length). From Gingerich and Uhen (1996). (B) Modern mysticete *Balaena mysticetus* (~15 m). (C) Modern odontocete *Lagenorhynchus obscurus* (~3 m). (B) and (C) are modified from Fordyce and Muizon (2001).

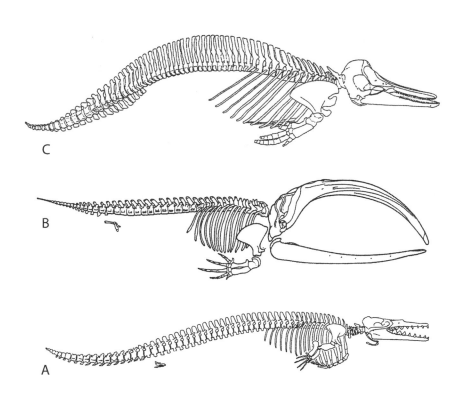

to be assigned to similarities and dissimilarities, for by one the resemblances are classed as convergences (adaptive) and by the other as a demonstration of blood relationship (non-adaptive).

Fortunately, new fossils and more critical consideration of previously known fossils have resolved this issue, and there is no longer any question of separate origins (Van Valen, 1968; Barnes et al., 1985; Fordyce and Barnes, 1994; Fordyce and Muizon, 2001). A classification of Cetacea is listed in Table 15.1.

Odontoceti

Odontocetes can be traced back in the fossil record to the late Oligocene, where they are represented by a range of primitive forms, including *Agorophius* Cope (1895), *Patriocetus* Abel (1914), *Archaeodelphis* Allen (1921), *Xenorophus* Kellogg (1923), *Waipatia* Fordyce (1994), and *Simocetus* Fordyce (2002). An undescribed skull evidently carries the record of Odontoceti to or near the Eocene/Oligocene boundary (Barnes and Goedert, 2000). The first-named genera *Agorophius*, *Patriocetus*, and *Archaeodelphis* have been compared to Archaeoceti at one time or another, if not actually classified as archaeocetes. Skulls of *Agorophius* and *Simocetus* are shown in Fig. 15.2 to illustrate not only the anteroposterior foreshortening of the frontal region of the skull seen in mysticetes but also the distinct telescoping of the maxillae over the frontals characteristic of odontocetes. Morphological characteristics of Odontoceti are listed in Table 15.2. In spite of their more derived skulls, such primitive odontocetes as *Agorophius* and *Simocetus* retain teeth that are basically archaeocete in cusp arrangement (see, e.g., the teeth of *Simocetus* in Fordyce, 2002).

Mysticeti

Similarly, mysticetes can be traced back in the fossil record to the late Oligocene, where they are represented by a range of primitive forms, including *Cetotheriopsis* Brandt (1871), *Kekenodon* Hector (1881), *Mauicetus* Benham (1939), *Mammalodon* Pritchard (1939), *Aetiocetus* Emlong (1966), *Chonecetus* Russell (1968), *Ashorocetus* Barnes et al. (1995), *Micromysticetus* Sanders and Barnes (2002a), and *Eomysticetus* Sanders and Barnes (2002b). Here again, the first-named genera *Cetotheriopsis*, *Kekenodon*, *Mauicetus*, *Mammalodon*, *Aetiocetus*, and *Chonecetus* were compared to, or in several cases actually classified as, Archaeoceti at one time or another. *Phococetus* Gervais (1876) and *Llanocetus* Mitchell (1989) extend the range of mysticetes back into the early Oligocene or even possibly the latest Eocene (Fordyce, 2003). Skulls of *Aetiocetus* and *Eomysticetus* are shown in Fig. 15.2 to illustrate anteroposterior foreshortening of the intertemporal region of the skull compared to archaeocetes. Morphological characteristics of Mysticeti are listed in Table 15.3.

Archaeoceti

The best known archaeocetes can be divided into two distinct grades, a middle-to-late Eocene fully aquatic grade, classified as Basilosauridae, and a middle Eocene semiaquatic grade, classified as Protocetidae. Pakicetidae range back into the early Eocene and represent a slightly more primitive, but still poorly known grade (see below), whereas middle Eocene Ambulocetidae and Remingtonocetidae are divergently specialized and seemingly unrelated to later cetaceans. Early archaeocetes are known only from the eastern Tethys Sea, in what is now India and Pakistan, but later

Table 15.1   Outline classification of Cetacea at the family level

Suborder Archaeoceti	Superfamily Physeteroidea
Superfamily Protocetoidea	Physeteridae (sperm whales)
Pakicetidae	Kogiidae (pygmy sperm whales)
Protocetidae	Superfamily Ziphioidea
Ambulocetidae(?)	Ziphiidae (beaked whales)
Superfamily Remingtonocetoidea	Superfamily Platanistoidea
Remingtonocetidae	Squalodontidae
Superfamily Basilosauroidea	Squalodelphinidae
Basilosauridae	Waipatiidae
Suborder Mysticeti	Dalpiazinidae
Superfamily Aetiocetoidea	Platanistidae (river dolphins)
Aetiocetidae	Superfamily Eurhinodelphinoidea
Llanocetidae	Eurhinodelphinidae
Mammalodontidae	Eoplatanistidae
Superfamily Eomysticetoidea	Superfamily Inioidea
Eomysticetidae	Iniidae (river dolphins)
Superfamily Balaenopteroidea	Pontoporiidae (river dolphins)
Cetotheriidae	Superfamily Lipotoidea
Balaenopteridae (fin whales)	Lipotidae (river dolphins)
Superfamily Eschrichtioidea	Superfamily Delphinoidea
Eschrichtiidae (gray whales)	Kentriodontidae
Superfamily Balaenoidea	Albireonidae
Neobalaenidae (pygmy right whales)	Delphinidae (dolphins)
Balaenidae (right whales)	Phocoenidae (porpoises)
Suborder Odontoceti	Monodontidae (white whales)
Superfamily Agorophioidea	Odobenocetopsidae
Agorophiidae	
Simocetidae	
Patriocetidae(?)	

Sources: Mysticeti follows Sanders and Barnes (2002b), Odontoceti follows Fordyce and Muizon (2001), with additions.

Notes: Common names are listed in parentheses for extant families. (?) indicates uncertainty of rank or placement.

Protocetidae and Basilosauridae are known from the eastern and western margins of the North Atlantic, and Basilosauridae are known from the South Pacific.

Basilosauridae include dorudontines, such as *Dorudon* Gibbes (1845), *Zygorhiza* True (1908), and *Saghacetus* Gingerich (1992), with normally proportioned vertebrae, and larger basilosaurines, such as *Basilosaurus* Harlan (1834) and *Basiloterus* Gingerich et al. (1997), with conspicuously inflated and elongated vertebrae. The best known basilosaurids are *Basilosaurus isis* (Gingerich et al., 1990) and *Dorudon atrox* (Uhen 1996, 2004), which are known from virtually complete skeletons. The skeleton of *Dorudon* is illustrated in Fig. 15.1A. Morphological characteristics of Basilosauridae are listed in Table 15.4.

Dorudon is the kind of generalized basilosaurid that might have given rise to later whales. It has multicusped cheek teeth that resemble those of primitive mysticetes and primitive odontocetes, and a well-developed fossa for the pterygoid sinus. Other salient characteristics of the skeleton are a vertebral formula (cervicals:thoracics:lumbars:sacrals:caudals) of 7:17:20:0:21, for a total of 65 vertebrae in the skeleton (Uhen 1996, 2004). Cervicals are short anteroposteriorly, neural spines on thoracics and lumbars are medium in length and robust, there is no sacrum, and the tail was evidently fluked (Uhen 1996, 2004). Forelimbs retain a mobile elbow joint, and the hindlimbs retain evidence of *Basilosaurus*-like feet and toes. The hindlimbs are greatly reduced in size in relation to the rest of the skeleton, and the innominates did not contact the vertebral column, meaning that the hindlimbs could not have borne weight on land or have played any substantial role in locomotion.

The best known of the early middle Eocene protocetids are *Protocetus* Fraas (1904), *Rodhocetus* Gingerich et al. (1994), *Artiocetus* Gingerich et al. (2001a), and *Qaisracetus* Gingerich et al. (2001b), and the best known late middle Eocene protocetid is *Georgiacetus* Hulbert et al. (1998). Postcranial remains attributed to *Indocetus* by Gingerich et al. (1993) are now known to belong to *Remingtonocetus*. Skulls of *Artiocetus* and *Protocetus* are shown in Fig. 15.2 to illustrate their generalized mammalian form, with none of the anteroposterior foreshortening in the intertemporal region of the skull seen in mysticetes and odontocetes. Morphological characteristics of Protocetidae are listed in Table 15.5.

The skeleton of the early protocetid *Rodhocetus* is illustrated in Fig. 15.3B, based on the skull and axial skeleton of *Rodhocetus kasranii* (Gingerich et al., 1994) and fore- and hindlimbs of *R. balochistanensis* (Gingerich et al., 2001a). As a whole, the early protocetid skeleton retains more generalized mammalian proportions compared to those of later basilosaurids, mysticetes, and odontocetes. The vertebral formula is known from several individuals to be 7:13:6:4:*x* (the number of caudal vertebrae in the tail is not yet known).

Fig. 15.2. Comparison of skulls. Frontal bones are stippled. Note the foreshortening of frontals in mysticetes and the telescoping of maxillae over frontals in odontocetes. All skulls are shown in dorsal view, drawn to the same length (not to scale). (A) Early middle Eocene archaeocete *Artiocetus clavis*. Redrawn from Gingerich et al. (2001a). (B) Early middle Eocene archaeocete *Protocetus atavus*. Modified from Fordyce and Muizon (2001). (C) Late Oligocene mysticete *Aetiocetus cotylalveus*. Modified from Emlong (1966). (D) Late Oligocene mysticete *Eomysticetus whitmorei*. Modified from Sanders and Barnes (2002b). (E) Late Oligocene odontocete *Agorophius pygmaeus*. Modified from Whitmore and Sanders (1977) and Fordyce (1981). (F) Late Oligocene odontocete *Simocetus rayi*. Redrawn from Fordyce (2002).

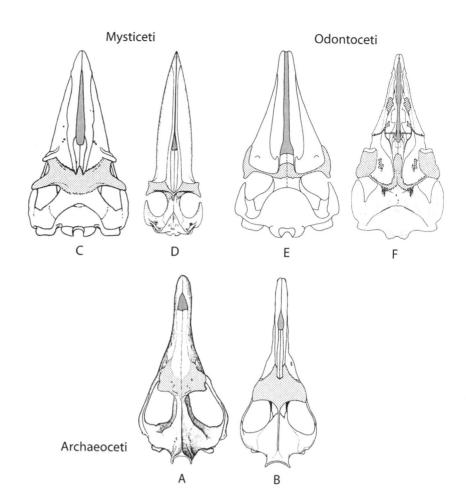

Table 15.2   Morphological characters of Oligocene-Recent Odontoceti as a group

Number	Character	Number	Character
1	Retention of heterodont dentition with some multicusped teeth (e.g., *Simocetus;* primitive) to homodont dentition of simple conical teeth lacking accessory cusps (derived)	8	Dentaries relatively straight
		9	Left and right sides of skull symmetrical in some taxa (primitive) to highly asymmetrical (derived)
2	Maxilla extends back over frontal, to or beyond orbit	10	Ability to echolocate inferred (primitive) to echolocation well developed (derived)
3	Anteroposterior shortening of intertemporal region of skull (primitive) to shortened intertemporal region of skull with broad anteroposterior telescoping (derived)	11	Cervical vertebrae short (primitive) to highly compressed anteroposteriorly (derived)
		12	Immobile elbow articulation
4	Dense tympanic bulla	13	Sacrum lost entirely
5	Pterygoid sinus present	14	Innominate retained within body wall; femur and lower leg lacking
6	Mandibular canal large		
7	Coronoid process retained on dentary (primitive) to coronoid process reduced (derived)	15	Small-to-medium body size (primitive) to large body size (derived)

Source: Abstracted from Fordyce and Muizon (2001), with additions.

The neck is relatively long for a cetacean, and the anterior thoracic vertebrae retain high neural spines, like those of land mammals (e.g., *Elomeryx* in Fig. 15.3A). There are distinct anticlinal and diaphragmatic vertebrae in the posterior thoracic series. The centra of the sacrum are solidly co-ossified in some early protocetids and less so in others. The vertebrae at the base of the tail are robust, but neither the number of caudals nor the length of the tail is known. The scapula is narrow, like that of land mammals. The humerus is relatively long, and bones of the forearm are relatively short. The manus retains five digits, of which the central three retain small, flattened, hooflike ungules (Gingerich et al., 2001a).

The hindlimb of *Rodhocetus* (Fig. 15.3B) has a large innominate articulating with the sacrum, but the ilium is short.

Table 15.3 Morphological characters of Oligocene-Recent Mysticeti as a group

Number	Character	Number	Character
1	Retention of multicusped teeth (e.g., *Aetiocetus;* primitive) to filter-feeding with baleen (derived)	8	Coronoid process retained on dentary (primitive) to coronoid process on dentary reduced (derived)
2	Maxilla extends back under frontal, toward orbit	9	Dentaries relatively straight (primitive) to dentaries bowed (convex laterally; derived)
3	Palatal surface of maxilla flat (primitive), to palatal surface of maxilla flat and grooved for nutrient supply to baleen (derived)	10	Cervical vertebrae short (primitive) to cervical vertebrae highly compressed anteroposteriorly (derived)
4	Anteroposterior shortening of intertemporal region of skull (primitive) to shortened intertemporal region of skull with conspicuous telescoping on midline (derived)	11	Immobile elbow articulation
		12	Sacrum lost entirely
		13	Innominate and femur retained within body wall
5	Dense tympanic bulla	14	Small-to-medium body size (primitive) to medium-to-large body size (derived)
6	Pterygoid sinus present		
7	Mandibular canal large (primitive) to mandibular canal reduced (derived)		

Source: Abstracted from Fordyce and Muizon (2001), with additions.

Table 15.4 Morphological characters of middle and late Eocene Basilosauridae as a group

Number	Character	Number	Character
1	Retention of heterodont dentition with multicusped cheek teeth	9	Coronoid process retained on dentary
		10	Dentaries relatively straight
2	Dental formula 3.1.4.2/3.1.4.3	11	Cervical vertebrae short anteroposteriorly
3	Maxilla-frontal contact largely in front orbit	12	Mobile elbow articulation
4	Long intertemporal region of skull; no telescoping	13	Sacrum lost entirely
5	Left and right sides of skull symmetrical	14	Innominate retained within body wall; femur, lower leg, and foot present
6	Dense tympanic bulla	15	Medium body size (primitive) to medium-to-large body size (derived)
7	Pterygoid sinus present		
8	Mandibular canal large		

Sources: Abstracted from Kellogg (1936), Luo and Gingerich (1999), and Uhen (1996, 2003), with additions.

Note: Based primarily on *Basilosaurus* and *Dorudon*.

Table 15.5 Morphological characters of middle-late Eocene Protocetidae as a group

Number	Character	Number	Character
1	Retention of heterodont dentition with multicusped cheek teeth	11	Cervical vertebrae medium in length (primitive) to cervical vertebrae short anteroposteriorly (derived)
2	Dental formula 3.1.4.3/3.1.4.3	12	Mobile elbow articulation
3	Maxilla-frontal contact in front of orbit	13	Sacrum of four vertebrae, with anterior centra solidly fused (primitive) to sacrum reduced to a single centrum or lost entirely (derived)
4	Long intertemporal region of skull; no telescoping		
5	Left and right sides of skull symmetrical	14	Innominate articulates with sacrum; ilium short; lower leg and foot present (primitive) to innominate retained within body wall; ilium short; lower leg and foot present (?) (derived)
6	Dense tympanic bulla		
7	Pterygoid sinus absent (primitive) to anterior pterygoid sinus present (derived)		
8	Mandibular canal large		
9	Coronoid process retained on dentary	15	Small-to-medium body size (primitive) to medium body size (derived)
10	Dentaries relatively straight		

Sources: Based primarily on *Rodhocetus* (Gingerich et al., 1994, 2001a), *Georgiacetus* (Hulbert, 1998; Hulbert et al., 1998), and *Eocetus* (Fraas, 1904; Uhen, 1999).

Note: (?) denotes a characteristic that has not been found in association with cranial or dental remains.

Fig. 15.3. Comparison of skeletons. (A) Early Oligocene anthracotheriid artiodactyl *Elomeryx armatus.* Redrawn from Scott (1894). (B) Early middle Eocene protocetid archaeocete *Rodhocetus balochistanensis* with an elongated skull, generalized mammalian vertebral formula with seven cervicals, 13 thoracics, six lumbars, and four sacrals, shortened neck, retention of long neural spines on thoracic vertebrae, robust proximal caudal vertebrae (length of tail is not known), short forelimb (ulna and radius), short ilium relative to ischium, short femur, and elongated pes (see Fig. 15.4). From Gingerich et al. (2001a).

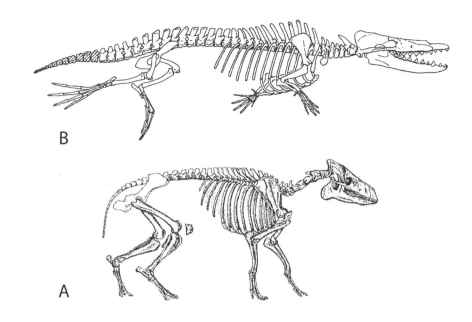

The femur is relatively short and robust, and the tibia and fibula are longer. The tarsus is known in *Artiocetus,* and the entire pes is known in *Rodhocetus.* The tarsal bones are interesting because of their general similarity to those of anthracotheriid artiodactyls. A comparison of the pes of *Rodhocetus* with that of the anthracothere *Elomeryx* is shown in Fig. 15.4. There are four notable points of resemblance: (1) the astragalus has a distal as well as a proximal trochlea, making it double-pulleyed; (2) the calcaneum has a prominent convex facet on its dorsal surface for articulation with the distal fibula; (3) the cuboid is distinctly stepped or notched to receive the distal process of the calcaneum; and (4) the entocuneiform is a relatively large, flat, platelike bone, with a small distal articular facet interpreted to indicate retention of metatarsal I. The first three tarsal traits are general artiodactyl characteristics (Gingerich et al., 2001a), but the fourth, conformation of the platelike entocuneiform, although possibly primitive, is a particular resemblance to anthracotheres. Scott (1894) described the entocuneiform of *Elomeryx* as resembling that of *Hippopotamus* (Scott, 1894: 485).

Comparison of the tarsus of *Rodhocetus* to that of *Elomeryx* shows many points of similarity, but proportions of the pes as a whole are very different in the two genera (Fig. 15.4). The metatarsals are longer and the proximal and medial phalanges are much longer in *Rodhocetus* than they are in *Elomeryx.* Both are drawn at the same calcaneum length in Fig. 15.4, and normalization to the same tarsus width would make these differences even more conspicuous.

Multivariate study of *Rodhocetus* and *Elomeryx* in a matrix of limb and trunk proportions for living semiaquatic mammals shows *Rodhocetus* to have been a desmanlike swimmer and *Elomeryx* to have been a more terrestrial, hippolike mammal (Gingerich, 2003b). More-aquatic semiaquatic mammals are distinguished from more-terrestrial semiaquatic mammals by the contrast between (1) long manual and pedal phalanges, combined with short femora and ilia in the former (more aquatic), versus (2) short manual and pedal phalanges, combined with long femora and ilia in the latter (more terrestrial). *Rodhocetus* has a more-aquatic suite of skeletal proportions, and *Elomeryx* has a more-terrestrial suite of proportions.

Pakicetid archaeocetes are known principally from cranial and dental specimens (Gingerich and Russell, 1981; Gingerich et al., 1983; Bajpai and Gingerich, 1998; Thewissen and Hussain, 1998; Thewissen et al., 2001). Some postcranial elements have been attributed to *Pakicetus,* including isolated vertebrae, a scapula, innominates, rare long bones and pieces of long bones, and astragali and calcanea (Gingerich, 1977; Thewissen et al., 1987, 2001). Some morphological characteristics of Pakicetidae are listed in Table 15.6. *Pakicetus* has generally been interpreted as a semiaquatic archaeocete (e.g., Gingerich et al., 1983; Thewissen and Hussain, 1993), but it has recently been reinterpreted as a "terrestrial mammal" with "running adaptations" that was "no more amphibious than a tapir" (Thewissen et al., 2001: 278). The presence of pointed anterior teeth in an elongated rostrum, dense auditory bullae, short cervical vertebrae, elongated caudal vertebrae, and an ilium no longer than the ischium all favor *Rodhocetus*-like semiaquatic habits for *Pakicetus* rather than terrestrial cursoriality. Definitive interpretation of *Pakicetus* will require associations of skeletal elements that have not yet been documented. The oldest pakicetid, early Eocene *Himalayacetus* Bajpai and Gingerich (1998), comes from marine strata, and it appears, as might have been expected, that the origin of whales goes hand-in-hand with aquatic adaptation.

ORIGIN OF CETACEA

Modern cetaceans, Mysticeti and Odontoceti, can be traced back in the fossil record to Oligocene times, when some genera in each group are seen to retain primitive characteristics,

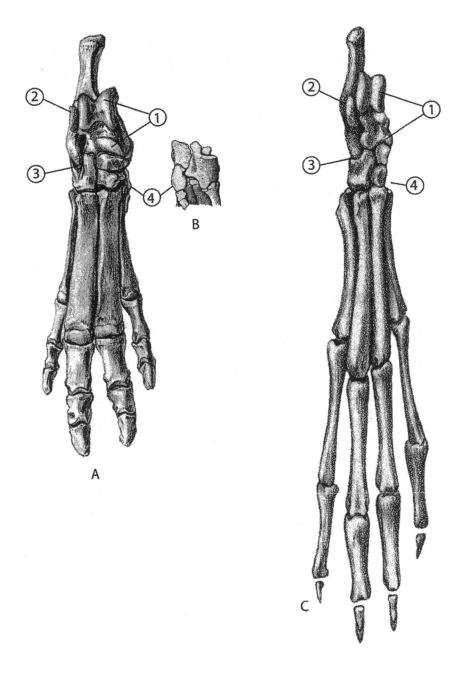

Fig. 15.4. Comparison of the pes. (A) Early Oligocene anthracotheriid artiodactyl *Elomeryx armatus.* Redrawn from Scott (1940). (B) Early Oligocene anthracotheriid artiodactyl *Elomeryx armatus.* From Scott (1894). (C) Early middle Eocene protocetid archaeocete *Rodhocetus balochistanensis.* From original specimen, drawn by B. Miljour. Both (A) and (C) are right feet, reduced to the same calcaneum length (52 and 54 mm, respectively; not to same scale). Characteristics of resemblance interpreted as synapomorphies of primitive Artiodactyla and early Cetacea are (1) double-pulley astragalus with a distal, as well as proximal, trochlea; (2) large convex fibular facet on the calcaneum; (3) stepped cuboid with distinct notch for calcaneum; and (4) large platelike entocuneiform with facet for metatarsal I. Note the much greater elongation of all metatarsals and proximal phalanges in the semiaquatic *Rodhocetus* relative to those of terrestrial *Elomeryx.*

Table 15.6 Morphological characters of middle-late Eocene Pakicetidae as a group

Number	Character	Number	Character
1	Retention of heterodont dentition with multicusped cheek teeth	9	Coronoid process retained on dentary
2	Dental formula 3.1.4.3/3.1.4.3	10	Dentaries relatively straight
3	Maxilla-frontal contact in front of orbit	11	Cervical vertebrae medium in length anteroposteriorly (?)
4	Long intertemporal region of skull; no telescoping	12	Mobile elbow articulation (?)
5	Left and right sides of skull symmetrical	13	Sacrum of four vertebrae, with anterior centra solidly fused (?)
6	Dense tympanic bulla	14	Innominate articulates with sacrum; ilium short; lower leg and foot present (?)
7	Pterygoid sinus absent	15	Small-to-medium body size
8	Mandibular canal small		

Sources: Based on Gingerich and Russell (1981), Gingerich et al. (1983), Thewissen and Hussain (1998), and Thewissen et al. (2001).

Notes: Based primarily on *Pakicetus.* (?) denotes a characteristic that has not been found in association with cranial or dental remains.

such as multicusped cheek teeth and a coronoid process on the dentary, linking them to late Eocene archaeocetes (Fordyce and Barnes, 1994; Fordyce and Muizon, 2001). Archaeoceti can, in turn, be traced back to middle Eocene times, when the protocetids *Artiocetus* and *Rodhocetus* are known from skulls and associated skeletons that retain such primitive characteristics as a double-pulley astragalus, convex fibular facet on the calcaneum, and a stepped cuboid, linking them to Artiodactyla among land mammals (Gingerich et al., 2001a).

Gingerich et al. (1990: 155) noted that a paraxonic pes in *Basilosaurus* "is consistent with serological evidence of relationship to Artiodactyla" but did not claim that this demonstrated such a relationship. Thewissen et al. (1998: 452) described an isolated "?pakicetid" astragalus and argued that the "absence of a trochleated astragalar head argues against . . . inclusion of Cetacea in Artiodactyla unless the flat head of the cetacean is interpreted as a secondary aquatic adaptation." A year later, Thewissen and Madar (1999: 23, 28) described the same astragalus as "pakicetid" (without a query) and concluded that "new evidence of Eocene cetacean tarsal morphology is . . . consistent with inclusion of cetaceans in artiodactyls, if one assumes that the wide arc of rotation of the trochleated head was lost during the origin of Cetacea." O'Leary and Geisler (1999) dismissed such claims about pakicetids, because the astragalus in question is fragmentary and was not associated with diagnostic cetacean material. After reading Gingerich et al. (2001a) in manuscript, Thewissen et al. (2001) acknowledged that the astragalus identified as "?pakicetid" and "pakicetid" was misidentified as cetacean, and Thewissen et al. (2001) then illustrated two new astragali as pakicetids that resemble astragali of *Artiocetus, Rodhocetus,* and artiodactyls. It is still true, as of this writing, that the only skeletons of archaeocetes that preserve associated ankle bones primitive enough and complete enough to demonstrate artiodactyl relationships are those of *Artiocetus* and *Rodhocetus* (Gingerich et al., 2001a). Now, knowing this association, it is possible to recognize that some astragali previously identified as representing artiodactyls are almost certainly from pakicetids (e.g., astragali described and illustrated as artiodactyls by Gingerich, 1977, and Thewissen et al., 1987, and those identified as *Pakicetus* and *Ichthyolestes* by Thewissen et al., 2001).

Fraas (1904) was clearly impressed by the upper molars of the early archaeocete *Protocetus,* when he removed Archaeoceti from Cetacea and included it as a subgroup of Creodonta (which, at the time, commonly included Mesonychidae). *Protocetus,* like *Pakicetus* and other primitive archaeocetes discovered later, retains much of the general mammalian tritubercular cusp pattern, with a large paracone and distinct remnants of a separate protocone and metacone positioned more or less like those of mesonychids. Van Valen (1966) transferred Mesonychidae to Condylarthra, and his inferences that archaeocetes evolved from mesonychid condylarths, while artiodactyls evolved from arctocyonid condylarths, were both based on dental resemblances

of primitive archaeocetes and artiodactyls, respectively, to earlier condylarths. Dental resemblances of archaeocetes and mesonychids were analyzed in more detail by O'Leary (1998), who cautioned that similar morphology here does not translate directly into similar toothwear patterns. One interpretation might be that the dental similarities evolved convergently.

The tarsal characteristics of early archaeocetes (Fig. 15.4) are very informative of relationships because of the presence of detailed similarities in three different bones representing three different articulations of the ankle that are otherwise known in combination only in living and fossil artiodactyls (Schaeffer, 1947, 1948; Rose, 2001):

1. The bodies of the astragalus and calcaneum have large, hemicircular dorsal articular surfaces (paired with an intervening trochlea on the astragalus) for contact with the tibia and fibula, respectively, restricting movement at the upper ankle joint to a parasagittal plane;
2. The astragalus has a plantar surface dominated by a sustentacular facet elongated parallel to the long axis of the bone, whereas the calcaneum has a distal process guiding the astragalus laterally, so movement at the lower ankle joint is stable but permits both folding of the ankle and shortening and elongation of the tarsus as a whole, as is characteristic of artiodactyls; and
3. The distal surface of the astragalus has a distinct trochlea for articulation with the navicular, whereas the calcaneum is elongated distally to contact and fit into a distinct step or notch in the cuboid—acting together, these stabilize the transverse tarsal joint and accommodate shortening and elongation at the lower ankle joint.

In mammals in general, this suite of characteristics is generally considered diagnostic of artiodactyls.

Rodhocetus and *Elomeryx* are illustrated here, but the evidence of relationship is based on more than this two-taxon comparison. The double-pulley astragalus, calcaneum with a convex fibular facet, and stepped cuboid are now known in all primitive pakicetid and protocetid archaeocetes for which such tarsal bones are known (Gingerich et al., 2001a; Thewissen et al., 2001; Zalmout et al., 2003), and these characteristics are known in all living and fossil artiodactyls for which the ankle is known (Schaeffer, 1947, 1948). It is doubtful that pakicetids or protocetids retained a primitive mesonychid-like remnant of an astragalar canal (suggested by Rose, 2001), as the depressions present on different astragali are all shallow and blind, and these vary in position more like ligamentous pits.

Primitive fossil archaeocetes with diagnostically artiodactyl ankle characteristics corroborate other, much earlier, comparisons of soft anatomy, suggesting a relationship of Cetacea to Artiodactyla (Hunter, 1787; Flower, 1883)—an interpretation that could not really be taken seriously in the absence of a more complete fossil record exhibiting more continuity through time (e.g., Simpson, 1945; also Langer,

2001). Fossils linking early Cetacea to Artiodactyla also corroborate more recent biochemical and molecular evidence for the association of these two groups (Boyden and Gemeroy, 1950; Sarich, 1985, 1993; Graur and Higgins, 1994; Irwin and Árnason, 1994; Gatesy et al., 1996, 1999; Montgelard et al., 1997; Shimamura et al., 1997, 1999; Gatesy, 1998; Milinkovitch et al., 1998; Ursing and Árnason, 1998; Liu and Miyamoto, 1999; Nikaido et al., 1999; Árnason et al., 2000; Shedlock et al., 2000).

Many of the molecular studies just cited group not only Cetacea with Artiodactyla, but group Cetacea with Hippopotamidae within Artiodactyla. This may or may not be consistent with the fossil record. Early paleontological studies (e.g., Falconer and Cautley, 1847; Lydekker, 1884; Andrews, 1906; Colbert, 1935a,b; Scott, 1940; Simpson, 1945) linked Hippopotamidae to Anthracotheriidae. Anthracotheres are unusual among artiodactyls in retaining appendages with a five-fingered mesaxonic manus and some remnant of a fifth toe on the four-toed pes (Kowalevsky, 1873; Scott, 1894, 1940). These conformations of the hand and foot are similar to those of the primitive artiodactyl *Diacodexis* (Rose, 1982; Thewissen and Hussain, 1990) and to the protocetid archaeocete *Rodhocetus* (Gingerich et al., 2001a). If hippopotamids are derived from anthracotheres, then it appears plausible that hippopotamids may be the closest living relatives of whales. However, other paleontologists, including Matthew (1929), Pickford (1983, 1989), McKenna and Bell (1997), and Kron and Manning (1998), grouped hippopotamids with pigs and peccaries rather than anthracotheres. More work needs to be done to achieve a consensus on hippopotamid ancestry and evolution. As it stands, the record provides a permissive and plausible case for a relationship of Cetacea to hippopotamids within Artiodactyla through intermediate anthracotheriids.

The stratigraphic ranges and relationships of artiodactyls and whales are illustrated diagrammatically in the phylogeny of Fig. 15.5. The fossil record of Artiodactyla on the northern continents (Europe, Asia, and North America) extends to the beginning of the Eocene, and artiodactyls are one of the key index taxa marking the beginning of Eocene time (Sudre et al., 1983; Estravís and Russell, 1989; Gingerich, 1989, 2003a; Smith, 2000; Bowen et al., 2001, 2002). As an order, Cetacea has a similarly rich and continuous fossil record, extending from the present back into the early Eocene. Early Eocene archaeocetes are rare but are known from at least two sites: Kuthar Nala in Himachal Pradesh, India (Bajpai and Gingerich, 1998), and Panandhro in Gujarat, India (Bajpai and Thewissen, 2002). Pakicetids from the Kuldana Formation of Pakistan, previously thought to be early Eocene in age (e.g., Gingerich et al., 1983; Thewissen and Hussain, 1998), are more likely to come from the earliest middle Eocene (Gingerich, 2003c). Thus, there is a million-year-plus gap between the first appearance of artiodactyls in the fossil record at the beginning of the Eocene and the first appearance of the oldest known archaeocete cetaceans later in the early Eocene. However, in spite of this,

Cetacea is the last of the major orders of mammals to appear in the fossil record. Hence, we can expect that diversification of whales was probably one of the more profound and rapid adaptive radiations.

The fossil record indicates that Cetacea evolved in a Tethyan aquatic adaptive zone that was new for Artiodactyla and artiodactyl-like mammals. It is almost certain that the ancestor of the earliest archaeocetes would itself have been an artiodactyl, and it is therefore possible that Artiodactyla is a paraphyletic clade. This is true whether cetaceans evolved from an anthracotheriid or other stem lineage leading to hippopotamids, the plausible scenario favored by recent molecular studies, or from some other early and as yet unknown lineage. On a broader scale, the origin of Artiodactyla (or the greater Artiodactyla-Cetacea clade) is uncertain, but origin of this larger group from arctocyonid Condylarthra is suggested by the evidence at hand (Rose, 1996).

TIMING OF ORIGIN AND DIVERSIFICATION

Consideration of the timing of origin of a taxonomic or evolutionary group necessarily involves comparison of the likelihoods of alternative hypotheses (Strauss and Sadler, 1989). For any group that is reasonably well known in the fossil record, as cetaceans are, the group's stratigraphic or temporal range has been established. This is the range into which all known samples fall. It is possible to hypothesize a larger total range for a group than that observed, but any such new hypothesis has a quantifiable and smaller likelihood relative to the range now known, and the question then becomes whether the smaller relative likelihood of the hypothesized range lies within the bounds of credibility. To appreciate this, it is useful to construct a geometric model of the sample space. A uniform model was employed by Strauss and Sadler (1989), triangular and composite exponential models, respectively, were illustrated by Gingerich and Uhen (1994, 1998), and I develop a new composite exponential model here.

A temporal range can be thought of as a line of known length, symbolized by K (for "known"), or, equivalently, as the area under a unit uniform distribution of length K (and hence, area K). Any hypothesized extension can be represented by some additional length E (for "extension"). The known range K is established by a set of samples drawn from K, including N independently discovered samples (where $N \geq 2$, because a minimum of two independent records is required to establish a finite range). When a total range is hypothesized to be larger than the known range by some extension E, it is necessary to explain how, by chance, all samples drawn from a larger sample space $K + E$ (represented as a length, area, volume, or hypervolume) fall only in the known range K. This is easy to explain when the hypothesized extension is small and the number of independent

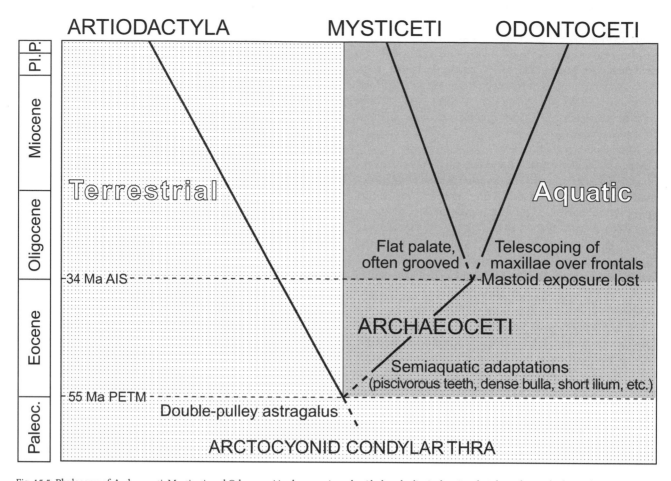

Fig. 15.5. Phylogeny of Archaeoceti, Mysticeti, and Odontoceti in the aquatic realm (darker shading), showing their hypothesized relationship to terrestrial Artiodactyla and arctocyonid Condylarthra. The common clade Artiodactyla-Cetacea is distinguished from earlier condylarths and other mammals by the presence of a double-pulley astragalus and other synapomorphies identified in Fig. 15.4. Archaeoceti and later Cetacea differ from Artiodactyla in sharing semiaquatic adaptations, including simple pointed anterior teeth, dense auditory bullae, and a short ilium. Mysticeti differ from other Cetacea in having a flat palate that is often grooved by nutrient canals associated with the presence of baleen. Odontoceti differ from other Cetacea in having the maxillae telescoped over the frontals. Note that semiaquatic-to-aquatic Archaeoceti appear in the fossil record following the Paleocene-Eocene thermal maximum at about 55 million years ago, and Mysticeti and Odontoceti appear in the fossil record with or following the reorganization of ocean circulation that led to Antarctic glaciation and the development of the Antarctic ice sheet (see Fig. 15.7). Removing the time axis, purely cladistic relationships in this figure could be expressed as: Arctocyonia-(*Artiodactyla*-(Archaeoceti-(*Mysticeti-Odontoceti*))). The phylogenetic history of early artiodactyls is poorly known, and it is possible that the branching order here was Arctocyonia-(*non-hippo Artiodactyla*-(*hippo Artiodactyla*-(Archaeoceti-(*Mysticeti-Odontoceti*)))). (Extant taxa are italicized.) Abbreviations: AIS, Antarctic ice sheet; PETM, Paleocene-Eocene thermal maximum.

records is also small, because the associated probability is large. The probability that each independent sample falls in the smaller known range K when it was really drawn from a larger hypothesized range $K + E$ is the ratio $K/(K + E)$. This can be illustrated by a simple example: if K is assumed to be one unit in length, then one hypothesis might be that E is as large as K, meaning $E = K = 1$. Now the probability of a sample falling in K when it was drawn from $K + E$ is $K/(K + E) = 1/2 = 0.5$. Compare this to the probability of a sample falling in K when it was drawn from K, which is $K/K = 1$ (maximum likelihood). The relative likelihood is the ratio of probabilities of the two hypotheses, which is $0.5/1 = 0.5$, in this example. Hypotheses with such large relative likelihoods, or ratios of probabilities, are well within the bounds of reason.

In general, the relative likelihood L of a stratigraphic or temporal range extension depends on K, E, and N:

$$L = (K/[K + E])^{N-1},\qquad\text{(Equation 1)}$$

which illustrates how the number of independent samples contributes to a calculation of relative likelihood. Any extension can be hypothesized, but the chance of it being reasonable or even credible, relative to the range we know, decreases as the size of the extension E becomes larger and decreases as the number of independent records N increases. A hypothesized extension can be one-tailed, meaning that E is an extension in one direction, or it can be two-tailed, meaning that E is partitioned to extend the range both forward and backward in time. Here we are only concerned with one-tailed range extensions backward in time.

An empirical example of such a comparison of likelihoods is shown graphically in Fig. 15.6A. The stratigraphic range of Archaeoceti is represented by the cross-hatched uniform distribution spanning much of the Eocene. The oldest archaeocete known to date is *Himalayacetus* (Bajpai and Gingerich, 1998) from low-latitude marine strata of the eastern Tethys, dated at about 53.5 million years ago (early

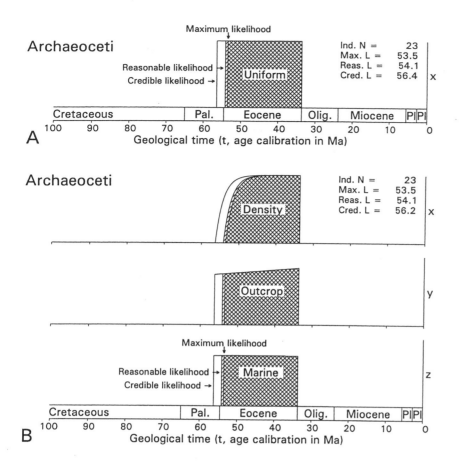

Fig. 15.6. Geometric models for understanding the relative likelihoods (*L*) of different times of origin of Archaeoceti. Abscissa in each model is time *t*. (A) Simple uniform *x* versus *t* model, with ordinate *x* a unitary constant (Strauss and Sadler, 1989). (B) More complex and hence more representative *x-y-z* versus *t* model, with increasing density of taxa or individuals during diversification on the *x*-axis, increasing outcrop area of sedimentary rocks in younger times (Blatt and Jones, 1975) on the *y*-axis, and changing marine area of the earth's surface (Smith et al., 1994) on the *z*-axis. Answers given by the two models are virtually identical, and it is not necessary to understand the complex model to appreciate the power of the statistical logic. Cross-hatched areas are a proportional representation of the difference on each axis over the known temporal range of the group. *K* is a simple $x \times t$ area in the uniform case (A), and an $x \times y \times z \times t$ four-dimensional hypervolume in the complex case (B). In this example, the number of independent records *N* is 23. Equation 1 is most easily solved iteratively, increasing *E* by small increments until reaching critical values of *L*.

Eocene). The youngest archeocetes are basilosaurids (e.g., *Saghacetus;* Gingerich, 1992) from low-latitude marine strata of the Tethys, extending near or to the end of the Eocene at 33.7 million years ago. Thus, archaeocetes have a known temporal range of about 19.8 million years. The number *N* of *independent* records can be estimated by counting the number of records coming from different geological formations, representing different temporal epochs in various geographic states or countries of the world. A survey of the Georef online database yields an estimate of *N* = 23 independent records. This *N* is conservative, because an ongoing compilation of published records now includes approximately twice as many independent records as are listed in the Georef database (see table 1 in Gingerich and Uhen, 1998).

The relative likelihood of successive incremental extensions of the range of archaeocetes back in time can be explored by adding small increments to the first record at 53.5 million years ago. Each added increment represents a new hypothesis, with its own associated probability. We can stop adding increments when the likelihood ratio *L* of the total hypothesized range extension *E* reaches one-half of the maximum likelihood. This can be considered as a landmark of *reasonable likelihood,* because such an extension is half as likely as the known range, given what we know about the number and distribution of empirical records. Using a uniform model, our estimate of reasonable likelihood for the origin of archaeocetes is about 54.1 million years ago.

Alternatively, we can continue adding increments until the likelihood of the total hypothesized range extension *E* reaches 5% of the maximum likelihood. This can be considered a landmark of *credible likelihood,* because it is equivalent to the 95% confidence limit, in the sense that this or any greater hypothesized range extension is 5% or less likely compared to the known range. We ordinarily accept such a 95% confidence limit as marking the bounds of credibility. The maximum likelihood estimate for the time of origin of archaeocetes is about 53.5 million years ago, whereas the credible limit by this standard, based on the uniform distribution, is 56.4 million years ago.

It is possible to make the same calculations using a *t:x:y:z* four-dimensional hypervolume sample space to represent (1) the density of archaeocetes during their diversification (dimension *x*); (2) the declining availability of outcrop area for older sedimentary rocks (dimension *y;* Blatt and Jones, 1975); and (3) the relative proportion of the earth's surface covered by oceans (dimension *z;* Smith et al., 1994). This more complicated and hence more representative model, shown graphically in Fig. 15.6B, gives a slightly younger credible limit for the time of origin of archaeocetes (and cetaceans) of 56.2 million years ago (late Paleocene).

The oldest known mysticete is *Llanocetus* (Mitchell, 1989) from high-latitude marine strata of the Eocene/Oligocene boundary in Antarctica (Fordyce and Muizon, 2001). A survey of the Georef online database yields an estimate of *N* = 34 independent records of Mysticeti. This, like the

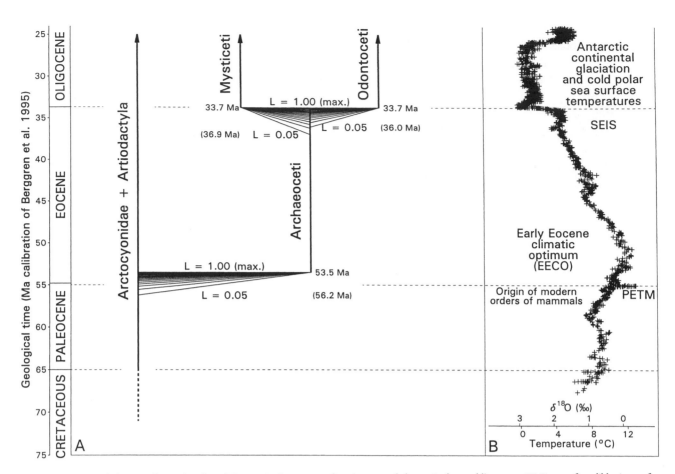

Fig. 15.7. Summary phylogeny of Artiodactyla and Cetacea in the context of environmental change in the world's oceans. (A) Range of credible times of origin of Archaeoceti, and credible times of divergence of Mysticeti and Odontoceti (calculated using the more complex model in Fig. 15.6). (B) Oxygen isotope record of temperature change preceding the onset of Antarctic ice sheets in the early Oligocene (data from Zachos et al., 2001). Note the coincidence of the origin of archaeocetes (and Cetacea) with the Paleocene-Eocene thermal maximum and the appearance of other modern orders at or near the Paleocene/Eocene boundary, and the divergence of Mysticeti and Odontoceti with the initiation of small ephemeral ice sheets in the southern oceans in the late Eocene, leading to full-scale Antarctic continental glaciation, cold polar sea-surface temperatures, and formation of cold bottom water in the oceans in the early Oligocene. Abbreviations: PETM, Paleocene-Eocene thermal maximum; SEIS, small ephemeral ice sheets.

survey of Archaeoceti described above, is undoubtedly conservative. Relative likelihoods calculated for Mysticeti, as in Fig. 15.6, yield a maximum likelihood estimate for the time of origin of about 33.7 million years ago and a credible limit of 37.0 million years ago.

The oldest known odontocete is an unnamed taxon from mid-latitude marine strata of the Eocene/Oligocene boundary in Washington state in western North America (Barnes and Goedert, 2000). A survey of the Georef online database gives an estimate of $N = 46$ independent records of Odontoceti; again, a conservative number. Relative likelihoods calculated for Odontoceti, as in Fig. 15.6, yield a maximum likelihood estimate for the time of origin of about 33.7 million years ago and a credible limit of 36.1 million years ago.

Ranges of possible divergence times for Archaeoceti from early Artiodactyla, Mysticeti from Archaeoceti, and Odontoceti from Archaeoceti are summarized graphically in Fig. 15.7A, based on these likelihood calculations. Credible estimates for the divergence of extant Cetacea from extant Artiodactyla are in the narrow range of 53.5–56.2 million

years ago, and credible estimates for the divergence of extant Odonticeti from Mysticeti within Cetacea lie in the restricted range of 33.7–36.9 million years ago.

This approach to estimating a time of origin requires that the whales included in a taxon be identifiable as such—in this case, identifiable to the suborders Archaeoceti, Mysticeti, or Odontoceti—but it does not require any assumption about the ancestral stock that gave rise to the taxon (i.e., the ancestral stock need not be known) or about the holophyly (non-paraphyly) of taxa in general. The alternative "ghost lineage" approach (O'Leary and Uhen, 1999; Gatesy and O'Leary, 2001) portrays all taxonomic groups as holophyletic, with sister taxa giving rise to one another only at the moment of their own conception. By this logic, the time of origin of a taxon of interest is not the time of its own origin, but at minimum, the time of appearance of its oldest sister taxon. Hence, the time of origin of Mysticeti and Odontoceti considered separately or the time of origin of the two considered together ("Neoceti") would be, at a minimum, the time of appearance of Archaeoceti; the time of origin of Archaeoceti would be that of Artiodactyla; and the time of origin of Artiodactyla would be that of Arctocyonidae plus Artiodactyla considered together.

All groups like Artiodactyla and Archaeoceti that were evolutionarily successful in the sense of giving rise to descendants outside the group itself are necessarily paraphyletic, and hence a ghost lineage approach to their times of origin is inappropriate. Furthermore, the limits of credible likelihood, calculated as in Fig. 15.6, show that many taxonomic groups (Mysticeti, Odontoceti, and Archaeoceti included) cannot be as old as their putative sister taxa, and that paraphyly must be common.

ENVIRONMENTAL CONTEXT FOR THE ORIGIN OF CETACEA

The stratigraphic and temporal records summarized in Fig. 15.7A provide a framework for discussion of the origin and diversification of Cetacea in a broader paleoenvironmental context. Temperature is one of the most important components of paleoclimate and paleoenvironment. A deep-sea temperature proxy, the oxygen isotope record ($\delta^{18}O$) for the world's oceans, is shown on a matching time scale in Fig. 15.7B (Zachos et al., 2001). These oxygen isotope data provide constraints on the evolution of deep-sea temperatures and continental ice volumes. Deep-sea waters are derived primarily from the cooling and sinking of water in polar regions, so the deep-sea temperature record doubles as a time-averaged record of high-latitude sea-surface temperatures. Three events in this record are of particular interest:

1. The abrupt spike of climatic warming, known as the Paleocene-Eocene thermal maximum (PETM) right at the Paleocene/Eocene boundary (this is the event initially regarded as the late Paleocene thermal maximum [LPTM] by marine stratigraphers);
2. The early Eocene climatic optimum (EECO), when temperatures were the warmest known for the entire Cenozoic; and
3. The abrupt spike of cooling in the late Eocene, marking the first appearance of small, ephemeral ice sheets (SEIS) and leading to the first full-scale Antarctic continental glaciation in the early Oligocene, with cold polar sea-surface temperatures (Zachos et al., 2001).

Many of the modern orders of mammals, including Artiodactyla, appear abruptly in the fossil record at the PETM, a brief 80,000- to 200,000-year interval of abrupt global warming that coincides with the Paleocene/Eocene boundary (this is not really a coincidence, of course, because the epochs of the geological time scale generally reflect such substantial differences in faunas and floras). There is a 1.0- to 1.5-million-year lag in the appearance of archaeocetes, but the Paleocene/Eocene boundary at 55.0 million years ago lies within the range of credible estimates for the time of origin of whales. Thus, Cetacea, like other orders, probably originated directly or indirectly in response to global

warming during the PETM. Warm climates during the EECO would have facilitated adaptation to the marine realm; such conditions probably prevailed on the margins of the relatively warm, low-latitude Tethys Sea (where *Himalayacetus* and all later pakicetid fossils are found).

SEIS, the late Eocene cooling event, occurred at about the earliest credible time of origin of both Mysticeti and Odontoceti, anticipating the onset of Antarctic glaciation and cold polar sea-surface temperatures in the early Oligocene, when it appears that both mysticetes and odontocetes were already present in middle-to-high latitude waters. The result of SEIS was a reorganization of ocean circulation, with formation of cold bottom water and enhanced nutrient-rich upwelling (reviewed in Fordyce, 1980, 1989). These changes may well have affected life at high latitudes more than they did life in more equatorial areas; hence, it is not unreasonable to expect some temporal overlap of archaic and modern whales, even if they did not overlap geographically. It appears doubtful that Archaeoceti survived the abrupt ocean cooling recorded at the Eocene/Oligocene boundary. The great number of different kinds of both mysticetes and odontocetes present in the late Oligocene provides evidence that these modern groups continued to diversify during early Oligocene glaciation, although the fossil record is poor during this interval.

The origin of Cetacea and their diversification are both correlated in time with important environmental changes in the sea and on land. The environment changed abruptly at both the Paleocene/Eocene and Eocene/Oligocene boundaries, and there was a substantial pulse of resulting evolutionary turnover at or near both epoch boundaries. This is consistent with the "turnover pulse" hypothesis that environmental change drives evolution, as articulated by Vrba (1985). Whales illustrate how evolution is both stimulated and channeled by the physical environment.

RETROSPECTIVE

It is encouraging to see how much has been learned about whale evolution from the fossil record in the past sixty years, and we can thank George Gaylord Simpson for articulating so clearly how little was known in 1945. There are now remarkable "annectent types" connecting Artiodactyla to Archaeoceti and linking Archaeoceti to Mysticeti and Odontoceti. There is general agreement on a unified phylogeny of whales, with support from both paleontology and molecular biology. And finally, the appearance of successive grades of structure can be tied explicitly to environmental changes on land and in the sea. Simpson would write very differently today.

SUMMARY

Cetaceans are interesting from an evolutionary point of view, because they were able to colonize a new aquatic adaptive zone markedly different from that of their terrestrial ancestors. Two suborders, Odontoceti and Mysticeti, are living today. Odontocetes can be traced back in the fossil record almost to the Eocene/Oligocene epoch boundary, and mysticetes can be traced back to the early Oligocene and possibly, to the latest Eocene. When they first appear in the fossil record, both modern groups resemble archaic whales, or Archaeoceti, in the form of their teeth and other characteristics. Archeocetes can be traced back from fully aquatic forms in the middle and late Eoene to amphibious, semiaquatic forms in the early and middle Eocene. Among early archaeocetes, *Artiocetus* and *Rodhocetus* are particularly important, because they preserve ankle bones, (calcaneum, astragalus, and cuboid) in direct association with protocetid skulls and skeletons. The ankle bones are diagnostically artiodactyl in their form both as individual bones and in their articulation as a unit. Cetacea is very closely related to Artiodactyla, and the common ancestor of the two would almost certainly be considered a generalized artiodactyl if it were known. Derivation from an early anthracothere-like artiodactyl is a distinct possibility, in which case, a sister-group relationship of Cetacea to hippos within Artiodactyla, suggested by molecular comparisons of living animals, is also plausible in terms of the fossil record. The known fossil record of archaeocetes indicates a time of origin of Cetacea in the range of 53.5–56.2 million years ago (late part of the late Paleocene to the early part of the early Eocene). This interval includes the PETM event of global climate change, and it is likely that Cetacea originated in concert with other mammalian orders as part of the Paleocene-Eocene transition in earth history.

ACKNOWLEDGMENTS

I thank the editors for the invitation to participate in a symposium honoring George Gaylord Simpson. The chapter has been improved as a result of careful readings by Ewan Fordyce, Kenneth Rose, and an anonymous reviewer. Field research on fossil cetaceans was supported by a series of grants from the National Geographic Society and National Science Foundation, most recently, EAR-9714923.

REFERENCES

Abel, O. 1914. Die Vorfahren der Bartenwale. Denkschriften der Mathematisch-Naturwissenschaftlichen Klasse der Kaiserlichen Akademie der Wissenschaften, Wien 90: 155–224.

Allen, G. M. 1921. A new fossil cetacean (*Archaeodelphis patrius* gen. et sp. nov.). Bulletin of the Museum of Comparative Zoology, Harvard University 65: 1–14.

Andrews, C. W. 1906. A Descriptive Catalogue of the Tertiary Vertebrata of the Fayum, Egypt. British Museum (Natural History), London.

Árnason, Ú., A. Gullberg, S. Gretarsdottir, B. Ursing, and A. Janke. 2000. The mitochondrial genome of the sperm whale and a new molecular reference for estimating eutherian divergence dates. Journal of Molecular Evolution 50: 569–578.

Bajpai, S., and P. D. Gingerich. 1998. A new Eocene archaeocete (Mammalia, Cetacea) from India and the time of origin of whales. Proceedings of the National Academy of Sciences USA 95: 15464–15468.

Bajpai, S., and J.G.M. Thewissen. 2002. Vertebrate fauna from Panandhro lignite field (lower Eocene), District Kachchh, western India. Current Science, Bangalore 82: 507–509.

Barnes, L. G., D. P. Domning, and C. E. Ray. 1985. Status of studies on fossil marine mammals. Marine Mammal Science 1: 15–53.

Barnes, L. G., and J. L. Goedert. 2000. The world's oldest known odontocete (Mammalia, Cetacea). (Abstract). Journal of Vertebrate Paleontology, Abstracts of Papers 20 (supplement to no. 3): 28.

Barnes, L. G., M. Kimura, H. Furusawa, and H. Sawamura. 1995. Classification and distribution of Oligocene Aetiocetidae (Cetacea; Mysticeti) from western North American and Japan; pp. 392–431 *in* L. G. Barnes, N. Inuzuka, and Y. Hasegawa (eds.), Evolution and Biogeography of Fossil Marine Vertebrates in the Pacific Realm. Island Arc 3.

Benham, W. B. 1939. *Mauicetus: A* fossil whale. Nature 143: 765.

Berggren, W. A., D. V. Kent, C. C. Swisher III, and M.-P. Aubry. 1995. A revised Cenozoic geochronology and chronostratigraphy; pp. 129–212 in W. A. Berggren, D. V. Kent, M.-P. Aubry, and J. A. Hardenbol (eds.), Geochronology, Time Scales and Global Stratigraphic Correlation. SEPM Special Publication 54. Society for Sedimentary Geology, Tulsa.

Blatt, H., and R. L. Jones. 1975. Proportions of exposed igneous, metamorphic, and sedimentary rocks. Geological Society of America Bulletin 86: 1085–1088.

Bowen, G. J., W. C. Clyde, P. L. Koch, S. Ting, J. Alroy, T. Tsubamoto, Y. Wang, and Y. Wang. 2002. Mammalian dispersal at the Paleocene/Eocene boundary. Science 295: 2062–2065.

Bowen, G. J., P. L. Koch, P. D. Gingerich, R. D. Norris, S. Bains, and R. M. Corfield. 2001. Refined isotope stratigraphy across the continental Paleocene-Eocene boundary on Polecat Bench in the northern Bighorn Basin; pp. 73–88 *in* P. D. Gingerich (ed.), Paleocene-Eocene Stratigraphy and Biotic Change in the Bighorn and Clarks Fork Basins, Wyoming. University of Michigan Papers on Paleontology 33. University of Michigan Museum of Paleontology, Ann Arbor, Michigan.

Boyden, A. A., and D. G. Gemeroy. 1950. The relative position of the Cetacea among the orders of Mammalia as indicated by precipitin tests. Zoologica, New York Zoological Society 35: 145–151.

Brandt, J. F. 1871. Bericht über den Fortgang meiner Studien über die Cetaceen, welche das grosse zur Tertiärzeit von Mitteleuropa bis Centralasien hinein ausgedehnte Meeresbecken bevölkerten. Bulletin de l'Académie des Sciences de Saint-Pétersbourg 16: 563–566.

Colbert, E. H. 1935a. Distributional and phylogenetic studies on Indian fossil mammals. IV: The phylogeny of the Indian Suidae and the origin of the Hippopotamidae. American Museum Novitates 799: 1–24.

———. 1935b. Siwalik mammals in the American Museum of Natural History. Transactions of the American Philosophical Society 26: 1–401.

Cope, E. D. 1895. Fourth contribution to the marine fauna of the Miocene period of the United States. Proceedings of the American Philosophical Society 34: 135–155.

Emlong, D. 1966. A new archaic cetacean from the Oligocene of north-west Oregon. Bulletin of the Museum of Natural History, University of Oregon 3: 1–51.

Estravís, C., and D. E. Russell. 1989. Découverte d'un nouveau *Diacodexis* (Artiodactyla, Mammalia) dans l'Eocène inférieur de Silveirinha, Portugal. Palaeovertebrata, Montpellier 19: 29–44.

Falconer, H., and P. T. Cautley. 1847. Fauna antiqua sivalensis, being the fossil zoology of the Sewalik Hills, in the north of India; pp. 69–71 *in* Suidae and Rhinoceratidae. Part 8. Smith, Elder, and Co., London.

Flower, W. H. 1883. On whales, past and present, and their probable origin. Notices of the Proceedings of the Royal Institution of Great Britain, London 10: 360–376.

Flower, W. H., and R. Lydekker. 1891. An Introduction to the Study of Mammals Living and Extinct. Adam and Charles Black, London.

Fordyce, R. E. 1980. Whale evolution and Oligocene southern ocean environments. Palaeogeography, Palaeoclimatology, Palaeoecology 31: 319–336.

———. 1981. Systematics of the odontocete *Agorophius pygmaeus* and the family Agorophiidae (Mammalia: Cetacea). Journal of Paleontology 55: 1028–1045.

———. 1989. Origins and evolution of Antarctic marine mammals; pp. 269–281 *in* J. A. Crane (ed.), Origins and Evolution of the Antarctic Biota. Special Publication 47. Geological Society, London.

———. 1994. *Waipatia maerewhenua,* new genus and new species (Waipatiidae, new family), an archaic late Oligocene dolphin (Cetacea: Odontoceti: Platanistoidea) from New Zealand; pp. 147–176 *in* A. Berta and T. A. Deméré (eds.), Contributions in Marine Mammal Paleontology Honoring Frank C. Whitmore, Jr. Proceedings of the San Diego Society of Natural History 29.

———. 2002. *Simocetus rayi* (Odontoceti: Simocetidae, new family): A bizarre new archaic Oligocene dolphin from the eastern North Pacific; pp. 185–222 *in* R. J. Emry (ed.), Cenozoic Mammals of Land and Sea: Tributes to the Career of Clayton E. Ray. Smithsonian Contributions to Paleobiology 93.

———. 2003. Early crown-group Cetacea in the southern ocean: The toothed archaic mysticete *Llanocetus.* (Abstract). Journal of Vertebrate Paleontology, Abstracts of Papers 23 (supplement to no. 3): 50A–51A.

Fordyce, R. E., and L. G. Barnes. 1994. The evolutionary history of whales and dolphins. Annual Review of Earth and Planetary Sciences 22: 419–455.

Fordyce, R. E., and C. de Muizon. 2001. Evolutionary history of cetaceans: A review; pp. 169–233 *in* J.-M. Mazin and V. de Buffrénil (eds.), Secondary Adaptation of Tetrapods to Life in Water. Verlag Dr. Friedrich Pfeil, Munich.

Fraas, E. 1904. Neue Zeuglodonten aus dem unteren Mitteleocän vom Mokattam bei Cairo. Geologische und Paläontologische Abhandlungen, Jena 6: 197–220.

Gatesy, J. E. 1998. Molecular evidence for the phylogenetic affinities of Cetacea; pp. 63–111 *in* J.G.M. Thewissen (ed.), The Emergence of Whales: Evolutionary Patterns in the Origin of Cetacea. Plenum, New York.

Gatesy, J. E., C. Hayashi, M. A. Cronin, and P. Arctander. 1996. Evidence from milk casein genes that cetaceans are close relatives of hippopotamid artiodactyls. Molecular Biology and Evolution 13: 954–963.

Gatesy, J. E., M. C. Milinkovitch, V. G. Waddell, and M. J. Stanhope. 1999. Stability of cladistic relationships between Cetacea and higher-level artiodactyl taxa. Systematic Biology 48: 6–20.

Gatesy, J. E., and M. A. O'Leary. 2001. Deciphering whale origins with molecules and fossils. Trends in Ecology and Evolution 16: 562–570.

Geisler, J. H. 2001. New morphological evidence for the phylogeny of Artiodactyla, Cetacea, and Mesonychidae. American Museum Novitates 3344: 1–53.

Geisler, J. H., and Z. Luo. 1998. Relationships of Cetacea to terrestrial ungulates and the evolution of cranial vasculature in Cete; pp. 163–212 *in* J.G.M. Thewissen (ed.), The Emergence of Whales: Evolutionary Patterns in the Origin of Cetacea. Plenum, New York.

Gervais, P. 1876. Remarques au sujet du genre *Phocodon* d'Agassiz. Journal de Zoologie, Paris 5: 62–70.

Gibbes, R. W. 1845. Description of the teeth of a new fossil animal found in the green-sand of South Carolina. Academy of Natural Sciences, Philadelphia, Proceedings 2: 254–256.

Gingerich, P. D. 1977. A small collection of fossil vertebrates from the middle Eocene Kuldana and Kohat formations of Punjab (Pakistan). Contributions from the Museum of Paleontology, University of Michigan 24: 190–203.

———. 1989. New earliest Wasatchian mammalian fauna from the Eocene of northwestern Wyoming: Composition and diversity in a rarely sampled high-floodplain assemblage. University of Michigan Papers on Paleontology 28: 1–97.

———. 1992. Marine mammals (Cetacea and Sirenia) from the Eocene of Gebel Mokattam and Fayum, Egypt: Stratigraphy, age, and paleoenvironments. University of Michigan Papers on Paleontology 30: 1–84.

———. 2003a. Mammalian responses to climate change at the Paleocene-Eocene boundary: Polecat Bench record in the northern Bighorn Basin, Wyoming; pp. 463–478 *in* S. L. Wing, P. D. Gingerich, B. Schmitz, and E. Thomas (eds.), Causes and Consequences of Globally Warm Climates in the Early Paleogene. Special Paper 369. Geological Society of America, Boulder, Colorado.

———. 2003b. Land-to-sea transition of early whales: Evolution of Eocene Archaeoceti (Cetacea) in relation to skeletal proportions and locomotion of living semiaquatic mammals. Paleobiology 29: 429–454.

———. 2003c. Stratigraphic and micropaleontologic constraints on the middle Eocene age of the mammal-bearing Kuldana Formation of Pakistan. Journal of Vertebrate Paleontology 23: 643–651.

Gingerich, P. D., M. Arif, M. A. Bhatti, M. Anwar, and W. J. Sanders. 1997. *Basilosaurus drazindai* and *Basiloterus hussaini,* new Archaeoceti (Mammalia, Cetacea) from the middle Eocene Drazinda Formation, with a revised interpretation of ages of whale-bearing strata in the Kirthar Group of the Sulaiman Range, Punjab (Pakistan). Contributions from the Museum of Paleontology, University of Michigan 30: 55–81.

Gingerich, P. D., M. Haq, I. H. Khan, and I. S. Zalmout. 2001b. Eocene stratigraphy and archaeocete whales (Mammalia, Cetacea) of Drug Lahar in the eastern Sulaiman Range, Balochistan (Pakistan). Contributions from the Museum of Paleontology, University of Michigan 30: 269–319.

Gingerich, P. D., M. Haq, I. S. Zalmout, I. H. Khan, and M. S. Malkani. 2001a. Origin of whales from early artiodactyls: Hands and feet of Eocene Protocetidae from Pakistan. Science 293: 2239–2242.

Gingerich, P. D., S. M. Raza, M. Arif, M. Anwar, and X. Zhou. 1993. Partial skeletons of *Indocetus ramani* (Mammalia, Cetacea) from the lower middle Eocene Domanda Shale in the Sulaiman Range of Pakistan. Contributions from the Museum of Paleontology, University of Michigan 28: 393–416.

———. 1994. New whale from the Eocene of Pakistan and the origin of cetacean swimming. Nature 368: 844–847.

Gingerich, P. D., and D. E. Russell. 1981. *Pakicetus inachus,* a new archaeocete (Mammalia, Cetacea) from the early-middle Eocene Kuldana Formation of Kohat (Pakistan). Contributions from the Museum of Paleontology, University of Michigan 25: 235–246.

Gingerich, P. D., B. H. Smith, and E. L. Simons. 1990. Hind limbs of Eocene *Basilosaurus isis:* Evidence of feet in whales. Science 249: 154–157.

Gingerich, P. D., and M. D. Uhen. 1994. Time of origin of primates. Journal of Human Evolution 27: 443–445.

———. 1998. Likelihood estimation of the time of origin of Cetacea and the time of divergence of Cetacea and Artiodactyla. Palaeontologia Electronica 1(2). http://palaeo-electronica.org/1998_2/ging_uhen/issue2.htm.

Gingerich, P. D., N. A. Wells, D. E. Russell, and S.M.I. Shah. 1983. Origin of whales in epicontinental remnant seas: New evidence from the early Eocene of Pakistan. Science 220: 403–406.

Graur, D., and D. G. Higgins. 1994. Molecular evidence for the inclusion of cetaceans within the order Artiodactyla. Molecular Biology and Evolution 11: 357–364.

Harlan, R. 1834. Notice of fossil bones found in the Tertiary formation of the state of Louisiana. Transactions of the American Philosophical Society 4: 397–403.

Hector, J. 1881. Notes on New Zealand Cetacea, recent and fossil. Transactions and Proceedings of the New Zealand Institute, Wellington 13: 434–436.

Hulbert, R. C. 1998. Postcranial osteology of the North American middle Eocene protocetid *Georgiacetus;* pp. 235–267 in J.G.M. Thewissen (ed.), The Emergence of Whales: Evolutionary Patterns in the Origin of Cetacea. Plenum, New York.

Hulbert, R. C., R. M. Petkewich, G. A. Bishop, D. Bukry, and D. P. Aleshire. 1998. A new middle Eocene protocetid whale (Mammalia: Cetacea: Archaeoceti) and associated biota from Georgia. Journal of Paleontology 72: 907–927.

Hunter, J. 1787. Observations on the structure and oeconomy of whales. Philosophical Transactions of the Royal Society of London 77: 371–450.

Irwin, D. M., and Ú. Árnason. 1994. Cytochrome *b* gene of marine mammals: Phylogeny and evolution. Journal of Mammalian Evolution 2: 37–55.

Kellogg, R. 1923. Description of an apparently new toothed cetacean from South Carolina. Smithsonian Miscellaneous Collections 76 (7): 1–7.

———. 1936. A review of the Archaeoceti. Carnegie Institution of Washington Publications 482: 1–366.

Kowalevsky, W. 1873. Sur l'*Anchitherium aurelianense* Cuv. et sur l'histoire paléontologique des chevaux. Mémoires de l'Académie Impériale des Sciences de St.-Pétersbourg, VII Série 20 (5): 1–73.

Kron, D. G., and E. M. Manning. 1998. Anthracotheriidae; pp. 381–388 in C. M. Janis, K. M. Scott, and L. L. Jacobs (eds.), Evolution of Tertiary Mammals of North America. Volume I. Terrestrial Carnivores, Ungulates, and Ungulate-like Mammals. Cambridge University Press, Cambridge.

Kükenthal, W. G. 1893. Vergleichend-Anatomische und Entwickelungsgeschichtliche Untersuchungen an Walthieren. Part II. Denkschriften der Medicinisch-Naturwissenschaftlichen Gesellschaft, Jena 3(2): 223–448.

———. 1922. Zur Stammesgeschichte der Wale. Sitzungsberichte der Preussischen Akademie der Wissenschaften, Physikalisch-mathematische Klasse, Berlin 1922 (9): 72–87.

Langer, P. 2001. Evidence from the digestive tract on phylogenetic relationships of ungulates and whales. Journal of Zoological Systematics and Evolutionary Research 39: 77–90.

Liu, F.-G.R., and M. M. Miyamoto. 1999. Phylogenetic assessment of molecular and morphological data for eutherian mammals. Systematic Biology 48: 54–64.

Luo, Z., and P. D. Gingerich. 1999. Terrestrial Mesonychia to aquatic Cetacea: Transformation of the basicranium and evolution of hearing in whales. University of Michigan Papers on Paleontology 31: 1–98.

Lydekker, R. 1884. Indian Tertiary and post-Tertiary Vertebrata. Siwalik and Narbada bunodont Suina. Palaeontologia Indica, Memoirs of the Geological Survey of India, Calcutta 3: 35–104.

Matthew, W. D. 1929. Reclassification of the artiodactyl families. Bulletin of the Geological Society of America 40: 403–408.

McKenna, M. C. 1975. Toward a phylogenetic classification of the Mammalia; pp. 21–46 in W. P. Luckett and F. S. Szalay (eds.), Phylogeny of the Primates. Plenum, New York.

McKenna, M. C., and S. K. Bell. 1997. Classification of Mammals above the Species Level. Columbia University Press, New York.

Milinkovitch, M. C., M. Bérubé, and P. J. Palsboll. 1998. Cetaceans are highly derived artiodactyls; pp. 113–131 in J.G.M. Thewissen (ed.), The Emergence of Whales: Evolutionary Patterns in the Origin of Cetacea. Plenum, New York.

Miller, G. S. 1923. The telescoping of the cetacean skull. Smithsonian Miscellaneous Collections 76: 1–70.

Mitchell, E. D. 1989. A new cetacean from the late Eocene La Meseta Formation, Seymour Island, Antarctic Peninsula. Canadian Journal of Fisheries and Aquatic Sciences 46: 2219–2235.

Montgelard, C., F. M. Catzeflis, and E. Douzery. 1997. Phylogenetic relationships of artiodactyls and cetaceans as deduced from the comparison of cytochrome *b* and 12S rRNA mitochondrial sequences. Molecular Biology and Evolution 14: 550–559.

Nikaido, M., A. P. Rooney, and N. Okada. 1999. Phylogenetic relationships among cetartiodactyls based on evidence from insertions of SINEs and LINEs: Hippopotamuses are the closest extant relatives of whales. Proceedings of the National Academy of Sciences USA 96: 10261–10266.

O'Leary, M. A. 1998. Phylogenetic and morphometric reassessment of the dental evidence for a mesonychian and cetacean clade; pp. 133–161 in J.G.M. Thewissen (ed.), The Emergence of Whales: Evolutionary Patterns in the Origin of Cetacea. Plenum, New York.

O'Leary, M. A., and J. H. Geisler. 1999. The position of Cetacea within Mammalia: Phylogenetic analysis of morphological data from extinct and extant taxa. Systematic Biology 48: 455–490.

O'Leary, M. A., and M. D. Uhen. 1999. The time of origin of whales and the role of behavioral changes in the terrestrial-aquatic transition. Paleobiology 25: 534–556.

Pickford, M. 1983. On the origins of the Hippopotamidae together with a description of two new species, a new genus, and a new subfamily from the Miocene of Kenya. Géobios 16: 193–217.

———. 1989. Update on hippo origins. Comptes Rendus de l'Académie des Sciences, Paris 309: 163–168.

Pritchard, G. B. 1939. On the discovery of a fossil whale in the older Tertiaries of Torquay, Victoria. Victorian Naturalist 55: 151–159.

Prothero, D. R., E. M. Manning, and M. S. Fischer. 1988. The phylogeny of the ungulates; pp. 201–234 in M. J. Benton (ed.), The Phylogeny and Classification of the Tetrapods. Clarendon Press, Oxford.

Rose, K. D. 1982. Skeleton of *Diacodexis,* oldest known artiodactyl. Science 216: 621–623.

———. 1996. On the origin of the order Artiodactyla. Proceedings of the National Academy of Sciences USA 93: 1705–1709.

———. 2001. The ancestry of whales. Science 293: 2216–2217.

Russell, L. S. 1968. A new cetacean from the Oligocene Sooke Formation of Vancouver Island, British Columbia. Canadian Journal of Earth Sciences 5: 929–933.

Sanders, A. E., and L. G. Barnes. 2002a. Paleontology of the late Oligocene Ashley and Chandler Bridge formations of South Carolina, 2: *Micromysticetus rothauseni,* a primitive cetotheriid mysticete (Mammalia: Cetacea); pp. 271–293 in R. J. Emry (ed.), Cenozoic Mammals of Land and Sea: Tributes to the Career of Clayton E. Ray. Smithsonian Contributions to Paleobiology 93.

———. 2002b. Paleontology of the late Oligocene Ashley and Chandler Bridge formations of South Carolina, 3: Eomysticetidae, a new family of primitive mysticetes (Mammalia: Cetacea); pp. 313–356 in R. J. Emry (ed.), Cenozoic Mammals of Land and Sea: Tributes to the Career of Clayton E. Ray. Smithsonian Contributions to Paleobiology 93.

Sarich, V. M. 1985. Rodent macromolecular systematics; pp. 423–452 in W. P. Luckett and J.-L. Hartenberger (eds.), Evolutionary Relationships among Rodents. A Multidisciplinary Analysis. Plenum, New York.

———. 1993. Mammalian systematics: Twenty-five years among their albumins and transferrins; pp. 103–114 in F. S. Szalay, M. J. Novacek, and M. C. McKenna (eds.), Mammal Phylogeny: Placentals. Springer-Verlag, New York.

Schaeffer, B. 1947. Notes on the origin and function of the artiodactyl tarsus. American Museum Novitates 1356: 1–24.

———. 1948. The origin of a mammalian ordinal character. Evolution 2: 164–175.

Scott, W. B. 1894. The structure and relationships of *Ancodus.* Academy of Natural Sciences, Philadelphia, Journal 9: 461–497.

———. 1940. The mammalian fauna of the White River Oligocene—Part IV. Artiodactyla. Transactions of the American Philosophical Society 28: 363–746.

Shedlock, A. M., M. C. Milinkovitch, and N. Okada. 2000. SINE evolution, missing data, and the origin of whales. Systematic Biology 49: 808–817.

Shimamura, M., H. Abe, M. Nikaido, K. Ohshima, and N. Okada. 1999. Genealogy of families of SINEs in cetaceans and artiodactyls: The presence of a huge superfamily of tRNA^{Glu}-derived families of SINEs. Molecular Biology and Evolution 16: 1046–1060.

Shimamura, M., H. Yasue, K. Ohshima, H. Abe, H. Kato, T. Kishiro, M. Goto, I. Munechika, and N. Okada. 1997. Molecular evidence from retroposons that whales form a clade within even-toed ungulates. Nature 388: 666–670.

Simpson, G. G. 1945. The principles of classification and a classification of mammals. Bulletin of the American Museum of Natural History 85: 1–350.

Slijper, E. J. 1936. Die Cetaceen, Vergleichend-Anatomisch und Systematisch. Capita Zoologica 6–7: 1–590.

Smith, A. G., D. G. Smith, and B. M. Funnell. 1994. Atlas of Mesozoic and Cenozoic Coastlines. Cambridge University Press, Cambridge.

Smith, T. 2000. Mammals from the Paleocene-Eocene transition in Belgium (Tienen Formation, MP7): Paleobiogeographical and biostratigraphical implications; pp. 148–149 in B. Schmitz, B. Sundquist, and F. P. Andreasson (eds.), Early Paleogene Warm Climates and Biosphere Dynamics. Geologiska Föreningens Förhandlingar, Stockholm 122.

Strauss, D. J., and P. M. Sadler. 1989. Classical confidence intervals and Bayesian probability estimates for ends of local taxon ranges. Mathematical Geology 21: 411–427.

Sudre, J., D. E. Russell, P. Louis, and D. E. Savage. 1983. Les artiodactyles de l'Éocène inférieur d'Europe (première partie). Bulletin du Muséum National d'Histoire Naturelle, Série 4, Paris 5: 281–333.

Thewissen, J.G.M. 1994. Phylogenetic aspects of cetacean origins: A morphological perspective. Journal of Mammalian Evolution 2: 157–184.

Thewissen, J.G.M., P. D. Gingerich, and D. E. Russell. 1987. Artiodactyla and Perissodactyla (Mammalia) from the early-middle Eocene Kuldana Formation of Kohat (Pakistan). Contributions from the Museum of Paleontology, University of Michigan 27: 247–274.

Thewissen, J.G.M., and S. T. Hussain. 1990. Postcranial osteology of the most primitive artiodactyl *Diacodexis pakistanensis* (Dichobunidae). Anatomia Histologia Embryologia, Zentralblatt für Veterinärmedizin, Reihe C 19: 37–48.

———. 1993. Origin of underwater hearing in whales. Nature 361: 444–445.

———. 1998. Systematic review of the Pakicetidae, early and middle Eocene Cetacea from Pakistan and India. Bulletin of the Carnegie Museum of Natural History 34: 220–238.

Thewissen, J.G.M., and S. I. Madar. 1999. Ankle morphology of the earliest cetaceans and its implications for the phylogenetic relations among ungulates. Systematic Biology 48: 21–30.

Thewissen, J.G.M., S. I. Madar, and S. T. Hussain. 1998. Whale ankles and evolutionary relationships. Nature 395: 452.

Thewissen, J.G.M., E. M. Williams, L. J. Roe, and S. T. Hussain. 2001. Skeletons of terrestrial cetaceans and the relationship of whales to artiodactyls. Nature 413: 277–281.

True, F. W. 1908. The fossil cetacean, *Dorudon serratus* Gibbes. Bulletin of the Museum of Comparative Zoology, Harvard University 52: 65–78.

Uhen, M. D. 1996. *Dorudon atrox* (Mammalia, Cetacea): Form, Function, and Phylogenetic Relationships of an Archaeocete from the Late Middle Eocene of Egypt. Ph.D. Dissertation, University of Michigan, Ann Arbor.

———. 1999. New species of protocetid archaeocete whale, *Eocetus wardii* (Mammalia: Cetacea) from the middle Eocene of North Carolina. Journal of Paleontology 73: 512–528.

————. 2004. Form, function, and anatomy of *Dorudon atrox* (Mammalia, Cetacea): An archaeocete from the middle to late Eocene of Egypt. University of Michigan Papers on Paleontology 34: 1–222.

Ursing, B., and Ú. Árnason. 1998. Analyses of mitochondrial genomes strongly support a hippopotamus-whale clade. Proceedings of the Royal Society of London, Series B 265: 2251–2255.

Van Valen, L. M. 1966. Deltatheridia, a new order of mammals. Bulletin of the American Museum of Natural History 132: 1–126.

————. 1968. Monophyly or diphyly in the origin of whales. Evolution 22: 37–41.

Vrba, E. S. 1985. Environment and evolution: Alternative causes of the temporal distribution of evolutionary events. South African Journal of Science 81: 229–236.

Weber, M. 1904. Die Säugetiere. Einführung in die Anatomie und Systematik der Recenten und Fossillen Mammalia. Gustav Fischer, Jena.

Whitmore, F. C., and A. E. Sanders. 1977. Review of the Oligocene Cetacea. Systematic Zoology 25: 304–320.

Winge, H. 1921. A review of the interrelationships of the Cetacea (Translated by Gerrit S. Miller). Smithsonian Miscellaneous Collections 72(8): 1–97.

Yablokov, A. V. 1964. Convergence or parallelism in the evolution of cetaceans. International Geological Review 7: 1461–1468.

Zachos, J. C., M. Pagani, L. C. Sloan, E. Thomas, and K. Billups. 2001. Trends, rhythms, and aberrations in global climate 65 Ma to present. Science 292: 686–693.

Zalmout, I. S., P. D. Gingerich, H. A. Mustafa, A. Smadi, and A. Khammash. 2003. Cetacea and Sirenia from the Eocene Wadi Esh-Shallala Formation of Jordan (Abstract). Journal of Vertebrate Paleontology, Abstracts of Papers 23(supplement to no. 3): 113.

INDEX

Taxa and terms listed solely in tables and figures are not included in this index.

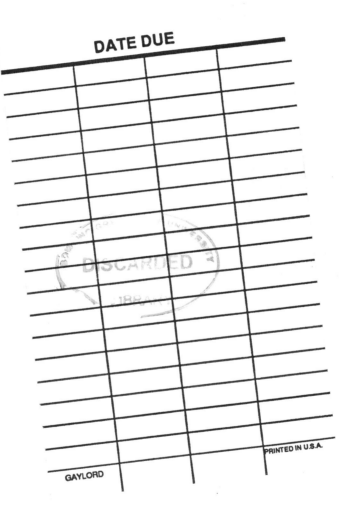